Practical Surface Analysis

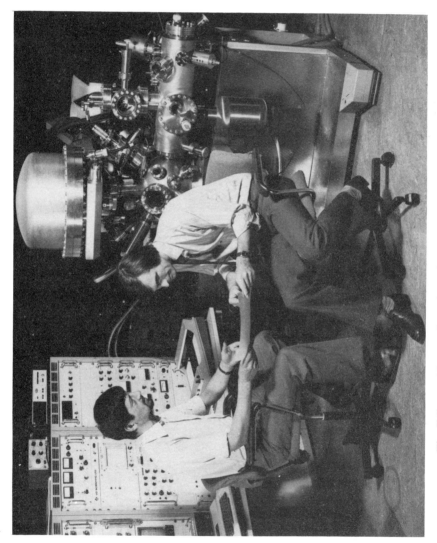

The Editors with a modern Auger/X-ray Photoelectron Spectrometer

Practical Surface Analysis

by Auger and X-ray Photoelectron Spectroscopy

Edited by

D. BRIGGS
ICI PLC, Petrochemicals and Plastics Division,
Wilton, Middlesbrough, Cleveland, UK

and

M. P. SEAH
Division of Materials Applications,
National Physical Laboratory,
Teddington, Middlesex, UK

JOHN WILEY & SONS
Chichester · New York · Brisbane · Toronto · Singapore

7/83-6792

PHYSICS

Library of Congress Cataloging in Publication Data:

Main entry under title:

Practical surface analysis.

Includes bibliographical references and index.
1. Surfaces (Technology)—Analysis. 2. Electron spectroscopy. I. Briggs, D. II. Seah, M. P.
TP156.S95P73   1983   620.1'1292   83-6618

ISBN 0 471 26279 X

British Library Cataloguing in Publication Data:

Practical surface analysis.
1. Auger effect 2. Photoelectron spectroscopy
3. Solids–Spectra
I. Briggs, D. II. Seah, M. P.
530.4'1   QC454.E4

ISBN 0 471 26279 X

Typeset by Preface Ltd, Salisbury, Wilts.
Printed by Page Bros. (Norwich) Ltd.

Contents

Preface

The amazing growth and diversification of surface analytical techniques began in the 1960s with the development of electron spectroscopy—first Auger electron spectroscopy (AES), closely followed by X-ray photo-electron spectroscopy (XPS or ESCA). Today these two complementary techniques still dominate the surface analysis scene with the total number of operational instruments well over one thousand. Since the 'heart' of either technique is an electron energy analyser, AES and XPS are frequently found in combination, with essentially no loss of performance in current fourth generation instruments. AES in the form of high resolution scanning Auger microscopy adds the surface compositional dimension to scanning electron microscopy. The addition of an ion source for sputter removal of surface layers allows either XPS or AES to perform composition depth profiling.

In recent years these techniques have rapidly matured with the generation of a large body of literature, the growth of 'user' groups in Europe, the United States and Japan, the systematic improvement of standards of operation and procedures for quantification, and so on. AES and XPS are now routinely used in a large number of industries; indeed, in some cases their development has been intimately connected with the growth of new industries (e.g. AES and the microelectronics industry). Many centres offering contract research facilities have appeared. Each year more and more people come into contact with the techniques through the need to solve problems relating to surface or interface composition and many people became practically involved in using the techniques without the advantage of relevant formal academic training. Although literature reviews of aspects of AES and XPS are legion, no text exists which reflects the new maturity of these techniques in a way which is useful *practically*, especially for newcomers to applied surface analysis.

The aim of this book is to correct this omission and to present, in one volume, all of the important concepts and tabulated data. A brief introduction gives the historical background to AES and XPS and sets them in the perspective of surface analytical techniques as a whole. The essentials of technique are covered in chapters on instrumentation, spectral interpretation, depth profiling and quantification. The remaining chapters are intended to give an insight into the major fields of application, both in terms of the special attributes of AES and XPS and the contribution they have made. These fields are microelectronics, metallurgy, catalysis, polymer technology and corrosion science. Throughout, the underlying electron spectroscopy link between AES and XPS is stressed. Aspects of technique which have fundamental importance in day-to-day operation such as instrument calibration, XPS binding energy referencing and XPS data processing (especially complex curve resolution) are discussed in the Appendices. Finally, there are full tabulations of major peak positions in AES and XPS, relative sensitivity factors for XPS and binding energy/Auger parameter data for elements and compounds.

Wilton   D. BRIGGS
Teddington   M. P. SEAH
February 1983

Practical Surface Analysis
by Auger and X-ray Photoelectron Spectroscopy
Edited by D. Briggs and M. P. Seah
© 1983, John Wiley & Sons, Ltd.

Chapter 1

A Perspective on the Analysis of Surfaces and Interfaces

M. P. Seah

*Division of Materials Applications, National Physical Laboratory,
Teddington, Middlesex, UK*

D. Briggs

*ICI PLC, Petrochemicals and Plastics Division,
Wilton, Middlesbrough, Cleveland, UK*

1.1 Introduction

Surface analysis and surface science are evolutionary disciplines. They are where they are at present as a result of a small number of steps by a very large number of researchers. Progress along the whole front has not been even and advance has been rapid in certain areas but non-existent in others. Some of the pressures for advance come from mission-orientated studies for industry and some from the curiosity-motivated studies of academe. These pressures have led to an amazing plethora of techniques for surface analysis, each with its bureaucratic acronym, and to a very wide range of studies and data. Not all of these techniques and studies are relevant to researchers who wish to understand and solve surface-related problems in applied science. Over the years, Auger electron spectroscopy (AES) and X-ray photo-electron spectroscopy (XPS) have been found to show the greatest applicability. In this chapter, therefore, we chronicle the evolution of these techniques and show how they fit into the jigsaw of modern surface spectroscopies.

We have already mentioned surface analysis without defining what we mean. In its simplest sense we require the elemental composition of the outermost atom layer of a solid. Having found that, there will be immediate requests for detailed knowledge of the chemical binding state, precise sites of atoms in relation to crystal structure, surface homogeneity and the state of adsorbates. For many people surface science still concerns the complete

1

Practical Surface Analysis

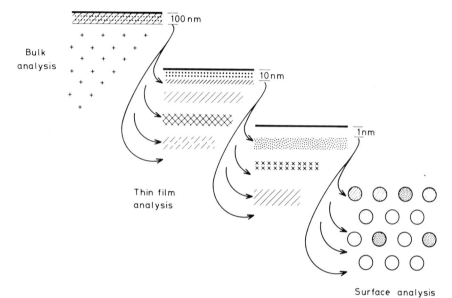

Figure 1.1   The regimes of surface analysis, thin film analysis and bulk analysis

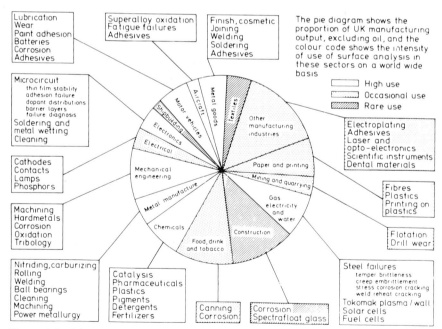

Figure 1.2   The intensity of application of surface analysis, illustrated by the manufacturing sectors of the United Kingdom. After Seah[1]

characterization of clean, low index, metal surfaces in vacuum; for others, the occasional adsorbed molecule of carbon monoxide may be added. However, in applied studies, the surfaces will be far more complex and the characterization will not generally be complete—simply adequate for the purpose. In applied studies all of the above will be required but, in addition, there will be requests for similar information for the atom layers below the surface as a function of depth, to depths of 1 μm or so, as shown in Figure 1.1. Each of the many surface analysis techniques approaches one or more of these aspects better than the others so that, in principle, each has a particular advantage. However, the value of a technique to the user depends not only on the theoretical advantage but also on the available experience in that technique, the back-up data and examples of similar approaches by previous workers. This books seeks to provide such information for the two principal surface analysis techniques, AES and XPS.

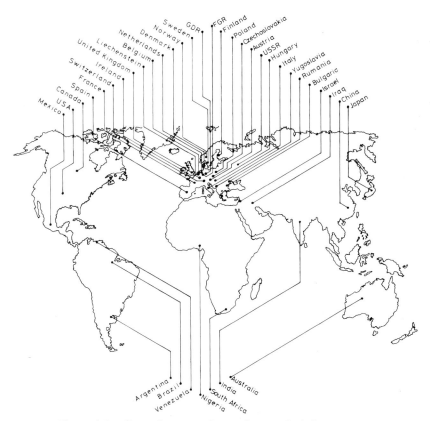

Figure 1.3   Countries operating surface analysis instruments

In the later chapters, examples are given of the use of AES and XPS in the major areas of microelectronics, metallury, catalysis, polymers and corrosion science. Figure 1.2 shows these and other areas in which surface analysis has made considerable contributions, ranging from the wear of cutting edges in strip metal production and the design of lubricant additives to opto-electronics and the architecture of integrated circuits. The ubiquity of surface analysis applications is matched by the range of countries now operating surface analysis instruments, as shown in Figure 1.3. This overall develop-ment has occurred in the years since 1970. Some developments took place before 1970 but most are concentrated into the last few years.

1.2   The Background to Electron Beam Techniques

1.2.1   Auger electron spectroscopy

The start of much of the work can be placed in the early 1920s when C. Davisson and H. E. Farnsworth were studying the secondary emission proper-ties of metal surfaces in high vacuum under bombardment by electrons with energies of up to a few hundred electronvolts. Davisson's interests at that time partly concerned the properties of the secondary electron emission of anodes in gas-filled and vacuum triode valves. During his experiments with L. H. Germer, an accident occurred to the glass vacuum system when the nickel target was being outgassed at a high temperaturc. This led to oxidation of the target which was then cleaned by prolonged heating in hydrogen, causing the originally fine grained target to recrystallize to a few large grains. Low energy electron diffraction (LEED) was observed from these crystals by Davisson and Germer in 1927[2] and was correctly interpreted in terms of de Broglie's publication on wave mechanics. Davisson, Germer, Farnsworth[3] and others then worked on LEED through the 1930s using glass vacuum systems and their own designs of Faraday cup detectors to monitor the LEED beams. The experiments were very difficult but these workers still managed to study many low index single crystal metal surfaces, under what were probably ultra-high vacuum (UHV) conditions, as a function of temperature and gas adsorption. The work had a very limited appeal until Germer[4] revived the display LEED concept of Ehrenberg[5] and the concept was marketed commercially in a metal UHV system by Varian in 1964. Within a few years other manufacturers were producing commercial systems and many laboratories were set up to study single crystal metal surfaces, by LEED, in relation to a whole range of notionally applied problems that today, with the benefit of hindsight, may appear a little far-fetched. Many of the laboratories were, unfortunately, not in a position to handle the theoretical requirements of LEED.

In 1967 Harris published two reports from GE, subsequently appearing in the regular literature,[6,7] which followed the notion of Lander[8] that Auger electrons from solids could be used for surface analysis. Briefly, the principle of the technique is that the sample is bombarded by an electron beam of 1–10 keV energy and this ejects core electrons from a level E_x in atoms in a region of the sample up to 1 μm or so deep. The core hole is then filled by an internal process in the atom whereby an electron from a level E_y falls into the core hole with the energy balance taken by a third electron from a level E_z. This last electron, called an Auger electron after Pierre Auger who first observed such events in a cloud chamber,[9] is then ejected from the atom with an energy E_a, given very approximately by

$$E_a = E_z + E_y - E_x \tag{1.1}$$

The electrons thus have energies unique to each atom and, if the energy spectrum from 0 to 2 KeV is measured, the energies of the Auger electron peaks allow all the elements present, except hydrogen and helium, to be identified. The reason that AES is a surface-sensitive technique lies in the intense inelastic scattering that occurs for electrons in this energy range, so that Auger electrons from only the outermost atom layers of a solid survive to be ejected and measured in the spectrum. Considerably earlier measurements on Auger electrons had been made, but not at these energies and not in relation to surface analysis.[10] Harris realized that the direct energy spectrum, with its small peaks on a large background, could be much easier to present and analyse if differentiated. He thus developed the potential modulation of the analyser to present the differential mode spectra used ever since. Harris also demonstrated the great sensitivity of the technique in measuring surface contaminants with a signal-to-noise ratio of over 200.

In 1967 Harris discussed his work with Peria who immediately realized that the standard LEED apparatus could be modified for AES work with the addition of a small amount of standard electronics. Weber and Peria's publication[11] appeared before those of Harris due to unfortunate refereeing problems.[12] Some of the early euphoria for LEED was fading and so standard LEED apparatuses were rapidly converted to AES work. Many early AES instruments were thus ideally suited for UHV work on single crystal surfaces but, because of the inherent differences of the LEED system as an electron spectrometer compared with the 127° deflection spectromer of Harris, the LEED system was generally used at poor energy resolution. The poor resolution led many workers to believe that Auger electron peaks were broad and that they contained little chemical information. Thus AES developed from the beginning as an elemental analysis technique but was well established for carefully prepared single crystal surfaces in UHV. This enabled Palmberg and Rhodin[13] to show, by depositing single atom layers on low index crystal faces, that AES is characteristic of the surface to a depth of only 5–10 Å. The next major step came with the introduction of the cylindrical mirror analyser by Palmberg, Bohm and Tracy[14] which significantly improved the signal-to-noise ratio available with AES and which opened the door to the wide spectrum of applications we see today.

1.2.2 Other techniques

Concomitant with these developments have been the developments of a number of parallel electron beam analytical techniques which have their own particular following but which do not have such general appeal as AES. Techniques such as appearance potential spectroscopy (APS),[15] in which the onset of excitation of the core levels is measured as the beam energy is increased, were popular as cheaper alternatives to AES but have poor sensitivity and so are only used now for solid-state studies. Ionization loss spectroscopy (ILS),[16] in which the electron losses associated with the onset of ionization of a core level are measured, similarly has poor sensitivity but now is being revived as a technique appropriate for very localized chemical analysis in the scanning transmission electron microscope (STEM).[17] Other variants on these techniques exist but the only other analytical electron beam technique to have established itself is high resolution low energy electron loss spectroscopy (HRLEELS).[18,19] In this technique the electron loss events from a highly monochromatic electron beam are measured with an energy resolution in the 5–20 MeV range. These losses are associated with the excitation of the vibrational modes of atoms and molecules in specific sites and orientations on metal surfaces. Reasonable quantities of data are now established so that the technique is becoming more general in application; however, polar compounds cannot be studied as the spectra become swamped by losses due to the easy excitation of optical phonons. A final technique which analyses structure rather than chemistry, but which is finding strong application today in the growth of epitaxial layered materials, is reflection high energy electron diffraction (RHEED). Like LEED, RHEED can be used to analyse the crystal structure of the outermost layers of a solid. To achieve a similar surface sensitivity to the 10–200 eV electrons used in LEED, the 10–30 keV RHEED electron beam is arranged to strike the surface at grazing angles of incidence. The analysis of diffracted intensities here is more straightforward than is the case for LEED and partly explains its current popularity.

1.3 The Background to Photon Beam Techniques

1.3.1 X-ray photoelectron spectroscopy[20]

XPS has a very illustrious pedigree in which the Auger effect plays a part. Its long history is intricately bound up with the developments of wave particle duality and the early days of atomic physics. For the more discerning reader, with a little more time to invest, this history is excellently and vividly chronicled by the workers at LaTrobe University[20-23] whose articles contain interesting historical perspectives and anecdotes.

XPS has its origins in the investigations of the photo-electric effect (discovered by Hertz in 1887) in which X-rays were used as the exciting photon source. For example, Innes described, in 1907,[24] experiments involving a Röntgen tube with a platinum anti-cathode and subsequent magnetic analysis of the emitted electrons by velocity selection using two Helmholz coils and photographic detection.

Rutherford's laboratory, in Manchester, was at this time ideally placed to develop the field, having much experience in the measurement of the β-ray spectra of radioactive materials by magnetic analysis and being at the forefront of the new field of X-ray spectroscopy. The early experiments were carried out by Moseley, Rawlinson and Robinson before the outbreak of World War I. Indeed, in 1914 Rutherford made a first stab[25] at stating the basic equation of XPS, which was subsequently modified to

$$E_K = h\nu - E_B$$

where E_K is the kinetic energy of the β-rays (photo-electrons), $h\nu$ the incident photon energy and E_B the electron binding energy. After the War, Robinson continued the work and Maurice de Broglie started research in France, both still using photographic detection. Understanding developed rapidly and in the early 1920s the photo-electron spectra of many elements, excited by a variety of high energy X-rays, had been obtained. Anomalous 'lines' (not obviously from core levels) were found to correspond with excitations by X-ray fluorescence photons characteristic of the atoms under study. The realization of the true nature of these electrons came only with Auger's demonstration of the radiationless transition,[9] which bears his name, in cloud chamber experiments. At this time XPS was seen as extending the study of atomic structure beyond the confines of X-ray spectroscopy (XES). However, it eventually became obvious that the limitations of the experimental XPS technique on the one hand and new advances in XES technique on the other had changed the situation heavily in favour of XES. Only Robinson continued research, though without any major advance in resolution or sensitivity, through the 1930s, and this effort too ended with the outbreak of World War II.

After the War, Steinhardt and Serfass at Lehigh University conceived the idea of reviving XPS as an analytical tool, particularly for studying surface chemical phenomena. Several instruments were constructed which, though failing to advance performance, introduced G-M electron counting and eventually a 127° electrostatic energy analyser. Despite the use of very high energy X-rays, surface effects on spectra were, however, discerned.[26] Steinhardt's PhD thesis had the prophetic title 'An X-ray photoelectron spectrometer for chemical analysis'.

Meanwhile decisive developments were underway in Uppsala, Sweden. Kai Siegbahn had developed, during the 1940s, β-ray spectroscopy to very high levels of precision and had realised that electron spectroscopy using X-ray excitation, rather than radioactive sources, might now be possible. An iron-free double-focusing spectrometer for accurate magnetic energy analysis was specially developed with a resolving power of 10^{-5}. In 1954 the first X-ray photo-electron spectrum from cleaved sodium chloride was obtained. Up to this point all previous spectra had consisted of a series of bands with a more-or-less well-defined edge followed by a tail. The Uppsala spectrum revealed for the first time a completely resolved line on the high kinetic energy side of each 'edge', representing electrons which had lost zero energy. The peak maximum (E_K) allowed the electron binding energy (E_B) to be measured accurately and the goal of using XPS for atomic structure investigation had been realized. Subsequently Siegbahn's group observed the chemical shift effect on core-level binding energies and went on to develop the whole field of electron spectroscopy during the period 1955–1970. Siegbahn coined the acronym ESCA (electron spectroscopy for chemical analysis) to underline the fact that both photo *and* Auger electron peaks appear in the 'XPS' spectrum.[27] A seminal book[28] on the subject was published with this title in 1967 which brought large-scale awareness of the potential of XPS. Commercial instruments started to appear around 1969–1970, although at this time, as a result of work with stearates by Siegbahn *et al.*,[28] it was thought that XPS characterized the outermost 100 Å of the surface. It was not until the work of Brundle and Roberts[29] in UHV that XPS truly became a surface technique whilst in 1972 Seah[30] pointed out that the thickness of the surface layer characterized must be very similar to that in AES.

1.3.2 Other techniques

Instead of using photons of kiloelectronvolt energy, much lower energy photons may be used to excite electrons in the solid. In ultra-violet photo-electron spectroscopy (UPS) the source of photons is a differentially pumped inert gas discharge 'lamp'. This produces discrete low energy resonance lines (e.g. He[I], 21.2 eV, and He[II], 40.8 eV) with an inherent width of a few millielectronvolts. Since there is only sufficient energy to emit electrons from the valence band, the technique has no analytical potential (in the sense of this book). UPS is, however, widely used in the study of the electron band structure of metals, alloys and semi-conductors and of adsorption phenomena.[31] A further source of photons for photo-electron spectroscopy is synchrotron radiation—the radiation emitted by accelerating electrons. Electrons circulated continuously in a storage ring at energies of ~ 1 GeV produce a continuous spectrum of photons with energies from a few electronvolts to several kiloelectronvolts. With a monochromator, therefore, a variable energy photon source is provided and this has several attractions for photo-electron spectroscopy.[32,33] The synchrotron source is also ideal for carrying out extended X-ray absorption fine-structure spectroscopy (EXAFS). Here, information about local structure and short-range order is obtained from the pattern of oscillations on the high energy side of X-ray absorption edges. Essentially this is a bulk analytical technique to complement X-ray diffraction (XRD) but applicable to amorphous materials. However, 'surface' information is obtained, for instance, in the study of very small metal crystallites in supported metal catalysts.[34] True surface information can also be obtained by a variety of methods, e.g. by monitoring the Auger electron yield over the same photon energy range in the region of an absorption edge. Such variants are referred to as surface EXAFS or S-EXAFS.[35]

1.4 The Background to Ion Beam Techniques

Over a similar period to that of AES has been the development of two major ion beam techniques: secondary ion mass spectroscopy (SIMS),[36,37] and ion scattering spectroscopy (ISS).[38,39] In SIMS a mass filtered beam of argon, oxygen or cesium ions strike the surface and remove surface atoms. Some of the emitted atoms are either positively or negatively ionized and these may be detected, very sensitively, by a mass spectrometer. For surface analysis a primary beam of argon, at current densities of the order of nanoamperes per square centimetre, is used with the beam spread out over some tens of square millimetres. In this mode, called static SIMS, the technique has a very high surface sensitivity, especially to electropositive elements, and can also give, in principle, detailed chemical state information and, with some effort, moderate quantification. Although now twelve years old, the technique has not yet found as much favour as AES and XPS. It is quite likely, however, that interest in the use of static SIMS will grow in the near future for samples already characterized by AES or XPS. The latter techniques rapidly provide an overall picture but SIMS is sometimes better at unravelling detailed aspects of chemistry, has a much higher sensitivity and furthermore can be used to observe hydrogen. In the alternative operating mode of dynamic SIMS the ion optics allow submicron images of the sample to be produced for different elements. The high primary ion flux densities of dynamic SIMS cause the surface to be eroded, as measurement occurs, in contrast to static SIMS where the top atom layer may remain for many hours before removal. In dynamic SIMS the alternative ion sources of oxygen and cesium are used as these enhance the yields for positive and negative ions, respectively, to increase the detectability. Here limits are given in parts per million (p.p.m.) or billion (US) instead of fractions of a percent as in AES and XPS. In dynamic SIMS the chemical information is lost and the instrument can be operated in one of two ways. In one mode the instrument behaves similarly to the electron probe X-ray microanalyser (EPMA) for bulk microanalysis and imaging or, in the other mode, by rastering the primary beam over an area of approximately 50×50 μm, high depth resolution concentration–depth profiles of dopants in semi-conductors can be attained.

The second ion beam technique, ISS, relies on a simple billiard ball collision concept in which the primary ion is scattered into the spectrometer with an energy defined by the mass of the struck target atom. The technique does not have the broad applicability of AES, XPS and SIMS but has the unique advantages that it gives information of the outermost atom layer only and also, by varying the polar and azimuthal angles of the interrogating ion beam with respect to a single crystal target, of the adatom site position.

Two other ion beam techniques should be mentioned, sputtered neutral mass spectroscopy (SNMS)[40] and Rutherford back-scattering spectroscopy (RBS).[41] SNMS uses the same principles as SIMS but, recognizing that most of the signal is lost because most of the emitted particles are neutralized on leaving the target surface, seeks to re-ionize them in a plasma just above the surface. This should greatly increase the sensitivity over that of SIMS but, due to the difficulty of operating the plasma and of obtaining spatial resolution, the technique has not yet gained popularity. On the other hand, RBS is more akin to ISS, using the billiard ball collisions and energy analysis of the reflected beam ions. By using ions of a few MeV energy the depth distributions of different elements can be measured non-destructively over depths up to 1 μm with a depth resolution of 20 nm. Heavy elements of adjacent atomic number are difficult to resolve but sensitivities below 1 per cent. of a monolayer are claimed. The equipment is simple and rugged and suitable Van der Graaff accelerators are becoming available as surplus to requirements from the nuclear measurements field, where higher energy machines are being purchased. Ion neutralization spectroscopy (INS),[42] in which low energy ions neutralized at the surface produce electrons which are characteristic of the valance band, is not discussed here since, like UPS, it has not general analytical capability.

1.5 Conclusions

The broad scope of some of the surface analysis techniques* is shown in Table 1.1, where some techniques are summarized in terms of their input and measured radiations. Fuller versions of such a table and a more detailed discussion of many of the techniques may be found in the review of Coburn and Kay,[43] articles in *Methods of Surface Analysis*[44] and the review of Honig.[45] Table 1.1 reminds us that for each input radiation there are many simultaneous pro-

---

*Little mention has been made of vibrational spectroscopy techniques which can provide surface analytical information, since they do not meet a basic requirement of identifying *elements*. However, in the study of adsorption of molecules or of organic and polymeric surfaces they have an important role to play. Some of these techniques are mentioned in Chapter 9.

Table 1.1   Easy reminder of the principles of surface techniques

| | | Incident Radiation | | |
		Electrons	Ions	Neutrals	X-rays
Radiation detected	Electrons	AES LEED HREELS	INS		UPS XPS
	Ions	ESD	SIMS ISS RBS	FABMS	
	Neutrals		SNMS		
	X-rays	(EPMA)	(PIXE)		S-EXAFS

cesses occurring in the sample. Thus AES instruments with high spatial resolution can usefully have added X-ray detectors, as in the electron probe X-ray microanalyser (EPMA), but electron stimulated desorption (ESD)[46] must also be recognized. Since a measurement in AES requires a certain dose of electrons and the fraction of surface atoms removed in ESD concerns the dose per unit area, it is clear that ESD sets limits on spatial resolution in AES.[47] Some extra acronyms are given in Table 1.1, where those in brackets are not true surface analysis techniques but are added for completion. Particle-induced X-ray emission (PIXE)[48] uses MeV α-particles, or protons, to irradiate the sample and, by analysing the emitted X-rays gives p.p.m. detectability for elements with $Z > 12$ over depths comparable with the EPMA. Fast atom bombardment mass spectrometry (FABMS)[49] replaces the ion beam of SIMS with energetic neutrals and so overcomes any charging effects of insulators.

The more important of these techniques are presented in some detail in Table 1.2 where some of the less easily quantified aspects are presented. Cost is not included since the bulk of the cost is often not with the optics of the particular technique but with the specimen handling arrangements, computer analysis, laboratory overheads and social security payments. Ease of specimen preparation and an understanding of the measurement are generally more important. Ease of specimen preparation is the main limitation in the application of the atom probe field ion microscope[50] although, for micro-analysis at interfaces, it is doubtful if alternative techniques can approach within an order of magnitude of its spatial resolution. Sharp tips of material must, of course, be prepared; however, the range of possible materials is continually increasing.

Before concluding this chapter it should be noted that the surface analysis instrumentation field is moving very rapidly and within a year Table 1.2 is likely to be out of date in relation to spatial resolution and sensitivity. AES and XPS will continue to have the greatest popularity since they are both

Table 1.2 Survey of the more popular techniques for surface and interface analysis

Technique	Information (E = elemental, C = chemical)	Spatial resolution	Sampling depth monolayers	Sensitivity (order of)	Quantification (√ = easy)	Elements not covered	Popularity	Specimen preparation (√ = easy)	Ease of use	Extent of support data	Effective take-off year
AES	E*	0.2 µm	3	0.3 %	√	H,He	****		****	*****	1968
Atom probe FIM	E†	1 nm	1	1 %	√		*	√	*	**	1968
HREELS	C	1 mm	1	1 %			*	√	**	**	1970
ISS	E	1 mm	1	1 %	√	H,He	*	√	***	***	1967
RBS	E	1 mm	100	1 %	√	H,He	*	√	***	***	1967
SIMS (static)	C	1 mm	2	0.1 %			**	√	**	**	1970
SIMS (dynamic, imaging)	E	0.5 µm	40	1 p.p.m.	√‡		**	√	***	****	1968
SIMS (dynamic, depth prof)	E	50 µm	40	1 p.p.m.	√‡		**	√	***	***	1975
UPS	C	3 mm§	3	1 %		N/A	**	√	***	****	1969
XPS	C,E	0.2 mm§	3	0.3 %	√	H,He	****	√	****	*****	1967

* C is available, but not with high spatial resolution due to electron stimulated desorption effects.[45]
† C may generally be deduced.
‡ When compared with a close standard.
§ 1 µm resolution has been reported in a new instrument concept using an He UPS source and resolutions below 100 µm are calculated for XPS energies.[51]

self-contained techniques which are available for any laboratory and for which there has consequently developed an enormous store of expertise and data. Both are limited in the sense that XPS has poor spatial resolution and AES tends to be rather destructive to weakly bound species. Count rates are likely to increase as spectrometer design improves and as multi-channel detectors become more popular. This will allow higher spatial resolution in XPS and reduce the damage associated with AES. Spatial resolutions of 100 μm should be possible in XPS using this approach although something below this figure may be possible using the axial magnetic field photo-electron spectromicroscope of Beamson, Porter and Turner.[51] This latter instrument is of a revolutionary rather than evolutionary nature and makes use of trapped electron orbits in the eight tesla field of a superconducting magnet.

The histories of many of our surface analysis techniques are measured in decades and yet we see from Table 1.2 that the effective take-off date for many of the techniques, when they mushroomed into many laboratories, is bracketed by the years 1968 to 1970. This is not a coincidence but most probably occurred through the maturing of UHV technology and the ready availability of complex UHV systems in the middle and late 1960s, so catalysing a general awareness of surfaces and their importance. From those days, with the careful handling of special vacuum systems by people who 'knew' about UHV, we have come today to systems of greater reliability, where samples may be inserted and removed through airlocks in minutes rather than days, and which may be operated by people no longer skilled in solving vacuum problems but skilled in solving surface problems.

In the chapters that follow the reader will find first the details of instrumentation followed by the interpretation of spectra generated from solid surfaces by both AES and XPS. A major proportion of the applied work currently involves not just the outer surface but also the composition through reaction layers or coatings. The basic problems concerned with such depth profiling are discussed together with the status of the quantification in the spectra. This part of the book, together with the tabulations of data and technique in the voluminous appendices, provides the reader with all that should be needed to get the most from the techniques. The second part of the book, Chapters 6 to 10, illustrates the power of AES and XPS in the solution of problems in the major technological areas, as shown in Figure 1.2. These chapters show how the different aspects of the techniques, discussed in the earlier part of the book, may be turned to particular advantage in the solution of important problems.

References

1. M. P. Seah, *Surface and Interface Anal.*, **2**, 222 (1980).
2. C. Davisson and L. H. Germer, *Phys. Rev.*, **30**, 705 (1927).

3. H. E. Farnsworth, *Phys. Rev.*, **33**, 1068 (1929).
4. E. J. Scheibner, L. H. Germer and C. D. Hartman, *Rev. Sci. Ins.*, **31**, 112 (1960).
5. W. Ehrenberg, *Phil. Mag.*, **18**, 878 (1934).
6. L. A. Harris, *J. Appl. Phys.*, **39**, 1419 (1968).
7. L. A. Harris, *J. Appl. Phys.*, **39**, 1428 (1968).
8. J. J. Lander, *Phys. Rev.*, **91**, 1382 (1953).
9. P. Auger, *J. Phys. Radium*, **6**, 205 (1925).
10. H. R. Robinson and A. M. Cassie, *Proc. Roy. Soc.*, **A113**, 282 (1926), and many later papers.
11. R. E. Weber and W. T. Peria, *J. Appl. Phys.*, **38**, 4355 (1967).
12. L. A. Harris, *J. Vac. Sci. Technol.*, **11**, 23 (1974).
13. P. W. Palmberg and T. N. Rhodin, *J. Appl. Phys.*, **39**, 2425 (1968).
14. P. W. Palmberg, G. K. Bohm and J. C. Tracy, *Appl. Phys. Lett.*, **15**, 254 (1969).
15. C. Webb and P. M. Williams, *Surface Sci.*, **53**, 110 (1975).
16. R. L. Gerlach, *J. Vac. Sci. Technol.*, **8**, 599 (1971).
17: R. D. Leapman, S. J. Sanderson and M. J. Whelan, *Metal Sci.*, **12**, 215 (1978).
18. H. Ibach and S. Lehwald, *J. Vac. Sci. Technol.*, **15**, 407 (1978).
19. H. Ibach and S. Lehwald, *Surface Sci.*, **76**, 1 (1978).
20. J. G. Jenkin, R. C. G. Leckey and J. Liesegang, *J. Electron Spectrosc.*, **12**, 1 (1977).
21. J. G. Jenkin, J. D. Riley, J. Liesegang and R. C. G. Leckey, *J. Electron Spectrosc.*, **14**, 477 (1978).
22. J. G. Jenkin, J. Liesegang, R. C. G. Leckey and J. D. Riley, *J. Electron Spectrosc.*, **15**, 307 (1979).
23. J. G. Jenkin, *J. Electron Spectrosc.*, **23**, 187 (1981).
24. P. D. Innes, *Proc. Roy. Soc.*, **A79**, 442 (1907).
25. E. Rutherford, *Phil. Mag.*, **28**, 305, (1914).
26. R. G. Steinhardt, *Anal. Chem.*, **23**, 1585 (1951).
27. K. Siegbahn, personal communication.
28. K. Siegbahn, C. N. Nordling, A. Fahlman, R. Nordberg, K. Hamrin, J. Hedman, G. Johansson, T. Bermark, S. E. Karlsson, I. Lindgren and B. Lindberg, *ESCA: Atomic, Molecular and Solid State Structure Studied by Means of Electron Spectroscopy*, Almqvist and Wiksells, Uppsala (1967).
29. C. R. Brundle and M. W. Roberts, *Proc. Roy. Soc.*, **A331**, 383 (1972).
30. M. P. Seah, *Surf. Sci.*, **32**, 703 (1972).
31. P. M. Williams, Chapter 9 in *Handbook of X-ray and Ultraviolet Photoelectron Spectroscopy* (Ed. D. Briggs), p. 313, Heyden and Son, London (1977).
32. W. E. Spicer, in *Electron and Ion Spectroscopy of Solids* (Eds L. Fiermans, J. Vennick and W. Dekeyser), p. 34, Plenum, New York (1978).
33. T. A. Carlson, *Surface and Interface Anal.*, **4**, 125 (1982).
34. D. C. Komingsberger and R. Prins, *Trends. Anal. Chem.*, **1**, 16 (1981).
35. A. Bianconi, *Appn. Surf. Sci.*, **6**, 392 (1980).
36. A. Benninghoven, *Z. Physik*, **230**, 403 (1970).
37. H. W. Werner, *Surface and Interface Anal.*, **2**, 56 (1980).
38. W. L. Baun, *Surface and Interface Anal.*, **3**, 243 (1981).
39. E. N. Haeussler, *Surface and Interface Anal.*, **2**, 134 (1980).
40. H. Oechsner and E. Stumpe, *Appl. Phys.*, **14**, 43 (1977).
41. I. V. Mitchell, *Phys. Bull.*, **30**, 23 (1979).
42. H. D. Hagstrum, in *Electron and Ion Spectroscopy of Solids* (Eds L. Fiermans, J. Vennick and W. Dekeyser), p. 273, Plenum, New York (1978).
43. J. W. Coburn and E. Kay, *CRC Critical Revs.*, in *Solid State Sciences*, **4**, 561 (1974).

44. A. W. Czanderna (Ed.), *Methods of Surface Analysis*, Elsevier, New York (1975).
45. R. E. Honig, *Thin Solid Films*, **31**, 89 (1975).
46. C. G. Pantano and T. E. Madey, *Appn. Surface Sci.*, **7**, 115 (1981).
47. M. P. Seah, in *Surface Analysis of High Temperature Materials—Chemistry and Topography* (Ed. G. Kemeny), Applied Science (to be published).
48. J. L. Campbell, B. H. Orr, A. W. Herman, L. A. McNelles, J. A. Thomson and W. B. Cook, *Anal. Chem.*, **47**, 1542 (1975).
49. D. J. Surman, J. A. Van der Berg and J. C. Vickerman, *Surf. Interface. Anal.*, **4**, 160 (1982).
50. M. K. Miller and G. D. W. Smith, *Metal Sci.*, **11**, 249 (1977).
51. G. Beamson, H. Q. Porter and D. W. Turner, *Nature*, **290**, 556 (1981).

Practical Surface Analysis
by Auger and X-ray Photoelectron Spectroscopy
Edited by D. Briggs and M. P. Seah
© 1983, John Wiley & Sons, Ltd.

Chapter 2

Instrumentation

J. C. Rivière

UK Atomic Energy Research Establishment, Harwell, Didcot, Oxfordshire

2.1   Introduction

Without the means of measuring and recording electron energy distributions, electron spectroscopy would be merely speculation. The spectrometer is central to the whole subject, and it is therefore worth discussing spectrometers in some detail because of their importance. In this context the term 'spectrometer' is taken to include all aspects of an instrument that have any bearing on the process of the collection of electron spectroscopic data. Thus not only must the design and performance of the electron energy analyser itself be discussed, but also the design of electron and of X-ray photon sources, the nature of the ambient required in a spectrometer and how it is to be achieved, the ways in which a specimen for analysis is handled, i.e. transferred or manipulated, within the spectrometer and the methods of specimen surface preparation and treatment. It is the purpose of this chapter to describe current state of the art and practice in modern spectrometers used for XPS and AES in a way that will allow their operation and principles to be understood, and that will also provide some useful working relationships.

2.2   Vacuum Conditions

There are two reasons why electron spectrometers used in surface analysis must operate under vacuum. In the first place, electrons emitted from a specimen should meet as few gas molecules as possible on their way to the analyser so that they are not scattered and thereby lost from the analysis. Another way of expressing this requirement is that their mean free paths should be much greater than the dimensions of the spectrometer. This by itself does not impose much stringency on the working vacuum to be achieved, since vacua in the range 10^{-5}–10^{-6} torr would be adequate. However, the second reason *does* impose stringent requirements on the working vacuum. It has been established[1] in Chapter 5 (see Figure 5.5, in which the observed relationship between electron kinetic energy and inelastic mean free path is plotted) that both XPS and AES are highly surface-specific techniques, with sampling depths typically of a few atom layers. Most of the electrons whose energies are analysed originate in fact in the first one or two atom layers. Such surface specificity, combined with elemental sensitivities of the order of 0.5 per cent. of an atom layer, means that the techniques are very sensitive to surface contamination, from whatever source. Since in many experiments it is necessary to start with a well-characterized surface, either atomically clean or in some other stable condition, and since even very small amounts of contaminant can affect the course of an experiment drastically, it is clearly necessary to operate under conditions in which the rate of accumulation of contamination is negligible compared to the rate of change in the experiment. The principal source of contamination is from the residual gas in the vacuum system. Any textbook[2] on gas kinetic theory will provide the information that, to a good approximation, a monolayer of gas will accumulate on a surface in 1.5 s at a pressure of 10^{-6} torr and at room temperature, if every molecule hitting the surface stays there on impact (a sticking probability of unity). If one takes as typical requirements that no more than 0.05 atom layers of contaminant should accumulate in the space of 30 min, then, again for unity sticking probability, gas kinetics rule that the residual gas pressure should be 4×10^{-11} torr. In practice, sticking probabilities are not often quite as high as unity, and for the great majority of surface experiments it is found that a base pressure of 10^{-10} torr is adequate. Such a pressure falls into the regime of ultra-high vacuum (UHV).

2.2.1 UHV systems

2.2.1.1 *Materials*

Having established that XPS and AES should be operated at pressures in the UHV region, it is necessary to discuss briefly what implications that has for the materials of construction of spectrometers. Returning to gas kinetics again, but without going into detail, it is an easy matter to show that for practicable dimensions of chambers, pipework and pump, the ultimate vacuum attainable is always governed by the rate at which adsorbed gases leave the internal surfaces at ambient temperature. This outgassing rate, as it is called, varies considerably according to the chemical nature of the surface and its pre-treatment or history and according to the nature of the adsorbed gas. However, even in the best case, i.e. the lowest outgassing rate, if the temperature of the internal surfaces never goes above ambient, then UHV in the spectrometer will not be achieved, or at least not for a very long time indeed. In order to reach UHV in a reasonable time it is necessary to remove the adsorbed gases at a much faster rate than normal, and this is accomplished by raising the temperature of the spectrometer, or baking it, to about 200 °C for a few hours. On cooling to ambient temperature again, the outgassing rate from the internal surfaces drops by several orders of magnitude and the pumps are then capable of producing and maintaining UHV.

The necessity of baking imposes restrictions on the materials that can be used in a spectrometer. Nothing must be used in construction that might disintegrate, outgas excessively or lose strength at the elevated temperature, and this applies throughout the life of the spectrometer, since bake-out is repeated at regular intervals. For these reasons it is not permissible, for example, to use elastomer materials such as Viton in seals between UHV and atmosphere since not only can elastomers oxidize and become brittle but at the baking temperature they also become porous and allow gases to pass through by diffusion. Thus the seals used in the joints between the components of the spectrometer must be of metal. The bake-out temperature precludes the use of indium (MP 156 °C), which is otherwise a useful sealing material, and in fact the two metals most commonly used for this purpose are gold and copper, with the emphasis heavily on the latter. Gold is used in the form of annealed wire compressed between flat flanges, but such seals, although completely reliable in use, suffer from the defects that it is difficult to position the wire correctly on a vertical flange, that the gold cannot always be removed from a flange after compression and that gold is expensive. The type of seal now employed almost universally is shown in Figure 2.1. A flat, annular, copper gasket fits exactly into the recess formed between two identical flanges when they are brought together. Each flange has a circular tapered knife-edge machined in its face. When the flanges are forced together by tightening the nuts on bolts placed through the bolt holes, the opposing knife-edges, which are of identical diameter, bite into the copper gasket and force the copper to flow away from them on each side. On the inside diameter there is no restriction, but on the outside the gasket is prevented from distorting by the walls of the recess into which it fits. Thus very large forces build up in the copper in that region and a very effective seal is formed between the gasket and the flange material on the outside of the knife-edge itself, capable of withstanding innumerable temperature cycles during bake-out.

The point of bringing in the above discussion about vacuum seals here is that it will be obvious by now that flanges of the type shown in Figure 2.1 (called, variously, 'knife-edge', 'copper-seal' or 'Conflat*' flanges) must be fabricated from a metal that is hard, retains its hardness indefinitely at baking temperatures and does not oxidize readily. Thus aluminium or its alloys, which are otherwise useful materials in UHV, cannot be used for UHV vacuum vessel construction since they are too soft for knife-edge flanges. As a result, the material used in the fabrication of the great majority of UHV

*Trade name of Varian Associates Ltd.

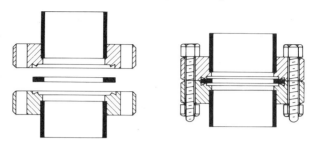

Figure 2.1   Knife-edge metal-to-metal UHV seal using an annular copper gasket. The left-hand diagram shows two identical flanges with circular knife-edges, and the gasket between them, before the seal is made. On the right the flanges have been clamped together and the gasket has been compressed against the inner walls of the recesses on the flanges by the action of the opposing knife-edges. The seal is made on the outer (sloping) surface of the knife-edge. This type of seal is also known as the Conflat* seal

vacuum vessels is stainless steel. The only exception occurs where magnetic screening is required, such as around energy analysers and in the vicinity of the specimen, in which case some manufacturers use mu-metal instead of stainless steel for the material of the vessel wall.

Although, as pointed out, the elastomer Viton should not be used between UHV and atmosphere, it is safe to use entirely within UHV if certain precautions are taken. These are that Viton rings should be de-greased and then pre-outgassed in an auxiliary vacuum system before insertion in the spectrometer and that those rings used in gate valve seats should never undergo bake-out with the valve in the closed position. Compressed Viton will take up a permanent set if baked repeatedly in that situation, leading to failure of the valve to close properly.

In addition to stainless steel and copper, there are several other materials that can safely be used between UHV and atmosphere, not as the principal constructional material but in components attached to the spectrometer for a variety of reasons. Before the advent of stainless steel vacuum systems, borosilicate glass was the constructional medium, and it is still used extensively in the form of windows mounted on flanges via graded glass-to-metal fused seals. For certain experiments quartz windows are also available. The other ceramic used widely is of course alumina; it is used in all places where electrical isolation is necessary to take current or voltage connections into the spectrometer.

Inside the spectrometer any material is permissible that does not cause an increase in the background pressure at either ambient or baking temperatures and that does not contain any constituents volatile at those temperatures that might settle on the inner surfaces and contaminate them. For instance, brass which contains a high proportion of volatile zinc must not be used, nor must any of the common plastic polymeric materials. However, it is permissible to use PTFE-covered wire for instrumental connections in UHV if good quality material is used. The commonly used metals for small-scale fabrication in UHV, e.g. for special sample mounts, sample heating, electrical connections, etc., are copper, nickel, platinum, molybdenum, tantalum and tungsten, and most workers in electron spectroscopy would reckon to have a stock of at least some of those metals in the forms of wire and foil kept near the spectrometer.

2.2.1.2 *Design and construction*

These two aspects are interdependent; the design of any component, large or small, for UHV cannot be carried out without knowing exactly what the constructional constraints and requirements are. Constraints that have to be borne in mind continually are, for example, the relationship of flange positions to each other and to the external surface of a vacuum vessel, so that there is always access provided for the insertion and tightening of bolts and the allowance of adequate space both for weld preparation and for the welding operation itself. Normally such matters concern the manufacturer rather than the scientist, but if one wishes to design a special vacuum vessel oneself, or modify an existing one, then they should not be overlooked.

Another important part of vacuum system design, which applies not just to UHV, is the provision of adequate pumping speed at *all* points in the vessels to be evacuated. It is unfortunately still a not uncommon fault, even amongst reputable manufacturers who should know better, to design a perfectly good vacuum system for a spectrometer, good that is when the vessels are empty or carry no attachments, and then to place inside or attach to the vessels devices of such sizes and shapes that clearances are reduced to the point where the pumping speeds around the devices are severely restricted. This is the principal reason why the achievement of UHV by bake-out can take so long in some spectrometers. The relative arrangement of components inside the vacuum should be considered at the design stage not only from the electron spectroscopic point of view but also from the vacuum standpoint as well, and that applies with equal force to attachments that have to be accommodated on side ports. If necessary, to improve the pumping speeds in the latter case, additional by-pass pumping lines to the main pumps should be provided, or even auxiliary pumps. The estimation of conductances, and therefore of local pumping speeds, to adequate accuracy can be made from the various well-known formulae based on gas kinetics in the pressure region of molecular flow. These can be found in any good textbook[2] on vacuum physics and will not be repeated here.

Fabrication of vessels and other components in stainless steel involves inert gas arc welding, a technique which is now standard. For UHV, however, there are certain practices in such welding which it is essential to follow otherwise the achievement of UHV will be difficult or impossible. These practices are aimed at avoiding trapped volumes with low-conductance paths to the vacuum side of the vessel wall; clearly such volumes could contain dirt or gases that would give rise to a constant source of contamination, i.e. act as 'virtual leaks'. To avoid their formation it is essential to specify that all joint welds must be internal *and* continuous. That is, enough filler rod must be on hand from the start to ensure that once the welding of, for example, a circular joint has begun it is continued without a break until the starting point is reached again. Only in this way can trapped volumes be avoided. Ideally, all such welding would be done by an electron beam in vacuum, but the latter method is slow and not suited to complex geometries such as are frequently encountered on the inside of UHV vessels.

The last operation to be carried out in the construction procedure is that of finishing and then cleaning the internal surfaces, and the effectiveness of this operation will govern the subsequent vacuum performance of the vessel or component. Nowadays the preferred finishing treatment is blasting with tiny glass beads, which not only removes scale, etc., from welds and surrounding areas but passivates the surface by introducing cold work into the surface region. in addition the topography of the surface is smoothed on a microscopic scale, so that potential traps for contamination are removed. Finishing by electropolishing, which was used widely not so long ago, is to be avoided at all costs, since the surface is physically removed to produce the polish and the weld regions are particularly susceptible to attack, allowing crevices and small holes to be uncovered. After glass-blasting it is essential to make sure that all the glass ball debris is removed, since if any remains it will eventually find its way into moving parts, with disastrous results.

De-greasing in clean pure solvent after finishing will produce a surface condition that would be usable without further treatment for high vacuum applications, but for UHV a final step is necessary, and that is pre-bake-out. By that is meant the baking of individual components on an auxiliary vacuum system, ideally to temperatures above those to be used during bake-out of the assembled spectrometer, in order to remove traces of the de-greasing solvent left on the internal surface. Following this final step, the components should be wrapped in clean plastic sheet or put into plastic bags, to remain clean until required for assembly. Needless to say, at no stage following glass-blasting is any component handled with bare hands; use of gloves is essential. In addition it is highly desirable that the pre-bake-out operation, the storage before use, and indeed the assembly, be carried out in dust-free enclosures, i.e. with filtered atmospheres.

The foregoing points of design and construction would not normally be ones that would be used by the individual research worker, since there are now very few laboratories that make their own vacuum equipment. Nor is it possible in the space available here to go into all the many other minor details of instrument design that have to be taken into account. However, in assessing the comparative standards of fabrication of potential instrument suppliers, the practices described above form a useful general yardstick as to the likely performance and reliability.

2.2.1.3 *Vacuum pumps*

Only UHV pumps will be discussed under this heading, and then only briefly with regard to their respective advantages and disadvantages, since descriptions of their modes of operation can be found in many other places.

The four types of pump found in various combinations on commercially produced spectrometers are diffusion, sputter ion, turbomolecular and titanium sublimation, and it seems that each instrument manufacturer has his own preference. About the only area of agreement between manufacturers seems to be in the use of titanium sublimation pumps as auxiliary and not main pumps for UHV.

Diffusion pumps can be used over a wide range of required ultimate vacua, according to the type of oil used; for UHV they need oil of a vapour pressure less than 10^{-9} torr, of high resistance to degradation leading to volatile products and of reluctance to creep over surfaces. The polyphenyl ethers have been used with success for some years, but other oils with improved characteristics are now becoming available. Although diffusion pumps themselves are relatively cheap, they need efficient liquid nitrogen traps situated between them and UHV, and such traps in fact cost more than the pumps. When the costs of liquid nitrogen, of cooling water and of power over a long period are also taken into account, the cheapness is only apparent. Against this is the major advantage of diffusion pumps compared to those below—that they are well behaved and not temperamental. They are prepared to pump almost any gas that is not reactive towards the hot oil and will work for very long periods without needing attention if they are not abused.

Sputter ion pumps have many attractions. They do not use fluids, do not need cooling water or liquid nitrogen, nor indeed much power, and can be connected directly to the vessel to be evacuated without any intervening traps. They can be brought into operation at the throw of a switch rather than having a long warm-up time and can be switched off again just as easily. Although inherently expensive, since a control unit is needed with them, no additional components are necessary and there are no cooling costs. It is likely that over a long period of time they are cheaper to run than diffusion pumps. In general use, however, as the main UHV pump these advantages are more than outweighed by their disadvantages. Principally, they are fussy about what they will pump, and how much of it. As long as the gas load consists of the normal residual gas constituents, such as nitrogen, oxygen and carbon dioxide, there are no problems. With the modern improved designs they will even pump noble gases with reasonable speeds, provided the noble gas pressure is kept below 10^{-6} torr and the pump is not exposed to it for long continuous periods. However, ion-pumping of helium is to be avoided, since helium diffuses into the titanium cathode and can cause cracking of the cathode if too great a volume of helium becomes incorporated. The same applies to hydrogen, with the additional complication that titanium forms a series of solid solutions and compounds with hydrogen, so that the end result of prolonged pumping of hydrogen could be disintegration of the cathodes. Furthermore, since water vapour is pumped by dissociation of the water molecule into hydrogen and hydroxyl ion in the pump discharge region, followed by reaction of the products with the titanium, excessive pumping of high pressures of water vapour will have the same eventual effect as pumping hydrogen. Another drawback of ion pumps is the memory effect. For example, since, as stated above, both helium and hydrogen are pumped by dissolution in the titanium, it follows that subsequent sputtering of the helium- or hydrogen-filled titanium will cause re-evolution of those gases. The problem can be overcome to a certain extent by a high temperature bake-out ($\sim500°C$) of the pump, but that operation is inconvenient, to say the least. In addition, although ion pumps will pump hydrocarbons effectively, the carbon remains associated with the titanium and can recombine with other gases to introduce contamination. Thus on an ion-pumped system there will always be molecules containing one and two, and often three, carbon atoms in the residual gas spectrum; methane in particular can be a principal component, along with hydrogen, at base pressures of less than 10^{-9} torr. For the same reason, admission of pure oxygen to an ion pump will cause an increase in the carbon monoxide partial pressure to the point where it interferes seriously with the course of an experiment. In summary, then, ion pumps are suitable, and highly convenient, for pumping atmospheric constituents, but cause many problems when their application differs from that.

Turbomolecular pumps are also attractive from the point of view of convenience. A turbomolecular pump is the only type of pump that can in principle take a vacuum system from atmospheric pressure to UHV (with bake-out, obviously), although in practice it is never used in that way since to do so regularly would impose too much strain on it. It will pump any gas, but the pumping efficiency decreases directly with the molecular weight of the gas, so that the efficiency for hydrogen and helium is not good; the residual gas at UHV will therefore be mainly hydrogen. The only limitations on the nature of the gas pumped will be either possible reaction with the lubricating oil on the low vacuum side or possible corrosive effects on the rotor blades. In normal practice neither of these possibilities causes any problems. Turbomolecular pumps do not need traps or baffles like diffusion pumps and, like ion pumps, are probably cheaper to run over a long period than diffusion pumps. There is some lingering doubt as to whether they are capable *on their own* of achieving pressures of less than 10^{-10} torr since they tend always to be used in conjunction with titanium sublimation pumps, but then the same can be said of diffusion pumps too. In fact most manufacturers will not guarantee ultimate vacua of 10^{-10} torr or better without the aid of a sublimation pump, whatever the type of principal pump used. Nowadays turbomolecular pumps run very quietly, one of the original objections being noise, and it seems that the only reasons that they are not used more often on spectrometers are the possibilities of vibration and an inherent distrust of their reliability, again based on past experience.

Titanium sublimation pumps are by far the simplest and cheapest of the UHV pumps and are used very widely indeed, but, as remarked earlier, never as the principal pump on a spectrometer. They can be, and are, operated at pressures as high as 10^{-6} torr, but in electron spectrometers and other UHV systems their main contribution is to achieve the last order of magnitude or so decrease in pressure necessary if ultimate vacua of 10^{-10} torr and below are to be reached. For maximum pumping speed the titanium should be evaporated onto a liquid nitrogen cooled surface, but no standard spectrometer offers such a facility, and indeed much of the benefit of their ease of operation would disappear if a liquid nitrogen feed had to be provided. At the least, however, they ought to be water cooled, but even that is unfortunately not as common as it should be. The first operation of a titanium filament for sublimation has to be carried out carefully, since considerable release of gas occurs on initial heating, but once outgassing is complete the heating current can be switched straight to the maximum on subsequent operations. Normally the pumping of a sublimation pump is timed by the control unit, so that the titanium is not used wastefully; as a rough guide the pump need be operated only every few hours for two minutes at a time, at pressures around 5×10^{-10} torr. Since sublimation pumps operate by 'gettering', i.e. by reaction of the evaporated titanium with reactive gases, they are excellent pumps for the major atmospheric constituents and for hydrogen and hydrocarbons, but do not pump noble gases at all. Furthermore, since the titanium film is not subsequently disturbed, as the cathode is in an ion pump, there are very few memory effects.

Based on the above remarks, the most trouble-free pumping systems would consist of diffusion or turbomolecular pumps acting in parallel with titanium sublimation pumps. Ion pumps are not recommended as the principal pumps, for the reasons given, although they should be used as appendage pumps for pumping small subsidiary volumes to which noble gases will not be admitted or occasionally as 'holding' pumps in the event of failure or maintenance shut-down of the principal pumps.

2.3 Sample Handling

Under this heading will be discussed all the steps necessary in the process of offering a sample to the energy analyser for analysis, from preparation of the sample outside the spectrometer to insertion in, and positioning inside, the spectrometer, and to further preparation *in vacuo* that might be required.

2.3.1 Sample preparation

Most of the preparation that occurs before insertion in the spectrometer is devoted to mounting a sample on a probe or a manipulator, or on a stub, in such a way that the surface to be studied is presented for analysis with minimal contamination. The wearing of gloves to avoid direct skin contact with the sample or any other component to go into UHV must therefore be routine. Similarly, all tools used to help with sample preparation must be de-greased, handled only with gloves and reserved for the purpose in a clean area. Ideally the clean area should be a separate room or cubicle into which the air supply is filtered, but few laboratories have space to spare for that, and it is generally adequate to use an enclosed bench or a glove box for the purpose, with a positive flow of filtered air through it.

Methods of sample mounting vary from sample to sample and from one laboratory to another and are often the subject of considerable ingenuity. Where samples are to be analysed in a routine fashion at room temperature, i.e. do not require any special heating or cooling arrangements, modern spectrometers provide standard sample carriers, whose design varies with the manufacturer, but all of which have two things in common. Firstly, they have features such as recesses or grooves or spring clips that allow them to be transferred from the insertion system to the manipulator and, secondly, they have a flat area on which the sample is actually placed. The ingenuity arises in the ways in which awkwardly shaped samples can be fixed to this area in such a way that the materials used for fixing do not interfere with analysis of the sample surface of interest. Flat samples are always the easiest; if the reverse side is not of interest, they can be stuck to the carrier with a blob of methanol dag (*not* water dag), which dries rapidly and holds them firmly even after drying. Otherwise they can be held by being slipped under small spring clips of beryllium copper or some other springy metal fastened to the carrier.

The mounting of irregularly shaped samples is a function of individual sample geometry. In some cases the sample can be held sufficiently firmly against the carrier with a simple strap of some malleable metal such as nickel which can mould itself to the sample shape; the ends of the strap can either be spot-welded to the carrier or fixed under screw heads. In other cases it may be necessary to wrap the sample in metal foil, often of platinum, leaving the surface of interest exposed; the foil wrap acts as a cup in which the sample is held, the ends of the foil again being either spot-welded or held under screwheads for fixing to the carrier. In whichever way it is mounted, care has always to be taken that the mounting material does not intrude into the analysed area or, if that is unavoidable, that the photo-electron spectrum of the mount does not interfere with that expected from the sample.

Powders pose some special problems. If they cannot be, or should not be, compressed or compacted, they can be placed loose in a recess either fixed to a carrier or part of a modified carrier, but there are obvious risks in so doing. Initial evacuation from atmospheric pressure, if carried out too quickly, will distribute much of the powder around the vacuum system; the powder is always likely to spill as the carrier is moved or transferred from one part of a spectrometer to another; and in some spectrometer configurations the analysed surface must be inclined to the horizontal at an angle of at least 30°, which must again lead to spillage into the system. A recommended way of 'fixing' a powder for analysis is to shake some of it over piece of indium foil to approximately uniform coverage and then to press gently down on the powder layer with a piece of clean, hard, metal such as molybdenum or tungsten. Enough of the powder then becomes embedded in the surface of the indium to give a continuous surface for analysis, and the excess powder can be shaken off. This method has the additional advantage that the proximity of the indium reduces surface charging effects, which can be severe in the analysis of loose powders. Another popular method of mounting powders is to use double-sided adhesive tape, but this is not recommended for AES.

In the more complicated situation where the sample has to be heated or cooled, either before or during analysis, the mounting of the sample needs greater precision. Generally, the experimental programmes involving such analyses will be of longer term than those in which samples are transferred on carriers, and samples are therefore mounted individually and directly on the manipulator rather than being transferred to it. Most manufacturers market either special probes or special manipulator attachments, for use with their spectrometers when heating or cooling is required, and an example of a combined stage is shown in Figure 2.2. Heating is provided indirectly by an insulated hot filament under the substrate supporting the sample and cooling by passing liquid nitrogen into a tank to which the substrate is connected by a thick copper braid. In general it is advisable to use a probe or attachment dedicated to one or the other, since the dual-purpose probe has inherent limitations on the temperatures it can achieve. Since the actual temperature reached by a sample during either heating or cooling depends very much on the extent of thermal contact of the sample and the substrate, it is always advisable, if it can be done without damage to the sample, to attach a thermocouple directly to it rather than rely on a thermocouple attached to the substrate. Alternatively, a dummy sample can be used at first, so that a calibration of the substrate temperature can be obtained in terms of the actual sample temperature.

One of the worrying aspects of the above method of sample heating, i.e. by conduction from a heated substrate, is that impurities present either on or in the substrate can become mobile at high temperature and contaminate the sample. Such an occurrence would clearly lead to spurious experimental results. It is therefore necessary, when the heating stage is first used, to run it

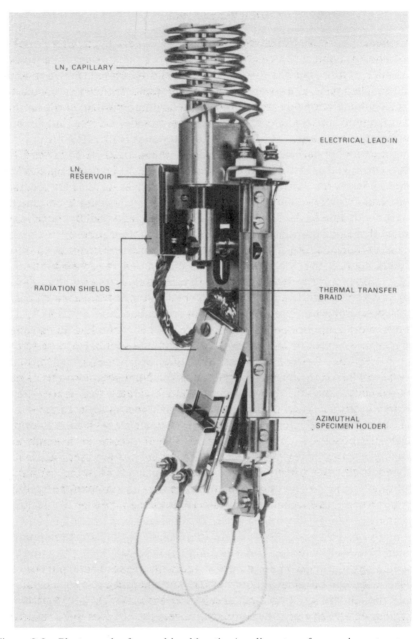

LN₂ CAPILLARY

ELECTRICAL LEAD-IN

LN₂ RESERVOIR

RADIATION SHIELDS

THERMAL TRANSFER BRAID

AZIMUTHAL SPECIMEN HOLDER

Figure 2.2   Photograph of a combined heating/cooling stage for specimen treatment. Heating is provided by conduction from an insulated hot filament under the substrate surface on which the specimen is mounted. Cooling is achieved by conduction along a copper braid fixed to a tank through which liquid nitrogen is passed. The stage is mounted on the shaft of a universal manipulator (see Figure 2.9). (Reproduced by permission of Vacuum Generators Ltd, Hastings)

for prolonged periods at the intended temperatures without any sample being present and to monitor the cleanliness of the substrate surface. If any contaminating species appears and persists, it should be removed by ion bombardment. The cycles of heating and ion bombardment should be continued until the contamination does not reappear on heating. In other words, during initial operation the substrate should be treated as if it were a sample, and cleaned in the same way.

In basic experimental work involving very pure materials, often in the form of single crystals, the above method of heating the sample is not regarded as adequate. There must be no possibility of contamination from any supporting material and the sample must be heated uniformly. According to the shape, thickness and nature of the sample, heating may be carried out by the passage of current through the sample, by conduction from heated supports, by radiation from an adjacent hot filament or by electron bombardment from a hot filament. Generally speaking, the supporting materials should either be the same and of the same purity as that of the sample or of a refractory metal such as tungsten or molybdenum that can be cleaned at a high temperature before being used as a support. For measurement of temperatures up to ~850 °C it is necessary to use a thermocouple either spot-welded very carefully to one edge of the sample or, if that is unacceptable, attached to a support as close to the sample as possible. Noble metal thermocouples, e.g. Pt–Pt/10% Rh, must always be preferred to the more common NiCr–NiAl thermocouples since the risk of contamination from the latter is much greater. For temperatures above 850 °C an optical pyrometer should be used, since at high temperatures there is an increased likelihood of reaction of the sample with thermocouple material; of course, it should be remembered that pyrometer readings are always subject to correction due to sample emissivity and to absorption in the glass window through which the sample is observed. With some mounting arrangements it may be possible to measure the sample temperature by using the temperature dependence of the resistance of wires supporting the sample.

2.3.2 Sample insertion

When a sample has had to be mounted individually on a manipulator, as discussed above, then clearly its insertion into the spectrometer involves bringing part of the spectrometer, usually the analysis chamber, to atmospheric pressure and bolting the manipulator directly to the chamber. This is of course a time-consuming procedure, since the chamber then has to be re-evacuated and the whole spectrometer baked in order to return to UHV. However, if the sample is one that is to be studied for a long period, then the time spent in regaining UHV is short in proportion. Until a few years ago this procedure was universal since no specimen transfer systems were available, which meant that for routine analyses at UHV sample turn-around was very slow. Some improvement could be achieved in routine analysis by multiple-sample mounting, e.g. on a carousel which could be rotated via the rotary drive on the manipulator, but in general there was little flexibility in sample queueing for analysis.

One of the principal directions in which the sample turn-around has been greatly improved in recent years has been that of the method of inserting and transferring samples from ambient to UHV. This method at present takes two different forms, depending on the individual manufacturer's philosophy. In one form the carrier holding the sample is placed at the end of a long shaft, or probe, with a very highly polished cylindrical surface. The shaft slides, entirely grease-free, in a series of close-fitting sealing rings of either Viton or Teflon*, depending on the manufacturer, from ambient through a fore-chamber into the UHV chamber. The fore-chamber is pumped to at least 10^{-6} torr before the gate valve to UHV opens. Once inside the UHV chamber, the shaft is moved forward in a controlled manner until the carrier docks with, and locks into, an empty position on the sample carousel, when the shaft can be retracted. Removal of a sample simply involves the reverse procedure.

The advantages of the shaft method of sample insertion are that it is fast, that it can be made semi-automatic and that only one transfer operation is necessary. The disadvantage is that part of the shaft surface intruding into the UHV chamber has arrived there directly from the outside ambient atmosphere without any intervening bake-out or outgassing apart from initial pump-down in the fore-chamber. This, of course, is always true of the sample itself, and of its carrier, but their total surface area is much less than that of the exposed shaft surface, and in any case the carriers are subject to bake-out at some stage in their use. It is usual, then, during sample insertion by shaft for the pressure in the UHV chamber not only to rise out of the UHV region but to take some time to drop below 10^{-9} torr again; in fact, after a series of sample insertions and removals in quick succession, it may be impossible to regain 10^{-9} torr without bake-out. This method, then, is suited to those applications which require routine analysis of relatively inert surfaces and in which the maintenance of pressures in the analysis chamber near 10^{-10} torr at all times is sacrificed to the speed of sample turn-around.

The other form of specimen insertion could be termed the successive transfer method. In it, the carrier with the sample is placed on a holder in a small fore-chamber that is then evacuated from atmospheric pressure to $\sim 5 \times 10^{-3}$ torr with a trapped rotary pump. A gate valve between the fore-chamber and a second, or preparation, chamber is opened and the carrier is moved into the second chamber on a trolley enclosed entirely within the vacuum. After lifting the carrier onto another trolley in the preparation chamber, by manual operation via a bellows-sealed fork, the first trolley is returned to the fore-chamber

---

*Registered trade name.

and the gate valve is closed. Since the preparation chamber has a much greater volume than that of the fore-chamber and is pumped by a large diffusion pump, its pressure returns quickly to $\sim 10^{-9}$ torr. A second gate valve between the preparation chamber and the analysis chamber is then opened, the carrier moved through into the analysis chamber on the second trolley and transferred to the manipulator, again by manual operation. Once the transfer is complete, the trolley is retracted and the second gate valve closed. Since the pressure in the preparation chamber during transfer is $< 10^{-9}$ torr, the UHV in the analysis chamber is hardly disturbed.

Clearly the above method is slower than the shaft insertion method, there is a not insignificant risk of dropping a carrier during transfer and it is difficult to see how it could be automated in any way. The overriding advantage from the experimental point of view is that the pressure in the analysis chamber always remains near the 10^{-10} torr achieved by bake-out, etc. If the sample itself were found to be a source of gas for any reason, it could be parked in the preparation chamber either until the outgassing had stopped or until it had been decided that it was unsuitable for analysis.

Not infrequently it is necessary to study samples that should not be exposed to atmospheric ambient at any stage, perhaps because the material of the sample itself would react too vigorously with the atmospheric constituent gases, or because there might be surface films of interest on the sample that would be destroyed or altered by reaction with the ambient, or even that the sample carries some chemical or radioactive contamination that should not be allowed to be dispersed. Various methods have been devised to minimize the exposure to the ambient during insertion of a sample into a spectrometer. The simplest consists merely of a large plastic bag surrounding either the insertion shaft or the entrance to the fore-chamber, as the case may be, in which is maintained a positive pressure of an inert gas such as argon. Sample mounting then takes place inside the bag by placing gloved hands through holes small enough to prevent back-streaming of atmospheric gases. A more elaborate, and more effective, version of this method is a glove box built to fit the end of the spectrometer into which insertion takes place, with proper sealed ports and sealed gloves. Again a positive pressure is maintained inside the box. Such a device has the advantages that many sensitive samples can be stored in it for future reference and that visibility of manipulation is very good.

More elaborate still for the protection of sensitive samples in transit to the spectrometer are the purpose-built transfer flasks, designed to pass through the standard port on a glove box and with integral flap valves. An example is shown in Figure 2.3. Inside the glove box, in an inert gas atmosphere, the valve is opened, the sample inserted and the valve closed again. The flask can then be removed, with the sample still surrounded by the inert gas, taken to the spectrometer and sealed to the fore-chamber. With the fore-chamber also filled with inert gas, the valve is opened and the flask and chamber pumped

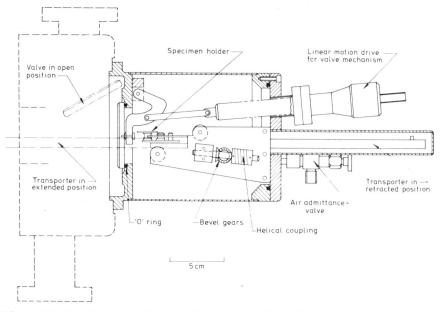

Figure 2.3   Specimen transfer flask for insertion of sensitive specimens into a spectrometer. The flask will pass through a standard port on a glove box and is designed to be attached to the entry port on ESCALAB. After the specimen has been loaded in an inert atmosphere, the valve is shut and the flask taken to the port, to which it is attached. The fore-chamber is then back-filled with inert gas, the valve opened and the chamber and flask evacuated together. Once a suitable vacuum has been achieved, the specimen is transported into the next chamber in the usual way. (Reproduced by permission of Vacuum Generators Scientific Ltd, East Grinstead)

out together to the normal vacuum required for sample insertion. After that, sample transfer follows the same procedure already described, using the transfer trolley in the flask rather than the one in the fore-chamber.

2.3.3   Sample treatment *in vacuo*

Treatments given to a sample after arrival in the spectrometer fall into three main categories, viz. preparation of a clean surface, depth profiling and surface reactions. Included in the latter are those treatments that consist of heating alone in order to produce surface segregation of impurities or to melt the sample if liquid metals are being studied.

2.3.3.1   *Preparation of a clean surface*

In the context of this volume, 'clean' is defined as the state of a surface in which the techniques of XPS and AES cannot detect characteristic spectral features from any elements, except those constituting the bulk composition oɪ the sample, above their limits of detection. Obviously if a surface analytical technique of greater sensitivity were to be applied to a surface judged clean by this definition, impurity elements would probably be detected. As always, cleanliness is a relative concept.

At first sight, the simplest cleaning technique is heat treatment. Several elemental materials, including silicon and the refractory metals, can be rendered clean according to the above definition by heating them to sufficiently high temperatures, generally a few hundred degrees Celsius below their melting points. That is, their surfaces will be clean while held *at* the high temperature. The problem is to maintain their cleanliness on cooling to room temperature, since as they cool their temperatures will pass through ranges in which impurities segregate quickly to the surface. The same problem arises with multielemental materials such as alloys and compounds whose temperatures cannot be raised sufficiently to achieve a clean surface while hot, but which have to undergo heat treatment as part of an experiment; in certain temperature ranges impurities present in the bulk at a low level can accumulate at the surface. The most common segregants are sulphur, oxygen and carbon, although in special cases other impurity elements such as nitrogen and phosphorus may appear in considerable amounts.

Heating a sample will thus hardly ever be effective by itself in producing a clean surface at room temperature. Another technique is needed to remove impurities physically from a surface, either those which have segregated as a result of heating or those present on the surface originally; by far the most commonly used is that of ion bombardment. In principle it is very simple. A beam of positive noble gas ions of energies between 500 eV and 5 keV, typically, is directed at the surface. As a result of the exchange of energy in the surface and subsurface regions, some atoms or clusters of atoms at the surface are given enough kinetic energy to leave the surface and be lost from it. In other words, the surface is eroded. The gas normally used is argon, chosen as a compromise between efficiency of removal of material, which increases with atomic weight, and vapour pressure at the temperature of liquid nitrogen traps, if in use; the heavier noble gases tend to condense at low temperatures and be difficult to pump away. The process of erosion of the surface is called sputtering and the source of ions is called an ion gun. Argon can be supplied to an ion gun in two ways: either by back-filling the system so that the pressure everywhere is that required to operate the gun or by passing the gas directly into the region of ion production so that only in that region is there a high pressure. The former method is a static one, in that it is used on ion pumped systems in which the ion pump must be valved off since it must not be allowed to pump argon continuously. The latter method is dynamic, since the exit apertures in the ion gun are sufficiently small for differential pumping to take place across them, allowing the pressure in the analysis chamber to be maintained at a low level by either diffusion or turbomolecular pumps. Sample cleaning is much more likely to be successful in a dynamically pumped ion gun, where impurities released by erosion are swept away, than in a static system where recontamination can occur. Description of the various types of ion gun in use and their modes of operation will be found in Section 2.4.3.

As remarked above, the technique of cleaning by ion bombardment is in principle simple. In practice it can be straightforward, too, for there are many published examples where surfaces have been cleaned by ion bombardment alone. In the majority of cases, however, it is found that although ion bombardment will remove the bulk of a layer of contamination, there is always left behind a small amount, of the order of a few atom per cent, of contaminant atoms. Carbon and oxygen are generally the most difficult to remove completely. In many applications and analyses the residual low level of contaminant may be acceptable, if all that was expected of the bombardment was the removal of sufficient masking contamination to be able to carry out an analysis of the sample. In other applications it will be essential to produce a clean surface, as defined above, in which case the most common procedure is to use a combination of techniques, viz. alternate cycles of heating and of ion bombardment.

If the sample is thin, e.g. in the form of foil, and already reasonably pure, it should be possible, by appropriate choice of temperature and ion dose in the above procedure, to denude the sample completely of those impurities that segregate to the surface on heating. Once the sample is in that condition, it is then easy to renew the cleanliness of its surface when desired by a single cycle of bombardment and/or heating, since no further contamination of bulk origin will appear. For thick samples, it is clearly impractical to attempt to remove all segregating impurities from the entire sample, but in many cases careful adjustment of temperature and ion dose can again achieve a situation in which the surface region, extending to perhaps $1-2$ μm, is denuded of impurities. Such a situation is obviously not stable under any subsequent heat treatment to temperatures near that used in the cleaning, but if there is to be no heat treatment, or only to a much lower temperature, then it represents a useful compromise.

Although ion bombardment is used so universally for surface cleaning, either by itself or in combination with heating, the potential user should be aware of the artefacts that can be introduced by it. These take the form of either or both chemical[3] or topographical[4] changes induced in the surface. Chemical changes here mean the alterations in the elemental composition or chemical state, or both together, of the major constituents that can occur at the surfaces of alloys and compounds during ion bombardment. Topographical changes are usually in the direction of increased surface roughness, or, occasionally, of statistically induced surface roughness. These effects are discussed in detail in Chapter 4, but it is useful to consider them briefly here in the context of surface cleaning.

Since the sputtering yields, i.e. the number of atoms sputtered per incident argon ion, vary substantially across the Periodic Table, as shown in Figure 2.4 from Seah,[5] it follows that in a material containing two or more elements the likelihood is that at least one of the elements will be removed preferentially due to its higher sputtering yield. Surface analysis after ion bombardment would therefore reveal a depletion in that element compared to the bulk or stoichiometric composition. Depletion would continue with further bombardment until an equilibrium situation was reached in which the relative atom populations of elements at the surface was balanced by the relative sputtering yields. This effect has been observed[6] frequently in alloy systems in which there is a significant disparity in sputtering yields between the constituents. If there are no matrix effects, and if the individual elemental sput-

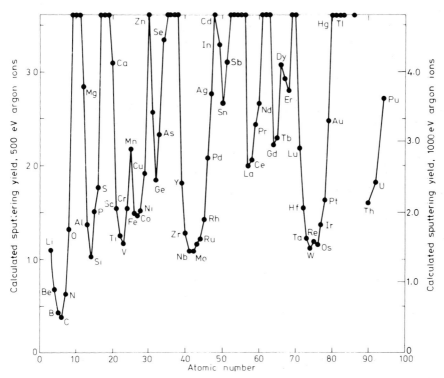

Figure 2.4   Predicted values of the sputtering yields of argon ions of energies 500 and 1000 eV across the Periodic Table. The yields of several elements with particularly high sputtering yields are not shown, for the sake of clarity, but may be found in the original paper by Seah[5]

tering yield are known, then it is possible to work back to a surface composition in the absence of sputtering; this will be discussed further in Chapter 5.

The other chemical changes induced by the bombardment itself can be equally serious. They include reduction of compounds[3] and persistence of elements due to 'knock-on'. In the former the more volatile elements such as oxygen in oxides, sulphur in sulphides, etc., are lost preferentially during bombardment even though their elemental sputtering yields may be similar to, or even lower than, those of the other constituents. Such loss is inevitably accompanied by a reduction in the oxidation state of, say, a metal in a metal oxide. In the effect of 'knock-on', as its name suggests, some elements, particularly light ones, can be driven further into the material by direct impact from a primary ion or by indirect impact via an energy cascade. The persistence of carbon in certain materials during prolonged bombardment, for example, is generally attributed to this effect.[7] It has not proved possible so far to take account in any systematic way of the artefacts of reduction and of 'knock-on' when attempting quantification in their presence.

The roughness of a surface can be increased by ion bombardment in several ways.[8] If the ions are directed at the sample surface from one direction only, i.e. from a single source, which is still the most common mode of operation, then in general any asperities will tend to be accentuated since sputtering efficiency is greater at higher incident angles. Eventually an equilibrium configuration will be reached in which the atom population in each different topographical position will be balanced by the sputtering yield in that position. A more serious situation can arise if there are impurity inclusions in the material whose sputtering yield is much less than that of the surrounding matrix, in which case the inclusion 'shadows' the material behind it from the incident ions and a conical pillar forms of progressively increasing height. Both these effects can be minimized, if not eliminated, by the use of two or more ion sources simultaneously from different directions, and an even better arrangement is to rotate the sample continuously during bombardment so that all angular effects are completely smoothed out. Unfortunately very few systems make provision for either of these arrangements.

Ion bombardment, then, although certainly effective in cleaning a surface, especially in conjunction with heating, brings in its train some highly undesirable side-effects, of which some are accountable but others as yet are not. Nevertheless, the technique is so widely used that it is clearly here to stay, and thus it is necessary to keep watch continually for intrusion of artefacts during the cleaning of any one sample.

In special cases other methods of producing clean surfaces are applicable. For example, soft metals such as lead, indium and tin have been cleaned[9] successfully by scraping their surfaces in vacuum with a tool, such as the edge of a razor blade, mounted on the end of a rod passing through a flexible bellows. Cleaning in this case obviously corresponds to the creation of a new surface by mechanical removal of the original surface and its associated contamination. Another specialized method is that of fracture in vacuum. Some materials such as alkali halides, silicon, germanium, etc., will cleave easily along well-defined cleavage planes, so that all that is required to produce a clean surface with a composition close to that of the bulk is some mechanical ingenuity in the method of holding the material, cleaving it and offering it to the energy analyser for analysis. In other materials such as polycrystalline metals and alloys, which are too ductile at room temperature to cleave, fracture may still be possible by lowering the temperature to the point at which they become brittle. The fracture path may then either pass through the individual grains (transgranular) or follow the grain boundaries (intergranular); in either case a surface free from contaminants is produced, although the intergranular surface may contain certain impurities segregated from the bulk, as described in Chapter 7. An example[10] of a fracture stage is illustrated in Figure 2.5. All such stages possess the facility of cooling the sample to be fractured to temperatures near that of liquid nitrogen.

Methods of preparing atomically clean surfaces on bulk samples have recently been collected in a useful review by Musket *et al.*[11] The most widely used method of producing a clean surface without having to resort to ion bombardment or heating of the sample is that of deposition of pure material onto a substrate in ultra-high vacuum by evaporation either from a heated filament or crucible or by electron bombardment. It is possible in principle to deposit *any* solid element as a thin film in this way, as well as many compounds, but in practice the requirement of a pure and atomically clean surface imposes stringencies that prevent the use of some elements. The two overriding criteria are (a) that the evaporated film should be of the same composition as the starting material and (b) that there should be no release of gas during evaporation. Difficulties in complying with the first of these arise if the material to be evaporated has to be supported in any way during evaporation, since then there is always the possibility of reaction by alloying, or if the starting material is itself multicomponent, since it is then possible for differential evaporation rates to produce a deposit of different composition. The second criterion can usually be obeyed by careful attention to evaporation procedure, i.e. a thorough outgassing of the sample and its support before the film to be used is actually deposited.

The most straightforward evaporators are those made from the sample material without any support; they usually consist of a tightly wound coil of wire of the material through which a current is passed to raise the temperature by ohmic heating to the point at which the material sublimes freely. Obviously the properties of the material must be such that the coil does not sag at the temperature of sublimation, and for this reason the method is limited to a few of the more refractory metals. The advantages are that there

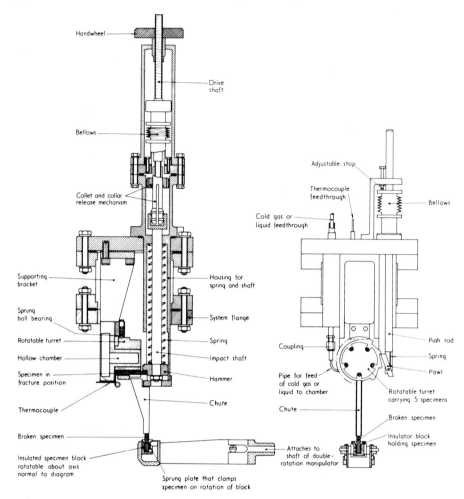

Figure 2.5   Impact fracture stage for fracture of brittle samples *in vacuo*. The rotating turret has holes for five specimens, which are clamped in place by grub screws. Rotation is achieved with a vertical push rod. To fracture a sample, the hammer is wound up against a powerful spring until at a predetermined point a release mechanism allows it to fly back and hit the free end of the sample. The broken piece then falls down a chute into a catcher with which it may be positioned for analysis. The sample may be cooled before fracture by passing liquid nitrogen into a tank on which the turret rotates. (After Coad *et al.*[10])

is no possibility of reaction with any support material and that the evaporator can be completely outgassed; the disadvantages are that the rate of deposition is slow and that the surroundings tend to become very hot unless proper radiation shielding is included.

Almost as straightforward are those evaporators that consist of a coiled filament supporting a sample material that wets but does not alloy with the filament material. Here the filament is invariably of ductile (i.e. annealed) tungsten or molybdenum. Metals that can be evaporated easily from such filaments include copper, silver and gold, typically, but others such as barium and magnesium are also possible. When using the noble metals it is desirable to pre-melt the sample onto the filament loops either in an auxiliary vacuum system or in a hydrogen stream, since the filament itself can then first be outgassed at a temperature much higher than that of subsequent evaporation. If that procedure is followed outgassing of the evaporator prior to evaporation merely involves re-melting the sample once or twice, following which evaporation can take place without any measurable rise in pressure. For more volatile non-alloying sample materials it may be difficult to outgas the evaporator to the same extent; the bare filament can always be pre-outgassed as above, but if attempts are made to melt the sample most of it may be lost prematurely. Instead it may be necessary to heat for long periods to temperatures at which the vapour pressure is still low. Some metals, such as cadmium, zinc, selenium, tellurium, the alkali metals and mercury, have vapour pressures at room or bake-out temperatures that are too high in any case to allow their presence in the bulk elemental form in an ultra-high vacuum system.

The majority of metals that one would like to be able to evaporate in order to produce a clean surface have the unfortunate property of alloying more or less rapidly with the filament material at the evaporation temperature. Included are all the transition metals, aluminium, the rare earths, platinum, zirconium, uranium and thorium, and some other less used metals. Clearly, if the sample alloys with the filament, with the formation of a liquid phase below the melting point of the filament material, not only will premature failure of the filament occur by breaking but if the vapour pressure of the alloy is also appreciable then the deposited film will be impure. In this situation it is necessary to design the filament-cum-sample configuration so that the possibility of disastrous alloying is minimized; what this means in practice is that the sample must be so distributed along the filament that at all points the local volume ratio of sample to filament materials is low—less then 1:5. If the evaporator can be arranged in that way, then it is possible to evaporate much of the sample in pure form before alloying causes breakage or contamination of the film. There are various ways that have been used to reduce the local material ratio, and ingenuity might well suggest others. One obvious one is to electroplate, or perhaps even ion plate, the sample material onto the filament as a uniform coating. The coating thickness, and therefore the local material ratio, can be controlled accurately. The disadvantage is the possibility of trapping impurities such as water in the plating, which might either be difficult to remove subsequently or cause poor adhesion to the filament. Another popular way of distributing the sample along the filament is to use the coiled coil. In this arrangement the sample is in the form of a fine wire with a diameter much smaller than that of the filament material; before the filament itself is wound, the sample is wound around the filament wire in a spiral whose pitch maintains the desired low local material ratio. Yet another way is to form the filament itself from several strands of fine wire, so that when the sample material, normally hung on the filament in small loops, is melted, it runs into the interstices of the strands by capillary action and distributes itself fairly uniformly. To be avoided is the method of hanging loops of sample material on *single* filamentary wire; on melting, blobs are formed in the filament turns, with consequent rapid alloying and failure.

In all the methods just described for the evaporation of metals that alloy with the filament (which is almost invariably of tungsten) the outgassing of the evaporator before the film is finally deposited presents some problems. Since the temperature of useful evaporation is also likely to be that at which some alloying with the filament may occur, clearly the prior outgassing must take place at a lower temperature, with the consequence that prolonged periods of heating are necessary. This is usually a matter of trial and error, for it is difficult to predict in any one case just what procedure will be successful; the published literature should be consulted for some guide to the most likely method.

Some metals present particular difficulties either because they will not wet the refractory metals used for filaments or because they are sufficiently volatile for evaporation only at temperatures much higher than their melting points. Amongst these are gallium, indium, lead, silicon, thallium, calcium, lithium, arsenic, antimony and bismuth. Because of the impossibility of evaporation from filaments, they are generally evaporated from crucibles of either alumina, beryllia, boron nitride or carbon, as appropriate, heated indirectly by a surrounding element of refractory metal. Such evaporators can rarely be outgassed sufficiently thoroughly for use in UHV, even after very prolonged heating, and indeed preparation of clean surfaces of the above metals by the evaporation method is hardly ever attempted.

The other popular method of evaporation is by electron bombardment, and this can often be successful for materials that are too difficult for one reason or another for filament evaporation. Electrons are focused from a hot filament onto the material to be evaporated, heating it to a temperature that can be controlled by the power input. Normally the electron emitter is near earth potential and the material at a high positive potential, but in some configurations the polarities may be reversed. According to the focusing conditions and the power dissipated, all or only part of the material may be heated to the temperature of evaporation. Commercially available electron bombardment evaporators have water-cooled hearths on which the sample material is placed, which avoids the problems described above of alloying with the support material. On the other hand, such evaporators cannot be regarded as

UHV-compatible since they are not usually bakeable and cannot be outgassed properly before use. Electron bombardment evaporation *has* been used[12] successfully in UHV, however, by the so-called pendant drop method, in which bombardment of the end of a wire or rod of the sample melts part of it, whereupon surface tension forces pull the molten volume into a blob that hangs from the end and acts as the evaporation source. Outgassing can be carried out by de-focusing the electron beam to heat the sample along its length without melting it.

Whatever the method of evaporation used, it is always necessary either to remove the substrate from the line of sight of the evaporator or to swing a mask in front of it, during the preliminary outgassing period, to avoid the deposition of contaminated or impure material. This is particularly vital when a significant proportion of the sample is likely to be lost by premature evaporation during the outgassing procedure.

2.3.3.2 Depth profiling

Ion bombardment has already been mentioned as a potential method of producing a clean surface on a sample; some of its pitfalls have also been mentioned. It is also used very extensively indeed, in conjunction with surface analysis, as a method of obtaining compositional information as a function of depth below the surface. When used with a technique in which data acquisition is slow (at the moment), such as XPS, the information is built up stepwise, i.e. by alternate cycles of bombardment and of analysis. When used with AES, however, which is a relatively fast technique, and if the partial pressure of argon (or whichever noble gas is being used) in the region of the energy analyser is less than 10^{-5} torr, then bombardment and analysis can be carried out simultaneously and a more continuous compositional profile is obtained. The technique is called depth profiling. All the artefacts that may be introduced by cleaning the surface by ion bombardment are of course present and are equally important during continued bombardment to produce a depth profile; indeed their effect may well increase with depth. With such widespread use of depth profiling, a great deal of attention has been and is being paid to all these effects and their influence on the composition, and in fact depth profiling is almost a subject in itself. For that reason it will not be discussed further here, since a full discussion is given in Chapter 4.

2.3.3.3 Surface reactions

Apart from the requirements of some experiments concerned solely with the physical properties of surfaces, the usual reason for wishing to prepare a surface in an atomically clean state (within the limits of detection) is to establish a reproducible, well-characterized, starting point for the study of the

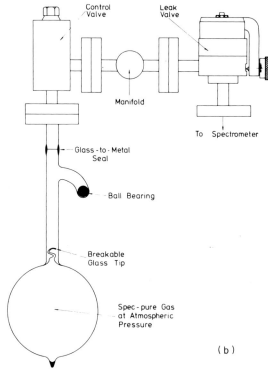

(b)

reactivity of that surface. The reactivity may be towards gases and vapours, either singly or together, towards liquids, towards other solids, or towards impurity or other atoms diffusing to the surface from the bulk. Since gases are in general easy to handle in a vacuum system, it will come as no surprise that by far the greatest number of surface reactions studied have been those between gases and solids.[13] This is not the place to discuss the reactions themselves—only the means to bring them about.

On all commercially available spectrometers a gas admission and control system can be installed as an optional extra. Usually it is separate from the argon feed to the ion bombardment guns. A gas admission system must have the following features:

(a) A series of containers of spectroscopically pure gases, as many as the number of different gases needed for the experiment. These containers consist in some cases of glass flasks filled to atmospheric pressure, in others of metal cans filled to a few atmospheres pressure. In both cases the gas can be released by breaking a seal after the container is attached to the system. Figure 2.6 shows examples of each type.

(b) Fixing points, or stations, for the gas containers, each consisting of a flange and an associated control valve. For glass containers the mating flange carries a glass-to-metal seal ending in a length of glass tubing, to which the tube on top of the container is fused when the container is connected to the station. Pressurized metal can containers normally carry their own valves as an integral part of the seal puncturing mechanism, but because of the high pressures involved it is essential not to do away with the second valve at the fixing point, since additional control will be necessary. Again Figure 2.6 shows the arrangement in each case.

(c) A manifold to which all the stations are connected, which has a volume large enough to allow sensible mixing of one or more of the gases if the experiment requires a gas mixture, but not so large that the gases are used wastefully. Its volume may also be governed by the total pressure of the gas or gases required for the experiment in the reaction chamber, and

Figure 2.6 (a) An arrangement for the mixing and admission of pure gases to a spectrometer in a controlled way, using pressurized metal can containers. When the valve on a can is first screwed onto it, a hollow plunger breaks a seal, thereby releasing the gas. However, since the pressure in the cans is about eight atmospheres it is essential to have additional valves between the can and the system as a precautionary measure and for proper control. (By courtesy of Messer Griesheim Ltd) (b) A typical gas-handling arrangement using glass bottles containing one litre of spec-pure gas at one atmosphere pressure. The gas is released by lifting the soft iron slug with a magnet and dropping it onto the breakable tip. Care has to be taken in the process that powdered glass does not get carried into the seats of any valves, and it is usual to place a wad of clean glass wool in the neck of the bottle to prevent this happening

the volume of that chamber. Generally speaking, however, since the gas can be stored in the manifold at a relatively high pressure, e.g. several torr if necessary, and since the pressures used in experiments are rarely greater than 10^{-4} torr, the manifold volume can be kept small. Obviously the pressure in the manifold must be measurable and over quite a large range, from UHV to the torr region, which implies that provision must be made to accommodate two gauges with overlapping ranges.

(d) A high quality leak valve for the admission of gas from the manifold to the reaction chamber. The valve should be one whose design and precision in construction are such that the rate of admission of gas can be controlled finely and smoothly and that the setting for any particular chosen gas flow is reproducible. An example of the construction of such a valve is given in Figure 2.7. Needless to say, the valve should also be capable of shutting completely against a high pressure on the manifold side.

To avoid contamination of the gas or gases admitted to it, the manifold and all the connecting pipework must be fabricated to the same UHV standards as the rest of the spectrometer, so that it can also be subject to the same bakeout procedure. On the other hand, the glass or metal gas containers clearly cannot be baked, which means that between the manifold and the containers

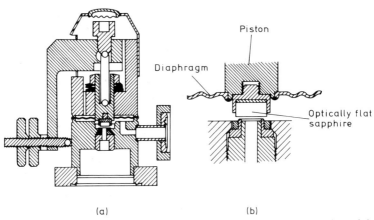

(a)   (b)

Figure 2.7   Leak valve for fine control of gas flow. The main construction of the valve is shown in the larger diagram and details of the seating mechanism in the smaller one. The valve seat consists of an optically flat sapphire disc pressing against a metal ring. The mechanical advantage of the drive mechanism is many thousand to one, so very fine control of the movement of the sapphire is possible. Provided the valve is not closed too tightly the setting for any particular flow rate is reproducible. (By courtesy of Varian Associates Ltd)

there is a region of temperature gradient during bake-out. Since the containers are positioned outside the bake-out zone, this region is accessible, and it is good practice to warm it carefully and judiciously with a hand-held heat gun once the bake-out temperature has been reached in the rest of the spectrometer in order to drive off any condensable impurities.

Even though a single gas stored in a manifold may be considered pure, or a gas mixture prepared in the manifold may have an accurately known composition, there is no guarantee that on admission to the reaction chamber the purity or the composition will remain unchanged. Exchange can take place with other gases adsorbed on the walls and other parts of the chamber, and hot filaments can cause dissociation followed by recombination to form different molecules. If the experiment is a dynamic one, i.e. with the reaction chamber open to the pumps so that the gases are flowing over the surface, there is always the risk of other reactions taking place in the pumps themselves. This is not particularly likely with diffusion or turbomolecularly pumped systems, but has certainly been observed in the presence of ion and sublimation pumps. It is therefore very advisable to have on the reaction chamber a means of checking the actual, as against the supposed, composition of the gas being admitted to the clean surface, and this check is normally performed by a small quadrupole mass spectrometer. Since such a device is also invaluable for diagnostic purposes, i.e. during a fault condition, and for leak testing, it can hardly be regarded as a luxury but as a necessity.

Reactions of clean surfaces with liquids have hardly ever been attempted in UHV; the problems associated with transferring a liquid to a surface in a controlled way in UHV seem too complex to be worth the effort. In recent years the study of solid–solid reactions, on the other hand, has been increasing steadily, stimulated by some of the requirements of the microelectronics industry for the production of surface layers with certain electronic properties. These experiments are in general quite straightforward, and consist of the deposition by evaporation of known quantities of one solid, usually a metal, onto the surface of another solid, sometimes a metal[14] but more often a semi-conductor.[15] The need for a completely hygienic deposition in terms of purity and contamination, as discussed earlier, is particularly stringent here since very small amounts of unwanted elements, either in the deposited film or at the film/surface interface, can cause large changes in the electronic properties of the layer. Reaction in some cases (e.g. palladium on silicon)[16] occurs spontaneously at room temperature; in other cases it is necessary to programme[17] the temperature of the substrate during the experiment either to encourage reaction to take place during deposition or to prevent it doing so until deposition is complete. Thus the basic surface preparation requirements for the majority of solid–solid reaction studies are a properly designed evaporator and a heating/cooling stage.

For the study of the adsorption at external surfaces of impurities segregating from the bulk the only basic surface preparation requirement is a heating stage capable of taking the material to the desired temperatures. If, however, it is the segregation to internal grain boundary surfaces that is to be studied, then the surfaces are prepared by fracture in UHV in one of the fracture stages already described and illustrated in Figure 2.5. In this case all heat treatments of the sample are carried out before insertion into UHV, although it will be necessary in most cases to *cool* the sample before fracture to increase the tendency to brittle fracture.

2.3.4 Sample positioning

There are several reasons why it is necessary to be able to manipulate the position of a sample in a finely controlled manner once it has been inserted into the spectrometer and brought into the proximity of the energy analyser.

Firstly, the shapes and sizes of samples can differ widely, so that the optimum position in terms of sensitivity and energy resolution for the surface analysis of one sample will not necessarily be optimum for the next. Secondly, it may be required during analysis to move the sample laterally in a reproducible way to allow different areas of it to be analysed; this of course applies more to XPS than AES since in the latter the electron beam can be deflected for the analysis of different areas and in any case the average area of analysis in AES is much smaller than in XPS. Thirdly, it is often desirable to be able to rotate the sample about an axis through its surface, again in a reproducible way, either to alter the angle of incidence of primary ions during ion bombardment or to alter the angle of take-off of electrons accepted by the analyser, for the purposes of profiling by variation of escape depth (see Chapters 3 and 4). Fourthly, if techniques other than electron spectroscopy are also available on the spectrometer, e.g. static SIMS that uses mass spectrometry rather than energy detection, it may be necessary for geometric reasons to have to move the sample into a position for analysis different from that for XPS and AES. Fifthly, as the reliability and degree of mechanical automation of surface analytical techniques improve, there will be increasing pressure from users for automatic sample change, which again involves accurate movement and positioning.

Accurate sample positioning is not an especially new requirement, or even one that is specific to XPS and AES. It has always, for example, been needed by LEED (low energy electron diffraction) from the earliest days of the study of surfaces, and many of the same reasons why it is needed by XPS and AES apply equally well to a whole range of other techniques. The considerations listed above suggest that at a minimum four degrees of freedom in movement should be available, viz. *X, Y* and *Z* linear motions and an axial rotational motion. For automatic sample change or for assuming analysis positions other than that for XPS and AES it would clearly be desirable, if not necessary, to have azimuthal rotation as well in which the samples would be mounted on a carousel. The positioning movements actually provided by manufactuers in their *standard* production equipment for surface analysis vary somewhat and are dependent on their philosophy in respect of sample insertion. Where a sample carousel is used to accept a sample on its carrier from the end of a sliding shaft, the sheer bulk of the carousel makes the facility for axial rotation very difficult to incorporate in the positioning degrees of freedom, making any variation of the ion incidence angle or of the electron take-off angle impossible. On the other hand, of course, azimuthal rotation of the carousel lends itself to automatic sample changing and to making any one sample available for analysis by a different technique. Where a sample is inserted by the successive transfer method, it is of course placed finally on the end of the shaft of a sample manipulator, where it can be translated along any of the three orthogonal directions and also be axially rotated. If the manipulator is fitted with a double rotation attachment, usually an optional extra, then azimuthal rotation of rather limited radius can be added to the movements, but the mechanism is not sufficiently precise to be usable in automatic sample changing.

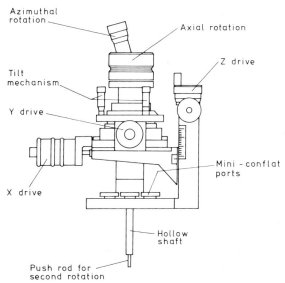

Figure 2.8 Diagram of high precision manipulator for reproducible positioning of specimens in a spectrometer. Movements available are shifts along the x, y and z axes, rotation about the central axis and about an axis at'right angles to it (azimuthal rotation), and tilt. Electrical and other feedthroughs can be taken into the vacuum via a series of mini ports set into the main flange, and the whole device is bakeable to 250 °C. (Reproduced by permission of Vacuum Generators Ltd, Hastings)

For experiments of a basic nature, in which the sample is not inserted by either of the two methods already described but is mounted individually, all manufacturers will provide sample manipulators with precise, reproducible movements. An example of one of these is shown in Figure 2.8. The translational movements are taken up in a highly flexible all-welded bellows, whose length can be increased considerably if required to provide for additional Z translation over the 50 mm or so of the standard device. Axial rotation is derived from an offset knuckle joint that also operates through a flexible bellows. If azimuthal rotation is needed, then an additional mechanism is included that drives a cable or rod inside the hollow shaft on a push/pull basis, the cable being attached to a pulley that rotates the sample mount. Earlier versions of such universal manipulators had to be partially dismantled for bake-out, but all current models are of such construction that no dismantling is needed and their precision of movement and reproducibility is unaffected by repeated bake-out up to 250 °C. With most of them it is possible to obtain a variety of attachments for heating, cooling, azimuthal rotation, offset axial rotation, or almost any combination of these, so that their flexibility of application is considerable.

2.4   Sources

2.4.1   X-ray sources

Devices for the production of high fluxes of X-ray photons of characteristic energies were in use long before the advent of XPS, principally in X-ray diffraction (XRD) instruments. For XRD, however, the characteristic energies are normally required to be high, typically many scores of kiloelectronvolts whereas reference to the description of the basics of XPS in Chapter 3 shows that the surface-specific property of XPS depends on using soft X-rays of characteristic energies of only a few kiloelectronvolts. In addition, sources used in XRD operate traditionally by the bombardment of an anode at earth potential by electrons from a hot filament at a very high negative potential, so that any other surrounding surfaces also at ground potential would also be bombarded. Since in XPS the X-ray photons have to pass from the vacuum in the source to the vacuum in the analyser chamber through a thin window which acts as a barrier to electrons and to possible contamination, it is clear that if the traditional configuration were to be maintained the window would also receive electron bombardment. Thus in XPS sources the configuration is reversed in that the filament is near earth potential and the anode at a high positive potential. Figure 2.9 shows two commonly used configurations.

When one comes to consider the choice of material for the anode for an XPS source, it is necessary to consider first the matter of energy resolution.

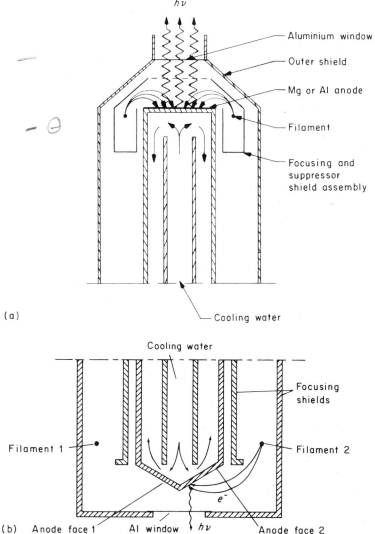

Figure 2.9 (a) Soft X-ray source with single anode of either magnesium or aluminium, deposited as a thick film on the flat end of a water-cooled copper block. The anode is surrounded by a cylindrical focusing shield at the same potential as the filament. An outer can acts as a radiation shield and carries the thin aluminium window that must be interposed between the target and sample. (Reproduced by permission of Perkin-Elmer, Physical Electronics Division) (b) Soft X-ray source with dual anode, allowing use of either magnesium or aluminium $K\alpha$ radiation by simple external switching without the need to break the vacuum in going from one to the other. The anode has a tapered end with two inclined faces on which films of magnesium and aluminium, respectively, are deposited. There are two semi-circular filaments, one for each face. The focusing arrangements are similar to those for the single anode of (a). (Reproduced from Barrie and Street[18] by permission of The Institute of Physics)

Practical Surface Analysis

The Einstein relation governing the interaction of a photon with a core level,

$$E_{KE} = h\nu - E_B - e\phi \qquad (2.1)$$

where E_{KE} = kinetic energy of ejected photo-electron
            $h\nu$ = characteristic energy of incident X-ray photon
            E_B = binding energy of core level electron
            ϕ = work function term,

shows that the line width of E_{KE} will depend, amongst other factors, on the line width of $h\nu$. (Line width means here the full width at half-maximum height of the exciting X-ray lines or emitted photo-electron lines, as the case may be.) Since in XPS one is continually seeking chemical information by detailed analysis of individual elemental spectra, it follows that the most accurate information will be provided by working at the best instrumental energy resolution compatible with sensitivity, i.e. signal-to-noise, consider-ations. In practice, at the time of writing, this means using energy resolutions in the range 1.0–2.0 eV, mostly nearer the lower than the higher figure. Clearly, then, to avoid limitation of the achievable resolution by the line width of the X-ray source, it is necessary to use materials for the source anode whose line width is less than 1.0 eV. Table 2.1 lists X-ray line energies and line widths for various characteristic lines from some materials that would be suit-able from other points of view, e.g. stability under prolonged electron bom-bardment. It can be seen that there are indeed very few materials whose

Table 2.1   Energies and widths of some charac-teristic soft X-ray lines

Line	Energy, eV	Width, eV
Y $M\zeta$	132.3	0.47
Zr $M\zeta$	151.4	0.77
Nb $M\zeta$	171.4	1.21
Mo $M\zeta$	192.3	1.53
Ti $L\alpha$	395.3	3.0
Cr $L\alpha$	572.8	3.0
Ni $L\alpha$	851.5	2.5
Cu $L\alpha$	929.7	3.8
Mg $K\alpha$	1253.6	0.7
Al $K\alpha$	1486.6	0.85
Si $K\alpha$	1739.5	1.0
Y $L\alpha$	1922.6	1.5
Zr $L\alpha$	2042.4	1.7
Ti $K\alpha$	4510.0	2.0
Cr $K\alpha$	5417.0	2.1
Cu $K\alpha$	8048.0	2.6

characteristic X-ray lines have sufficiently small widths. The yttrium and zirconium $M\zeta$ lines are narrow and have been used successfully in special applications, but their energies are much too low for general application since the number of photo-electron lines that could be excited would be insufficient for unambiguous analysis. The only other lines that fit the requirements are the $K\alpha$ lines of magnesium, aluminium and silicon. Although silicon has also been used[19] for special applications, it is not a metal, so that its heat transfer characteristics are poor and it is difficult to apply to an anode surface. Thus one is left with the magnesium and aluminium $K\alpha$ lines, and in fact it is these two that are used so universally in XPS.

Although the characteristic $K\alpha$ line energies of magnesium and aluminium are in the region of 1250–1500 eV, it is necessary to use exciting electron energies about an order of magnitude higher for efficient production of X-rays, according to the dependence[20] of emitted flux on bombarding energy shown in Figure 2.10. In all commercial spectrometers the maximum available accelerating potential is 15 kV, which is adequate. It is also necessary, for purposes of optimizing sensitivity at a given energy resolution, to be able to use as high an electron bombarding current as the source will stand since the photon flux will be directly proportional to that current. Against this design requirement must be set the necessity of making the physical size of the source near the sample analysing position as small as possible since the source must be placed as near as possible to the sample. The flux irradiating the sample will of course vary as the inverse square of the distance of the anode from the sample surface. All these requirements have led designers inexorably to the types of configuration shown in Figure 2.9, in which the necessary compactness of the source means that the maximum power dissipation that can be achieved is 1 kW, provided that adequate water cooling can be provided. The necessity of forced water cooling to remove heat from the anode before it melts also implies that the anode block must be of high heat conductance, which in turn means fabrication of the block and the integral water tubes from copper. Thus the anode material itself, i.e. the emitting surface, is normally deposited on the copper block as a thick film, typically ~10 μm in thickness, representing a compromise between being thick enough to exclude copper $L\alpha$ radiation and thin enough to allow adequate heat transfer. In most source control units the bombarding current can be selected from a set of switched fixed values, e.g. 5 mA, 10 mA, 20 mA, etc., the precise values varying with the manufacturer. The accelerating voltage, on the other hand, can be chosen anywhere in the continuous range from 0–15 kV, although in general the source will be unstable below ~2 kV.

Most commercially available X-ray sources are of the type shown in Figure 2.9(b), in which there are *two* filaments for electron bombardment and *two* anode surfaces, one of magnesium and one of aluminium. It is thus possible by simple external switching to go from Mg $K\alpha$ to Al $K\alpha$ radiation in a few

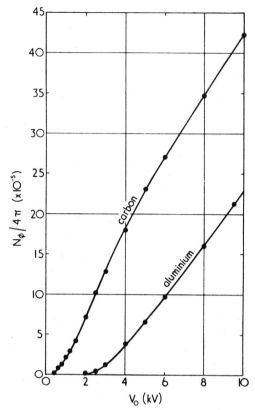

Figure 2.10   Dependence of efficiency of production of Al $K\alpha$ and C $K\alpha$ characteristic radiation on the energy of the bombarding electrons. (Reproduced from Dolby[20] by permission of The Institute of Physics)

seconds. There are two reasons why it is desirable to have the double-anode facility. One is that the two characteristic $K\alpha$ lines have different line widths, as can be seen by reference to Table 2.1, and although the Al $K\alpha$ line is of slightly greater general usefulness than that of magnesium, there are frequent occasions when the inherently better resolution obtained with the magnesium line is needed. The other reason is that in any X-ray-excited ejected electron spectrum both photo-electron and Auger peaks appear, and interferences can result. Since photo-electron energies are linked directly to the energy of the exciting photon, whereas Auger energies are fixed, a change in the X-ray line energy will resolve the interference. For special purposes, other combinations of anode material have been prepared, amongst which Mg/Zr and Al/Zr are popular, the Zr $L\alpha$ radiation being useful[21] when increased sensitivity for the detection of aluminium and silicon is needed.

As mentioned earlier, it is necessary to interpose a thin window between the anode and the sample to screen the latter from stray electrons, from heating effects and from any possible contamination originating in the source region. The material chosen for the window should obviously be reasonably transparent to the X-radiation being used, and for the Mg $K\alpha$ and Al $K\alpha$ lines it is both convenient and practicable to use aluminium foil about 2 μm thick. For that thickness the flux attenuation is about 24 per cent for Mg $K\alpha$ and 15 per cent for Al $K\alpha$. When other exciting radiation is used care must be taken that the appropriate window material is substituted for the standard aluminium. For example, if the $M\zeta$ radiation at 151.4 eV from a zirconium source were required, aluminium would be opaque and instead either beryllium or carbon windows would be suitable.

The description so far of X-ray sources for XPS refers for the most part to 'natural' or, more accurately, unmonochromatized radiation. It must be realized that the emission spectrum from *any* material is complex and consists of a broad continuous background (called Bremsstrahlung) on which are situated the more or less narrow characteristic lines. An example is shown in Figure 2.11 for aluminium. The Bremsstrahlung continuum is a function of the energy of the bombarding electrons, and after passing through the window will have a maximum between 20 and 40 per cent of that energy, according to the thickness of the window, so that it extends beyond the energy of the principal characteristic line and is therefore useful for exciting Auger transitions from atomic levels too deep to be ionized directly by the characteristic radiation. Complexities in the line spectrum arise from satellite emission associated with the principal lines and also from the multiplicity of lines as the Periodic Table is ascended, as discussed in Chapter 3.

2.4.1.1   *X-ray monochromatisation*

Removal of satellite interference, improvement of signal-to-background by eliminating the Bremsstrahlung continuum and selection of an individual line from the unresolved principal line doublet can be achieved by monochromatization of the emitted X-rays. There are several methods of doing this but they all depend on dispersion of X-ray energies by diffraction in a crystal, which is of course governed by the well-known Bragg relation:

$$n\lambda = 2d \sin \theta \qquad (2.2)$$

where n = diffraction order
λ = wavelength of X-rays
d = crystal spacing
θ = Bragg angle

For first-order diffraction of Al $K\alpha$ X-rays, for which λ = 0.83 nm, it is found that quartz crystals are very suitable, since the crystal spacing of the

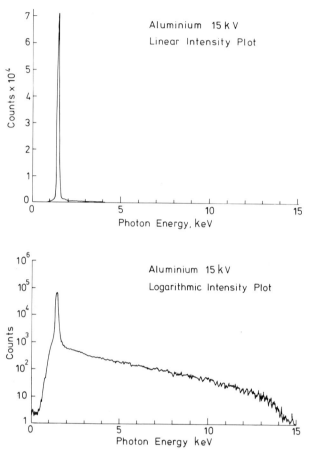

Figure 2.11   X-ray emission spectrum of an aluminium target under bombardment by 15 kV electrons, recorded by a lithium-drift detector through a beryllium window of thickness 7.5 μm. Upper curve, photon intensity plotted on a linear scale, on which little is evident except the intense characteristic $K\alpha$ line. Note that the energy broadening of the solid-state detector attenuates the peak by a factor of about 100. Lower curve, the same plotted on a logarithmic scale, that reveals more clearly the broad Bremsstrahlung background extending to energies much higher than the characteristic line. The background intensity at very low energies will have been reduced by absorption in the beryllium window. (Measurements by courtesy of Mr R. W. M. Hawes, Materials Development Division, Harwell)

$10\bar{1}0$ planes is 0.425 nm and the Bragg angle is thus 78.5°. Quartz has many advantages, since it can be obtained in perfect crystals of very large size which can easily be bent elastically or ground, and can be baked to high temperatures without damage or distortion.

The principle of the method of monochromatization used most estensively

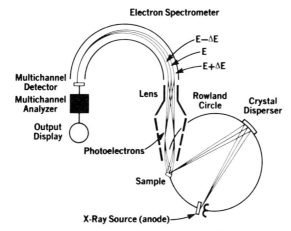

Figure 2.12 Principle of monochromatization of X-radiation. Selection of a particular wavelength is achieved by diffraction in a crystal of suitable lattice spacing, and for the Al $K\alpha$ lines quartz is a convenient crystal to use. The quartz crystal is placed on the surface of a Rowland circle, being accurately ground to match the curvature, and the X-ray source is placed at another point on the circle. Photons of the required wavelength will be focused at a third point on the circle, where the sample is placed. For the Al $K\alpha_1$ line the angle between incident and diffracted beams is 23°. A typical circle diameter would be 0.5 m. (From Kelly and Tyler[22])

in commercial instruments is that of fine focusing,[22] shown in Figure 2.12. The quartz crystal is placed on the surface of a Rowland or focusing sphere, having been ground accurately to the shape of the sphere, i.e. in both directions of curvature. With the anode itself lying on the Rowland circle, irradiated with a sharply focused electron beam, the $K\alpha$ X-rays are dispersed by diffraction and re-focused at another point on the Rowland circle where the sample is placed. If the $K\alpha_1$ component of the principal Al $K\alpha$ line is required, then the included angle between incident and diffracted rays will be 23°. According to the diameter chosen for the Rowland circle the dispersion around the circumference will vary; for the popular 0.5 m monochromator the dispersion is 1.6 mm/eV and for the larger 1 m instrument, 3.2 mm/eV. To achieve, therefore, line widths substantially less than the natural widths of the $K\alpha$ radiation, e.g. ~0.4 eV, involves focusing of the bombarding electron beam on the anode sufficiently finely that the width of the irradiated area on the sample in the dispersion plane is less than ~0.6 mm, using the 0.5 m monochromator.

Since the function of a monochromator is to allow only a selected small portion of the total $K\alpha$ emission to fall on the sample, it follows that the photon flux available for the same power dissipation is considerably less than with the unmonochromatized sources. For an aluminium anode, for example, the flux will be lower by a factor of about 40 when using a 1 m monochromator compared to that from a standard source. Thus count rates will be much reduced, but against that must be set increased detectability in terms of signal-to-background since the background has been substantially removed, the absence of satellites resulting in a 'clean' spectrum, and excellent resolution. Most manufacturers can provide monochromators as optional extras, but they are inevitably both bulky and expensive. Very recently one manufacturer has also described the monochromatisation of Ag $K\alpha$ radiation.[23]

2.4.1.2 Synchrotron radiation

The soft X-ray sources discussed up to now have all been discrete line sources. Useful as these are, they have inherent limitations as far as XPS is concerned, the most obvious one being that of photo-ionization cross-section, since for any fixed line source the cross-section for one group of core levels may be near maximum, while for another group it may be near minimum. All that can be done to change the cross-sections in conventional equipment is to go to another fixed line source, which would not necessarily be of the right energy, and in any case the number of available line sources of any use has been seen to be very limited. Ideally one would like to have a continuously tunable source of high intensity at all the required energies. A close approximation to such a source exists in the form of synchrotron radiation.

It has been known for a long time that when a charged body is accelerated it emits radiation; the process is particularly efficient for an electron because of its very small mass. This effect is used in a synchrotron in which electrons are accelerated around an approximately circular torus (or 'doughnut') under the influence of pulsed magnetic fields. As they are accelerated around the curved path, the electrons emit light in a continuous spectrum whose intensity maximum is proportional to the radius of curvature and inversely proportional to the cube of the electron energy. Acceleration is taken to near-relativistic velocities, when the emitted radiation is concentrated in a very narrow cone tangential to the electron orbits, with an angular divergence of the order of 0.1 mrad (0.006°). It is therefore ideal for passing into a monochromator for energy selection. The general shape of the radiation spectrum of an electron moving in a curved orbit[24] is shown in Figure 2.13. Typically λ_c, the critical wavelength at the intensity maximum, is of the order of 0.1 to 0.4 nm, and the maximum energy would be between 1 and 10 keV. However, Figure 2.13 shows that towards longer wavelengths, i.e. lower energies, the intensity falls off quite slowly, so that there is still plenty of usable intensity at a few tens of electronvolts.

Typically a synchrotron radiation source would have a radius of many metres and the electron beam would have an energy of several GeV and current up to 1A, so it is not a device to be purchased ready-made from a

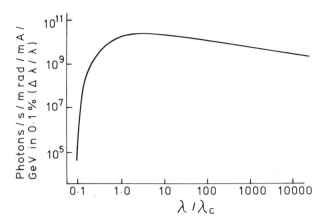

Figure 2.13 Shape of the radiation spectrum of an electron travelling in a curved orbit. The vertical scale of photon flux is a function only of the electron energy and current, while the horizontal scale is defined by λ_c, the so-called critical wavelength. If the maximum intensity is required to be at an energy near 1000 eV, then λ_c should be 1 °A or less. To achieve that the orbiting electrons must be accelerated to several gigaelectronvolts, and the radius of the orbit should be of the order of a few metres. (After Farge and Duke[24])

manufacturer and installed in one's laboratory. They are so large that they can only be nationally owned and would-be users have either to take their spectrometers to one or use the spectrometers available on site.

Despite the continuous nature of the spectrum of the emitted radiation, the synchrotron is not yet quite as continuously tunable as one would like because selection of the required energy or range of energies has to be performed by monochromators. So far monochromator design has enabled the energy region from about 20 eV to about 500 eV to be tunable continuously, but there is still quite a big gap to the higher energies of the discrete line sources. It is expected that this gap will be filled, but design constraints include operation in UHV and bake-out to reduce surface contamination of reflecting surfaces in the monochromator. Pressure to extend the range to 1000 eV or more comes from the fact that the X-ray photon flux from a typical synchrotron in that part of the spectrum is at least two orders of magnitude higher than from a conventional Al $K\alpha$ source. Even with losses in the monochromator, the flux will still be greater than available now.

2.4.2 Electron sources

In conventional AES the only function of the incident electron beam is to produce ionization in core levels in order to initiate the Auger transitions. For that purpose alone the energy in the beam is unimportant. If, however, electron loss spectra are to be studied then the line width, by analogy with X-ray sources, should be minimized. Without energy selection the width will be limited by the thermal spread, according to the Boltzmann distribution, to 0.3–0.5 eV. Such a width would be quite unacceptable in low energy electron loss spectroscopy (LEELS) in which vibrational spectra are studied, but a discussion of LEELS is outside the scope of this volume. For AES, then, no steps need be taken to reduce the primary beam width.

The two types of electron source used in AES are thermionic and field emission, and of these the former is by far the most common. All materials possess a work function, which is simply the energy necessary to remove an electron completely from the material. Work functions are usually[25] of several electronvolts magnitude and thus represent a barrier at the surface to the escape of electrons; at room temperature, and without any applied electric fields, electrons cannot escape and there is no emission. The two types of source use different physical methods to allow sufficient numbers of electrons to overcome the barrier and thus produce emission from the material. In the thermionic method the material is simply heated to a temperature high enough to give some of the electrons sufficient additional energy to pass over the barrier into the vacuum, so that the higher the temperature the greater is the thermionic current. The emission is governed by the Richardson equation:

$$J = A(1 - \bar{r})T^2 \exp(-e\phi/kT) \qquad A/cm^2 \qquad (2.3)$$

where J = current density
A = Richardson constant, ideally equal to 120 A/(cm^2 deg^2) but varies with material
$\bar{r}$ = zero-field reflection coefficient for incident electrons
T = absolute temperature
ϕ = work function

Field emission operates rather differently. Basically it does not set out to give the electrons in the material additional energy, but to reduce the height of the work function barrier so that even at room temperature electrons can escape. Since most metallic work functions are between 4 and 5 eV[25] and the work function barrier extends over only a few nanometres, it can be calculated at once that the field at the surface due to the barrier is very large. To attempt to eliminate it completely by the application of an equal and opposite field would be quite impracticable, but that is not necessary since the application of even a moderate field will cause sufficient reduction in the barrier to produce adequate emission. This is because electrons have a finite probability of passing through the barrier by 'tunnelling'; the probability is too low at room temperature and under field-free conditions to be measurable, but when the height and shape of the barrier are altered sufficiently by the application of a high electric field, then the probability becomes useful. To achieve the high field required the material is fashioned into a sharp point with a tip radius of typically 50nm, and a positive electrostatic field of many kilovolts is applied between it and an extraction electrode. The process is called field emission and is governed by the Fowler–Nordheim expression:

$$J = (1.55 \times 10^{-6} E^2/\phi)\exp[-6.86 \times 10^7 \phi^{3/2}\theta(x)/E] \qquad A/cm^2 \quad (2.4)$$

where J = current density
  E = field strength
  $\theta(x)$ = Nordheim elliptic function
  x = $3.62 \times 10^{-4} E^{1/2}/\phi$

It will be noted from the expression (2.4) that the current density in field emission is strongly dependent on the value of the work function of the emitting surface, so that it is standard practice to use for the emitter a needle-shaped single crystal oriented with a crystal plane of low work function at the tip. Since the adsorption of residual gases in a vacuum system onto a clean metal surface almost always increases the work function, it is necessary, if the current density is not to decrease as a result, to maintain a particularly good level of UHV around the emitter. Before use any accumulated adsorption is removed by 'flashing' the emitter, i.e. raising its temperature briefly to a very high value. Because of the flashing necessity, and also because the emitter material must withstand the high electrostatic field without disintegrating, tungsten is always used—oriented so that a low work function plane such as the (310) is parallel to the tip.

The two types of source, thermionic and field emission, each have their respective advantages and disadvantages. Because the emitting area of a field emission source is so small and the emission is concentrated in a small angle, the current density per unit solid angle obtainable from it is much higher than from a thermionic source, and for this reason field emission sources are known as 'high brightness' sources. Figures quoted[26] are in excess of 10^7 A/(cm² sr) compared to a typical figure of 10^4 A/(cm² sr) for a tungsten thermionic filament. By using an emitter of lower work function, higher brightnesses can be obtained[26] for thermionic emitters, e.g. of the order of 10^6A/(cm² sr) for an LaB$_6$ crystal, but against these high brightness figures must be set both the greater difficulty of fabricating both field emission and LaB$_6$ sources and their variable reliability. A conventional tungsten thermionic filament may have a lower brightness, but it is relatively easy to make and to replace, its emission properties are entirely reproducible, it can be used over a wide range of emission currents and it is robust. With progressively improving technology, the emphasis will be placed more and more on the higher brightness sources, but at the time of writing the great majority of electron sources still use tungsten filaments in the thermionic mode.

The principal direction of development of electron sources for AES has always been, and still is, towards the production of ever-smaller irradiated areas, or spot sizes, on the sample from electron guns whose construction must be compatible with UHV. Until recently the latter requirement has meant that the electron optics of a gun have been entirely electrostatic, since there were difficulties in designing and fabricating the pole pieces and coils for electromagnetic optics that would withstand bake-out for UHV. However, these difficulties have been largely overcome in the high spatial resolution AES equipment available in the last few years, as will be seen below.

In purely electrostatic guns using tungsten thermionic filaments, the available spot size has decreased progressively from the original 400 μm at $\sim$20 μA and 2.5 keV obtained with a converted oscilloscope gun, to 40 μm at $\sim$5 μA and 5 keV in the guns used in the mid 1970s, to 5 μm at $\sim$10^{-7}A and 5 keV in the late 1970s and to 0.5 μm at $\sim$10^{-8} A and 10 keV typical of the present. With particularly careful alignment of filament and optics, with accurate fabrication of apertures and with attention to the elimination of stray magnetic fields, it is possible to achieve a spot size of $\sim$0.2 μm in current electrostatic guns, but for general purposes 0.5 μm is more usual. It is probable that for a simple electrostatic system operating at a voltage that does not need bulky insulators or expensive high-voltage supplies, the performance has now been virtually optimized.

One of the commercial designs for such an electrostatic gun is shown schematically in Figure 2.14. The assembly carrying the V-shaped tungsten filament and the first electrode, the Wehnelt cylinder, is mounted on a flexible bellows and has alignment screws to align the filament and Wehnelt aperture with the rest of the optics. Alignment of the filament itself with respect to the Wehnelt is achieved by a jigging arrangement, but it is difficult to make such an arrangement completely reproducible and optimum alignment is still a matter of experience.

Focusing and shaping of the electron beam is performed by two condenser lenses and an objective lens, and in addition any astigmatism (focusing in the X and Y directions not coincident) can be corrected by a set of stigmator poles before the objective lens. Finally, scanning of the beam is carried out by X and Y scanning poles at the exit from the gun. The gun is pumped through the final aperture and since that is small, with a consequently low pumping speed to the body of the gun, and particularly to the filament region, care has to be taken in outgassing a new filament so that an excessive pressure rise does not occur around it.

As remarked above, for the achievement of much smaller spot sizes, a change has to be made from electrostatic to electromagnetic optics. It is also necessary to operate at higher incident energies, in the range of 10–30 keV, since for electron optical reasons smaller spot sizes can be reached only if the energy in the electron beam is increased. For these reasons, reduction of a

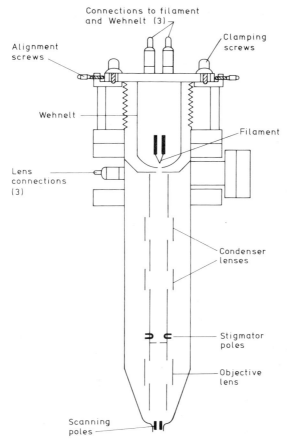

Figure 2.14 Electrostatically focused electron gun designed to produce beams of electrons of energies up to 10 keV. At beam currents between 10^{-9} and 10^{-8}A the optimum spot size on the specimen is about 0.5 μm. The filament, which is a tungsten hairpin, can be aligned with the Wehnelt cylinder by a jigging arrangement, while alignment with the optical axis of the gun is achieved by adjustment of the whole assembly via screws and a flexible bellows. (Reproduced by permission of Vacuum Generators Scientific Ltd, East Grinstead)

further order of magnitude in the spot size, from ~0.5 μm, involves a corresponding increase in the complexity of the electron gun and in the cost. One has to be quite sure that the additional information to be gained is worth the effort!

Just as for electrostatic guns, various designs of electromagnetic gun have appeared, but they are of necessity rather similar in basic layout. An example is given in Figure 2.15. There are two electromagnetic condenser lenses for 'squeezing' the electron beam to a small diameter, and an electromagnetic

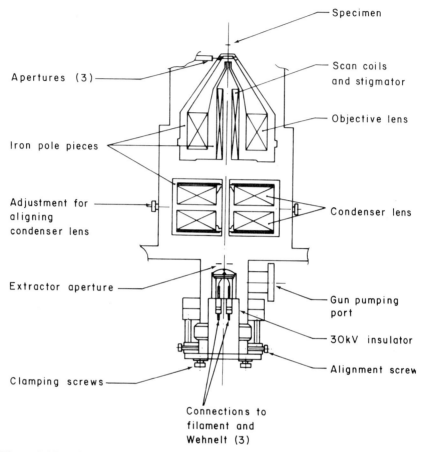

Figure 2.15   Electromagnetically focused electron gun designed to produce beams of electrons of energies up to 30 keV. At beam currents of a few nanoamperes the optimum spot size on the specimen is about 50 nm. As for the gun illustrated in Figure 2.14, the filament assembly can be aligned on the axis of the gun via alignment screws operating through a flexible bellows. Since the volume of the gun is much greater than that in Figure 2.14, a port is provided for separate pumping. The pole pieces and windings are bakeable to 160 °C. (Reproduced by permission of Vacuum Generators Scientific Ltd, East Grinstead)

objective lens for focusing. The windings for the lens are all totally enclosed, thus reducing the problems resulting from bake-out, although because of the lenses the maximum bake-out temperature has to be restricted to 160 °C. Filament mounting is similar to that in the electrostatic gun, and the assembly can be aligned as before with alignment screws operating on a bellows. Since the maximum voltage of operation is now 30 kV, however, the whole assembly is necessarily bulkier because of the additional insulation requirements.

Further alignment of the electron beam with the electron optical axis of the gun can be performed by adjusting the position of the condenser lenses themselves with set screws. Again the corrections for astigmatism and the scanning of the beam are carried out before the beam passes through the final aperture, and in this case the stigmator and the scanning coils are built into the objective lens, which is itself shaped in such a way as to allow reasonable access to the specimen. Instead of a fixed final aperture there is a sliding plate perforated with three apertures of different sizes, any of which can be selected by moving the plate into a predetermined position. Because of the sheer volume of the gun and the very low pumping speed through any of the final apertures, the gun is in this case separately pumped through a side port.

In the long run, the ability to achieve and to maintain over a useful length of time a spot size of the order of a few tens of nanometres depends not so much on the basic design, which tends inevitably to be fixed, but on the precision with which the lenses can be fabricated, particularly on the machining and alignment of the iron pole pieces. Also, it is important to construct the gun with very good mechanical stability and to be able to eliminate stray alternating magnetic fields completely from the gun and specimen region. The differences in performance between different manufacturers' guns will depend principally on those factors rather than on the possibility of producing a radically new design.

Field emission sources have also been used in electromagnet electron optics of the type illustrated in Figure 2.15, although rarely. The combination of the field emitter operated at high voltages (20–60 kV) and electromagnetic optics has pushed the progressive reduction in spot size to its present minimum of around 20 nm, but as yet very few spectrometers of that type are used in AES. However, field emission sources have been used more often with electrostatic optics, and this combination has achieved spot sizes comparable to those obtained with the thermionic source used with electromagnetic optics. A schematic example[27] is shown in Figure 2.16, along with the dimensions, from which it can be seen that the overall size is not large. The performance of this particular gun is that at a sample current of 10^{-9} A and a beam energy of 21 keV the spot size is 50 nm, at a working distance of 55 mm. An additional feature is the filament F located close to the first anode A1 used for outgassing the latter; one of the limitations in operation was found to be noise generated at the field emitter tip by ions released from the surface of A1 by electron bombardment.

Direct comparisons of the three emitters, viz. thermionic tungsten, LaB_6 and field emitting tungsten, indicate[26] that at beam currents above about 10^{-8} A, the LaB_6 source is superior to both thermionic and field emitting tungsten from the points of view of the spot size obtainable and the signal-to-noise ratio. At beam currents below 5×10^{-9} A, however, the field emission source is superior on both counts.

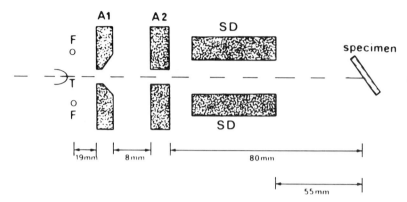

Figure 2.16   Schematic diagram of an electron gun using a field emission electron source. The field emitting tip T is that of a tungsten single crystal wire oriented [310] along its axis. F is a filament located close to the first anode A1, intended for outgassing the latter. A2 is the second anode and SD an eight-pole stigmator/deflector. At a beam energy of 21 keV and current of 10^{-9}A the gun focuses to 50 nm spot size at a distance of 55 mm from the SD. (After Prutton *et al.*[27])

2.4.3   Ion sources

Although it is certainly possible to excite Auger emission with primary ions, and there are several groups of workers whose principal study is ion-excited AES, the intention in this section is to discuss the devices used for ion erosion of surfaces, as described in Section 2.3.3. There are now several different types of ion source available which have been designed with surface cleaning and ion profiling in mind, each with its own characteristics and mode of operation, and the choice of type in any one situation is still partly a matter of personal preference and partly what is offered by the supplier of one's spectrometer.

The simplest type, now not used very much, is rather similar in construction and operation to a standard ionization gauge, and is shown schematically in Figure 2.17. As in the gauge a hot filament provides electron emission to a positively biased grid, the electrons making several oscillations within the grid space before being collected. Argon to pressures between 10^{-5} and 10^{-4} torr is admitted to the system, and the argon gas in the grid space is ionized. Unlike the ionization gauge the ion gun has no positive ion collector, but instead the positive argon ions are accelerated from the grid space towards the cylindrical electrode surrounding the filament and grid through a negative potential gradient of about 200 V. They pass through meshed apertures in that electrode and in an outer earthed screening can, the apertures being offset from the gun axis to avoid possible contamination effects from the hot filament. The entire assembly of filament, grid and accelerating electrode can be

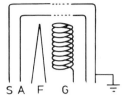

S A F G

Figure 2.17   Design of an ion gun based on an ionization gauge geometry. The hot filament F emits electrons to a positively biased grid G, within which the electrons make several oscillations before being collected. With argon admitted to the gun at pressures between 10^{-5} and 10^{-4} torr, argon ions are produced by ionization in the grid space. The positive ions are then accelerated from the grid space towards a cylindrical electrode A surrounding the filament and grid, through a potential gradient of about 200 V. The ions pass out into the system through meshes in A and S, an outer screening can at earth potential. The mesh apertures are offset to avoid any contamination from the filament striking the specimen. By floating the filament–grid–accelerator with respect to earth, the ion energy can be varied from 200 to 500 eV. At an argon pressure of 10^{-4} torr, the ion current density on the specimen is about 1 μA/cm^2

floated on a positive potential with respect to the screen so that the energies of the emerging ions in the beam can be varied from 200 to 500 eV.

The 'ion gauge' type of ion source is not very powerful, producing a maximum ion current density of only about 1 μA/cm^2 at the maximum working pressure of 10^{-4} torr. Its advantages are cheapness, reliability and relative purity of ion beam, since few energetic neutral atoms are produced in the ionizing region.

Probably the most popular ion sources in current use employ gas discharges to produce ionization of the gas in the gun. The discharge can be initiated either by Penning discharge or by a high-voltage discharge. In the former the combined action of magnetic and electrostatic fields on argon at a high pressure in a confined space produces a cold-cathode discharge from which the positive ions are extracted. A typical arrangement is illustrated diagrammatically in Figure 2.18. Argon gas is piped directly into the back of the gun and thence to the discharge chamber rather than being present from back-filling of the entire analysis chamber, as happens with the 'ion gauge' type. After extraction the positive ion beam can be focused to a greater or lesser extent, depending on the sophistication of the lens system used. The simpler guns would be incapable of producing an ion spot size of less than 3 mm diameter; the more complex ones can focus to diameters of about 0.1 mm if the sample is at the correct distance from the final aperture. With the finer focused beams there are many advantages to be gained by rastering the beam on the surface, and so X and Y deflection plates are added after the focusing lens system. Guns of this type operate at accelerating voltages between 2 and 10 kV and, since the pumping speed through the gun to the discharge region is low, at

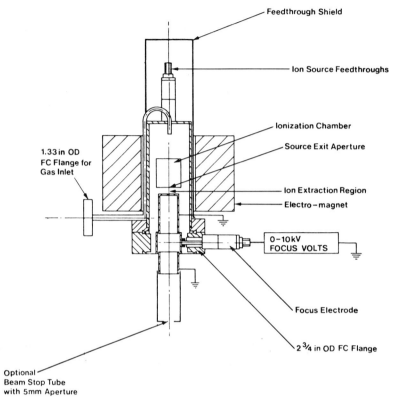

Feedthrough Shield

Ion Source Feedthroughs

Ionization Chamber

Source Exit Aperture

1.33 in OD
FC Flange for
Gas Inlet

Ion Extraction Region

Electro-magnet

0-10kV
FOCUS VOLTS

Focus Electrode

2 ¾ in OD FC Flange

Optional
Beam Stop Tube
with 5mm Aperture

Figure 2.18   Design of an ion gun using a Penning discharge for production of positive ions. Argon gas is fed directly into the back of the gun in this case, rather than being admitted to the whole system, and since its exit speed is restricted the pressure in the interior of the gun is much higher than the pressure of argon in the system. During operation the pressure in the system is between 3×10^{-7} and 3×10^{-6} torr. A high voltage and an axial magnetic field are applied to an ionization chamber in which a discharge is induced and positive ions are extracted from the discharge into a focusing region. Ion energy can be varied betwen 500 eV and 10 keV, and the gun can deliver between 10 and 100 μA into an area of diameter about 5 mm, according to the pressure of argon in the source. (Reproduced by permission of Vacuum Generators Ltd, East Grinstead)

pressures in the analysis chamber between 3×10^{-7} and 3×10^{-6} torr. The ion current on the sample varies with both voltage and pressure, and as an example of performance it would provide about 20 μA into a spot of diameter 5 mm at an accelerating voltage of 5 kV and a chamber pressure of 4×10^{-7} torr. The focusing voltage would then be about 3.6 kV. Under the extreme conditions of 10 kV and 3×10^{-6} torr it would deliver 200 μA.

The greatly increased current density of such ion gun types over the 'ion gauge' type is somewhat offset by one or two disadvantages. Ions inside the gun can strike other internal surfaces as well as being extracted into the beam, and hence sputtered material gradually accumulates; when the accumulation on insulating surfaces becomes sufficient breakdown or arcing occurs and the action of the gun becomes both unreliable and unreproducible. At that stage there is no option but to dismantle the gun and clean it according to the manufacturer's instructions, and it is generally desirable to keep a stock of spare insulating parts for replacement. The other disadvantage is that the beam emerging from the gun is not pure, in that as well as positive ions there is an appreciable proportion of energetic neutral atoms. Clearly it is only the current of positive ions striking the sample that can be measured, so the additional erosion effect of the neutrals is at the moment a little-known quantity. However, all the discharge-type ion guns suffer from this same drawback to a greater or lesser extent.

Ion sources that produce ion beams by high-voltage discharge depend on the geometry of their structure to increase the path lengths of the oscillating electrons and hence the ionization probability per electron. There are various possible geometries, but they all aim to shape the electrostatic field so that electrons are trapped for a significant length of time in one or more potential wells in the field. Because of the field configuration usually employed, this type of source is known as the 'saddle field' source, and a diagram of one of the smaller versions is shown in Figure 2.19. A pair of tungsten wires, surrounded for most of their length by shielding sleeves, protrude into the body of the gun, which in this case is spherical. A high positive voltage between 5 and 10 kV is applied to the wires, the body being earthed. Argon is admitted directly to the gun at a pressure in the region of 10^{-4} torr, whereupon a discharge is set up between the anode wires and the body, the latter acting as the cathode. The field configuration then constricts the electron paths to the 'figure-of-eight' shown in Figure 2.19, most of the ionization of the argon taking place in the loops at each end of the field. Positive ions are extracted through a slit cut in the body in the same plane as that of the field; the ion beam maintains the same width as that of the slit in the plane of the figure, but diverges progressively from the slit width in the directions above and below the plane. By adjusting the distance of the sample from the slit, the shape of the bombarded area can thus be altered from rectangular to square as desired. Since the only pumping path to the system is through the slit, pressures between 10^{-6} and 10^{-5} torr can be maintained in the system, with adequate pumping speed.

Saddle field sources are capable of providing very high ion current densities. Even a small one like that of Figure 2.19 can produce a density of 200 μA/cm^2 at 6–7 kV, while the larger ones are so powerful that they tend to be used not so much for ion bombardment in UHV as for what is called 'ion milling', in which, as the name suggests, large quantities of material are

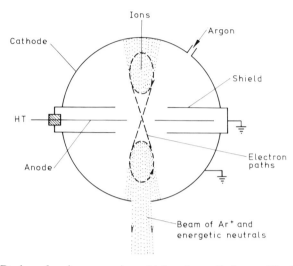

Figure 2.19   Design of an ion gun using a high-voltage discharge. The high voltage is applied to a pair of tungsten wires protruding into the body of the gun, which in this case is spherical. Argon gas is fed directly into the gun from a pipe at the back, and as shown in the design, the pumping speed for argon in the gun is low, so that the pressure inside is much higher than the pressure outside. A discharge is set up between the wires and the body and the electron paths are then constricted to the 'figure-of-eight' paths shown, with most of the ionization of the argon taking place in the loops at each end. Positive ions emerge through a slit cut in the body in the same plane as that of the field. Because of the shape of the field configuration this type of source is also known as the 'saddle field' source. At ion energies of 5–7 keV such a source can produce 100–200 μA of positive ions into an area of about 1 cm². (Reproduced by permission of Ion Tech Ltd, Teddington)

removed. As with the other sources, however, they have their disadvantages as well as their advantages. Unlike the sources using a Penning discharge, they tend to be unstable at voltages much below 5 kV and difficult to control, and are therefore restricted in general application to the high-voltage end of their operating regime. In addition, as can be seen from Figure 2.19, positive ions are not only extracted through the slit but bombard the interior of the body as well, so that it is possible for sputtered material also to pass through the slit and be deposited on the sample. It is therefore necessary either to fabricate the body from a metal that does not sputter easily, e.g. aluminium or titanium, or to insert an internal liner of such material. Finally, these sources do produce a rather high proportion of energetic neutral atoms in their beams, of the order of 50 per cent under maximum voltage operation, a higher proportion than from the types using Penning discharges.

The most recently developed[28] type of ion source is the field emission liquid-metal ion source, whose principle is illustrated in Figure 2.20. A needle whose tip has a radius in the range 1–10 μm is placed with its shank passing

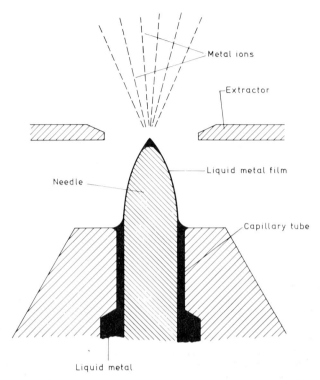

Figure 2.20   Schematic of the principle of operation of a liquid-metal field emission ion source. A needle with a tip of radius 1–10 μm dips into a reservoir of liquid metal through a close-fitting capillary tube. The liquid metal is drawn along the needle and over its tip by a capillary action, provided the liquid wets the material of the needle. A high voltage of between 4 and 10 kV is then applied to an extractor electrode placed a short distance in front of the tip, whereupon the liquid metal is drawn out into a cusp-like protrusion from which positive metal ions are extracted by field emission. Since high currents, of the order of 100 μA, can be drawn from the very small volume of the cusp, the source has a high brightness and the ion beam can thus be focused into a small spot while retaining usefully high currents. (Reproduced from Prewett and Jeffries[28] by permission of The Institute of Physics)

through a close-fitting capillary tube into a reservoir of liquid metal. The material of the needle is chosen so that the liquid wets it, but does not react with it. On wetting, the liquid metal is drawn up over the protruding needle and over its tip by capillary action, any loss being made good by flow from the reservoir. A high voltage between 4 and 10 kV is then applied between the needle and an extractor electrode positioned a short distance in front of the needle, whereupon the film of liquid metal covering the tip is distorted into a cusp-shaped protrusion. This protrusion is formed by the balance of forces due to electrostatic stress and surface tension. Due to the consequent very high electric field at the tip of the cusp, a beam of positively charged metal ions is formed, which is then extracted through a circular aperture in the extractor. Although the source is described as a field emission source, it is in fact not yet clear whether the ionization mechanism includes field desorption or field ionization, or both, or indeed other contributing processes. Metals used as liquids in these sources have included to date cesium, indium, gallium, tin, bismuth, copper and gold; of these cesium has received most attention due to its potential applications in ion milling and in SIMS.

For reasons similar to those for which the field emission electron source was described as a 'high brightness' source, the liquid metal ion source is also a high brightness source. The high currents obtainable, ranging up to 100 μA, coupled with the very small emitting volume imply very high brightness values; unfortunately it is difficult to be precise because the volume is not known with any certainty. The high brightness, however, just as in the case of the electron source, leads to the possibility of producing useful currents in a finely focused beam, enabling ion erosion to be performed at much higher spatial resolution in principle than can be achieved with the gas discharge types. Preliminary experiments have shown[29] that ion spot sizes of less than 0.5 μm can be achieved without much effort, and it seems likely that resolutions of 0.1–0.2 μm should soon be possible.

There are at the moment several unknown quantities in respect of liquid-metal sources, although much research and development is being carried out and it will not be long before much more is known about them. They do seem to suffer from instabilities in certain regions of operating parameters, and this may be a function of the mechanisms of formation of the liquid cusp. The composition of the ion beam, in terms of proportions of singly and doubly charge ions, etc., varies with the metal being used and to a certain extent with the applied voltage. Molecular clusters may also be formed. There is a considerable divergence of the ion beam from the aperture, so that a carefully designed focusing lens system is required to make optimum use of the high brightness. However, these drawbacks are not necessarily insuperable and it is expected that liquid-metal ion sources will play an increasing part in surface analytical technology.

2.5   Electron Energy Analysers

The device which measures the energies (or, more correctly, the velocities) of electrons emitted or scattered from a surface is the heart of any spectrometer. In this section the device will be called the 'analyser' to distinguish it from the term 'spectrometer', which has been taken in this chapter to mean the complete instrument, as defined in Section 2.1. However, it should be noted that many authors use the term 'spectrometer' when they intend it to apply only to the analyser; neither usage is entirely wrong or entirely right, and it seems usually to be a matter of personal preference and definition.

Over the years of development of XPS and AES various types of analyser have been put forward as being eminently suited to the techniques, but there now seems to be a general consensus of agreement as to those that are most suitable. Because of the low kinetic energies of the electrons to be analysed in the techniques, and because magnetic fields are difficult to produce and handle in UHV, the analysers are all of the electrostatic type. The two principal ones are the cylindrical mirror analyser (CMA) and the concentric hemispherical analyser (CHA), which are both dispersive analysers, that is to say, in each the action of a deflecting electrostatic field disperses the electron energies so that for any given field only those energies in a certain narrow range are measured. Also still in use, and not always to be despised, is the retarding field analyser (RFA) based on the spherical electron optics used for low energy electron diffraction (LEED), in which a retarding potential is imposed in front of a collector allowing only those electrons of energies greater than the potential barrier to reach the collector. The descriptions of these three analysers will be given in the approximate chronological order in which they appeared on the surface analysis scene. Before that can be done, however, the requirements of XPS and AES in terms of energy resolution must be established.

2.5.1 Energy resolution requirements

There are two definitions of energy resolution in current use and it is necessary to be clear about the differences between them. The first is the *absolute* resolution, ΔE, usually measured as the full width at half-maximum (FWHM) height of a chosen observed peak. An alternative measure sometimes used is the width ΔE_B of the peak at its base, called the base width, and obviously in an ideal situation $\Delta E_B = 2\Delta E$. The second is the *relative* resolution defined as

$$R = \Delta E/E_0 \qquad (2.5)$$

where E_0 is the kinetic energy at the peak position. The resolution R is usually expressed as a percentage, i.e. $(\Delta E/E_0) \times 100$. It is also common to find the relative resolution expressed as the *resolving power*, which is simply the reciprocal of R, i.e.

$$\rho = 1/R = E_0/\Delta E \qquad (2.6)$$

Thus the absolute resolution can be specified independently of peak position in a spectrum, but the relative resolution can be specified only by reference to a particular kinetic energy.

Now in XPS it is necessary for identification of possible differences in chemical states of elements to be able to apply the same absolute resolution to any photo-electron peak in the spectrum, i.e. at any kinetic energy. As has been shown, the natural (i.e. unmonochromatized) line widths of the commonly used soft X-ray sources are 0.70 eV for Mg $K\alpha$ and 0.85 eV for Al $K\alpha$. To match the absolute resolution to these natural line widths at the maximum available photo-electron energies (i.e. 1253.6 eV using magnesium, 1486.6 eV using aluminium), would require a relative resolution of $\sim 6 \times 10^{-4}$ (or, if preferred, a resolving power of ~ 1700). Such a relative resolution is not too difficult to achieve, but would be accompanied by an unacceptable loss in sensitivity. In addition, if monochromatized radiation were to be used, the absolute resolution needed, of about 0.2 eV, would imply a relative resolution of $\sim 10^{-4}$ (resolving power of 10000); such a relative resolution is very difficult to achieve without construction of an exceptionally large and expensive analyser.

For the above reasons, it is standard practice to retard the kinetic energies of the photo-electrons *either* to a chosen analyser energy, called the pass energy, *or* by a chosen ratio. In either case the pass energy or the ratio is kept fixed during the recording of any one spectrum. Thus, for example, if retardation to a pass energy of 50 eV were chosen an absolute resolution of 0.7 eV would require a relative resolution of only $\sim 10^{-2}$. Retardation therefore enables the same absolute resolution to be obtained for a lower relative resolution, but, as will be seen later, brings certain penalties with it.

In AES, the natural line width of the exciting electron energy is irrelevant since the Auger line width is independent of that of the incident energy. (If electron energy loss spectroscopy is to be performed as well, the situation is of course different.) The energy resolution requirements are set by the inherent widths of the Auger peaks themselves. Since most Auger peaks are several electronvolts wide, there is not the need for the same absolute resolution as in XPS, and an average absolute resolution of 2–3 eV is adequate. In practice it is desirable to have high sensitivity at high Auger energies (e.g. >1000 eV) and high absolute resolution at low Auger energies (e.g. <150 eV); this is achieved by operation at a constant relative resolution rather than constant absolute resolution.

A typical relative resolution for an Auger analyser would be 4×10^{-3} (resolving power 250). Under those conditions the absolute resolution at an Auger energy of 50 eV would be 0.2 eV, which is more than adequate, and at 1000 eV it would be 4 eV. Although the absolute resolution at high energies is degraded, the sensitivity is much higher at 1000 eV than at 50 eV.

If high resolution Auger spectra are to be studied, for those materials or situations in which the Auger widths happen to be narrow,[30] then the spectra are recorded by X-ray excitation with an analyser operated in the XPS mode, i.e. with pre-retardation. In this case only selected regions of the total spectrum are of interest, and any variations in sensitivity or in absolute resolution over an extended energy range are irrelevant. This variation of AES is sometimes called XAES (X-ray-excited Auger electron spectroscopy) or HRAES (high resolution Auger electron spectroscopy).

2.5.2 Retarding field analyser (RFA)

Long before XPS and AES were established the technique of LEED was using electron optics of spherical symmetry to display surface diffraction patterns by acceleration of the diffracted electron beams onto a phosphor-coated collector or screen. The original arrangement consisted of two grids concentric with the screen, the inner grid being grounded, the second grid held at a negative potential close to that of the primary energy and the screen held at a

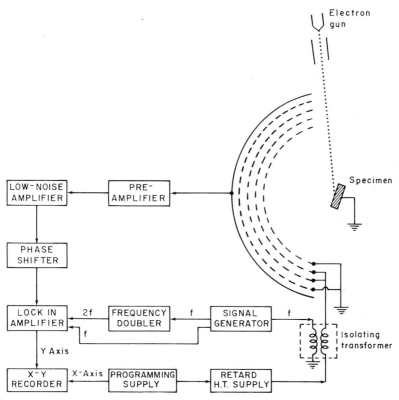

Figure 2.21   Schematic arrangement of a retarding field analyser (RFA) for AES based on four-grid LEED optics. The LEED electron gun would normally be shown protruding through the collector (screen) and grids, but is omitted here because for most analytical purposes it is unsuitable, since the electron energy range it can produce is too restricted and the spot size is too large. The retarding potential applied to the two inner grids is modulated with a small potential from a signal generator, via an isolating transformer, and the amplified signal from the collector is compared in the lock-in amplifier with either the fundamental or the first harmonic of the modulation frequency. The first comparison gives the energy distribution $N(E)$ according to expression (2.8) and the second the differential distribution $dN(E)/dE$ according to expression (2.9)

high positive potential. After it was realized[31] in about 1967 that the LEED optics could be used as an energy analyser the number of grids increased quickly to three and then four, with the final (and present) arrangement being that of Figure 2.21.

In the four-grid RFA of Figure 2.21 the incident electrons come either from the LEED gun projecting through the optics, to strike the sample at normal incidence, or from an auxiliary gun outside the optics, for grazing incidence excitation. The inner and outer grids are grounded, in order to provide a uniform field in the sample region and to shield the collector (i.e. the LEED screen) from capacitative coupling with the modulation on the other grids, respectively. The middle two grids are connected together and have the retarding potential applied to them; the reason for using two grids there is to reduce the serious lens effects suffered by electrons passing through the holes in a mesh. Superimposed on the retarding potential is a small modulating voltage from a signal source. The modulation is usually sinusoidal but need not be so.

The modulated collector current is then compared in a phase-sensitive detector (otherwise known as a lock-in amplifier or synchronous detector) with the reference signal from the signal source. Amplification of the component of collector current at the fundamental frequency ω then gives the energy distribution $N(E)$, while amplification of the component at the frequency 2ω gives the differential distribution $N'(E)$. These relationships can be derived quite easily by expansion of the signal as a Taylor series and algebraic rearrangement. Thus the collected current I is a function of the retarding potential V and, if the modulation is of the form $k \sin \omega t$, I can be written $I(V + k \sin \omega t)$. The Taylor expansion is then

$$I(V + k \sin \omega t) = I + k \sin \omega t (dI/dV) + \frac{k^2 \sin^2 \omega t}{2!} (d^2I/dV^2) + \cdots \quad (2.7)$$

When this is rearranged algebraically, the coefficient of $\sin \omega t$ is found to be

$$k(dI/dV) + (k^3/8)(d^3I/dV^3) + \ldots \quad (2.8)$$

and the coefficient of $\cos 2\omega t$ to be

$$(k^2/4)(d^2I/dV^2) + (k^4/48)(d^4I/dV^4) + \ldots \quad (2.9)$$

Sinse (dI/dV) is the energy distribution $N(E)$ and (d^2I/dV^2) the differential distribution $N'(E)$, it can be seen that the amplitude of the first harmonic is proportional to the former and of the second harmonic to the latter, *provided* the higher order derivatives can be neglected in each case. This will be so only if k is small enough. In practice this criterion will be obeyed if the modulation amplitude is less than half the width of an Auger peak; since the widths of Auger peaks vary, the modulation should obviously be chosen to resolve without distortion the narrowest peak in the spectrum.

Enhanced sensitivity can be obtained by using large modulating voltages, but at the cost of a distorted spectrum.

With sufficient care taken in construction and with some modifications to the grid spacings normally used for LEED, a relative resolution of about 0.3 per cent has been achieved for an RFA. For most RFA systems, though, the usually obtainable resolutions are in the range 0.5–1.0 per cent, which are just about adequate for AES. Where the RFA tends to score over the dispersive analysers is in the very low kinetic region, below 100 eV, since there a relative resolution of 0.5 per cent implies an absolute resolution of less than 0.5 eV, resulting in high resolution spectra.

Despite its inherent simplicity, the RFA is not suitable for general AES applications because of one overriding disadvantage, and that is that at any one retarding potential on the grids, *all* electrons with energies greater than that potential are allowed to reach the collector. Shot noise is thus generated by many electrons of energies other than those being analysed, with the result that the signal-to-noise characteristics are poor. There are also other problems such as uniformity of grid structure and the parallelism of the grids, for minor deviations from strict geometry can have devastating effects on the resolution obtainable. However, for those whose main research effort is devoted to the application of LEED to surface structural studies, the RFA presents a convenient means of checking the cleanliness of a surface at any stage in an experiment.

2.5.3 Cylindrical mirror analyser (CMA)

The drawbacks inherent in the RFA led the early users of AES to seek other types of energy analyser that would be better suited to the technique, and it was soon realized[32] that the CMA had many properties that made it almost ideal for the purpose. Over the years various commercial and home-made versions of the CMA have appeared, but in general the differences between the versions have been in details of construction, the basic form and mode of operation remaining unchanged.

The CMA is shown diagrammatically in Figure 2.22. Two cylinders of radii r_1 (inner) and r_2 (outer) are positioned accurately coaxially. The potential of the inner cylinder is at earth; that of the outer cylinder is $-V$. Electrons emitted from a source on the axis at an angle α pass through an aperture in the inner cylinder and those of a particular energy E_0 are deflected by the outer cylinder potential through another aperture to a focus on the axis. The general relationship between V and E_0 is

$$\frac{E_0}{eV} = \frac{K}{\ln(r_2/r_1)} \qquad (2.10)$$

CMAs can and have been built for a variety of entrance angles α, but for

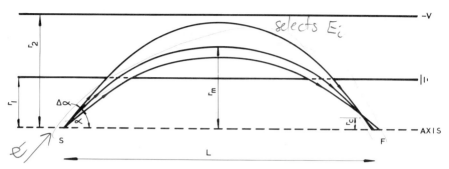

Figure 2.22   Diagrammatic arrangement of a cylindrical mirror analyser (CMA). The radii are r_1 for the inner cylinder and r_2 for the outer cylinder. The inner cylinder is earthed and a potential $-V$ is applied to the outer cylinder. Electrons emitted from a source S on the axis with a kinetic energy E_0 are re-focused at F according to the expression (2.10). The entrance angle α is chosen to be $42°\ 18'$, since at that angle the CMA becomes a second-order focusing device. A typical angular aperture $\Delta\alpha$ would be $6°$. L is the distance between S and F, r_c is the position of the minimum trace width and r_m the maximum distance off the axis for electrons entering the analyser at $42°\ 18'$. (After Bishop, Coad and Rivière[36])

the special case where $\alpha = 42°\ 18'$, the CMA becomes a second-order focusing instrument.[33] For that case $K = 1.31$, and the above relationship between E_0 and V becomes

$$\frac{E_0}{eV} = \frac{1.31}{\ln(r_2/r_1)} \qquad (2.11)$$

Apertures are in practice of finite width, and in any case it is desirable to achieve maximum sensitivity compatible with adequate resolution by operating the CMA with apertures that accept a spread of angles about α, typically with $\Delta\alpha = 6°$. Under these conditions the electron paths through the analyser, depicted in Figure 2.22, consist of annuli rather than an infinitely narrow trace, and it can be shown[33,34] that the minimum trace width ω_m is given, for $\Delta\alpha < 6°$, by

$$\frac{\omega_m}{r_1} = 7.76(\Delta\alpha)^3 \qquad (2.12)$$

The important feature of the minimum trace width is its position, which is off-axis and slightly ahead of the focal position. Its distance r_c off the axis is given by

$$\frac{r_c}{r_1} = 5.28(\Delta\alpha)^2 \qquad (2.13)$$

The importance lies in the fact that if an additional aperture is placed[35] at the position of minimum trace width, and made equal in size to it, a useful improvement in resolution is obtained with little sacrifice in sensitivity. Some designs incorporate[36] the possibility of varying the width of the additional aperture, but this is not an operation that is easy to arrange for remote control, i.e. from outside the vacuum system, and therefore the tendency is to fix that aperture size at some convenient compromise and to allow variation instead of the sizes of the principal apertures.

For the unique angle 42° 18' the distance on the axis between the source and focus is $6.1r_1$, and the other important constructional parameter is the maximum distance off the axis reached by electrons that are to pass through the exit apertures. This distance is given by[29]

$$\frac{r_m}{r_1} = \exp(1.3 \sin^2 \alpha) \tag{2.14}$$

Thus all the CMA parameters scale with the radius r_1 of the inner cylinder, and once that is chosen the rest follows more or less automatically.

Not shown in Figure 2.22 for reasons of clarity are some other important operational requirements, namely the fringe field plates. For practical reasons, i.e. access to the specimen position, the source of ejected electrons must lie *outside* the end of the CMA, and it can be seen at once that such an arrangement involves terminating the analyser rather abruptly. In fact, all commercial CMAs are cut back from their axes, so that their ends are cone shaped. However, the electrons passing through the analyser must see only uniform radial fields, otherwise there will be severe aberrations, so the field in the analyser must be terminated at each end without introducing field distortion in the aperture regions. In some CMAs this is achieved[36] by a series of ring-shaped electrodes whose potentials are adjusted via dividing resistors to ensure a smooth transition from the deflecting potential V to earth, in others by ceramic discs coated with a film whose radial resistance performs the same function as the dividing resistors, but in a more continuous fashion.

As in all dispersive analysers, simply scanning the potential $-V$ applied to the outer cylinder of a CMA gives the energy distribution of electrons passing through it directly, but it is important to note that because the transmission of the CMA varies as E_0 the distribution derived is not $N(E)$ but $EN(E)$. Thus modulation of V by a small sinusoidal signal, as described for the RFA, can still be used to extract a differential distribution, in this case by tuning the phase-sensitive detector to the fundamental frequency ω, and not to the second harmonic, but again the recorded distribution will be $EN'(E)$ rather than $N'(E)$.

The energy resolution of the CMA and CHA can be written [36, 37] in terms of the base resolution ΔE_B as

$$\frac{\Delta E_B}{E_0} = Aw + B(\Delta \alpha)^n \tag{2.15}$$

where w = slit width (equal at the entrance and exit) and A, B and n are constants. A more approximate expression[38] involving the half-width ΔE is

$$\frac{\Delta E}{E_0} = \tfrac{1}{2}Aw + \tfrac{1}{4}B(\Delta \alpha)^n \qquad (2.16)$$

which is valid only if $B(\Delta \alpha)^n \approx Aw/2$. For the CMA and CHA the values of the constants are given in Table 2.2.

Insertion of the values for the CMA into the two expressions above gives, for the base resolution,

$$\frac{\Delta E_B}{E_0} = \frac{0.36w}{r_1} + 5.55(\Delta \alpha)^3 \qquad (2.17)$$

and, for the half-width resolution,

$$\frac{\Delta E}{E_0} = \frac{0.18w}{r_1} + 1.39(\Delta \alpha)^3 \qquad (2.18)$$

In the case of the CMA the relatively wide angular apertures used can limit the resolution obtainable; in effect the source size replaces the slit widths in the above expressions. As an example, a CMA built with an inner radius r_1 of 3 cm and a semi-angular aperture $\Delta \alpha$ of 6° will have a theoretical half-width resolution $\Delta E/E_0$ of 0.15 per cent for a source size of 5 μm. However, for greater source sizes, the theoretical resolution degrades as follows: 10 μm, 0.16 per cent; 50 μm, 0.18 per cent; 100 μm, 0.21 per cent; 500 μm, 0.48 per cent; 1 mm, 0.80 per cent. The practical resolution obtainable will of course depend on the accuracy of fabrication and on stray magnetic fields, but in a reasonably well-constructed analyser, the degradation from these effects might cause the half-width resolution to deteriorate from 0.16 to 0.25 per cent for a 10 μm spot size.

The simple (single-pass) CMA as described up to now has been most effective in AES because of its very high transmission. This is the most important property for an AES analyser because the source area in AES using a highly focused electron beam is invariably much smaller than the acceptance area of the analyser. In other words, luminosity, which is the product of acceptance area and transmission, is not particularly relevant in AES. On the other hand, for XPS luminosity is all-important because, as described, standard unmono-

Table 2.2 Values of coefficients giving the base and half-width energy resolutions of the CMA and CHA

	A	B	n
CMA (42.3°)	$0.36/r_1$	5.55	3
CHA (180°)	$1/R_0$	1	2

chromatized X-ray sources are not focused but flood a large area of the sample with X-rays. In addition, XPS needs in general a greater absolute resolution than does AES. For these reasons, the single-pass CMA is not suitable for XPS. Another reason is that the precise energetic position of a photo-electron or Auger peak is so dependent on sample position in front of a single-pass CMA that reproducibility adequate for XPS would be totally lacking. (For example, Sickafus and Holloway[39] have shown that at 1500 eV a shift of 1 mm in the axial sample position causes a peak shift of 17.5 eV.)

The high transmission characteristics of the CMA are attractive for XPS, but higher luminosity should be achieved while maintaining adequate resolution. By using two CMAs in series, i.e. a two-stage or, as it is better known, a double-pass CMA, with pre-retarding grids before the entrance slit, an analyser suitable for XPS has been constructed.[40] It is shown diagrammatically in Figure 2.23. The two pre-retarding grids are spherical and centred on the source position of the specimen.

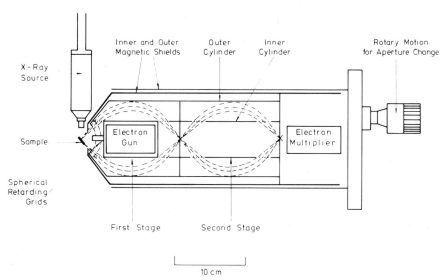

Figure 2.23 Diagrammatic arrangement of a double-pass CMA, used for both AES and XPS. The exit aperture from the first stage is the entrance aperture to the second stage. At the front end of the analyser are two spherical retarding grids centred on the source area of the sample that retard photo-electrons to a constant pass energy for XPS, For AES the grids are at earth potential, as is the inner cylinder. An externally operated rotary motion allows the entrance and exit apertures to the second stage to be changed remotely, from large sizes for XPS to small sizes for AES. The electron gun is situated on the axis of the CMA internally, but the X-ray source, of the type seen in detail in Figure 2.9(a), is external and positioned as close to the sample as the geometry will allow. (Reproduced from Palmberg[40] by permission of Elsevier Scientific Publishing Company)

The effect of pre-retardation is that by going from a kinetic energy of E_0 to a fixed analyser band-pass energy E_p, the energy resolution is improved (i.e. $\Delta E/E_0$ is reduced) by a factor (E_p/E_0). At the same time, the effective source volume is collapsed by a factor $(E_p/E_0)^{1/2}$. Thus, to maintain overall luminosity at a level suitable for XPS, the entrance and exit apertures to the *second* stage must be increased as far as possible, which means in practice to the point where the overall energy resolution is not quite degraded. To do this, the apertures can be altered by an external control so that either small apertures for AES or large apertures for XPS can be selected. In the double-pass CMA, therefore, the effective energy resolution is set by the first stage and the sensitivity by the second stage, although the two functions are clearly not inseparable.

For AES, the double-pass CMA is operated in the 'normal' mode, i.e. with the retarding grids grounded. If pre-retardation were attempted with AES, grid scattering would reduce the transmission undesirably. In XPS, however, although grid scattering reduces the transmission when using pre-retardation, the effective acceptance area is increased proportionately so that the luminosity is unaffected.

The opposing effects of collapse of the source area on retardation coupled with the effective source area increase due to grid scattering are functions both of kinetic energy and pass energy, and are somewhat difficult to estimate. The combined effect for any chosen pass energy will not necessarily be zero, or even linear with kinetic energy. In general, for absolute resolution ΔE worse than about 1.4 eV, the high luminosity of the double-pass CMA means that the signal intensity, for identical sample and operating conditions, is greater than for the CHA. However, as the pass energy is decreased to achieve absolute resolution of $\leqslant 1.0$ eV, the signal intensity drops off more rapidly than in the CHA, and it is likely that the retardation grid effects mentioned above are largely to blame. Simply increasing the overall size of the CMA, i.e. scaling with the inner radius r_1, serves to reduce the effect for a given pass energy, and this increase is in fact the direction in which the manufacturers are currently going.

2.5.4 Concentric hemispherical analyser (CHA)

At roughly the same time (1969) as the CMA was established as highly suitable for AES, XPS was being recognized as a valuable surface-specific analytical technique, and the type of analyser used by its founder, Siegbahn,[41] was being adapted to operation in UHV. This was the concentric hemispherical analyser (CHA), also called the spherical deflector analyser, and it is still the most widely used energy analyser for XPS, being supplied in the spectrometers from at least four manufacturers, whereas the double-pass CMA is supplied from only one.

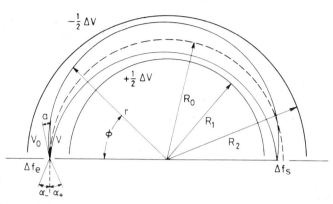

Figure 2.24 Principle of operation of the concentric hemispherical analyser. Two hemispherical surfaces of inner radius R_1 and outer radius R_2 are positioned concentrically. A potential ΔV is applied between the surfaces so that the outer is negative and the inner positive with respect to ΔV. R_o is the median equipotential surface between the hemispheres, and the entrance and exit slits are both centred on R_o. If E is the kinetic energy of an electron travelling in an orbit of radius R_o, then the relationship between E and V is given by expression (2.21). ϕ and r are the angular and radial coordinates, respectively, of an electron of energy E_o ($= e\Delta V_o$) entering the analyser at an angle α to the slit normal. If this electron is to pass through the exit slit, its path in the analyser is governed by the conditions of expression (2.22). (Reproduced from Roy and Carette[43] by permission of the National Research Council of Canada)

The CHA is shown schematically in Figure 2.24. Two hemispherical surfaces of inner radius R_1 and outer radius R_2 are positioned concentrically. A potential ΔV is applied between the surfaces so that the outer sphere is negative and the inner positive. There will be a median equipotential surface between the hemispheres of radius R_0, and the entrance and exit slits are each centred on a distance R_0 from the centre of curvature and lie on a diameter. In an ideal situation $R_0 = (R_1 + R_2)/2$, and that relationship is usually obeyed closely.

The coordinate position of an electron of initial kinetic energy E_0 travelling between the hemispheres is governed by two focusing conditions. Firstly,[42] if an electron enters the analyser at an angle α to the normal through the slit centre, its position in terms of r, its radial distance from the centre of curvature, and ϕ, its angular deflection, is given by

$$y = -y_0 \tan \alpha \sin \phi + (y_0 - c^2)\cos \phi + c^2 \qquad (2.19)$$

where $y = R_0/r$
y_0 = initial coordinate of electron
$c^2 = Ey_0^2/[E_0 \cos^2 \alpha - 2E(1 - y_0)]$

Here E corresponds to the kinetic energy which an electron would have if travelling on the circular orbit of radius R_0; for the 180° CHA, the first condition reduces to

$$y_f = 2c^2 - y_0 \qquad (2.20)$$

where y_f is the final coordinate of the electron. The relationship between the deflecting potential ΔV and the ideal energy E is

$$e\Delta V = E(R_2/R_1 - R_1/R_2) \qquad (2.21)$$

The second focusing condition arises[43] from the limits that must be imposed on the greatest and least radii of the electron trajectories. The maximum and minimum excursions can be expressed with respect to the median trajectory radius R_0 by

$$y_\pm = c^2 \pm (c^4 + 2k)^{1/2} \qquad (2.22)$$

where $k = \frac{1}{2} y_0^2 (\tan^2 \alpha + 1) - c^2 y_0$

This second condition merely says that if the electron is to be transmitted by the analyser its final position y_f must be contained within the aperture of the exit slit and that the $y_\pm$ are not outside the limits imposed by the hemispherical surfaces.

In the earlier spectrometers, the mid-potential of the analyser was tied to the ground potential, and the scanning through the kinetic energy range was performed by varying the specimen potential. Nowadays, however, the situation is reversed in that the specimen is grounded and the whole analyser floats isolated from ground so that scanning is now carried out by varying the potential (or 'pedestal') on which the analyser sits.

The base and half-width resolutions can be written down from expressions (2.15) and (2.16), as well as from Table 2.2. They are

$$\frac{\Delta E_B}{E} = \frac{w}{R_0} + \alpha^2 \qquad (2.23)$$

and

$$\frac{\Delta E}{E} = \frac{w}{2R_0} + \frac{\alpha^2}{4} \qquad (2.24)$$

(Note that the entrance angle α into the CHA is equivalent here to the semi-angular aperture $\Delta\alpha$ of the CMA.) In general it is desirable to work with a large entrance angle for increased sensitivity, and since a compromise must be reached between sensitivity and resolution it is common to choose α so that $\alpha^2 \approx w/2R_0$. Then expressions (2.23) and (2.24) become

$$\frac{\Delta E_B}{E} = \frac{1.5w}{R_0} \qquad (2.25)$$

and

$$\frac{\Delta E}{E} = \frac{0.63w}{R_0} \qquad (2.26)$$

Thus, if R_0 has been chosen and a certain absolute resolution is required, there is an inverse relationship between the pass energy E and the slit width.

There are two retarding modes in which CHAs are currently used. In one, the electrons are decelerated by a constant factor, or ratio, from their initial kinetic energies. Clearly this mode operates at constant *relative* resolution; it is termed either the CRR or the FRR mode. In the other, the electrons are decelerated to a constant pass energy, and it is obvious that the mode is that of constant *absolute* resolution. It is termed the CAT or FAT mode, standing for constant or fixed analyser transmission. Each mode has its adherents and each its advantages and disadvantages. The CAT mode is generally easier for quantification, since the absolute resolution is the same at all parts of the spectrum, but the signal-to-noise characteristics become progressively worse towards low kinetic energies. The CRR mode is less easy to quantify, but small peaks at low kinetic energies can be detected without difficulty.

Retardation when used with a CHA is usually accomplished with planar grids across the entrance slit, but may also be achieved with a lens system. Lens systems are used in other ways, too, to improve the overall efficiency of analysis. By removing the sample from the entrance to the analyser, and instead projecting the analysed area of the sample onto the entrance slit, a much greater working distance is available below the lens, allowing closer approach of the X-ray source, electron gun and ion gun. In the cases of the two latter, this close approach is necessary to be able to achieve minimum beam spot sizes. According to whether either or both CRR and CAT modes of retardation are required, the lens is more or less complex, for in the CRR mode transmission between the sample and the analyser is constant, whereas in the CAT mode the transmission in the lens is varying. The former could be performed with a simple unipotential lens, but the latter needs variable ratios for which three or more elements are necessary.

Various criteria or figures of merit have been applied to the CMA and CHA to establish their respective ranking, including transmission,[44] luminosity and étendue[45] (the product of the entrance area and entrance solid angle). According to Roy and Carette,[46] who have summarized all these criteria for various analysers including these two, the CMA in the second-order focusing arrangement (i.e. $\alpha = 42.3°$) is consistently the best. However, they also make the very pertinent point that the criteria take no account of the way in which a particular analyser is to be applied, and that point is certainly relevant in the present context. The choice must depend on practical considerations. This is precisely the reason why the CMA has up to now been the prime choice for AES, because of its high transmission at moderate energy resolution, while the CHA is always preferred for XPS, because, particularly with the addition of a lens, it can maintain adequate luminosity at high-energy resolution. Also, of course, the CHA, with its narrow acceptance angle, is suited to angular dependence measurements, while the CMA is in general not so suited because it collects over 360° (although in the new models there is an internal shutter which can be used to select variable sectors; see Chapter 4.) Development of the CMA will probably go in the direction of increased size only, since it is difficult to see how it could accommodate a lens, whereas it is unlikely that the development of the input lens to the CHA has reached its optimum.

REFERENCES

1. M. P. Seah and W. A. Dench, *Surf. Interface Anal.*, **1**, 2 (1979).
2. For example, M. Pirani and J. Yarwood, *Principles of Vacuum Engineering*, Chapman and Hall, London (1961).
3. R. Kelly, *Surface Science*, **100**, 85 (1980).
4. G. W. Lewis, G. Kiriakides, G. Carter and M. J. Nobes, *Surf. Interface Anal.*, **4**, 141 (1982).
5. M. P. Seah, *Thin Solid Films*, **81**, 279 (1981).
6. P. H. Holloway and R. S. Bhattacharya, *J. Vac. Sci. Tech.*, **20**, 444 (1982).
7. W. O. Hofer and V. Littmark, *Phys. Lett.*, **71**, 6 (1979).
8. R. Kelly and O. Auciello, *Surface Science*, **100**, 135 (1980).
9. A. C. Parry-Jones, P. Weightman and P. T. Andrews, *J. Phys. C: Solid State Phys.*, **12**, 1587 (1979).
10. J. P. Coad, J. C. Rivière, M. Guttmann and P. R. Krahe, *Acta Met.*, **25**, 161 (1977).
11. R. G. Musket, W. McLean, C. A. Colmenares, D. M. Makowiecki and W. J. Siekhaus, *Applic. Surface Science*, **10**, 143 (1982).
12. C. Benndorf, H. Seidel and F. Thieme, *Rev. Sci. Inst.*, **47**, 778 (1976).
13. H. H. Madden, *J. Vac. Sci. Tech.*, **18**, 677 (1981).
14. M-G. Barthès and C. Pariset, *Thin Solid Films*, **77**, 305 (1981).
15. C. A. Crider and J. M. Poate, *Appl. Phys. Lett.*, **36**, 417 (1980).
16. S. Okada, K. Oura, T. Hanawa and K. Satoh, *Surface Science*, **97**, 88 (1980).
17. M. Eizenberg and R. Brener, *Thin Solid Films*, **88**, 41 (1982), and **89**, 355 (1982).
18. K. Yates, A. Barrie and F. J. Street, *J. Phys. E: Sci. Inst.*, **6**, 130 (1973).
19. J. E. Castle, L. B. Hazell and R. D. Whitehead, *J. Electron. Spec.*, **9**, 247 (1976).
20. R. M. Dolby, *Brit. J. Appl. Phys.*, **11**, 64 (1960).
21. J. E. Castle, L. B. Hazell and R. H. West, *J. Electron. Spec.*, **16**, 97 (1979).
22. M. A. Kelly and C. E. Tyler, *Hewlett-Packard Journal*, **24**, 2 (1972).
23. K. Yates and R. H. West, *Surf. Interface. Anal.* (in press).
24. Y. Farge and P. J. Duke (Eds), *European Synchrotron Radiation Facility*, Supplement I, European Science Foundation (1979).
25. J. C. Rivière, *Solid State Surface Science*, **1**, 79 (1969); H. B. Michaelson, *J. Appl. Phys.*, **48**, 4729 (1977).
26. A. Christou, *J. Appl. Phys.*, **47**, 5464 (1976).
27. M. Prutton, R. Browning, M. M. El Gomati and D. Peacock, *Vacuum*, **32**, 351 (1982).
28. P. D. Prewett and D. K. Jefferies, *J. Phys. D; Applied Phys.*, **13**, 1747 (1980).

29. P. D. Prewett and D. K. Jefferies, *Second International Conf. Low Energy Ion Beams*, Bath (1980).
30. J. C. Fuggle, in *Electron Spectroscopy* (Eds C. R. Brundle and A. J. Baker), Vol. IV, Academic Press, London (1982).
31. P. W. Palmberg and T. N. Rhodin, *J. Appl. Phys.*, **39**, 2425 (1968).
32. P. W. Palmberg, G. K. Bohn and J. C. Tracy, *Appl. Phys. Letters*, **15**, 254 (1969).
33. V. V. Zashkvara, M. I. Korsunskii and O. S. Kosmachev, *Sov. Phys. Tech. Phys.*, **11**, 96 (1966).
34. H. Z. Sar-El, *Rev. Sci. Inst.*, **41**, 561 (1970).
35. H. Hafner, J. A. Simpson and C. E. Kuyatt, *Rev. Sci. Inst.*, **39**, 33 (1968).
36. H. E. Bishop, J. P. Coad and J. C. Rivière, *J. Electron Spec.*, **1**, 389 (1972–1973).
37. J. A. Simpson, in *Methods of Experimental Physics* (Eds V. W. Hughes and H. L. Schultz), Vol. 4A, pp. 124–135, Academic Press, New York (1967).
38. M. E. Rudd, in *Low Energy Electron Spectrometry* (Ed. K. D. Sevier), pp. 17–32, Wiley-Interscience, New York (1972).
39. E. Sickafus and P. H. Holloway, *Surface Science*, **51**, 131 (1975).
40. P. W. Palmberg, *J. Electron. Spec.*, **5**, 691 (1974), and *J. Vac. Sci. Tech.*, **12**, 379 (1975).
41. K. Siegbahn, C. Nordling, A. Fahlman, R. Nordberg, K. Hamrin, J. Hedman, G. Johansson, T. Bergmark, S. Karlson, I. Lindgren and B. Lindberg, *ESCA, Atomic, Molecular and Solid State Structure Studied by Means of Electron Spectroscopy*, Almqvist and Wiksells Boktryckeri AB, Uppsala (1967).
42. E. M. Purcell, *Phys. Rev.*, **54**, 818 (1938).
43. D. Roy and J-D. Carette, *Canadian J. Phys.*, **49**, 2138 (1971).
44. M. P. Seah, *Surf. Interface Anal.*, **2**, 222 (1980).
45. D. W. O. Heddle, *J. Phys. E: Sci. Inst.*, **4**, 589 (1971).
46. D. Roy and J-D. Carette, *Topics in Current Physics*, Vol. 4, *Electron Spectroscopy for Chemical Analysis*, pp. 13–58, Springer-Verlag, Berlin (1972).

Practical Surface Analysis
by Auger and X-ray Photoelectron Spectroscopy
Edited by D. Briggs and M. P. Seah
© 1983, John Wiley & Sons, Ltd

Chapter 3

Spectral Interpretation

D. Briggs
*ICI PLC, Petrochemicals and Plastics Division,
Wilton, Middlesbrough, Cleveland, UK*

J. C. Rivière
*UK Atomic Energy Research Establishment,
Harwell, Didcot, Oxfordshire*

3.1   Introduction

In this chapter are considered the features which may be found in secondary electron and in X-ray photo-electron spectra, from the viewpoints both of their physical origin and of their information content. Since Auger features are contained in both spectra, while X-ray-excited spectra contain in addition photo-electron features, it is logical to describe the secondary electron spectrum first. Before doing either, however, it is necessary to discuss the derivation of the nomenclature used for the description of the two principal types of spectral feature.

3.2   Nomenclature

For a proper discussion of the nomenclature used in the two techniques being described here, one must go back to details of the momenta associated with the orbiting paths of electrons around atomic nuclei. Since an electron is a charged particle, its orbit around a nucleus induces a magnetic field whose intensity and direction depend on the electron velocity and on the radius of the orbit, respectively. Clearly the two latter quantities can be characterized by an angular momentum, called the *orbital* angular momentum, which of course is quantized since the electron can travel only in certain discrete orbitals. The characteristic quantum number is l, and l can take the values 0, 1, 2, 3, 4, . . . Another property of an orbiting electron is the electron spin, i.e. positive or negative, which also induces an inherent magnetic field; the latter in turn has an associated *spin* momentum, characterized by a spin quantum number s, which can take either of the values $\pm\frac{1}{2}$. Thus the *total* electronic angular momentum is a combination of the orbital angular and spin momenta, and this combination is in fact simply the vector sum of the two momenta. However, it is most important in the present context to note that the vector summation can be carried out in *two ways*, rejoicing in the names of j–j coupling and L–S (also called Russell–Saunders) coupling, respectively.

3.2.1 j–j coupling

In this summation, the total angular momentum of a single isolated electron is obtained by summing vectorially the individual electronic spin and angular momenta. For the particular electron the *total* angular momentum is then characterized by the quantum number j, where $j = l + s$. Obviously j can take the values $\frac{1}{2}, \frac{3}{2}, \frac{5}{2}$, etc. To arrive at the total angular momentum for the whole atom, summation is then performed for all electrons, the result being the total atomic angular momentum with an associated quantum number J, where $J = \Sigma j$. This description of the summation is known as j–j coupling.

Strictly speaking, j–j coupling is the best description of electronic interaction in elements of high atomic number, i.e. $Z > \sim 75$, but in fact the nomenclature based on it has been used for both Auger and spectroscopic features for all parts of the Periodic Table. This does not matter for the features arising from photo-electron production, since the final state of the atom is singly ionized, but in the Auger process, where the final state is doubly ionized, interactions between the two holes in the final state can lead to situations where the j–j description is inadequate. This will be discussed below.

Under the j–j coupling scheme the nomenclature is based on the principal quantum number n and on the electronic quantum numbers l and j mentioned above. In the historical X-ray notation, states with $n = 1, 2, 3, 4, \ldots$ are designated K, L, M. N, $\ldots$, respectively, while states with various combinations of $l = 0, 1, 2, 3, \ldots$ and $j = \frac{1}{2}, \frac{3}{2}, \frac{5}{2}, \frac{7}{2}, \ldots$ are given conventional suffixes, $1, 2, 3, 4, \ldots$, according to the listing in Table 3.1. The X-ray notation is almost always used for Auger transitions, so that, for example, in j–j coupling there would be six predicted KLL transitions, i.e. KL_1L_1, KL_1L_2, KL_1L_3 KL_2L_2, KL_2L_3 and KL_3L_3. This is still the most universally used nomenclature in AES.

The spectroscopic nomenclature is directly equivalent to the X-ray, and is more obviously related to the various quantum numbers. In it the principal quantum number appears first, then states with $l = 0, 1, 2, 3, \ldots$ are designated $s, p, d, f, \ldots$, respectively, and follow the first number, and finally the j values are appended as suffixes. Thus the state written L_3 in the X-ray notation, in which $n = 2$, $l = 1$ and $j = \frac{3}{2}$, would be written $2p_{3/2}$ in the spectro-

Table 3.1   X-ray and spectroscopic notation

Quantum numbers					Spectroscopic level
n	l	j	X-ray suffix	X-ray level	
1	0	$\frac{1}{2}$	1	K	$1s_{1/2}$
2	0	$\frac{1}{2}$	1	L_1	$2s_{1/2}$
2	1	$\frac{1}{2}$	2	L_2	$2p_{1/2}$
2	1	$\frac{3}{2}$	3	L_3	$2p_{3/2}$
3	0	$\frac{1}{2}$	1	M_1	$3s_{1/2}$
3	1	$\frac{1}{2}$	2	M_2	$3p_{1/2}$
3	1	$\frac{3}{2}$	3	M_3	$3p_{3/2}$
3	2	$\frac{3}{2}$	4	M_4	$3d_{3/2}$
3	2	$\frac{5}{2}$	5	M_5	$3d_{5/2}$
	etc.		etc.	etc.	etc.

scopic notation. In Table 3.1 the spectroscopic terms are listed opposite their X-ray equivalents. It is conventional to identify a photo-electron feature in terms of the spectroscopic name of the atomic level from which the photo-electron was ejected.

3.2.2   L–S coupling

The other way of carrying out the vectorial summation is first to sum all the individual electronic angular momenta and then all the individual electronic spin momenta. These two total momenta are then characterized by two quantum numbers, the total atomic *orbital* angular momentum quantum number L, which is equal to Σl, and the total atomic *spin* quantum number S, which is equal to Σs. Coupling of the two total momenta to give the total atomic angular momentum can then be characterized as before by the quantum number J, which is now, however, equal to $|L \pm S|$. Since L and S can take the values 0, 1, 2, 3, . . ., J can take any integral value between $|L-S|$ and $|L+S|$. The origin of the name 'L–S coupling' is obvious.

L–S coupling has been found to apply to elements of low atomic number, i.e. $Z < \sim 20$. In this coupling scheme the nomenclature is that of term symbols of the form $^{(2S+1)}L$ describing the electron distribution in the final state. By analogy with the spectroscopic notation of Table 3.1, states with $L = 0, 1, 2, 3, \ldots$ are designated in capitals $S, P, D, F, \ldots$, while the total spin quantum number S enters as the prefix $(2S + 1)$. (The state S corresponding to $L = 0$ should not here be confused with S, the total spin quantum number.) As for j–j coupling, L–S coupling predicts six possible components in the *KLL* series, listed in Table 3.2, but one of these is forbidden through the

Table 3.2   Notation in L–S coupling

Transition	Configuration	L	S	Term
KL_1L_1	$2s^02p^6$	0	0	^1S
$KL_1L_{2,3}$	$2s^12p^5$	$\begin{cases}1\\1\end{cases}$	0 , 1	^1P , ^3P
$KL_{2,3}L_{2,3}$	$2s^22p^4$	$\begin{cases}0\\[1\\2\end{cases}$	0 , 1 , 0	^1S , $^3P]^*$, ^1D

*Forbidden.

principle of conservation of parity[1]. Also shown in Table 3.2 is a frequently used way of describing the final-state configuration following an Auger transition ABC by writing down the electron populations in the levels B and C, e.g. $2s^12p^5$ for $KL_1L_{2,3}$.

The L–S classification and notation has been used mostly by those recording Auger spectra at high energy resolution in order to provide data for comparison with theoretical calculations, and is not normally encountered in everyday use. The same is true of the mixed, or intermediate, coupling scheme which must be used in the region of the Periodic Table where neither L–S nor j–j coupling is adequate to describe the final-state configuration. In intermediate coupling each L–S term is split into multiplets of different J values, so that the term symbols are now of the form $^{(2S+1)}L_J$. As can be seen from Table 3.3, ten possible final states are predicted in the KLL series, but one is forbidden for the same reasons as before. There have been several examples of experimental evidence for the existence of nine lines in a KLL spectrum.

Table 3.3   Notation in intermediate coupling

Transition	Configuration	L–S term	L	S	J	IC term
KL_1L_1	$2s^02p^6$	^1S	0	0	0	$^1S_0$
$KL_1L_{2,3}$	$2s^12p^5$	^1P	1	0	1	$^1P_1$
		^3P	1	1	0	$^3P_0$
			1	1	1	$^3P_1$
			1	1	2	$^3P_2$
		^1S	0	0	0	$^1S_0$
$KL_{2,3}L_{2,3}$	$2s^22p^4$	^3P	1	1	0	$^3P_0$
			[1	1	1	$^3P_1]^*$
			1	1	2	$^3P_3$
		^1D	2	0	2	$^1D_2$

*Forbidden.

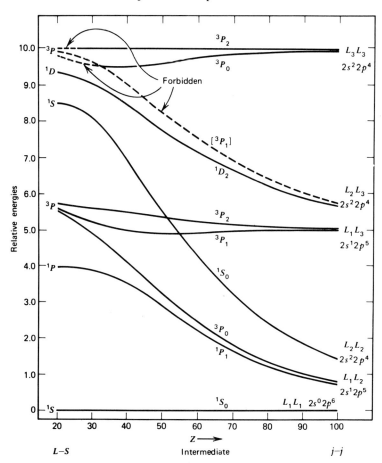

Figure 3.1 The transition from pure $L-S$ through intermediate to pure $j-j$ coupling across the Periodic Table, and its effects on relative energies in the KLL Auger series. (Reproduced from Sevier[2] by permission of John Wiley & Sons Inc.)

As Table 3.3 implies, it is customary to use a mixed notation, so that the KLL series, in approximate order of increasing energy, would be $KL_1L_1({}^1S_0)$, $KL_1L_2({}^1P_1)$, $KL_1L_2({}^3P_0)$, $KL_2L_2({}^1S_0)$, $KL_1L_3({}^3P_1)$, $KL_1L_3({}^3P_2)$, $KL_2L_3({}^1D_2)$, $KL_2L_3({}^3P_1)$ (forbidden), $KL_2L_3({}^3P_0)$ and $KL_3L_3({}^3P_2)$. The transition in the KLL series across the Periodic Table from pure $L-S$ through intermediate to pure $j-j$ coupling is shown in Figure 3.1.

3.3 The Electron-excited Secondary Electron Spectrum

When the energies of secondary electrons produced as a result of irradiation of a solid surface by an incident electron beam are analysed, the energy

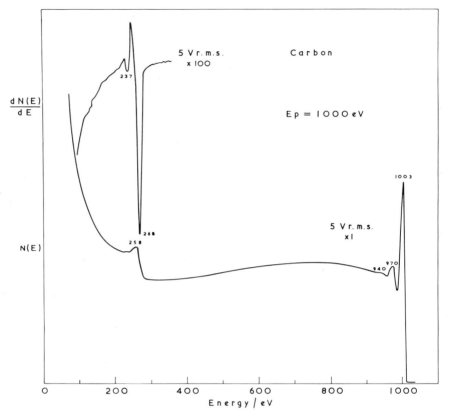

Figure 3.2   Lower curve: distribution of energies of secondary electrons ejected from a graphite surface by incident electrons of energy 1000 eV. Upper curve: differential distribution over the energy range containing the carbon *KLL* Auger peaks. In the differential distribution the peak 'position' is taken to be that of the high energy minimum, by convention

distribution of the ejected electrons looks typically like that in the lower part of Figure 3.2. At the energy of the primary electrons, there is a large, narrow, peak in intensity corresponding to those electrons that have been reflected from the surface with virtually no loss in energy, i.e. elastically scattered electrons. If the surface is that of a single crystal then these electrons contain structural information by virtue of diffraction from the regular atomic array, and are in fact used in LEED and RHEED. Associated with the elastic peak at lower energies is a series of smaller, broader, peaks whose intensity decreases successively away from the elastic peak. These are the plasmon loss peaks. At the other end of the energy distribution can be seen another large peak, very broad, peaking at a few electronvolts and extending many tens of electron-volts up the energy scale. In that peak are the so-called 'true' secondary electrons; of course, *all* electrons ejected from a surface on primary electron irradiation are secondaries, but conventional terminology distinguishes the true secondaries from the others, mainly for historical reasons. The many investigations carried out under the heading of 'secondary electron emission' have all been concerned with the low-energy secondaries, which are produced by a cascade process.

Between the elastic and the true secondary peak, and often actually situated on the slope of the latter, are other features, some distinguishable easily as peaks, and some not, due mostly to Auger emission. Those few that are not Auger peaks are either associated plasmon loss peaks or ionization loss peaks. In conventional AES, where the highest spatial resolution is not sought and in which therefore the current in the incident electron beam can be maintained at 10^{-6}–10^{-7} A, it is normal to record the spectrum as the differential with respect to energy. Thus the energy distribution in the lower half of Figure 3.2 would appear as the differential distribution shown in the upper half of the figure. Not only does differentiation help to make weak features more readily identifiable, but it also removes the background to a large extent; it is particularly valuable in revealing more clearly features situated on the steep slopes of the 'true' secondary electron peak.

In the following section the features and their origins, arising from Auger emission and from various loss processes, will be described. Most of the description will be based on the differential spectrum, since it is still the one most commonly encountered.

3.3.1 Auger electron spectra

3.3.1.1 *The Auger process*

The left-hand side of Figure 3.3 shows a schematic energy level diagram of a solid, energy being measured downwards from an assumed zero of energy at the Fermi level. More formally, the zero should be taken as that at the vacuum level at an infinite distance, but in both XPS and AES it is normal to measure binding energies with respect to the Fermi level.

In the centre of Figure 3.3 is shown the sequence of events following ionization of a core level. For this example the K level is shown as being ionized by an incident electron, whose energy E_p must obviously be greater than the binding energy E_K of an electron in K. Due to the way in which the ionization cross-section depends on E_p it is in fact necessary in practice for E_p to be greater than about $5E_K$ for efficient ionization, as will be shown in Chapter 5. Following creation of a hole in the level K, the atom relaxes by filling the hole via a transition from an outer level, in this example shown as L_1. As a result of that transition the energy difference $(E_K - E_{L_1})$ becomes available as excess kinetic energy, and this excess energy can be used by the

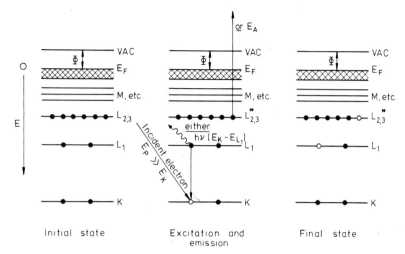

Figure 3.3   Schematic diagram of the process of Auger emission in a solid. The ground state of the system is shown at the left. In the centre an incident electron of energy E_p has created a hole in the core level K by ionization; for this to occur efficiently E_p should be $\geqslant \sim 5E_K$. The hole in the K shell is filled by an electron from L_1, releasing an amount of energy $(E_K - E_{L_1})$, which can appear as a photon of energy $h\nu = (E_K - E_{L_1})$ or can be given up to another electron. In this example the other electron is in the $L_{2,3}$ shell, and it is then ejected with energy $(E_K - E_{L_1} - E^*_{L_{2,3}})$; $E^*_{L_{2,3}}$ is starred because it is the binding energy not of $L_{2,3}$ in its ground state, but in the presence of a hole in L_1. The doubly ionized final state is shown on the right

atom in either of two ways. It can appear as a characteristic X-ray photon at that energy *or* it can be given to another electron either in the same level or in a more shallow level, whereupon the second electron is ejected. The first process is that of X-ray fluorescence, the second that of Auger emission. Clearly both cannot take place from the same initial core hole, so that they compete. However, as Figure 3.4 indicates for K shell ionization, the probability of relaxation by Auger emission is favoured overwhelmingly over that of X-ray fluorescence for relatively shallow core levels, i.e. with binding energies below about 2 keV. The same is true for the L, M, N, etc., atomic levels. If this were not so, signal strengths in AES would be much smaller and it would not therefore be as useful a technique as it is.

The Auger transition depicted in Figure 3.3 would be named in the conventionally used j–j coupling $KL_1L_{2,3}$. Obviously in the same notation other transitions are possible in the atom depicted in that figure, e.g. KL_1L_1, $KL_{2,3}L_{2,3}$, $L_1L_{2,3}L_{2,3}$, etc. The electrons taking part in the Auger process might also originate in the valence band of the solid, in which case the convention writes the transitions as, for example, $KL_{2,3}V$ if one electron comes from the valence band and, for example, KVV if both do.

The energy of the ejected Auger electron in the example of Figure 3.3 is

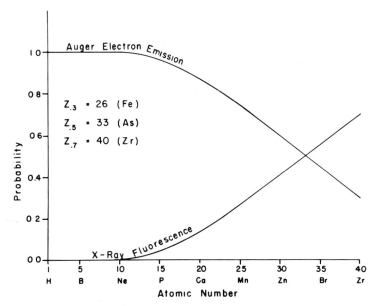

Figure 3.4   Relative probabilities of relaxation by emission of an Auger electron and by emission of an X-ray photon of characteristic energy, following creation of a core hole in the K shell

$$E_{KL_1L_{2,3}} = E_K - E_{L_1} - E^*_{L_{2,3}} \qquad (3.1)$$

where E_i are the binding energies of the ith atomic energy levels. $E^*_{L_{2,3}}$ is starred because it is the binding energy of the $L_{2,3}$ level in the presence of a hole in level L_1, and is therefore different from $E_{L_{2,3}}$. The various contributions to the difference between E^*_i and E_i, in the general case, will be discussed below. It is sufficient for the moment to note that the Auger energy expressed by equation (3.1) is a function only of atomic energy levels, so that for each element in the Periodic table there is an unique set of Auger energies, there being no two elements with the same set of atomic binding energies. Thus analysis of Auger energies immediately leads to elemental identification. Even when either or both electrons in the Auger process originate in the valence band, the analysis provides elemental identification, since the dominant term in equation (3.1) is always the binding energy of the initially ionized core level.

For heavy elements it will be clear from equation (3.1) that the number of energetically possible Auger transitions, for a given primary electron energy, can become very large indeed, as the number of atomic levels proliferates. Fortunately for AES, the transition probabilities favour only a few of the very many, so that even for the heaviest elements the problem is not intractable.

3.3.1.2   Auger energies

An approximation to the energy E_{ABC} of the Auger transition ABC in an atom of atomic number Z, that has proved sufficiently accurate for most practical purposes, is the one derived empirically by Chung and Jenkins.[3] It is

$$E_{ABC}(Z) = E_A(Z) - \tfrac{1}{2}[E_B(Z) + E_B(Z + 1)] - \tfrac{1}{2}[E_C(Z) + E_C(Z + 1)]$$

$$(3.2)$$

$E_i(Z)$ are the binding energies of the *i*th levels in the element of atomic number Z and $E_i(Z + 1)$ of the same levels in the next element up the periodic table, various compilations of E_{ABC}, such as that of Coghlan and Clausing[4], are available, based on equation (3.2).

A more physically acceptable expression for the Auger energy is

$$E_{ABC} = E_A - E_B - E_C - \mathscr{F}\,(BC : x) + R_x^{in} + R_x^{ex} \qquad (3.3)$$

where $\mathscr{F}\,(BC{:}x)$ is the energy of interaction between the holes in B and C in the final atomic state x and the R_x are relaxation energies. The latter energy terms arise from the additional screening of the atomic core necessary when there is a hole in a core level, and it is achieved by an inward collapse or 'relaxation' of outer electronic orbitals towards the core. As the relaxation takes place, there is an increased interaction of the electrons in the outer orbitals with electrons in the inner orbitals, via Coulomb and exchange integrals. The term R_x^{in} is the *intra-atomic* relaxation energy, i.e. the relaxation energy appropriate to an isolated atom. In a molecule or in a solid, where additional screening electrons are available from the environment of the ionized atom, i.e. from other atoms or from the valence band, respectively, then the additional R_x^{ex} term appears, called the *extra-atomic* relaxation energy. The magnitudes of the $\mathscr{F}$ and R terms are often considerable. Table 3.4 gives examples calculated for the $L_3M_{4,5}M_{4,5}$: ^3P transitions of metallic nickel, copper and zinc by Kim, Gaarenstroom and Winograd.[5]

In the calculation of Auger energies using equation (3.3) it is customary to use experimentally determined binding energies for the E_i and calculated values for the other terms, so that the approach is semi-empirical. Various

Table 3.4   Calculated values of the $\mathscr{F}$ and R energies in equation (3.3) for the $L_3M_{4,5}M_{4,5}$: ^3P transition in Ni, Cu and Zn. (From Kim, Gaarenstroom and Winograd[5])

Metal	$\mathscr{F}\,(M_{4,5}M_{4,5} : {}^3P)$	R_{3P}^{in}	R_{3P}^{ex}
Ni	26.6	9.9	18.2
Cu	26.3	10.6	11.0
Zn	29.4	12.3	9.6

authors have provided tables of such semi-empirically determined Auger energies, the most extensive tabulation being that of Larkins,[6] who has covered the Auger series KLL, KLM, L_3MM, L_3MN, M_5NN, M_5NO, and N_5OO, all within the intermediate coupling system.

This identification of features in the secondary electron spectrum as arising from Auger emission could be made, in principle, by checking the observed kinetic energies against the energies calculated from expressions (3.2) or (3.3), the latter being more accurate and also providing energies of the individual final-state terms. On the other hand, there are by now several atlases of recorded Auger spectra available, in which the principle Auger peaks are labelled with their kinetic energies, making elemental identification easy in most cases. There are slight differences between these compilations, due to different recording conditions and different energy analysers used in each case, but these differences are not usually large enough to matter. It must be remembered that the spectra are in all cases shown in the differential $E\,dN(E)/dE$ mode, in which by *convention* the energetic position of the Auger peak is taken as that of the minimum in the high-energy negative excursion. Clearly such a conventional energy does not correspond to the actual peak energy in the undifferentiated, or direct, spectrum, and will differ from it by an amount of energy that depends on the peak width. Thus close agreement between Auger energies obtained from the differential distribution and those calculated should not be expected, but such a comparison is not usually attempted. In practise, the Auger atlases are used on a day-to-day basis in applied AES, in which the differential mode is still by far the more common, while the results of the detailed energy calculations are used by those working in the undifferentiated mode, generally derived from X-ray rather than electron excitation.

3.3.1.3 *Auger intensities*

Inspection of any one spectrum, or comparison of the spectra from any two adjacent elements in the above-mentioned atlases, reveals at once that the relative intensities of different Auger transitions within the same element, or of the same Auger transition in different elements, show large variations. The reasons for these variations are in general well known, but, unlike the situation for the calculation of Auger energies, that for the calculation of Auger intensities (or transition probabilities) is not nearly so good. There is no simple semi-empirical formula that will produce a set of usable relative intensities, and even the complex formulations that have to be used by the theoreticians have not yet been particularly successful in the accurate prediction of intensities, especially in the solid state. The relationship between the intensity of an Auger peak from a given element and its concentration, when present with other elements on a surface, will be discussed in Chapter 5, but in the present context it is sufficient merely to note that intensities cannot be predicted with the same certainty as can energies. Where comparison of intensities is important, recourse must be had once again either to the published Auger atlases or to sets of elemental spectra compiled in one's own laboratory.

3.3.1.4 Characteristic Auger series

Since, as shown in Chapter 5, the dependence of the cross-section for ionization by electrons on the primary electron energy passes through a maximum at about 4–5 times the primary energy, and since a typical primary energy in current use would be 10 keV, there are necessarily limitations in any one part of the Periodic Table on the Auger transitions that can be excited with optimum efficiency. Thus Auger transitions based on K shell ionization are limited in usefulness to the range from lithium (in the solid state) to silicon, on L_3 shell ionization to the range from magnesium to rubidium, on M_5 ionization to the range from gallium to osmium, and so on. The practical effect of these limitations is that in each region of the Periodic Table there is a characteristic set, or series, of Auger transitions which is the most prominent under the standard experimental conditions normally employed. Some examples of

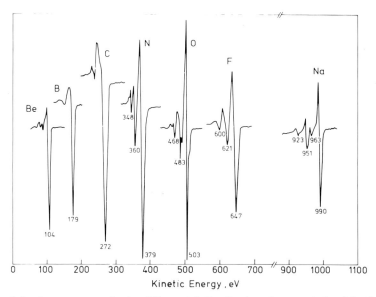

Figure 3.5 Auger spectra in the differential distribution characteristic of the lightest elements. The principal peak is the $KL_{2,3}L_{2,3}$. The relative intensities are not to scale. (Reproduced from Davis *et al.*[7] by permission of Perkin-Elmer Corporation)

these series are shown in Figures 3.5 to 3.9, taken from the *Handbook of Auger Electron Spectroscopy* by Davis *et al.*[7]

· In Figure 3.5 it can be seen that for the light elements of the Auger transition $KL_{2,3}L_{2,3}$ (KL_1L_1 for beryllium) dominates for most of the range. For magnesium, aluminium and silicon, which are not illustrated, the low-energy $L_{2,3}VV$ transition also becomes strong and for the elements from phosphorus to calcium it is sufficiently intense to be used for identification, K shell ionization no longer being available as mentioned above. In the series of $3d$ transition metals, from scandium to copper, the characteristic 'fingerprint' is the LMM triplet, consisting, in order of decreasing energy, of the transitions L_3VV,

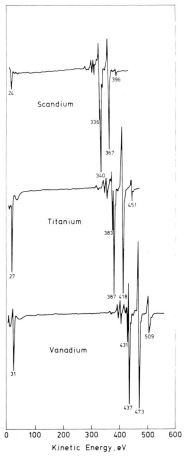

Figure 3.6 Auger spectra in the differential distribution for scandium, titanium and vanadium. The characteristic features are the LMM triplet and the low-energy $M_{2,3}VV$ peak. (Reproduced from Davis *et al.*[7] by permission of Perkin-Elmer Corporation)

$L_3M_{2,3}V$ and $L_3M_{2,3}M_{2,3}$. Figures 3.6 to 3.8 illustrate the progression of this triplet through the series, and it can be seen clearly how the relative magnitudes of the individual components change, particularly that of the L_3VV to the other two (note, also, Figure 5.7 in Chapter 5). The gradual increase in the intensity of the latter transition reflects, of course, the progressive filling of the valence band to completion at copper. Another striking and characteristic Auger series can be found amongst the 4d transition metals, and is shown in Figure 3.9. The principal feature there is the closely spaced $M_{4,5}N_{4,5}N_{4,5}$ doublet, the spacing being that between M_4 and M_5. In other regions of the

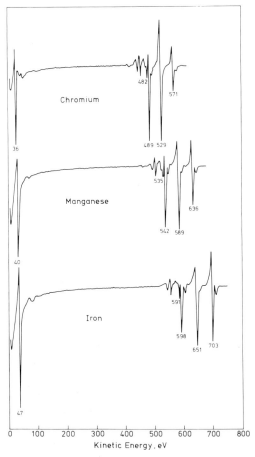

Figure 3.7   Auger spectra in the differential distribution for chromium, manganese and iron. The *LMM* triplet seen at the start of the first transition series in Figure 3.6 is now well developed, and the $L_{2,3}VV$ peak is also stronger. (Reproduced from Davis *et al.*[7] by permission of Perkin-Elmer Corporation)

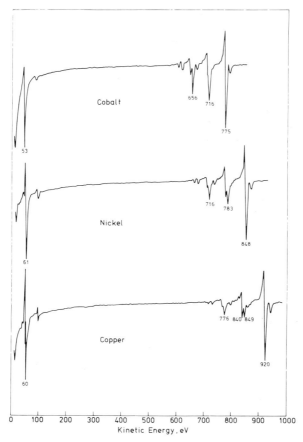

Figure 3.8 Auger spectra in the differential distribution for cobalt, nickel and cop-per. In the *LMM* triplet the L_3VV peak at the highest indicated energies is now predominant due to the filling of the metal *d* bands, the filling being complete at copper. (Reproduced from Davis *et al.*[7] by permission of Perkin-Elmer Corporation)

Periodic Table there are other series characteristic of the elements in those regions, but none is quite as immediately recognizable as any of those described above.

The Auger spectra in Figures 3.5 to 3.9 also reveal many minor features, mostly due to weak Auger transitions, and obviously in alloys or compounds or other multielement materials there is always present the possiblity of over-lap of the Auger features of one or more of the constituent elements. Where, as for example amongst the 3*d* transition metals, there are several prominent transitions, it is always possible to find one that is clear of interference, but problems can arise where there is overlap between the position of the only Auger peak available for a light element and a prominent peak belonging to a

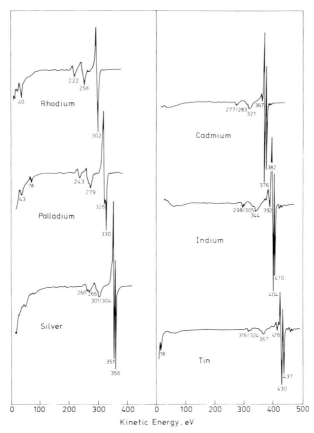

Figure 3.9 Auger spectra in the differential distribution for metals in the second transition series. Apart from rhodium and palladium, where it is poorly resolved, the characteristic feature is the sharp, closely spaced $M_{4,5}N_{4,5}N_{4,5}$ doublet, the separation being that between M_4 and M_5. (Reproduced from Davis *et al.*[1] by permission of Perkin-Elmer Corporation)

heavy element. Examples are chlorine and argon with molybdenum, boron with niobium, sulphur with zirconium and carbon with ruthenium. Where in doubt, it is best to use spectrum subtraction techniques, in which a standard 'clean' spectrum is subtracted from the suspect one.

3.3.1.5 *Auger fine structure*

Fine structure in Auger spectra can be seen frequently from both metals and non-metals, and originates either in chemical effects or in final-state effects. In general, structure due to chemical effects can usually be observed in conventional electron-excited AES, but that due to final-state or hole localization effects must be recorded at much higher energy resolution, in the direct or undifferentiated mode, using X-ray excitation.

Chemical effects can arise from the same source as the so-called 'chemical shift' that is the strength of XPS, i.e. the binding energy of electrons in the core level being ionized can alter with chemical environment, leading to a

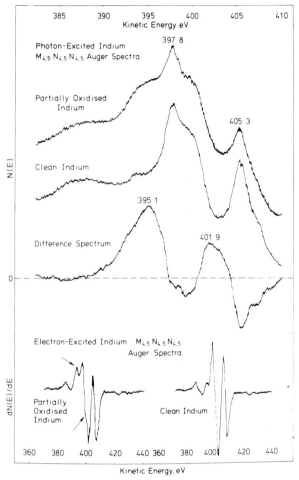

Figure 3.10   Example of fine structure in Auger spectra, due to a chemical shift in the binding energy of the ionized core level. The electron-excited differential Auger spectra of indium in the clean and partially oxidized states shown at the bottom reveal fine structure after partial oxidation, not seen in the clean state. Analysis of the photon-excited and undifferentiated Auger spectra in the top part shows by subtraction that the fine structure originates in shifts of 2.7 and 3.4 eV in the M_4 and M_5 levels on oxidation. The Auger spectrum in the partially oxidized state is thus a superposition of the clean and 'oxidized' spectra

similar shift in the consequent Auger transition. Where atoms of a particular element are present on a surface in more than one chemical state, then there is the consequent possibility of Auger fine structure by this mechanism. An example shown in Figure 3.10 is that of indium where the fine structure arrowed in the conventional differential spectrum at the bottom is seen to be due to a shifted Auger peak when the spectrum is integrated numerically, as shown at the top. In this case it is the shift of 2.8 and 3.4 eV in the M_4 and M_5 core levels, respectively, on oxidation that causes the shift in the $M_{4,5}N_{4,5}N_{4,5}$ Auger transitions and hence the appearance of the additional structure.

The more commonly observed fine structure due to chemical effects, however, arises in the Auger spectra of non-metallic elements present on surfaces in one or more of various possible chemical states. When, for example, such an element is part of a chemisorbed layer on a metallic substrate, the most

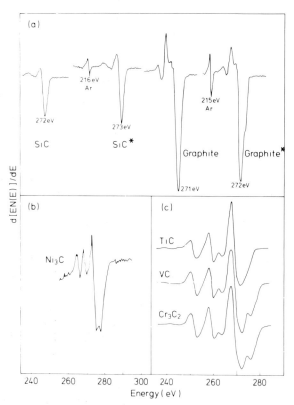

Figure 3.11   The *KLL* Auger spectra of carbon in different chemical situations. In the spectra in the upper part the asterisk * signifies recordings made after ion bombardment. (Reproduced from Kny[8] and Kleefeld[8] by permission of the American Institute of Physics and Elsevier Sequoia)

intense Auger transitions characteristic of the element will be those in which ionization of a core level in the element is followed by transfer of valence electrons *from the substrate* for the subsequent Auger relaxation. In other words, although the average energetic position of the resultant peak in the Auger spectrum will be characteristic of the element, which can thus still be identified, the shape of the peak, including the fine structure, will be related to the local density of the valence states adjacent to the chemisorption site. Since the local density of states (LDOS) will vary from one substrate to another, and from one type of site to another on any one substrate, it follows that variations in the shape and fine structure associated with the Auger peak(s) of the chemisorbed species will be observed.

The earliest example to be observed of these variations, and still one of the most striking, is that of the KVV Auger spectrum of carbon in different chemical situations. Figure 3.11, taken from various authors,[8] shows that both the overall appearance of the Auger peak shape, and the associated fine structure change markedly on going from graphite to various carbides, and even after ion bombardment of solid carbon. Note the particularly sharp fine structure present in the carbide spectra. Another example is given in Figure 3.12 for the sulphur $L_{2,3}VV$ Auger peak on three transition metals; the differences are not quite so great as for carbon, but are significant. Similar variations have been recorded for nitrogen, phosphorus and even oxygen,[10] which is more electronegative than the others mentioned here. Such valence spectra can be used either to derive information about the LDOS at the ionized surface site, by self-deconvolution, or, at a lower but more immediately useful level, to act as 'fingerprints' of the particular chemical form or compound present. In the present context of spectral interpretation it is sufficient to point out that the fine structure due to chemical effects is not uncommon and that it must therefore be recognized for what it is.

The other type of fine structure, originating in correlations between the two holes that are left in the atom at the end of the Auger process, is intrinsic to certain metals. In these metals Auger transitions of the type CVV, where C is a core level, produce line shapes whose widths even before deconvolution are much narrower than those of the valence bands V, and thus there would be no possibility whatsoever of deriving information about the valence density of states in the metals. The widths are in fact similar to those expected from the same transition in the free atom, allowing for solid-state broadening, and the Auger spectra are thus labelled 'quasi-atomic'. The physical explanation is that in such metals the effective interaction energy between the two holes in the final state is much larger than the width of the valence band, and therefore the holes cannot decay rapidly as they would be able to do if their interaction energy were lower. As a result the holes became localized near the ionization atom for a significant time, leading to a situation formally similar to that in an isolated atom. Under these circumstances the electronic correlation between

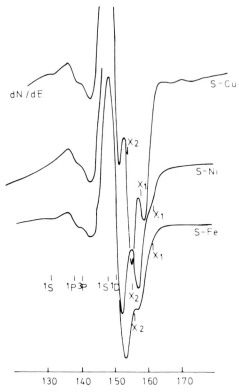

Figure 3.12   The $L_{2,3}VV$ Auger spectra of sulphur on the surfaces of three transition metals, showing chemical effects due to the differences in the bonding and therefore in the local densities of electronic states near the sulphur atoms. Calculated positions of spectral terms are indicated. The peaks labelled X_1 and X_2 are due to so-called 'inter-atomic' transitions. (After Salmeron, Baco and Rojo[9])

the holes gives rise to a splitting of the CVV spectrum into multiplets that can be described by the term symbols given in Section 3.2.2 on L–S and intermediate coupling.

The metals for which the 'quasi-atomic' type of behaviour has definitely been established are those from nickel to selenium, from silver to tellurium and from thallium to bismuth; it is possible there may be others. Some metals for which the 'quasi-atomic' Auger character is not normally observed can show that character when present in a dilute alloy.[11] Since some of the multiplet splittings are small, it is necessary to record the Auger spectra at higher energy resolution than that normally used; this is generally only possible by recording the undifferentiated spectrum by pulse counting in an XPS system, i.e. using X-ray rather than electron excitation. Examples of Auger spectra obtained in this way, showing well-developed fine structure arising from the

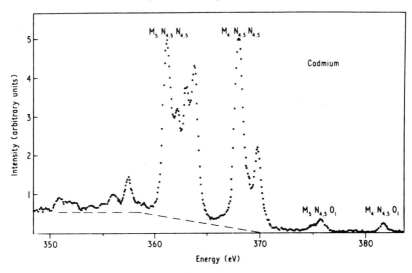

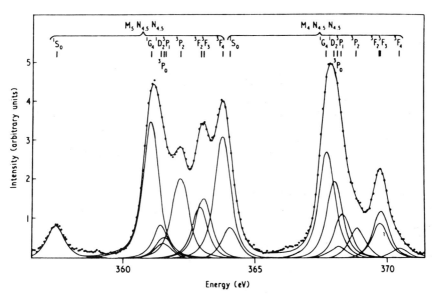

Figure 3.13 Fine structure in the $M_{4,5}N_{4,5}N_{4,5}$ Auger spectrum of cadmium due to multiplet splittings in the final state of the Auger process. The upper part is the observed spectrum with the assumed inelastic background shown as the dashed line. The lower part shows the calculated positions and intensities of the various multiplet components for each group, with the resultant theoretical envelope (solid line) compared with the experimental points. (Reproduced from Aksela and Aksela[12] by permission of the Institute of Physics)

above 'quasi-atomic' behaviour, are given in Figure 3.13. The nomenclature based on j–j coupling labels the two groups of peaks as $M_5N_{4,5}N_{4,5}$ and $M_4N_{4,5}N_{4,5}$, respectively, taking no account of the multiplet splittings. Underneath the experimental spectra are shown the calculated positions and relative intensities of the multiplets corresponding to the nine permitted final states; it is not easy to distinguish all nine since the energies of the $^3P_0$ and $^3P_1$ states are almost the same, but it can be seen that the energetic positions based on atomic structure calculations agree well with experiment.

3.3.1.6 Plasmon loss features

Any electron of sufficient energy passing through a solid can excite one or other of the modes of collective oscillation of the sea of conduction electrons. These oscillations have frequencies characteristic of the material of the solid, and therefore need characteristic energies for excitation. An electron that has given up an amount of energy equal to one of these characteristic energies, in the course of excitation, is said to have suffered a plasmon loss. Within the solid the loss is said to be that of a 'bulk' plasmon, and if the fundamental characteristic frequency of the bulk plasmon is ω_b, then the plasmon energy loss is clearly $\hbar\omega_b$. Since electrons that have suffered a plasmon loss in energy can themselves suffer further losses of this kind in a sequential fashion, then a series of losses, all equally spaced by $\hbar\omega_b$ but of decreasing intensity, will occur.[13]

At a surface the regular atomic lattice of the solid terminates and the conditions for the setting up of oscillations of frequency ω_b are no longer satisfied. Instead, a rather localized type of collective oscillation can be excited, of fundamental frequency ω_s, referring to the surface, where ω_s is less than ω_b. For a free electron metal ω_s can be shown[13] theoretically to be equal to $\omega_b/\sqrt{2}$, or more generally[14] to $\omega_b/(1 + \varepsilon)^{1/2}$, where ε is the dielectric constant.

When, therefore, the energies of electrons that have been ejected from a solid are analysed, there will be found associated with prominent peaks in the spectrum a series of plasmon loss peaks. The fundamental or 'first' plasmon loss will always be visible, and, dependent on the material and the experimental conditions, several multiple plasmon losses of decreasing size may also be visible. According to the surface condition of the material and to the kinetic energy of the parent peak, surface plasmon loss peaks may be discernible, but will always be of a lower intensity than the adjacent bulk plasmon peak.

A very clear example of an Auger spectrum containing all the above-mentioned features is given in Figure 3.14, in which the various losses are associated with the *KLL* peaks of aluminium. In the figure the *KLL* transitions are shown as excited by both electrons and soft X-ray photons, demonstrating that, for the same energy resolution, the shape of an Auger peak is the

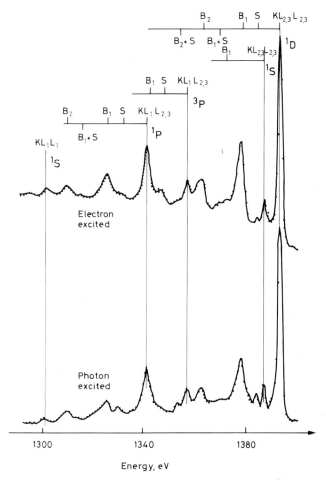

Figure 3.14 Electron-excited and photon-excited plasmon loss fine structure associated with the aluminium *KLL* Auger spectrum. The parent Auger peaks and their spectral terms are indicated as $KL_{2,3}L_{2,3}(^1D)$, $KL_{2,3}L_{2,3}(^1S)$, $KL_1L_{2,3}(^3P)$, etc. The surface plasmons are indicated as S and the successive bulk plasmons as B_1 and B_2; coupling between bulk and surface plasmons is thus $(B_1 + S)$ and $(B_2 + S)$. (Reproduced from Dufour *et al.*[16] by permission of The Royal Swedish Academy of Sciences)

same for either method of excitation and also that the same associated plasmon losses occur.

In an AES spectrum the largest peak is generally that of the elastically scattered electrons, and if the primary electron beam has a narrow spread in energy, then the elastic peak is also narrow when adequate energy resolution is used in the energy analyser. Under these conditions the plasmon loss structure associated with the elastic peak is both intense and well resolved, and the surface plasmon loss $h\omega_s$ in particular has proved useful as a surface cleanliness diagnostic. By reducing the primary energy to 50–300 eV, typically, where the penetration of the beam into the solid is very low and all loss processes are occurring at or near the surface, the surface plasmon loss can be considerably enhanced. The level of detectability of the resultant loss peak, coupled with the sensitivity of the surface plasmon loss to surface contamination and reaction, is then such that the state of cleanliness of a surface can be judged at least as sensitively from observation of that peak as that of the Auger peaks characteristic of contaminants.

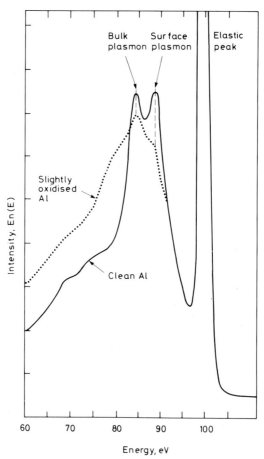

Figure 3.15 Bulk and surface plasmon losses from clean and slightly oxidized aluminium surfaces associated with the elastically scattered peak at a primary energy of 100 eV. The magnitude of the surface plasmon loss peak is a sensitive monitor of surface cleanliness. (Reproduced from Massignon *et al.*[16] by permission of the American Institute of Physics)

The surface plasmon loss peak has been used in this way by, for example, Massignon *et al.*[16] Figure 3.15, from their paper, shows how sharply the surface plasmon loss peak from aluminium is reduced due to slight oxygen contamination.

3.3.1.7 Ionization loss features

As depicted earlier in Figure 3.3, the first step in the electron-excited Auger process is the ionization of a core level by an incoming primary electron. Once an electron has entered a solid and lost energy by one or more loss processes any information about its origin is lost and it becomes indistinguishable from the other electrons in the solid. Thus on ionization two electrons are created in the solid, and since an electron can give up any fraction of its kinetic energy either of the two electrons can have energy from $(E_p - E_i)$ to zero. For ionization of E_i, kinetic energies above $(E_p - E_i)$ are not possible since all levels are filled up to E_F, from which E_i is measured, and there is thus nowhere for an excited electron to go except to empty, allowed, states beyond E_F.

In the undifferentiated secondary electron spectrum, therefore, the ionization loss feature would appear as a shallow step at a kinetic energy $(E_p - E_i)$, followed by a decreasing tail stretching all the way to zero energy. Such a step is very difficult to observe in the undifferentiated spectrum, but may be revealed more clearly, as for Auger features, by the usual differentiation with respect to energy.[17]

3.4 The X-ray Photoelectron Spectrum

3.4.1 Primary structure

Figure 3.16 depicts a *wide scan* spectrum of a clean silver surface obtained using Mg $K\alpha$ radiation and an analyser operating in the constant pass energy mode (constant ΔE). A series of peaks are observed on a background which generally increases to low kinetic energy (high binding energy) but which also shows step-like increases on the low kinetic energy side of each significant peak. When, as in this case, the X-ray source is non-monochromatic, the output consists of a broad continuous distribution (Bremsstrahlung radiation upon which are superimposed lines characteristic of the target (anode) material. All these X-rays give rise to photo-electron emission according to the equation

$$E_K = h\upsilon - E_B - \phi \qquad (3.4)$$

here E_K is the measured electron kinetic energy (KE), $h\upsilon$ the energy of the exciting radiation, E_B the binding energy (BE) of the electron in the solid and

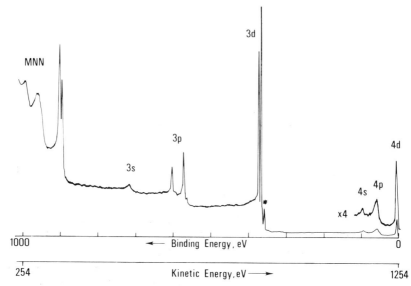

Figure 3.16   X-ray photo-electron spectrum of silver excited by Mg $K\alpha$ (essentially Mg $K\alpha_{1,2}$) and recorded with a constant analyser energy of 100 eV (*Ag $3d$ 'satellite' excited by Mg $K\alpha_{3,4}$)

ϕ the 'work function' (a catch-all term whose precise value depends on both sample and spectrometer, see Appendix I). Equation (3.4) assumes that the photo-emission process is *elastic*. Thus each characteristic X-ray (Mg $K\alpha_{1,2}$, etc., see below) will give rise to a series of photo-electron peaks which reflect the discrete binding energies of the electrons present in the solid.

The photo-emission process is *inelastic* if the photo-electron suffers an energy change (usually an energy loss) between photo-emission from an atom in the solid and detection in the spectrometer. Characteristic energy loss processes are discussed later; for the moment it is sufficient to note that the background 'step' to low kinetic energy of the photo-electron peaks is due to inelastic photo-emission (energy loss within the solid). Photo-emission by Bremsstrahlung radiation gives rise to a general background which is dominant in the low binding energy region of the spectrum. Secondary electrons resulting from inelastic photo-emission increasingly dominate the background at lower kinetic energy. The peaks observed in Figure 3.16 can be grouped into three basic types: peaks due to photo-emission from *core levels* and *valence levels* and peaks due to X-ray excited Auger emission (*Auger series*).

3.4.1.1   Core levels

The core-level structure in Figure 3.16 is a direct reflection of the electron structure of the silver atom. Mg $K\alpha$ radiation (1253.6 eV) is only energetic

Table 3.5 Spin-orbit splitting parameters

Subshell	j values	Area ratio
s	$\frac{1}{2}$	—
p	$\frac{1}{2}$, $\frac{3}{2}$	$1:2$
d	$\frac{3}{2}$, $\frac{5}{2}$	$2:3$
f	$\frac{5}{2}$, $\frac{7}{2}$	$3:4$

enough to probe the core levels of silver up to the $3s$ shell. It is immediately clear that the core levels have variable intensities and widths and that non-s levels are doublets.

The doublets arise through spin-orbit $(j-j)$ coupling which has previously been discussed in Section 3.2.1 Two possible states, characterized by the quantum number j $(j = l + s)$ arise when $l > 0$ (see Table 3.1). The difference in energy of the two states, ΔE_j, reflects the 'parallel' or 'anti-parallel' nature of the spin and orbital angular momentum vectors of the remaining electron. The magnitude of this energy separation is proportional to the spin-orbit coupling constant ξ_{nl} which depends on the expectation value $\langle 1/r^3 \rangle$ for the particular orbital. The separation can be many electron volts. ΔE_j is therefore expected to increase as Z increases for a given subshell (constant n, l) or to increase as l decreases for constant n (e.g. Figure 3.16 shows for silver that the splitting of $3p > 3d$).

The relative intensities of the doublet peaks are given by the ratio of their respective degeneracies $(2j + 1)$. Thus the area ratios and designations (nl_j) of spin-orbit doublets are given in Table 3.5.

Relative intensities The basic parameter which governs the relative intensities of core-level peaks is the atomic photo-emission cross-section, α. Values of α have been directly calculated and also derived from X-ray mass absorption coefficients. The calculated values of Scofield[18] are presented in Figure 5.14 of Chapter 5 to show the trends. Variation in the exciting energy also affects values of α, although for the commonly used Mg $K\alpha$ and Al $K\alpha$ radiations the effect on *relative* photo-emission cross-sections is very small. The transmission characteristics of the electron analyser, being a function of electron energy, apply an important modulation. These effects are thoroughly discussed in the context of quantification in Chapter 5.

Peak widths The peak width, defined as the full width at half-maximum (FWHM) ΔE, is a convolution of several contributions:

$$\Delta E = (\Delta E_n^2 + \Delta E_p^2 + \Delta E_a^2)^{1/2} \qquad (3.5)$$

where ΔE_n is the natural or inherent width of the core level, ΔE_p is the width of the photon source (X-ray line) and ΔE_a the analyser resolution, all expressed as FWHM. Equation (3.5) assumes that all components have a Gaussian line shape.

The analyser contribution is discussed in detail in Chapters 2 and 5; it is the same for all peaks in the spectrum when the analyser is operated in the constant analyser energy (CAE) mode, but varies across the spectrum when the analyser is operated in the constant retard ratio (CRR) mode (since in this case $\Delta E/E$ is a constant).

The inherent line width of a core level, i.e. the range in KE of the emitted photo-electron, is a direct reflection of uncertainity in the lifetime of the ion state remaining after photo-emission. Thus from the uncertainity principle we obtain the line width (in energy units)

$$\Gamma = \frac{h}{\tau} = \frac{4.1 \times 10^{-15}}{\tau} \quad \text{eV} \tag{3.6}$$

with Planck's constant (h) expressed in electronvolt-seconds and the lifetime expressed in seconds. Clearly the narrowest core levels (e.g. Ag $3d$) have lifetimes between 10^{-14} and 10^{-15} s whilst the broader core levels (e.g. Ag $3s$) have lifetimes close to or even slightly less than 10^{-15} s.

Core-hole lifetimes are governed by the processes which follow photo-emission, in which the excess energy of the ion is dissipated or decays. Any of three mechanisms may be involved (Figure 3): emission of an X-ray photon (X-ray fluorescence), or emission of an electron either in an Auger process or in a Coster–Kronig process.

The Coster–Kronig process is a special type of Auger process in which the final doubly charged ion has one hole in a shell of the same principle quantum number (n) as that of the original ion. The relevant point at this juncture is that Coster–Kronig processes are heavily favoured for the *initial* decay of holes in core levels with low angular momentum quantum numbers (e.g. $2s$, $3s$ and $3p$, $4s$ and $4p$, etc.) and in addition they are extremely fast. These core levels, therefore, are broader than those which decay by normal Auger processes.

It can be noted that the line widths of the principle light element core levels ($1s$, $2p$) systematically increase with an increase in atomic number. The increase in valence electron density enhances the probability of the relevant Auger process (see below), decreasing the lifetime of the core hole. This effect can even be seen in high resolution spectra from one core level in different chemical environments. For example, the chemically distinct N atoms in the linear azide ion (N_3^-) give rise to two N $1s$ peaks (due to the chemical shift effect, see below), of which the peak due to the central atom, with the lowest electron density, is the narrower.

3.4.1.2   Valence levels

Valence levels are those occupied by electrons of low binding energy (say 0–20 eV) which are involved in de-localized or bonding orbitals. The spectrum in this region consists of many closely spaced levels giving rise to a *band* structure. Two situations can be distinguished as in Figure 3.17, viz. insulators and conductors (metals). Figure 3.17 illustrates the density of electron states (per unit energy in unit volume) in these two cases. In the case of an insulator the occupied *valence band* is separated from the empty *conduction band*, whilst in the case of a metal these bands overlap and the uppermost occupied state is termed the *Fermi level* (E_F). Note that E_F is not the true zero point of the electron energy scale, although BEs are often referenced to this point. The true zero is the vacuum level (E_v) and, to a first approximation, $E_F - E_v = \phi$, where ϕ is the work function of the material.

Figure 3.17 shows that at low energies the unoccupied states have structure. Consequently, if photo-electrons ejected from filled levels have KEs which fall within this structured region the *observed* intensity will be a convolution of the filled and empty density of states together with the matrix of transition probabilities. This is the case in ultra-violet photo-electron spectroscopy (UPS) and gives rise to the strong dependence of valence-level spectra on photon energy. In the case of XPS the KE of the valence photo-electrons is such that the final states are quite devoid of structure; thus the observed density of states closely reflects the initial filled density of states. In Figure 3.16 the peak labelled '4*d*' is in fact the conduction band spectrum which is dominated by 4*d* states. The complete spectrum at high resolution, Figure 3.18, shows the band structure and the sharp cut-off in electron density at E_F.

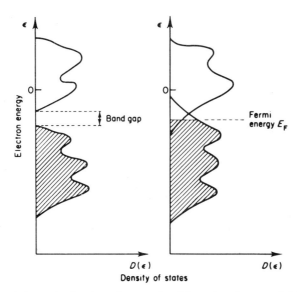

Figure 3.17 Schematic density of states for (a) an insulator and (b) a metal. The shading indicates the extent to which the energy levels are occupied. (After Orchard[19])

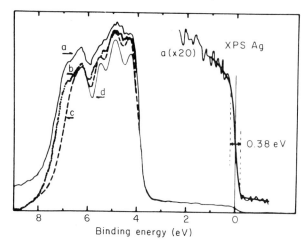

Figure 3.18   XPS valence spectrum for polycrystalline silver excited by monochromatic Al $K\alpha$ radiation in comparison with the theoretical density of states: (a) raw XPS data; (b) data after subtraction of smooth inelastic background; (c) and (d) total theoretical density of states after two different lineshape broadenings to include lifetime and shake-up effects. The Fermi edge has been expanded to reveal the instrumental resolution. (After Barrie and Christensen[20])

Cross-sections for photo-electron emission from valence levels are much lower than for core levels, giving rise to low intensities in general. However, valence-level spectra have their analytical value as discussed in Section 3.4.2.2.

3.4.1.3   Auger series

As explained in Section 3.3.1.1 the dominant decay mechanism for core-hole states is via radiationless transitions resulting in electron emission. The KEs of electrons resulting from Coster-Kronig transitions are very low and these are not normally detected; however, Auger electrons can give rise to dominant peaks in the 'photo-electron' spectrum as seen in Figure 3.16. The nomenclature for Auger peaks is described fully in Section 3.2.

Four principle Auger series can be observed in XPS. The *KLL* series is first observed for boron and can be exited through to sodium (with Mg $K\alpha$) or magnesium (Al $K\alpha$). With Ag $L\alpha$ (2984 eV) the whole series through to chlorine can be observed. The dominant line is KL_2L_3, as illustrated in Figure 3.19.

The *LMM* series is first observed with sulphur and can be followed through to germanium (Mg $K\alpha$), selenium (Al $K\alpha$) or ruthenium (Ag $K\alpha$). This series is very complex since lines due to initial ionization in L_2 or L_3 can then result

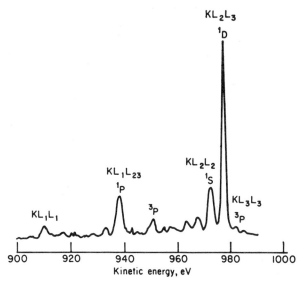

Figure 3.19 *KLL* Auger spectrum of sodium vapour. (After Hillig, Cliff and Mehlhorn[21])

in transitions involving M_{23} or M_{45}. The relative intensities of these groups changes with the atomic number (Z). An example is given in Figure 3.20.

The *MNN* series can be split into two types. The $M_{45}N_{45}N_{45}$ series probably first observed with molybdenum can be followed right through to neodymium with Mg $K\alpha$. The $M_{45}N_{67}N_{67}$ series is only accessible with higher energy photons (e.g. Ag $L\alpha$) because the theoretically observable lines from rare earth elements are weak and ill-defined until hafnium is reached (M_{45}

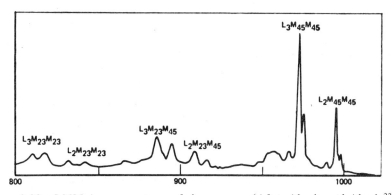

Figure 3.20 *LMM* Auger spectrum of zinc vapour. (After Aksela and Aksela[22])

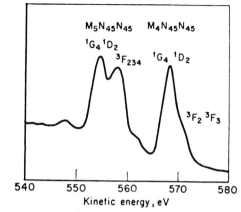

Figure 3.21   Cesium $M_{45}N_{45}N_{45}$ Auger spectrum from Cs_2SO_4. (After Wagner[23])

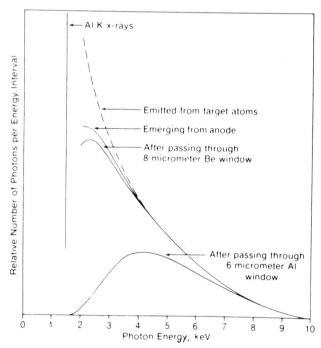

Figure 3.22   Bremsstrahlung radiation intensity from a conventional aluminium X-ray source. (Reproduced from Wagner and Taylor[24] by permission of Elsevier Scientific Publishing Co.). Note that use of a 2 μm aluminium window is roughly equivalent to using the beryllium window

shell $\simeq$ 1660 eV and therefore beyond Mg $K\alpha$ or Al $K\alpha$). A typical $M_{45}N_{45}N_{45}$ spectrum is shown in Figure 3.21.

The positions of the principle Auger peaks on the BE scale are included in Appendix 7. The KE of an Auger electron is given in equation (3.2); thus, in contrast to the photo-electron (equation 3.4) the Auger electron KE is independent of the exciting radiation. If, for example, the X-ray source is changed from Mg $K\alpha$ to Al $K\alpha$ all photo-electron peaks increase in KE by 233 eV whilst the Auger peaks remain unmoved. For spectra plotted directly in BE the reverse is true. This allows the two types of peak in the electron spectrum to be easily distinguished. Note also that the width of X-ray excited Auger peaks, as a consequence of equation (3.2), does not contain a contribution from the line width of the X-ray source.

In the above discussion it has been indicated which X-ray sources have possibilities for exciting the high energy Auger series (i.e. where the initial state energy is greater than that of Al $K\alpha$). However, for conventional single-target instruments the observation of principal photo-electron lines and these Auger lines in one experiment (see Section 3.4.2.3) is impossible, and attention has recently turned to the use of the Bremsstrahlung radiation for this purpose. As shown in Figure 3.22 useful intensities over a wide range of photon energy can be obtained from the normal aluminium target, particularly if a beryllium window is used in the X-ray source. Wagner and Taylor[24] have explored this alternative for twenty-four elements with Auger lines having KEs in the range 1400–2200 eV.

3.4.2 Information from primary structure

3.4.2.1 *Core-level chemical shifts*

The discovery, during the early days of XPS, that non-equivalent atoms of the same element in a solid gave rise to core-level peaks with measureably different BEs had a stimulating effect on the field. This BE difference was dubbed the 'chemical shift' by analogy with nuclear magnetic resonance (n.m.r.) spectroscopy. Non-equivalence of atoms can arise in several ways: difference in formal oxidation state, difference in molecular environment, difference in lattice site and so on. The physical basis of the chemical shift effect is illustrated by a relatively simple model used with much success (in appropriate cases) to interpret chemical shift data—the charge potential model:[25]

$$E_i = E_i^0 + kq_i + \sum_{i \neq j} \frac{q_i}{r_{ij}} \qquad (3.7)$$

where E_i is the BE of a particular core level on atom i, E_i^0 is an energy reference, q_i is the charge on atom i and the final term of equation (3.7) sums the potential at atom i due to 'point changes' on surrounding atoms j.

If the atom is considered to be an essentially hollow sphere on the surface of which the valence charge q_i resides, then the classical potential inside the sphere is the same at all points and equal to q_i/r_v, where r_v is the average valence orbital radius. A change in the valence electron charge (density) of Δq_i changes the potential inside the sphere by $\Delta q_i/r_v$. Thus the BE of all core levels will change by this amount. Moreover, as r_v increases, the BE shift for given Δq_i will decrease. In practice it is indeed found that for all cases where core–valence interaction is small the observable core levels from a given atom do suffer a similar BE shift and that BE shifts between equivalent compounds decrease as one descends a column of the periodic table (similar Δq_i, r_v increasing). If the summation term in equation (3.7) is abbreviated to V_i then the shift in BE for a given core level of atom i in two different environments is

$$E_i^{(1)} - E_i^{(2)} = k(q_i^{(1)} - q_i^{(2)}) + (V_i^{(1)} - V_i^{(2)}) \qquad (3.8)$$

The first term ($= k \, \Delta q_i$) clearly ensures that an increase in BE accompanies a decrease in valence electron density on atom i. The second term is not to be underestimated since it has the opposite sign to Δq_i. In molecular solids the atoms j are basically the atoms bonded to atom i, but in ionic solids the summation extends over the whole lattice; this is closely related to the Madelung energy of the solid and for this reason V is often referred to as a Madelung potential.

Equations (3.7) and (3.8) involve a number of simplifications. A major simplification of the charge potential model is that it neglects relaxation effects, i.e. no account is taken of the polarizing effect of the core hole on the surrounding electrons, both intra-atomic (on atom i) and extra-atomic (on atoms j). This is one example of a final-state effect, a term used to indicate the possibility that the electronic state of the photo-ionized atom may be different from that of the initial state (usually the atomic ground state). Other final-state effects are discussed later in this chapter.

In comparing equation (3.8) with equation (3.4) it will be realized that equation (3.8) is only valid if the work functon of the material incorporating $i_{(1)}$ and $i_{(2)}$ are the same. When conductors are involved 'ϕ' is the spectrometer work function and the problem does not arise, but in the case of insulators $\phi_{(1)}$ and $\phi_{(2)}$ may not be identical. Moreover, insulators acquire a surface charge under X-irradiation and correction of non-identical charging is required before chemical shifts can be accurately measured. The problem of charge correction or BE referencing is dealt with in Appendix 2.

Figure 3.23 illustrates the chemical shift effect for three different types of nitrogen atom in the N 1s spectrum of a single compound. In this case the relative shifts are easily measured but the absolute N 1s binding energies require some calibration procedure to be adopted before they can be measured. The clear trend in this example is that the N 1s binding energy increases with an increase in the formal oxidation number of the nitrogen atom. Table 3.6 shows this to be generally true for nitrogen in both inorganic and organic

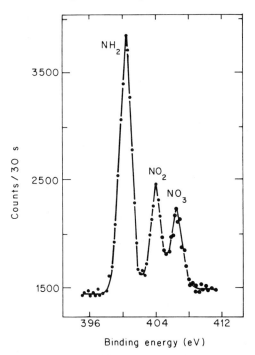

Figure 3.23 Nitrogen 1s spectrum from *trans*- $[Co(NH_2CH_2CH_2NH_2)_2(NO_2)_2]NO_3$. (Reprinted with permission from Hendrickson[26]. Copyright 1969 American Chemical Society)

compounds. In situations where the formal oxidation state is the same the general rule is that the core-level BE of the central atom increases as the electro-negativity (electron withdrawing power) of attached atoms or groups increases. Many examples of core-level chemical shifts appear in later chapters and Appendix 4 gives an extensive compilation of reported values for most elements.

Many cases obviously arise in which distinct chemical states give rise to photo-electron peaks whose relative chemical shift is sufficiently small for the peaks to overlap considerably. Whilst in principle such a shift should be measurable if ≈ 0.2 eV this is impossible when the contributory peaks have a line width typically of 1–2 eV. However, if the profile of the contributory peaks is known then an attempt can be made to deconvolute the envelope. This procedure is discussed fully in Appendix 3.

3.4.2.2   Valence band structure

As mentioned in Section 3.4.1.2 valence band spectra relate closely to the occupied density of states structure. This is very useful in the study of the

Table 3.6   N 1*s* binding energies. (Reproduced from Wagner *et al.*[27] by permission of Perkin-Elmer Corporation)

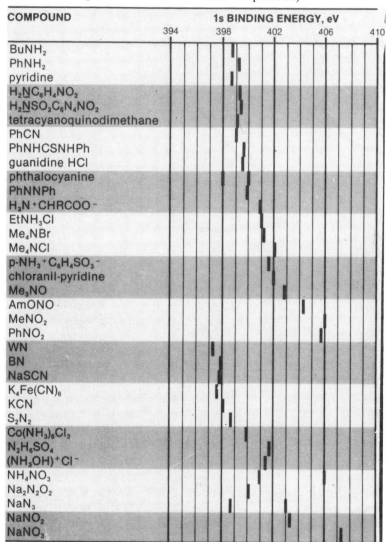

COMPOUND	1s BINDING ENERGY, eV
BuNH$_2$	
PhNH$_2$	
pyridine	
H$_2$NC$_6$H$_4$NO$_2$	
H$_2$NSO$_2$C$_6$N$_4$NO$_2$	
tetracyanoquinodimethane	
PhCN	
PhNHCSNHPh	
guanidine HCl	
phthalocyanine	
PhNNPh	
H$_3$N$^+$CHRCOO$^-$	
EtNH$_3$Cl	
Me$_4$NBr	
Me$_4$NCl	
p-NH$_3$$^+C_6H_4SO_3$$^-$	
chloranil-pyridine	
Me$_3$NO	
AmONO	
MeNO$_2$	
PhNO$_2$	
WN	
BN	
NaSCN	
K$_4$Fe(CN)$_6$	
KCN	
S$_2$N$_2$	
Co(NH$_3$)$_6$Cl$_3$	
N$_2$H$_6$SO$_4$	
(NH$_3$OH)$^+$Cl$^-$	
NH$_4$NO$_3$	
Na$_2$N$_2$O$_2$	
NaN$_3$	
NaNO$_2$	
NaNO$_3$	

electronic structure of materials, fundamental to many aspects of device applications, and in checking the accuracy of band structure calculations. As shown in Figure 3.18, very good agreement has been obtained in the case of silver and this is, in fact, quite typical of work on all types of metals. The applicability of various models for the electronic structure of alloys has been extensively studied in this way.

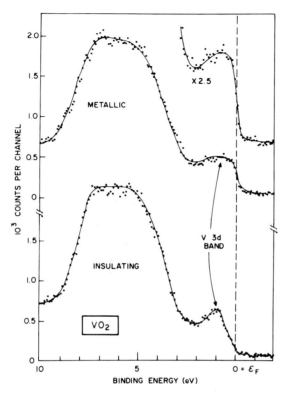

Figure 3.24   XPS valence bands of VO_2 at temperatures above and below the metal-insulator transition temperature. (Reproduced from Wertheim[28] by permission of Pergamon Press Ltd)

Further examples of the direct information obtained from valence band spectra are provided by the following. Vanadium dioxide (VO_2) is an insulator at room temperature but undergoes a structural change at 65 °C and becomes a metallic conductor. Figure 3.24 clearly shows the gap created at the lower temperature which results in the loss of conductivity. The tungsten bronzes are the family of materials with composition Na_xWO_3 ($0.32 < x < 0.93$). The metallic properties of these materials contrasts with the insulating behaviour of WO_3. Figure 3.25 shows that increasing the sodium content results in electron density increasing in a 'd' band. This is absent in WO_3 (d^0) but is clearly seen in ReO_3 (the d^1 counterpart) which has metallic conductivity.

In the case of polymers, valence bands reveal structural information often unobtainable from core-level studies. This fingerprinting capacity of polymer valence band spectra is discussed in Chapter 9.

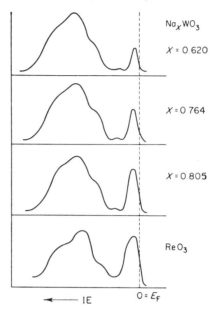

$Na_x WO_3$

$X = 0.620$

$X = 0.764$

$X = 0.805$

$Re O_3$

$\longleftarrow$ IE $0 = E_F$

Figure 3.25   XPS valence bands of the tungsten bronzes $Na_x WO_3$ and ReO_3. (After Wertheim *et al.*[29])

3.4.2.3   *Auger chemical shifts*

For the Auger processes in which the final vacancies arise in core levels it is clear that a change in chemical state giving rise to a chemical shift in the photo-electron lines will also produce a chemical shift in the Auger lines. The magnitude of the Auger chemical shift is, however, often significantly greater than that of the photo-electron chemical shift. For instance, Wagner and Biloen,[30] in an early study of this effect, found the Auger chemical shift betwen metals and their oxides exceeded the photo-electron chemical shift by the following: Mg, 5.0; Zn, 3.7; Ga, 4.2; Ge, 3.5; As, 3.4; Cd, 4.9; In, 1.8; and Sn, 2.5 eV. Chemical shifts in Auger lines are discussed more fully in Section 3.3 but the reason for this effect can be appreciated from the follow-ing simplified argument.[31] As discussed in Section 3.4.2.2 the difference in BE between two chemical states depends on the change in core electron energy and the change in intra-atomic and extra-atomic relaxation energies. Thus, for a $1s$ (K) electron:

$$\Delta E_B(K) = \Delta\varepsilon(K) - \Delta R(K^+) \qquad (3.9)$$

where $\Delta\varepsilon(K)$ is the $1s$ energy change and $\Delta R(K^+)$ is the change in relaxation energies (predominantly extra-atomic) in the final, singly ionized state. As explained above (Section 3.4.2.1):

$$\Delta\varepsilon(K) \approx \Delta\varepsilon(L) \tag{3.10}$$

and the kinetic energy change is given by

$$\Delta E(K) = -\Delta E_{\text{B}}(K) = \Delta R(K^+) - \Delta\varepsilon(K) \tag{3.11}$$

For the *KLL* Auger process the kinetic energy change between chemical states is given by

$$\Delta E\ (KLL) = \Delta\varepsilon(K) - \Delta R(K^+) - 2\Delta\varepsilon(L) - \Delta R(L^+L^+) \tag{3.12}$$

where (L^+L^+) is the doubly ionized final state. It can be argued that $R(L^+L^+) \simeq 4R(K^+)$ so from equations (3.10) and (3.12):

$$\Delta E\ (KLL) = 3\Delta R(K^+) - \Delta\varepsilon(K) \tag{3.13}$$

and combining equations (3.11) and (3.13):

$$\Delta E\ (KLL) - \Delta E(K) = 2\Delta R(K^+) \tag{3.14}$$

The difference between Auger and photo-electron chemical shifts therefore results from the difference in final-state relaxation energies between chemical states. It is possible to define a parameter, α, such that, for example:

$$\alpha = E(KLL) - E(K) \tag{3.15}$$

For insulators α is independent of any static charging and is a parameter, characteristic of a particular chemical state, which can be measured with greater accuracy than core-level BE or Auger peak KE. Wagner termed α the Auger parameter[32] and defined it practically as the difference between the KE of the most intense Auger peak and the KE of the most intense photo-electron line. The disadvantage of this procedure is that it can result in negative values. A 'modified' Auger parameter has been defined[33] as, for example:

$$\alpha' = \alpha + h\upsilon = E(KLL) + E_{\text{B}}(K) \tag{3.16}$$

which overcomes this problem. A good example of the use of this parameter comes from the work of West and Castle[34] on silicate minerals, in which a Zr $L\alpha$ source was used to excite the Si $1s$ and Si *KLL* levels. Many Auger parameters are given in Appendix 4.

3.4.2.4 Auger lineshapes

When an Auger process results in at least one vacancy in the valence levels then the intensity distribution of the lines in the Auger series can vary strongly from one compound to another. In effect the valence band or molecular orbital structure is convoluted into the Auger peak structure. This is nicely shown by the F *KLL* spectra in Figure 3.26. In fluorine the L_1 ($2s$) level is

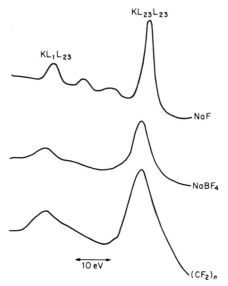

Figure 3.26   Fluorine *KLL* spectra for solids of different covalent character. (After
Wagner[35])

essentially core-like, whilst the L_{23} ($2p$) levels are involved in bonding. In
sodium fluoride the fluorine environment is similar to that of the free ion and
the *KLL* structure appears to be similar to the Na *KLL* structure where all
levels are core-like. In the complex ion and in the organic polymer, where the
covalency of fluorine is increasing and the L_{23} levels are increasingly involved
in bonding orbitals, the component lines in the *KLL* series become less
sharply defined.

With oxygen the *KLL* group takes on a greater variety of forms, as shown
in Figure 3.27. Note in particular how the separation between the KL_1L_{23} and
$KL_{23}L_{23}$ lines varies by up to 5 eV and how the $KL_{23}L_{23}$ profile changes from
an essentially single to a double peak. The extra variation in this peak is
expected because both vacancies are now in the valence shell, whereas the L_1
shell is again semi-core-like. Once again the O *KLL* line shape shows characteristic variations from one type of compound to another.

3.4.3   Secondary structure

3.4.3.1   *X-ray satellites and ghosts*

Before discussing true secondary structure features in the spectrum it is
necessary to identify the sources of 'spurious' low intensity peaks. These arise
in two ways.

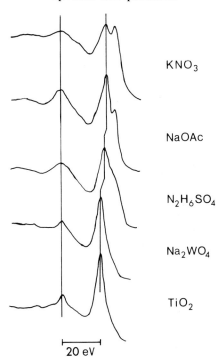

KNO$_3$

NaOAc

N$_2$H$_6$SO$_4$

Na$_2$WO$_4$

TiO$_2$

20 eV

Figure 3.27   Oxygen *KLL* Auger spectra from several materials in the solid state. (Reprinted with permission from Wagner, Zatko and Raymond[36] Copyright 1980 American Chemical Society)

Standard X-ray sources are not monochromatic. Besides the Bremmsstrahling radiation mentioned in Section 3.4.1 and the principle $K\alpha_{a,2}$ line, magnesium and aluminium targets also produce a series of lower intensity lines, referred to as X-ray satellites. The transitions giving rise to $K\alpha_{1,2}$ (an unresolved doublet) are $2p_{3/2,1/2} \rightarrow 1s$. Satellites arise from less probable transitions (e.g. $K\beta$; valence band $\rightarrow 1s$) or transitions in a multiply ionized atom (e.g. $K\alpha_{3,4}$). The observed line positions and intensities are given in Table 3.7.

X-ray ghosts are due to excitations arising from impurity elements in the X-ray source. The most common ghost is Al $K\alpha_{1,2}$ from a Mg $K\alpha$ source. This arises from secondary electrons produced inside the source hitting the thin aluminium window (present to prevent these same electrons hitting the sample). This radiation will therefore produce weak ghost peaks 233.0 eV to higher KE of those excited by the dominant Mg $K\alpha_{1,2}$. Old or damaged targets can give rise to ghost peaks excited by Cu $L\alpha$ radiation, the main line from the exposed copper base of the target. These peaks appear 323.9 eV (556.9 eV) to lower KE of Mg $K\alpha_{1,2}$ (Al $K\alpha_{1,2}$) excited peaks. In dual-anode X-ray sources, which are commonly being used with one anode capable of yielding

Table 3.7   High-energy satellite lines from Mg and Al targets. (From data of Krause and Ferreira in Carlson[37])

X-ray line	Separation from $K\alpha_{1,2}$(eV) and relative intensity ($K\alpha_{1,2}$ = 100)	
	Mg	Al
$K\alpha'$	4.5 (1.0)	5.6 (1.0)
$K\alpha_3$	8.4 (9.2)	9.6 (7.8)
$K\alpha_4$	10.0 (5.1)	11.5 (3.3)
$K\alpha_5$	17.3 (0.8)	19.8 (0.4)
$K\alpha_6$	20.5 (0.5)	23.4 (0.3)
$K\beta$	48.0 (2.0)	70.0 (2.0)

high-energy X-rays (e.g. zirconium, silver and titanium), misalignments inside the source can lead to cross-talk between filaments and anodes which is a further source of ghost peaks.

A ghost source which has not previously received much attention is O $K\alpha$ arising from the oxide layers present on aluminium and magnesium target surfaces. The possible importance of this interference when recording weak signals has been pointed out by Koel and White[38] as a result of their study of X-ray excited C KVV and O KVV Auger spectra from adsorbed species. O $K\alpha$ peaks appear 728.7 eV (961.7 eV) to lower KE of Mg $K\alpha_{1,2}$ (Al $K\alpha_{1,2}$) excited peaks.

3.4.3.2   Multiplet splitting

Multiplet splitting (also referred to as exchange or electrostatic splitting) of core-level peaks can occur when the system has unpaired electrons in the valence levels. As an example, consider the case of the $3s$ electron in the Mn^{2+} ion. In the ground state the five $3d$ electrons are all unpaired and with parallel spins (denoted ^6S). After ejection of the $3s$ electron a further unpaired electron is present. If the spin of this electron is parallel to that of the $3d$ electrons (final state ^7S) then exchange interaction can occur, resulting in a lower energy than for the case of anti-parallel spin (final state ^5S). Thus the core level will be a doublet and the separation of the peaks is the exchange interaction energy:

$$E = (2S + 1) K_{3s3d} \qquad (3.17)$$

where S is the total spin of the unpaired electrons in the valence levels ($\frac{5}{2}$ in this case) and K_{3s3d} is the $3s$–$3d$ exchange integral. The intensity ratio for the two peaks is given by

$$\frac{I(S + \frac{1}{2})}{I(S - \frac{1}{2})} = \frac{S + 1}{S} \qquad (3.18)$$

Cr(CO)$_6$

Cr(C$_5$H$_5$)$_2$

Cr(hfa)$_3$

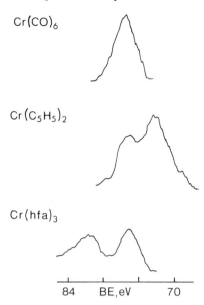

84   BE,eV   70

Figure 3.28   Multiplet splitting of the 3s peak from chromium compounds: Cr(CO)$_6$ is diamagnetic (no unpaired electrons); Cr(C$_5$H$_5$)$_2$ has two unpaired electrons in the e_{2g} level; Cr(hfa)$_3$ has three unpaired electrons in the t_{2g} level (hfa = CF$_3$COCH-COCF$_3$). Note also how the mean BE varies with the electronegativity of attached ligands. (Reproduced by Clark and Adams[41] by permission of North-Holland Publishing Co.)

For the 3s in Mn^{2+} this ratio is $(\frac{5}{2} + 1)/\frac{5}{2}$ or 1.4 and E is calculated to be ≈ 13 eV. In the real case of MnF$_2$ the intensity ratio is ≈ 1.8 and the separation ≈ 6.5 eV.[39] This discrepancy is due to relaxation and configuration interaction effects which complicate the simple model described above (see Ref. 40 for a detailed discussion). However, the data for chromium compounds shown in Figure 3.28 illustrates the fact that this model predicts the correct trends for changes in spin state.

Multiplet splitting for non-s levels is more complex because of the additional involvement of orbital–angular momentum coupling. Thus in Mn^{2+} the 3p level is split into four levels. Multiplet splitting is strongest when both levels involved are in the same shell (as with 3s, 3p–3d), but the effects are still apparent in transition metal systems for 2p–3d interaction. A good example is for cobalt which can exist in a variety of spin states. Multiplet splitting causes broadening (with asymmetry) in both 2p$_{1/2}$ and 2p$_{3/2}$ peaks, leading to apparent variation in the separation of the peak maxima. Table 3.8 gives some examples, showing that the doublet separation can be used diagnostically.[42] Similar variation has been reported for chromium compounds.[43]

Multiplet splitting is found strongly in the 4s levels of rare earth metals

Table 3.8 Variation in $2p$ spin-orbit splitting with spin state in Co complexes. (Reproduced from Briggs and Gibson[42] by permission of North-Holland Publishing Co.)

	Binding energy, eV			Unpaired electrons
	Co $2p_{1/2}$	Co $2p_{3/2}$	$2p_{1/2}-2p_{3/2}$	
Co(acac)$_2$	800.2	784.2	16.0	3
Co(acac)$_3$	799.0	784.0	15.0	0
(PEt$_2$Ph)$_2$Co(C$_6$Cl$_5$)$_2$	797.4	782.1	15.3	1
(PEt$_2$Ph)$_2$Co(2-methyl-1-naphthyl)$_2$	797.0	781.5	15.5	1

(unpaired electrons in the $4f$ valence levels) and can also be seen in the spectra of organic radicals in which the free electron is relatively localized.

3.4.3.3 Shake-up satellites

To the valence electrons associated with an atom the loss of a core electron by photo-emission appears to increase the nuclear charge. This major perturbation gives rise to substantial reorganization of the valence electrons (referred to as relaxation) which may involve excitation of one of them to a higher unfilled level ('shake-up'). The energy required for this transition is not available to the primary photo-electron and thus the two-electron process leads to discrete structure on the low KE side of the photo-electron peak (shake-up satellite(s)).

Most detailed assignment work concerns atomic systems (e.g. neon gas) and shows that the transitions obey monopole selection rules. In the solid state the examples are found to be much more complex, except for organic systems. Conjugated and, especially, aromatic systems show shake-up satellites with intensities up to 5–10 per cent of the primary peak. In aromatic systems the satellite structure has been shown[44] to be due to $\pi \rightarrow \pi^*$ transitions involving the two highest filled orbitals and the lowest unfilled orbital (see Chapter 9).

Very strong satellites are observed for certain transition metal and rare earth compounds which have unpaired electrons in $3d$ or $4f$ shells respectively. The satellite intensity is usually much higher than would be expected for atomic-like shake-up (typically $\simeq 10$ per cent) and the reason for this has proved a great source of discussion.[40] A convincing explanation is that in the final state there is significant ligand–metal charge transfer such than an extra $3d$ or $4f$ electron is present compared with the initial state. This immediately explains why closed-shell systems (e.g. Cu$^+$, $3d^{10}$) do not exhibit shake-up satellites, whereas open-shell systems (e.g. Cu$^+$, $3d^9$) do.[45] However, shake-up is a poor description of this process, as Fadley[40] has pointed out; a preferable

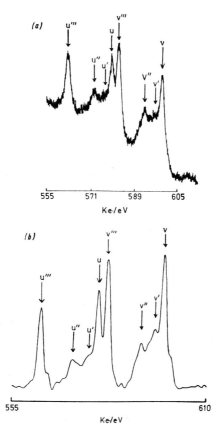

Figure 3.29 Cerium $3d$ signal from CeO_2: (a) raw Al $K\alpha$ excited spectrum, (b) deconvoluted spectrum. (After Burroughs *et al.*[46] by permission of the Royal Society of Chemistry)

description is that strong configuration interaction occurs in the final state due to relaxation. The complexity of such a process is illustrated in Figure 3.29 for cerium.

As in the case of multiplet splitting, shake-up satellites have diagnostic value. The distinguishing of copper (II) from copper (I) has already been mentioned. Tetrahedral nickel (II) gives satellites, square planar (diamagnetic) nickel (I) does not.[47] Cobalt (II) gives more intense satellites in the high spin (^4F) state than in the low spin (^2D) state, while cobalt (III) is diamagnetic and has no satellites.[42,48]

3.4.3.4 Asymmetric metal-core levels

In a solid metal there exists a distribution of unfilled one-electron levels above the Fermi energy (as shown in Figure 3.17) which are available for

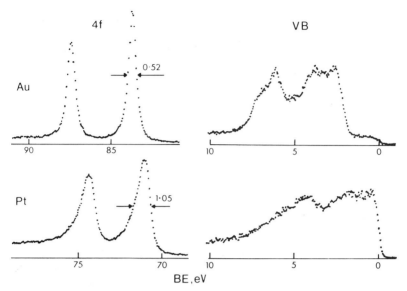

Figure 3.30   Core (4*f*) and valence (VB) photo-electron spectra of gold and platinum recorded using monochromatic Al $K\alpha$ radiation. Note the relationship between the degree of core-level asymmetry and the density of states at the Fermi level (BE = 0 eV). Reproduced from Barrie, Swift and Briggs[49])

shake-up-type events following core electron emission. In this case, instead of discrete satellites to low KE of the primary peak, the expectation is for the core level itself to be accompanied by a low KE tail. The higher the density of states at the Fermi level the more likely is this effect to be observed. This is confirmed by Figure 3.30. Such asymmetries must be borne in mind when attempting to deconvolute complex envelopes involving peaks from bulk metals (see Appendix 3).

3.4.3.5   Shake-off satellites

In a process similar to 'shake-up' valence electrons can be completely ionized, i.e. excited to an unbounded continuum state. This process, referred to as 'shake-off', leaves an ion with vacancies in both the core level and a valence level. Discrete shake-off satellites are rarely discerned in the solid state because: (a) the energy separation from the primary photo-electron peak is greater than for shake-up satellites, which means the satellites tend to fall within the region of the broad inelastic tail, and (b) transitions from discrete levels to a continuum produce onsets of increased intensity (i.e. broad shoulders) rather than discrete peaks. However, a weak structure on the inelastic tail of C l*s* peaks from polymers has been identified[50] (following double differentiation) as a shake-off structure which images the polymer valence band structure.

Plasmon loss processes have already been discussed in Section 3.3.1.6. Loss peaks of exactly the type described are found associated with core levels in exactly the same way as with Auger peaks.[51]

3.4.4 Angular effects

Two types of angular effect are important in XPS. The first involves the increase in surface sensitivity obtained at low angles of electron exit to the surface, whilst the second is found in single-crystal systems giving rise to 'photo-electron diffraction'.

3.4.4.1 Enhancement of surface sensitivity

The reason for this effect is simply demonstrated by reference to Figure 3.31. If λ is the inelastic mean free path (IMFP) of the emerging electron then 95 per cent of the signal intensity is derived from a distance 3λ within the solid. However, the vertical depth sampled is clearly given by

$$d = 3\lambda \sin \alpha \qquad (3.19)$$

and this is a maximum when $\alpha = 90°$. In the case of a substrate (s) with a uniform thin overlayer (o) the angular variation of intensities is given by

$$I_s^d = I_s e^{-d/\lambda \sin \alpha} \qquad \frac{2l}{\lambda \sin \theta} \qquad (3.20)$$

and

$$I_o^d = I_o(1 - e^{-d/\lambda \sin \alpha}) \qquad (3.21)$$

where λ is the appropriate value for the observed photo-electron. In the ideal

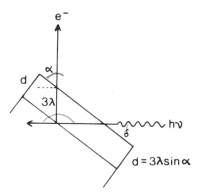

Figure 3.31   Surface sensitivity enhancement by variation of the electron 'take-off' angle

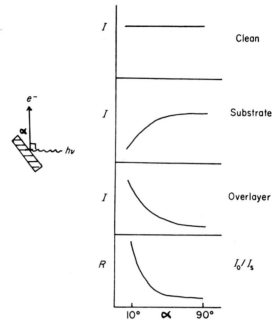

Figure 3.32   Theoretical angular dependent curves for a clean flat surface and a flat overlayer/substrate system

situation these equations lead to curves of the type shown in Figure 3.32. However, the real case is usually complicated by the fact that the system geometry imposes a response function also dependent on α. A useful summary of these effects for one particular spectrometer has been given by Dilks.[52] This complication is avoided by measurements of relative values of I_o/I_s so

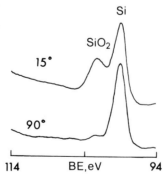

Figure 3.33   Effect of variation of take-off angle on the Si 2p spectrum from silicon with a passive oxide layer. Note the relative enhancement of the (surface) oxide signal at low angle (measured with respect to the surface). (After Wagner et al.[27])

that the instrument response function is cancelled. Thus, as shown in Figure 3.32, at low values of α, I_o/I_s increases significantly. An example of the effect is given in Figure 3.33.

The major requirement for surface sensitivity enhancement is that the surface is flat. Surface roughness leads to an averaging of electron exit angles and also to shadowing effects (both of the incident X-rays and emerging electrons) such that in most cases the surface enhancement effect cannot be observed.[53] In most commercial spectrometers the angle between the incident X-rays and detected electrons $(180 - \alpha - \delta)$ is fixed. However, it has been shown that at *very* low δ there is also a surface sensitivity enhancement effect due to the rapid fall-off in X-ray penetration depth as δ tends to grazing incidence.[54]

3.4.4.2 Photo-electron diffraction

Measurement of core electron intensity from a single crystal surface as a function of electron take-off angle gives rise to plots of the type shown in Figure 3.34 (and for any such angle similar plots will arise by rotation in the plane of the surface, i.e. by variation of the azimuthal angle). The pronounced fine structure superimposed on the instrument response function is primarily

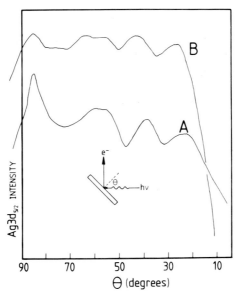

Figure 3.34 Angular variation of Ag $3d_{5/2}$ intensity from a clean (110) surface: (a) chemically polished, (b) not chemically polished after orientation. In both cases final cleaning was by argon ion etching and annealing cycles. (Reproduced from Briggs, Marbrow and Lambert[55] by permission of Pergamon Press Ltd.)

due to diffraction of photo-electrons from crystal planes. Studies of the related phenomena of high-energy electron channelling in crystals and Kikuchi band formation in LEED indicate[33] that photo-electron emission parallel with each set of planes (*hkl*) to within $\pm\theta_{hkl}$ (the first-order Bragg angle) will give rise to enhanced emission intensity. Here *hkl* are the usual Miller indices and θ_{hkl} is defined by

$$\lambda_e = 2d_{hkl} \sin \theta_{hkl} \tag{3.22}$$

where λ_e is the de Broglie wavelength of the core electron in question and d_{hkl} is the spacing between adjacent planes.

Photo-electron diffraction effects obviously complicate the quantification of peak intensity data from single crystal surfaces, but they have already been put to positive use analytically. Figure 3.35 shows how Evans and Scott[56] used X-ray photo-electron diffraction (XPD) patterns (the plot of relative core-

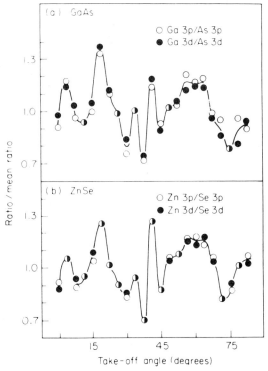

Figure 3.35   XPD patterns for (a) (100) GaAs wafer and (b) ZnSe layer grown on an identical GaAs substrate by hydrogen transport. The axis of rotation lay parallel to <110> and the take-off angle is given with respect to the surface normal. (After Evans and Scott[56])

level intensity as a function of take-off angle) to show that ZnSe grown onto (100) GaAs was expitaxial. The identical form of the plots immediately reveals identical crystal symmetry of substrate and overlayer. Evans, Adams and Thomas[57] have also carried out structural analyses of complex layer–silicate minerals using this technique.

3.4.5 Time-dependent spectra

Although XPS is generally regarded as a 'non-destructive' technique (particularly cf. AES and SIMS) this cannot be taken for granted. A fairly obvious problem concerns the exposure of materials to the high or ultra-high vacuum of the spectrometer. A recent systematic study[58] of several materials containing water of crystallization and some hydroxides has shown that their behaviour under vacuum can be predicted from the relationship between water vapour pressure and temperature (the Clausius–Clapeyron equation). Water loss is a particularly serious problem for intended studies of biological systems.

Less easy to predict is damage from the incident X-rays. In several cases the time dependence of XP spectra has been interpreted structurally and kinetically (e.g. reduction of platinum (IV) compounds, decomposition of $NaClO_4$).[59] Some radiation effects are essentially instantaneous at room temperature and therefore give rise to confusion. Perhaps the first example of this to be studied was the surface isomerization of the compound $(Ph_3P)_2Pt(C_2Cl_4)$ to $(Ph_3P)_2PtCl(C_2Cl_3)$.[60] Cooling the sample slows down these changes and often allows them to be followed *in situ*. X-radiation-induced changes are probably much more common than is often realized. A long literature argument about the appearance of a high BE (oxidized) S $2p$ peak in spectra of biologically important copper–sulphur proteins seems recently to have been resolved by the theory[61] that a radiation-induced redox reaction at surface copper–sulphur sites involving adsorbed water takes place, producing sulphone groups. Such examples indicate that radiation effects in XPS should be taken much more seriously than they have been hitherto.

References

1. See, for example, R. M. Eisberg, *Fundamentals of Modern Physics*, p. 436, John Wiley and Sons, New York (1963).
2. K. D. Sevier, *Low Energy Electron Spectrometry,* p. 55, John Wiley and Sons, New York (1972).
3. M. F. Chung and L. H. Jenkins, *Surface Science,* **21**, 253 (1970).
4. W. A. Coghlan and R. E. Clausing, *Atomic Data*, **5**, 317 (1973).
5. K. S. Kim, S. W. Gaarenstroom and N. Winograd, *Chem. Phys. Letters*, **41**, 503 (1976).
6. F. P. Larkins, *Atomic Data and Nuclear Tables*, **20**, 311 (1977).

7. L. E. Davies, N. C. MacDonald, P. W. Palmberg, G. E. Riach and R. E. Weber, *Handbook of Auger Electron Spectroscopy*, Physical Electronics Division, Perkin–Elmer Corporation, USA (1978). See also G. E. McGuire, *Auger Electron Spectroscopy Reference Manual*, Plenum, New York (1979).
8. E. Kny, *J. Vac. Sci. Tech.*, **17**, 658 (1980); J. Kleefeld and L. L. Levenson, *Thin Solid Films*, **64**, 389 (1979); M. A. Smith and L. L. Levenson, *Phys. Rev.*, **B16**, 1365 (1977).
9. M. Salmeron, A. M. Baro and J. M. Rojo, *Phys. Rev.*, **B13**, 4348 (1976).
10. See, for example, H. H. Madden, *J. Vac. Sci. Tech.* **18**, 677 (1981).
11. P. Weightman, *Rep. Prog. Phys.*, **45**, 753 (1982).
12. H. Aksela and S. Aksela, *J. Phys. B: Atom Mol. Phys.*, **7**, 1262 (1974).
13. R. H. Ritchie, *Phys. Rev.*, **106**, 874 (1957).
14. E. A. Stern and R. A. Ferrel, *Phys. Rev.*, **120**, 130 (1960).
15. G. Dufour, J. M. Mariot, P. E. Nilsson-Jatko and R. C. Karnatak, *Physica Scripta*, **13**, 370 (1976).
16. D. Massignon, F. Pellerin, J. M. Fontaine, C. le Gressus and T. Ichinokawa, *J. Appl. Phys.*, **51**, 808 (1980).
17. H. E. Bishop and J. C. Rivière (to be published).
18. J. H. Scofield, *J. Electron Spectrosc.*, **8**, 129 (1976).
19. A. F. Orchard, in *Handbook of X-ray and Ultraviolet Photoelectron Spectroscopy* (Ed. D. Briggs), p. 49, Heyden and Son, London (1977).
20. A. Barrie and N. E. Christensen, *Phys. Rev.*, **B14**, 2442 (1976).
21. H. Hillig, B. Cleff and W. Mehlhorn, *Z. Phys.*, **268**, 235 (1974).
22. S. Aksela and H. Aksela, *Phys. Lett.*, **A48**, 19 (1974).
23. C. D. Wagner, in *Handbook of X-ray and Ultraviolet Photoelectron Spectroscopy* (Ed. D. Briggs), p. 256, Heyden and Son, London (1977).
24. C. D. Wagner and J. A. Taylor, *J. Electron Spectrosc.*, **20**, 83 (1980).
25. K. Siegbahn *et al.*, *ESCA Applied to Free Molecules*, North Holland, Amsterdam (1969).
26. D. N. Hendrickson, J. M. Hollander and W. L. Jolly, *Inorg. Chem.*, **8**, 2642 (1969).
27. C. D. Wagner *et al.*, in *Handbook of X-ray Photoelectron Spectroscopy* (Ed. G. E. Muilenburg), Perkin–Elmer Corporation, USA (1979).
28. G. K. Wertheim, *J. Franklin Institute*, **298**, 289 (1974).
29. G. K. Wertheim, M. Campara, M. R. Shanks, P. Zuonsteg and E. Banks, *Phys. Rev. Lett.*, **34**, 738 (1975).
30. C. D. Wagner and P. Biloen, *Surf. Sci.*, **35**, 82 (1973).
31. C. D. Wagner, in *Handbook of X-ray and Ultraviolet Photoelectron Spectroscopy* (Ed. D. Briggs), p. 262, Heyden and Son, London (1977).
32. C. D. Wagner, *Farad. Discuss. Chem. Soc.*, **60**, 291 (1975).
33. C. D. Wagner, L. H. Gale and R. H. Raymond, *Anal. Chem.*, **51**, 466 (1979).
34. R. H. West and J. E. Castle, *Surf. Interface Anal.*, **4**, 68 (1982).
35. C. D. Wagner, in *Handbook of X-ray and Ultraviolet Photoelectron Spectroscopy* (Ed. D. Briggs), p. 266, Heyden and Son, London (1977).
36. C. D. Wagner, D. A. Zatko and R. H. Raymond, *Anal. Chem.*, **52**, 1445 (1980).
37. T. A. Carlson, *Photoelectron and Auger Spectroscopy*, Plenum, New York (1976).
38. B. E. Koel and J. M. White, *J. Electron Spectrosc.*, **22**, 237 (1981).
39. S. P. Kowalczyk, L. Ley, R. A. Pollak, M. R. McFeely and D. A. Shirley, *Phys. Rev.*, **B7**, 4009 (1973).
40. C. S. Fadley, in *Electron Spectroscopy: Theory, Techniques and Applications* (Eds C. R. Brundle and A. D. Baker), Vol. 2, Academic Press, London (1978).
41. D. T. Clark and D. B. Adams, *Chem. Phys. Lett.*, **10**, 121 (1971).

42. D. Briggs and V. A. Gibson, *Chem. Phys. Lett.*, **25**, 493 (1974).
43. G. C. Allen and P. Tucker, *Inorg. Chim. Acta*, **16**, 41 (1976).
44. D. T. Clark and A. Dilks, *J. Polym. Sci., Polym. Chem. Ed.*, **15**, 15 (1977).
45. D. C. Frost, C. A. McDowell and A. Ishitani, *Mol. Phys.*, **24**, 861 (1972).
46. P. Burroughs, A. Hamnett, A. F. Orchard and G. Thornton, *J. Chem. Soc. Dalton Trans.*, **1976**, 1686.
47. L. J. Matienzo, W. E. Swartz and S. O. Grim, *Inorg. Nucl. Chem. Lett.*, **8**, 1085 (1972).
48. D. C. Frost, C. A. McDowell and I. S. Woolsey, *Mol. Phys.*, **27**, 1473 (1974).
49. A. Barrie, P. Swift and D. Briggs, Preprints, 27th Pittsburgh Conference on Analytical Chemistry and Applied Spectroscopy (Cleveland, USA, 1976).
50. J. J. Pireaux, R. Caudano and J. Verbist, *J. Electron Spectrosc.*, **5**, 267 (1974).
51. See, for example, K. Siegbahn, *J. Electron Spectrosc.*, **5**, 3 (1974).
52. A. Dilks, in *Electron Spectroscopy: Theory, Techniques and Applications* (Eds C. R. Brundle and A. D. Baker), Vol 4, Academic Press, London (1981).
53. C. S. Fadley, *Prog. Solid State Chem.*, **11**, 265 (1976).
54. M. Mehta and C. S. Fadley, *Phys. Lett.*, **A55**, 59 (1975).
55. D. Briggs, R. A. Marbrow and R. M. Lambert, *Solid State Comm.*, **26**, 1 (1978).
56. S. Evans and M. D. Scott, *Surf. Interface Anal.*, **3**, 269 (1981).
57. S. Evans, J. M. Adams and J. M. Thomas, *Phil. Trans. Roy. Soc. Lond.*, **292**, 563 (1979).
58. K. Hirokawa and Y. Danzaki, *Surf. Interface Anal.*, **4** (1982).
59. R. G. Copperthwaite, *Surf. Interface Anal.*, **2**, 17 (1980).
60. D. T. Clark and D. Briggs, *Nature Phys. Sci.*, **237**, 15 (1972).
61. M. Thompson, R. B. Lennox and D. J. Zemon, *Anal. Chem.*, **51**, 2260 (1979).

Practical Surface Analysis
by Auger and X-ray Photoelectron Spectroscopy
Edited by D. Briggs and M. P. Seah
© 1983, John Wiley & Sons, Ltd

Chapter 4

Depth Profiling

S. Hofmann

*Max-Planck-Institut für Metallforschung, Institut für Werkstoffwissenschaften,
Seestrasse 92, D 7000 Stuttgart 1, FRG*

4.1 Introduction

In-depth distribution analysis of chemical composition is a special case of micro local analysis, where the third dimension perpendicular to the surface of a sample is of primary interest. In principle, this task requires the compositional analysis of thin sections (in the ultimate dimension of a monoatomic layer) defined on a depth scale. It can be obtained either by (a) non-destructive or (b) destructive techniques.[1-7]

4.1.1 Non-destructive techniques

Non-destructive techniques are based on an analytical signal parameter (e.g. intensity, energy) which has a well-defined dependence on its depth of origin. For example, the energy loss of high-energy charged particles is proportional to their path length in a given material. This property is extensively used in RBS (Rutherford back-scattering of primary ions, typically He^+, 100 keV–5 MeV, 0.1–30 μm depth).[8,9] Here the atomic number dependent elastic scattering cross-section contains the analytical information whereas the energy loss yields the depth information. In NRA (nuclear reaction analysis), the resonance of nuclear reaction cross-sections with energy of the impinging particles is used for profiling: by increasing the primary beam energy the reaction products originate from progressively deeper regions in the sample. These techniques are excellent with respect to quantitative results of both concentration and depth scale, with sensitivities more or less comparable to electron spectroscopy. Because of low scattering cross-sections for light elements, RBS is only appropriate for elements of $Z > 10$ and is of limited use for light elements in a heavy element matrix. NRA is defined to the detection of low Z elements ($Z < 20$) due to the presently used accelerator energies ($\leq$5 MeV). The depth resolution obtainable with RBS and NRA, mainly due to range straggling and useful energy resolution, is limited to between 5 and 200 nm depending on the material and total layer thickness.[8,9] For non-destructive techniques using X-ray analysis such as IIX (ion induced X-ray) analysis or EPMA (electron probe microanalysis), the depth resolution is even worse (typically fractions of 1 μm), which prevents their use in thin film depth profiling.

In electron spectroscopy, the intensity of a specific signal is dependent on the energy and the depth of origin of the Auger electrons or photo-electrons. The inelastic mean free path is the decisive parameter which determines the escape depth. For a given electron energy, a variation of the emission angle allows the variation of the effective escape depth between its full value perpendicular to the surface and almost zero.[4,10,11] This technique, which is limited to 50 nm total depth, is discussed in Section 4.5.1 in more detail.

4.1.2 Destructive techniques

Destructive techniques of depth profiling comprise the more classical methods of mechanical sectioning (tapered sectioning with recent improvements like ball cratering in conjunction with high lateral resolution techniques) and chemical sectioning with subsequent chemical analysis of the removed material. Apart from problems with specific materials (e.g. ceramics) depending on mechanical and chemical properties, these techniques suffer from the lack of control of surface reactions between subsequent steps, giving severe limitations in depth resolution, although recent progress in electrochemical polishing[12] has shown that depth resolution can be achieved in the 10 nm region.

A universally applicable 'sectioning' method is surface erosion by ion sputtering. The case of *in situ* combination with any surface analysis method has enhanced its now widespread application for depth profiling. A number of review articles have been published on this subject.[1-7]

Sputtering is a destructive method: the sample is bombarded with ions accelerated in an ion gun to an energy above 100 eV (typically 0.5–5 keV). A small fraction of the energy is transferred to surface atoms and causes them to leave the sample; they are sputtered away. Thus, the sample is successively decomposed in an abraded part which can be analysed, e.g. by SIMS, and in the residual surface which is analysed by AES and/or XPS.

The principle advantages of depth profiling by AES and XPS combined with ion sputtering are:

(a) The information depth is of the order of 1 nm.
(b) The analysis is independent of the sputtering yield.
(c) The influence of the matrix on the elemental detection sensitivity is small.

(d) The analysed area is small compared to the sputtered area, thus minimizing crater edge effects.

Point (d) is generally not fulfilled in XPS, as discussed later in Section 4.4.2.

4.2 Practice of Depth Profiling with AES and XPS

Since both AES and XPS analyse the residual surface left after a certain sputtering time, the basic implications with respect to depth profiling apply for both techniques. There are, however, a number of practical differences, e.g. detection speed, background and spatial resolution, which generally are more advantageous in Auger profiling. Therefore, we refer primarily to this technique if not otherwise stated. Any commercial equipment for AES/XPS comprises an electron energy analyser, an electron and/or X-ray source, an ion gun and the sample stage all mounted in a stainless steel chamber in which an ultra-high vacuum ($\leq 10^{-8}$ Pa) can be maintained.

4.2.1 Vacuum requirements

Considering the mean free path of electrons in gases, a vacuum better than 10^{-2} Pa would be sufficient for proper operation of the electron analysis. However, contamination of the surface during analysis has to be avoided.[13-15] Assuming a sticking coefficient of one, an adsorption monolayer is built up at a partial pressure of 10^{-4} Pa ($\approx 10^{-6}$ torr) in one second. To avoid an increase of 1 per cent of a monolayer for a typical measurement time (AES) of 100 seconds, the partial pressure of reactive gases such as CO, H_2O, C_xH_y should be below 10^{-8} Pa (10^{-10} torr), i.e. UHV. Inert gases (e.g. argon) which do not chemisorb can be tolerated up to a pressure several orders of magnitude higher.

4.2.2 Ion gun

The most frequently used ion guns in Auger spectrometers are simple electrostatic devices where the inert gas ions (Ar) are generated by collisional excitation with electrons of typically 100 eV energy from a hot filament. The positive ions are accelerated to between 0.5 and 5 keV and focused on the sample, creating a sputtering spot diameter of about 1–5 mm diameter. To achieve an ion current density of the order of 100 $\mu A/cm^2$, the pressure in the ion formation section should be about 5×10^{-3} Pa (5×10^{-5} torr). The general operation in ion pumped systems is to back-fill the whole chamber with argon with the ion pumps off. The sputtering gas purity required should meet the partial pressure ratio $10^{-8}/5 \times 10^{-3} = 2 \times 10^{-6}$. To avoid further contamination, a titanium sublimation pump with a cryopanel usually serves as an efficient scavenger for the reactive gases. Compared to such 'static' systems, by using the more modern type of ion gun operating in a 'dynamic' mode with constant throughput of the sputtering gas (pumped away by the main chamber pump) the overall contamination is reduced. A further improvement is differentially pumping the ion gun by a getter pump or an added turbomolecular or diffusion pump. The decreased pressure in the main chamber ($<10^{-2} \times$ ion gun pressure) which is due to the pressure stage action of the small opening (≈ 1 mm) at the end of the ion gun offers an additional advantage: the sputter damage of the electron gun cathode filament caused by accelerated positive ions is reduced and/or the X-ray source is prevented from arcing which is frequently encountered at a total pressure $>10^{-4}$ Pa.

Simple ion guns using a static spot must be aligned mechanically so that the analyser focus (together with the electron beam impinging on the sample) is close to the centre of the generally Gaussian ion current beam intensity distribution to avoid severe degradation of depth resolution.[14] The lack of uniformity of the ion beam intensity over millimetre dimensions precludes their application in XPS profiling.

For precise depth profiling, an ion gun with x/y beam deflection capability should be used. This enables an exact matching of analysed and sputtered areas to be made (e.g. by optimizing for the secondary electron emission monitored by a CMA). Rastering of the well-focused ion beam over a larger area (up to 10×10 mm) greatly improves the uniformity of the ion beam intensity,[15] leading to a flat bottom of the sputtering crater which is necessary for optimum depth resolution. Furthermore, at constant beam current the raster area is inversely proportional to the total primary ion density so that the sputtering rate is easily controlled. A rastered ion beam is a prerequisite for XPS profiling where the analysed area is some square millimetres.

To prevent shadowing effects in working with rough surfaces the angles between the ion beam , the electron beam and the electron take-off should be as small as possible. In particular, the ion beam should be directed close to the normal of the sample surface.[16] To avoid cone formation, sputtering with two guns at different angles is desirable.[17]

4.2.3 Analysing mode

Auger or photo-electron analysis can be done either discontinuously after subsequent sputtering or by continuous sputtering and simultaneous electron spectroscopy. In Auger profiling the latter technique is most frequently used together with multiplexing for different elemental peaks and a print-out of the respective Auger peak-to-peak heights (APPHS). In XPS, the strong increase in secondary electron emission during sputtering is unfavourable for simultaneous profiling. Furthermore, XPS peak area detection generally needs longer measurement times compared to AES, thus limiting the achievable

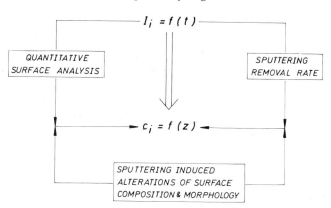

Figure 4.1 Principles of sputtering profile evaluation: conversion of a measured sputtering profile $I = f(t)$ to a true concentration profile $c = f(z)$[3]

depth resolution Δz_a for a given sputtering rate $\dot{z}$. If Δt_m is the time necessary for the measurement of a peak with a desired signal-to-noise ratio, $\Delta z_a = \dot{z} \Delta t_m$. Contrary to SIMS, where $\dot{z}$ and Δt_m are inversely proportional,[18] these quantities are independent in profiling by electron spectroscopy. Therefore Δz_a can be made negligibly small compared to the many other factors limiting depth resolution by reducing the primary ion current density or by discontinuous sputtering. In general, this procedure has to be used in XPS profiling because of the higher Δt_m. In AES profiling with a CMA, continuous sputter profiling with sputtering rates up to 5 nm/min is frequently applied. An additional advantage of a high sputtering rate is a reduced surface contamination (caused by adsorption from the residual gas atmosphere) which is due to increased sputter desorption. (cf. Section 4.4.2.1)

4.3   Quantification of Sputtering Profiles

The data primarily obtained during depth profiling experiment consist of signal intensities of the detected elements, I (e.g. Auger peak-to-peak heights), as a function of sputtering time, t, i.e. the 'measured sputtering profile', $I = f(t)$. The principal task is to obtain the original distribution of concentration c with depth z, $c = f(z)$, by an appropriate conversion of the measured data.[3-7]

The main steps of such an evaluation are visualized in Figure 4.1. First the sputtering time scale must be calibrated in terms of the main eroded depth, $z = f(t)$, and the intensity of the Auger signal must be calibrated in terms of local elemental concentration, $c = f(I)$. Thus a 'real' concentration profile is established which would be identical to the original profile if sputtering proceeded homogeneously in an ideal 'layer-by-layer' microsectioning. However, profile distortions mainly due to sputtering-induced topographical and compositional changes of the instantaneous sample surface must be taken into account in a second step in order to reveal the true original profile, $c = f(z)$. The precision by which the shape of an evaluated profile resembles that of the original profile can be described by the concept of depth resolution,[1-7] which is the key parameter in profiling. A knowledge of the depth resolution or, more precisely, the depth resolution function is the prerequisite for a deconvolution of the measured concentration profiles.

The steps in profile evaluation are outlined in the following sections.

4.3.1 Calibration of the depth scale ($z = f(t)$)

The velocity of surface erosion is described by an instanteneous sputtering rate $\dot{z} = \mathrm{d}z/\mathrm{d}t$ which determines the mean eroded depth z as a function of sputtering time t according to

$$z(t) = \int_0^t \dot{z}\, \mathrm{d}t \qquad (4.1)$$

The sputtering rate $\dot{z}$ (in units of metres per second) is determined by

$$\dot{z} = \frac{M}{\rho N_A e} S j_p \qquad (4.2)$$

where M = atomic mass number
ρ = density (kg/m^3)
N_A = 6.02×10^{26} (Avogadro number)
e = 1.6×10^{-19} A s (electron charge)
S = sputtering yield (atom/ion)
j_p = primary ion current density (A/m^2)

For a constant sputtering rate, equation (4.1) shows a direct proportionality between sputtered depth and time:

$$z = \dot{z}t \qquad (4.3)$$

In this case the z axis is linear with the sputtering time and (besides the zero point) only one point is necessary to determine the calibration factor $\dot{z}$. With equation (4.2), $\dot{z}$ can be calculated taking the sputtering rate S from the literature data[19-21] and measuring j_p with a Faraday cup. Figure 4.2 shows the orders of magnitude which may be achieved. However, the sputtering rate S depends on a variety of parameters including energy, mass and angle of the incident ions[19-21] and surface composition,[13] Therefore only a rough estimate can be expected from such a calculation.

A better method to obtain the sputtering rate $\dot{z}$ is to measure (for given ion gun operation parameters) the time required to sputter through a layer of

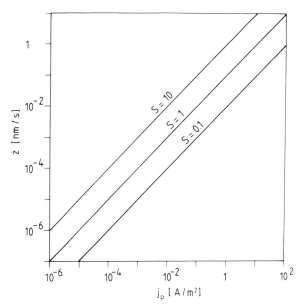

Figure 4.2 Sputtering rate $\dot{z}$ as a function of the primary ion density j_p for a mean $M/\rho = 10^{-2}$ m^3/mol and three different sputtering yields S according to equation (4.2)

known thickness, e.g. a metallic evaporation layer or an oxide layer. Anodized tantalum foils[22,23] have proved very useful for this purpose, since the thickness is easily controlled by the formation voltage.[24] In addition, due to the sharp oxide–metal interface they provide a quick test method for the instrumental depth resolution (see Section 4.4). Thus $\dot{z}$ is determined with the knowledge of the Ta_2O_5 sputtering rate[23] as compared to the material under study.

Another possibility is to measure the crater depth z_0 after sputtering by conventional interferometry or stylus methods.[25,26]

In general, however, $\dot{z}$ varies with composition because according to equation (4.2) M, ρ and S are a function of composition, and the relation between sputtering time and depth becomes non-linear.[27,28] Even for a constant composition, the sputtering rate may change within the first few layers due to the gradual build-up of structural and compositional changes until a saturation value is obtained after the sputter depth is of the order of the mean projected range of the primary ions.[29]

For these reasons it is obvious that a precise determination of the true depth scale requires a measurement of the instantaneous sputtering rate during profiling. This can be done by *in situ* measurement of the residual thin film thickness, e.g. by direct mass measurement with a quartz microbalance,[30] by X-ray emission analysis[29] or by *in situ* laser optical interferometry,[31,32] as developed recently by J. Kempf. Unfortunately, the latter methods are not available in standard equipment.

A first-order approximation of the correction for composition-dependent non-linearity in the time–depth relation has been demonstrated in the sputtering of Ni/Cr evaporation multilayers,[3,28] assuming a change of the sputtering rate proportional to composition. For a binary system A/B, the total instantaneous sputtering rate $\dot{z}$ is given by the mole fractions X_A, X_B and the sputtering rates of the pure components, $\dot{z}_A$, $\dot{z}_B$:

$$\dot{z} = X_A \dot{z}_A + X_B \dot{z}_B \qquad (4.4)$$

According to equations (4.1) and (4.4):

$$z_1(t_1) = \int_0^{t_1} [X_A(t)\dot{z}_A + X_B(t)\dot{z}_B] \, dt \qquad (4.5)$$

In this approximation, X_A and X_B may be taken from the normalized Auger signals I_A/I_A^0, I_B/I_B^0.

Limiting conditions and—in special cases—an estimation of the sputtering rate can be derived from the consideration of the electron spectroscopic signal quantification, as shown in the next section.

4.3.2   Calibration of the concentration scale ($c = f(I)$)

Quantitative AES and XPS are discussed in Chapter 5. With respect to depth profiling the important feature is the relation between intensity I_i and concentration c_i of an element i and can be described by an explicit function of the electron escape depth and—in the case of AES—an electron back-scattering factor:[33–38]

$$I_i = \frac{I_i^0}{\lambda_i} \int_0^\infty r_i(z)c_i(z)\exp\left(\frac{-z}{\lambda_i}\right) dz \qquad (4.6)$$

where I_i^0 is the intensity for an elemental bulk standard, λ_i is the 'effective' escape depth of the Auger electrons or photo-electrons perpendicular to the surface, $c_i(z)$ is the local concentration (in mole fractions) at depth z and $r_i \geq 1$ is the back-scattering (cf. Chapter 5 of this book). The effective escape depth λ_i is determined by the inelastic mean free path[35] of the electrons for a given energy and material, λ_i^0, and by the angle of emission φ of the detected electrons with respect to the normal to the sample surface, according to:

$$\lambda_i = \lambda_i^0 \cos \varphi \qquad (4.7)$$

Values for λ_i^0 are typically between 0.4 and 4 nm.[35] The angle φ is clearly defined for hemispherical analysers. For a cylindrical mirror analyser (CMA), a mean angle of emission, φ, can be defined.[33] Its dependence on the angle α

between the normal to the sample surface and the CMA axis is shown in Figure 4.3. For $\alpha > 47.7°$, part of the CMA acceptance cone is shadowed by the sample which gives a deviation from the cosine dependence, as shown in Figure 4.3.

The back-scattering factor is difficult to correct in the presence of concentration gradients. An approximation for binary systems has been given by Morabito and others.[6,37] In principle, the depth distribution of the excitation density must be known which is dependent on the concentration distribution. For a binary system, $r_{M,A} = (1 - r_B/r_A)c_A + r_B/r_A$. It has been shown[4,39] that for the intensity profile at a binary interface A/B a certain characteristic length for back-scattering (l_b) leads to an amplification factor for the A intensity given by $[1 + r_B/r_A - 1) \exp - (z_0 - z)/l_b]$ (where z_0 is the position of the interface), which is in accordance with experimental results.[4] In general it is easier to apply the back-scattering correction after the escape depth correction. With the exception of extreme cases (sharp interfaces of elements with large difference in atomic number) the electron back-scattering in AES is

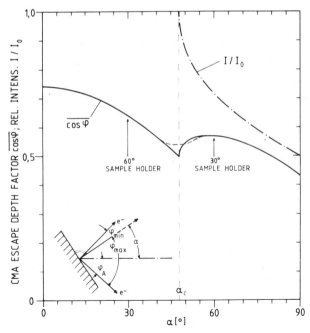

Figure 4.3 Mean effective escape depth factor $(\overline{\cos \varphi})$ as a function of the tilt angle α between the CMA axis and the sample normal. For $\alpha > \alpha_c = 47.7°$, the transmitted intensity I/I_0 decreases due to shadowing. The dashed line accounts for an angular acceptance $\Delta\varphi = 12°$

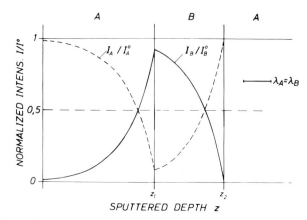

Figure 4.4   Shape of the measured profile of a thin sandwich layer A/B/A due to the escape depth influence after equations (4.8a,b), assuming ideal microsectioning[3,7]

only a second-order effect compared to the escape depth influence which is discussed in the following.

Neglecting r_i, equation (4.6) is valid for both AES and XPS. Depth profiling means a shift of the lower integration boundary from zero to an instantaneous sputter depth. Thus, for a given profile $c(z)$, integration of equation (4.6) as a function of the parameter z gives $I(z)$ for the case of ideal, continuous sputter erosion.[3,7,39] The expected intensity profile for the typical case of a thin sandwich layer of an element B in a matrix A is shown in Figure 4.4. Integration of equation (4.6) for this case gives

$$\left(\frac{I_B}{I_B^0}\right)_z = c_B\left[\exp\left(-\frac{z_1 - z}{\lambda_{B,A}}\right) - \exp\left(-\frac{z_2 - z}{\lambda_{B,B}}\right)\right]   \qquad (4.8)$$

for $z_1 - z \geq 0$ and $z_2 - z \geq 0$. Note that $\lambda_{B,A}$ is the electron escape depth of element B in A and $\lambda_{B,B}$ that of B in B. For $\lambda_{BA} = \lambda_{BB} = \lambda$ the intensity–depth relations are, for $z \leq z_1$:

$$\left(\frac{I_B}{I_B^0}\right)_z = c_B\left[1 - \exp\left(-\frac{z_1 - z_2}{\lambda}\right)\right]\exp\left(-\frac{z_1 - z}{\lambda}\right)   \qquad (4.8a)$$

and for $z_1 < z < z_2$

$$\left(\frac{I_B}{I_B^0}\right)_z = c_B\left[1 - \exp\left(-\frac{z_2 - z}{\lambda}\right)\right]   \qquad (4.8b)$$

Since any concentration profile can be approximated by a successive number of thin layers of thickness $z_2 - z_1$, equation (4.8) provides a method to receive the intensity at any depth z as a summation of these layers contributions.[7,39-42] In practice, it is sufficient to replace the upper integration limit (∞) by $z_u = z + 5\lambda$ (equation 4.6), because the maximum contribution (≤ 0.7 per cent) approaches the detection limit of AES or XPS.

A general solution of equation (4.6) for an arbitrary profile can be derived from equation (4.8) and is given by[3,7,43]

$$c_i(z) = \left(\frac{I_i}{I_i^0}\right)_z - \frac{d(I_i/I_i^0)}{dz}\lambda_i \qquad (4.9)$$

Using equation (4.9), any measured sputtering profile can be corrected for the escape depth influence.[29,39,44,45] As seen from Figure 4.4, a shift of the real profile of 0.7λ to lower depth with respect to the measured profile is obtained, which is broadened by $\Delta z_\lambda \approx 1.6\lambda$.[5,7,39] It is interesting to note that a maximum obtainable gradient follows from equation (4.9):[4,7]

$$\frac{dI_i}{dz} \leq \frac{I_i}{\lambda_i} \qquad (4.10)$$

This means a lower limit for the time–depth calibration factor $\dot{z}$, because $dz = \dot{z}\,dt$ gives

$$\dot{z} \geq \frac{\lambda_i}{I_i}\left(\frac{dI_i}{dt}\right) \qquad (4.10a)$$

Frequently, calibration of the intensities in a depth profile is made using relative sensitivity factors[34,46] and taking the concentration to be proportional to the instantaneous intensity—that means neglecting the λ effect. In this case, the relative error $(\Delta c_i/c_i)_\lambda$ depends on the slope of the measured profile and is of the order of the relative change of the intensity within a depth λ:

$$\left(\frac{\Delta c_i}{c_i}\right)_\lambda \approx \frac{I_i(z) - I_i(z + \lambda)}{I_i(z)} \qquad (4.11)$$

Applying equation (4.9) to the measured sputtering profile, a knowledge of the effective electron escape depth (and the back-scattering factor in AES) of the normalized intensities and of the instantaneous sputtering rate allows a quantitative evaluation of the measured depth profile (cf. Figure 4.1). For the special case of the detection of two peaks of the same element at different kinetic energies and therefore different escape depths λ_1, λ_2 (e.g. 60 and 920 eV copper AES peaks), equation (4.9) offers a possibility to derive the sputtering rate in terms of the two escape depths[47] if a measurable intensity slope with sputtering time exists. Since the right-hand side of equation (4.9) is the same for $I_1(\lambda_1)$ and $I_2(\lambda_2)$, it follows with $dz = \dot{z}\,dt$ that

$$\dot{z} = \frac{\lambda_1\,d(I_1/I_1^0)/dt - \lambda_2\,d(I_2/I_2^0)/dt}{(I_1/I_1^0) - (I_2/I_2^0)} \qquad (4.12)$$

Although the determination of $\dot{z}$ by equation (4.12) is of limited accuracy, it allows a quick test of the order of magnitude of the sputtering rate.[47]

At this stage we have not yet considered the sputtering-induced distortions of the original concentration profile which must be taken into account for its evaluation.

4.4 Sputtering Profile Compared to the Original Concentration Profile: The Concept of Depth Resolution

4.4.1 Depth resolution

4.4.1.1 Definition of depth resolution

The influence of the mean escape depth of the Auger electrons or photo-electrons on the measured profile results in a certain broadening with respect to the true, original profile as discussed in Section 4.3.2. So far, we still have considered sputtering to proceed as an ideal microsectioning, continuous with eroded depth. Experimentally this cannot be realized because ion bombard-ment inevitably induces changes in the topography and in the composition of the surface region of a sample. These sputtering-induced effects are recog-nized in the measured profile as an additional apparent broadening which can be mathematically described by a resolution function $g(z,z')$,[48,49] as visualized in Figure 4.5. Any infinitesimal thin segment of the true concentration–depth distribution $c(z')$ is apparently broadened and the sum of all these contribu-

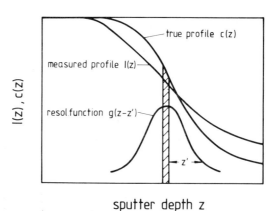

Figure 4.5   Visualization of the convolution process during sputter profiling: each thin layer of the true profile $c(z)$ is broadened into $g(z - z')$. The measured profile $I(z)$ is obtained by summation of the contributions from all such layers. (After Ho and Lewis[48])

tions at a sputtered depth z gives the normalized signal intensity at that depth when the resolution function is normalized:

$$\int_{-\infty}^{+\infty} g(z - z')\, dz' = 1 \qquad (4.13)$$

The normalized intensity $I(z)$ can be expressed as a convolution integral:[3,48,49]

$$I_i(z) = \int_{-\infty}^{+\infty} c_i(z')g(z - z')\, dz' \qquad (4.14)$$

If the resolution function $g(z - z')$ is known, the integral equation (4.14) can be deconvoluted and the true profile $c_i(z')$ is then determined. The procedures necessary for this step are outlined in Section 4.6.

It should be noted, that equation (4.14) also gives the convolution integral for quantitative AES/XPS, if $g(z - z') = \exp[-(z - z')/\lambda_i]$ (equations 4.5 and 4.6). In this case the solution is equation (4.9) and the shape of the resolution function for thin sections of a sample is similar to the curve depicted in Figure 4.4.

In many experiments of sputter profiling through sharp interfaces it has been observed that the measured profile can be approximated by an error function.[1-7,26,39,48-52] This gives a Gaussian function for $g(z - z')$, which is defined by a single parameter, the standard deviation σ. The most common definition of the depth resolution is $\Delta z = 2\sigma$, which corresponds to the difference of the depth coordinate z between 84 and 16 per cent of the intensity change at an interface.

The definition of depth resolution is shown in Figure 4.6(a), together with

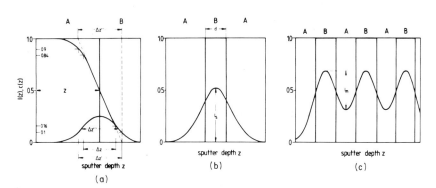

Figure 4.6   Determination of the depth resolution Δz for The Gaussian resolution function approximation: (a) step function true profile, $\Delta z = 2\sigma$ (0.84–0.16), $\Delta z'$ (0.9–0.1) = 2.564σ, $\Delta z''$ (inverse maximum slope) = 2.507σ. $\Delta z'''$ (FWHM of resolution function) = 2.355σ;[5,53] (b) single sandwich layer A/B/A where Δz is determined by I_s/I_0 according to equation (4.15a);[3,51] (c) multilayer sandwich structure A/B/A/B where Δz is determined by I_m/I_0 according to equation (4.15b)[3]

other definitions sometimes used. The relative depth resolution is defined by $\Delta z/z$ where z is the depth coordinate at the interface. For constant sputtering rate $\dot{z}$, the relative depth resolution corresponds to the respective value in sputtering time t coordinates, $\Delta z/z = \Delta t/t$.

The depth resolution can either be experimentally determined or estimated by calculations based on appropriate models.[3]

4.4.1.2 Measurement of depth resolution

The resolution function is directly obtained by sputtering through a very thin sandwich layer as visualized in Figure 4.5. A more frequently applied method is the sputtering through a rectangular profile and take the derivative.[3,48,49,50] In the Gaussian approximation a quantitative measure of the depth resolution Δz can be directly taken from the sputtering profile, as indicated in Figure 4.6(a).

For the case of a thin sandwich layer of thickness d, it has been shown[51] that the decrease of the maximum intensity I_0 (for ideal profiling) to a value I_s (Figure 4.6b) determines the depth resolution:

$$\frac{I_s}{I_0} = \mathrm{erf}\left(\frac{d}{\sqrt{2}\,\Delta z}\right) \tag{4.15a}$$

For multilayer sandwich structures (e.g. A/B/A/B/ ...) with constant single layer thickness d a similar expression can be derived, which relates the amp-

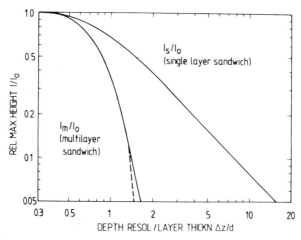

Figure 4.7   Diagram for the determination of Δz of single and multilayer structures of known thickness d shown in Figure 4.6(b) and (c) according to equations (4.15a,b). The approximations for the superposition of the two adjacent layers (——) and of four adjacent layers (– – –) is shown for the case of Figure 4.6(c)

litude of the signal intensity I_m (as defined in Figure 4.6c) for the normalized intensity of a non-degraded layer I_0[3,16,51,53] giving

$$\frac{I_m}{I_0} = \frac{1}{2} \sum_{k=-N}^{N} (-1)^k \left[\text{erf}\left(\frac{(2k-1)\,d}{\sqrt{2}\,\Delta z}\right) - \text{erf}\left(\frac{2k+1}{\sqrt{2}\,\Delta z}\right) \right]$$

$$= -2\,\text{erf}\left(\frac{d}{\sqrt{2}\,\Delta z}\right) + 2\,\text{erf}\left(\frac{3d}{\sqrt{2}\,\Delta z}\right) - 2\,\text{erf}\left(\frac{5d}{\sqrt{2}\,\Delta z}\right) + \cdots \quad (4.15b)$$

For $\Delta z \leq d$, the first two terms give a sufficient approximation. Figure 4.7 shows I_s/I_0 and I_m/I_0 as a function of $\Delta z/d$.[3,53]

A fundamental question concerning the experimental depth resolution is its dependence on the sputtered depth. Multilayer sandwich structures[28,54-58] are particularly suited for this study because depth dependence is obtained in a single experiment applying the approximate Δz evaluation according to equation (4.15b).[3,4,16,28] Results of such an analysis for different systems are shown in Figure 4.8.

4.4.2 Factors limiting depth resolution

Table 4.1 shows a summary of the effects which distort sputtering profiles. They can be divided into instrumental factors, into factors depending on the characteristics of the sample and into effects depending on the fundamental interactions of the ion beam (and/or electron beam in AES) with the sample.[3,14,59] In general we have no choice with regard to the sample, but the beam interactions and, above all, the instrumental conditions can be substantially optimized.

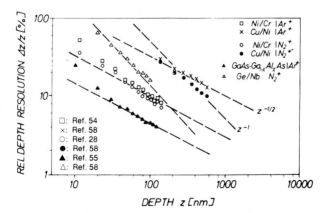

Figure 4.8 Dependence of the relative depth resolution $\Delta z/z$ on depth z for different multilayer sandwich structures obtained by sputtering with Ar^+ and/or N^+ ions and AES analysis

Table 4.1 Survey of the detrimental effects in sputter profiling and their contributions Δz_j to depth resolution

	Origin of profile distortion	Effect on profile and depth resolution	References
Instrumental factors	Adsorption from residual gas atmosphere	See equations (4.16a, b) (C, O)	1, 7, 14, 15, 50, 60
	Re-deposition of sputtered species	'Memory' effect of original surface composition	18
	Crater wall effects		
	Impurities in ion beam	Analysed as sample constituents	15
	Neutrals from ion gun		
	Non-uniform ion beam intensity	Non-uniform erosion: $\Delta z_i \propto z$	4, 5, 9, 62, 63, 108, 109
	Time-dependent ion beam intensity	Profile distortion: $\dot{z} = f(t)$	50
	Information depth	Profile shift: -0.7λ; $\Delta z_\lambda = \text{constant} \simeq 1.6\lambda$	3–5, 7, 29, 39, 44, 45, 51
Sample characteristics	Original surface roughness	$\Delta z_r \propto z^{1/2} \cdots z$	16, 17, 54, 56
	Crystalline structure and defects	$\Delta z_c \propto z$	64, 65, 110
	Alloys, compounds, second phases	→ Preferential sputtering-induced roughness	69, 70
	Insulators	→ Distortions by charging	77, 107, 111
Radiation-induced effects	Primary ion implantation	Initial $\dot{z}$ change, composition dependent $\dot{z}$	28, 29, 59, 112
	Atomic mixing	$\Delta z_k \rightarrow \text{constant} (\simeq R_p)$ for $z \gg R_p$	71, 78–83
	Sputtering-induced roughness	$\Delta z_s \rightarrow \text{constant}, \propto z^{1/2} \cdots z$	5, 50, 72, 75, 110
	Preferential sputtering and decomposition of compounds	$\Delta z_p \rightarrow \text{constant} (z \gg R_p)$	14, 47, 83–91
	Enhanced diffusion and segregation	$\Delta z_T = f(D_b, D_s, \dot{z}, t, \ldots)$	57, 76, 92–94
	Electron-induced desorption	$\Delta z_e = f(\sigma_e, j_p, z, t, \ldots)$	95, 96, 109
	Charging of insulators	Analysis distortion; electromigration	92, 111

4.4.2.1 *Instrumental factors*

An important condition in sputter profiling is to avoid chemical reactions of the sample with the ambience during sputtering.[50] This comprises adsorption, re-sputtering and impurities in the ion beam. With respect to adsorption of reactive gases the sample must be in UHV ($<10^{-9}$ torr or $<10^{-7}$ Pa). In a simultaneous AES sputtering experiment, assuming a sticking probability of one and a sputtering probability equal to that of the matrix, a simple calculation shows that the mean surface coverage of the impurity, θ, is given (for $\theta \ll 1$) by the ratio of adsorption rate and sputter desorption rate:[14,60]

$$\theta \simeq 10^5 \frac{p}{\dot{z}} \quad \text{(Pa s/nm)} \tag{4.16a}$$

where $\dot{z}$ is the sputtering rate in nanometres per second, p is the partial pressure in pascals and θ is the ratio of adsorbed atoms per sample surface atoms. Equation (4.16a) shows that θ will be well below the detection sensitivity (10^{-2}) for a sputtering rate of 1 Å/s if the partial pressure of the reactive gas is below $p < 10^{-8}$ torr (10^{-6} Pa). Mathieu and Landolt have demonstrated[15] that CH_4 partial pressures above 10^{-5} Pa can severely alter AES depth profiles. However, it must be pointed out that for an ion gun operating in a closed chamber back-filled at 5×10^{-5} torr with argon, the contaminant gas is also ionized and implanted in the sample surface. These impurities in the primary ion beam can be avoided by the use of a differential pumped ion gun. Another source of surface contamination is re-deposition of sputtered species (due to re-sputtering at construction parts close to the sample surface). This effect is less important in AES or XPS as compared to SIMS because of the lower detection sensitivity of the former techniques.

It should be mentioned that in the case of subsequent sputtering and surface analysis, as generally used in XPS profiling, the conditions of avoidance of adsoprtion are more stringent because the coverage is then

$$\theta \simeq 10^4 p t_m \quad \text{(Pa s)} \tag{4.16b}$$

where t_m is the measurement time in seconds at which the ion beam is switched off.

A non-uniform ion beam intensity over the analysed area means that the sample is eroded to different depths which contribute to the measured signal. The result is a depth resolution Δz_1 which increases proportionally to the sputtered depth z. For the case of a static ion beam of Gaussian intensity distribution with diameter (FWHM) d_1, Δz_1 depends on the mis-match distance b from the probing electron beam (in AES) of diameter d_e which is approximately given by[4,61,62]

$$\Delta z_1 \approx 4 \frac{d_e}{d_1^2} \left(\frac{d_e}{2} + b \right) z \tag{4.17}$$

A more rigorous treatment can be found elsewhere.[62] Equation (4.17) demonstrates the advantage of a small electron beam diameter in AES depth profiling as used in SAM ($d_e/d_1 < 0.01$) and the much worse situation in XPS, where d_e is equivalent to the diameter of the analysed area (typically of the order of a few millimetres, i.e. $d_e/d_1 \approx 1$). Therefore, rastering of the primary ion beam is mandatory in XPS profiling.[63] It should be noted, however, that neutrals generated before the electro-static raster deflection may still give rise to a small non-uniform ion intensity effect, which can only be excluded by an additional deflection after the rastering plates.

The mean escape depth of the Auger electrons or photo-electrons is an inherent instrumental property of electron spectroscopic analysis and has been discussed in Section 4.3.2.

4.4.2.2   Sample characteristics

Composition and structure of the sample cause certain limitations on the quality of a sputtering profile. Experimentally, it has been shown that the original surface roughness has a substantial influence on depth resolution.[54,56] A mathematical treatment has recently been given by Seah and Lea[16] which predicts an increase of Δz_r proportional to the sputtered depth z and increasing with the deviation from normal incidence of the primary ions. For normal incidence, they predict

$$\Delta z_r = 1.66\alpha_0^2 fz \qquad (4.18)$$

where α_0 is given by the standard deviation of the distribution of the angles between the average surface and that of the differently inclined elements of a rough surface. The factor f is defined by the ion incidence angle dependence of the sputtering yield[20,21] and is between 0.5 and 1.5.[16] A result of Seah and Lea's calculations is shown in Figure 4.9 for $f = 0.65$. The general conclusion is that we may expect optimum profiles only for flat polished samples and normal incidence of the primary ions.

The dependence of the sputtering yield on surface orientation[19,21] is responsible for the development of a sputtered surface with highest steps at the grain boundaries of the poly-crystalline material,[64] and a depth resolution Δz proportional to the sputtered depth is expected. It is generally observed that sputtering with oxygen ions (or flooding the surface with oxygen during argon sputtering) leads to a smoother sputtered surface. This is probably due to the build-up of a quasi-amorphous oxide layer which is sputtered more homogeneously.[65]

Walls and others[66,67] have shown how the geometry of a surface influences the development of a surface topography during sputtering. On the other hand, even in the case of flat, amorphous surfaces, cone formation may develop under ion bombardment due to the angle dependence of the sputtering

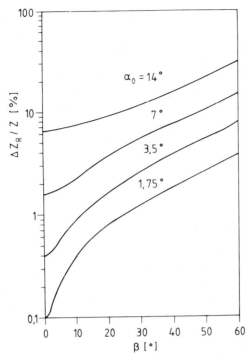

Figure 4.9 The dependence of the relative depth resolution $\Delta z_R/s$ on ion beam incidence angle β for several values of the roughness angle distribution given by α_0 (standard deviation) for $f = 0.65$ (Cu, Pd, Ag, Au). (After Seah and Lea[16])

yield.[21,68] These topographical effects can be diminished by the use of sample rotation or of sputtering with two ion guns at different incidence angles.[17]

Multiphase materials generally lead to increased roughness due to variations of the sputtering yield with local composition. Surface contamination may act in similar way.[69,70]

In view of the statements made above, sputtering of non-crystalline single-phase materials which amorphize during sputtering (e.g. semi-conductors and oxides) should yield much better profiling results than polycrystalline metallic materials, which is generally observed. Similarly, sputtering of the latter with reactive ions often show an improvement in depth resolution (see Figure 4.8).

4.4.2.3 Radiation-induced effects

The phenomena inherent to the sputtering process, as described by the Sigmund theory of sputtering,[71] seem to cause the most fundamental limitations of depth resolution. They comprise changes of the surface microtopography due to the statistical nature of surface atom emission and composition changes due to atomic mixing within the collision cascade. A first attempt to describe the contribution of microtopography development by the sequential layer sputtering model[5,50], could explain the often-found $z^{1/2}$ dependence of Δz.[72] However, experiments with reactive ion sputtering[73] and particularly sputtering of semi-conductors and oxides[74] have shown that a constant depth resolution can be achieved. Recent model calculations predict that a site-dependent sputtering probability,[75] as well as a certain surface mobility[76] within the framework of the sequential layer sputtering model, will lead to a constant Δz at larger depths. Therefore it is concluded that the profile broadening due to sputtering statistics is confined to a few monolayers. The often-found $z^{1/2}$ dependence of depth resolution[4,72] may be explained by a superposition of constant depth resolution parameters and those resulting in a linear depth dependence[14,77] (see Figure 4.11).

A fundamental contribution to the broadening of depth profiles is caused by the build-up of a collisional cascade.[71] Primary recoils are displaced into deeper layers ('knock-on' effect) and secondary recoils give rise to a more random distribution of relocated target atoms ('atomic mixing' or 'cascade mixing').[78-83] Calculations by Littmark and Hofer[80] have shown that a broadening of an interface occurs which is skewed towards the interior of the sample. This was experimentally confirmed by Etzkorn, Littmark and Kirschner.[82] The development of the broadening increases with ion dose until a steady state is achieved.[82] For larger depth, the atomic mixing contribution is constant with z but is approximately proportional to the square root of the primary ion energy[78] and is less for a higher sputtering yield.[83] The influence of atomic mixing on depth resolution is of the order of the projected range of the primary ions. Consequently the disturbing effect of knock-on and atomic mixing can be minimized by using heavy ions (e.g. Xe^+) at low energy (≤ 1 keV).

Due to different atomic sputtering yields, the composition of the surface layer of a multicomponent sample is generally changed during sputtering.[83-86] If the sputtering yield of the components is independent of their bulk concentration c_b, the surface composition c_s is inversely proportional to the respective sputtering yields S:[84]

$$\frac{c_{sA}}{c_{sB}} = \frac{S_B}{S_A}\frac{c_{bA}}{c_{bB}} \tag{4.19}$$

where the subscripts A and B denote the two components. This means that the element with the lower sputtering yield is enriched in the surface. The build-up of the altered layer during sputtering until the steady state, equation (4.19), is reached causes a transient in surface composition which has been derived by Ho *et al.*:[86]

$$c_{sA}(t) = (c_{bA} - c_{sA})\exp\left(-\frac{t}{\tau}\right) + c_{sA} \tag{4.20}$$

where $c_{sA}(t)$ is the instantaneous surface composition of component A be
tween the beginning of sputtering ($t = 0$) and $t \rightarrow \infty$ ($c_{sA}(t) = c_{sA}$ in equation
(4.19)). The parameter τ is the characteristic time to establish the altered
surface layer and is of the order of the projected range of the primary
ions.[47,83,86,87] The transient in surface composition (equation 4.20) superposes
the intensity–sputtering time profile and therefore affects the depth resolution.
More important, however, is the assymmetry of profiles due to the changing
sputtering rate with composition, discussed in Section 4.3.1. The most distor-
tional effect of preferential sputtering is the generation of an apparent profile
when there is no concentration gradient (see Figure 4.12).

In compounds, preferential sputtering of one component is linked with the
change of the chemical state of the sample constituents. Many oxides are
reduced to lower oxidation states.[47,87–89] This can be disclosed by XPS profiling.[14,47] For example, Figure 4.10 shows the influence of Ar^+ bombardment
on the 'chemical' sputtering profile of a Ta_2O_5/Ta layer of 30 nm thickness.
Figure 4.10(a) depicts the Ta $4f$ XPS peaks as a function of sputtering time. A
deconvolution of the peaks with respect to the different valencies of the
tantalum atoms[90] and an integration over the peak areas allows a quantification of the decomposition. This evaluation is shown in Figure 4.10(b) where
the intensities of Ta^{4+}, Ta^{2+} and Ta^0 are proportional to the suboxides TaO_2,
TaO and to the pure tantalum generated in the altered layer.[47] The remaining
Ta^{5+} is due to the contribution of the unchanged Ta_2O_5 beneath the oxygen-
depleted layer. A dynamic equilibrium is established after the initial transient
caused by preferential sputtering of oxygen. Decomposition of compounds
leads to difficulties in the interpretation of XPS sputtering profiles. Fortunately, oxides of the light elements (SiO_2, MgO, Al_2O_3) show negligible
reduction by ion bombardment.[45,47,88,91]

Sputtering may also give rise to enhanced diffusion in combination with
other driving forces like electrotransport[92] (in the case of insulators charged
by electron irradiation) and segregation.[93,94] An induced surface diffusion
may reduce the sputtering induced surface roughness and consequently
improve depth resolution,[76] but enhanced bulk diffusion broadens sputtering
profiles.[28,57] More systematic studies of temperature-dependent profiling are
needed to clearly distinguish between contributions of thermally activated
transport phenomena and atomic mixing and their interactions.

Whereas the X-rays in XPS rarely cause observable decomposition, the
electron beam in AES has a detrimental effect to many compounds. Electron
irradiation induced desorption of chlorine and oxygen is often observed.[95,96]
This can lead to an enhanced sputtering in the area of the electron beam, e.g.
in AES profiling with a loss in depth resolution[96] as a consequence. Since the
effect of electron beam induced desorption and sputtering compete at the
surface, this effect is suppressed by lowering the electron beam current
density (e.g. rastering) and/or increasing the ion beam current density.

The normal energy dissipation in the sample surface is negligible for X-ray

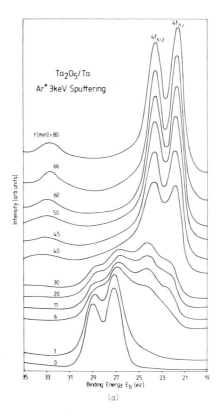

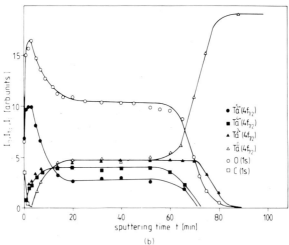

Figure 4.10 Decomposition of an anodic Ta_2O_5/Ta layer (30 nm thickness) during sputtering with 3 keV Ar^+ ions studied with XPS.[47] (a) Ta $4f$ peaks as a function of sputtering time. (b) XPS sputtering profile corresponding to a decrease of Ta_2O_5 and an increase of TaO, TaO_2 and Ta in the altered layer until a steady state is attained

irradiation and—with the exception of extreme cases—in ion sputtering for depth profiling. However, electron beam heating of the sample can lead to a temperature increase of several hundred degrees, particularly in thin films on glass substrates. A calculation of the temperature rise ΔT was given by Röll:[97]

$$\Delta T = \frac{p}{\pi r_0 \beta} \left[1 + 1.67 \left(\frac{\alpha}{\beta} \right) \left(\frac{h}{r_0} \right) \right]^{-1} \qquad (4.21a)$$

where p is the input power, r_0 is the radius of the electron beam, h is the thickness of the film, α and β are the thermal conductivities of the film and the substrate, respectively. For bulk samples equation (2.21a) can be replaced by

$$\Delta T = \frac{p}{\pi r_0 \beta} \qquad (4.21b)$$

Lowering the electron beam current density reduces ΔT. A test for the effect of electron beam heating on the depth profile can be made by sputtering the same sample with different current densities.[26,46]

4.4.3 Superposition of different contributions to depth resolution

If the simple contributions to depth profiling are independent of each other and can be approximated by Gaussian resolution functions of $2\sigma_j = \Delta z_j$, the total depth resolution Δz measured experimentally is described analogously to an error propagation law by an addition in quadrature:[72]

$$\Delta z = [\Sigma(\Delta z_j^2)]^{1/2} \qquad (4.22)$$

Equation (4.22) is only exact for Gaussian resolution functions, but it can be shown that it is a sufficient approximation, even for exponential functions (e.g. $\Delta z_\lambda = 1.6\lambda$).[49,53]

The different contributions may be characterized by their dependence on the total ion dose or the sputtered depth. Table 4.1 shows how the most important Δz_j's can be divided into constant, depth-dependent and sputter rate or time-dependent ones. For larger sputter depth, those Δz_j which increase with depth will be prevalent, whereas for small sputter depth the constant contributions will limit the observed sputter broadening.[14,72] Figure 4.11 shows the generally predictable shape of the depth dependence for the super-position of an instrumental effect ($\Delta z \propto z$, e.g. $\Delta z_1/z = 10$ per cent) and a constant contribution (e.g. atomic mixing, $\Delta z_k = \text{constant} = 2$ nm). According to an argument by Werner,[77] due to this type of superposition a dependence $\Delta z \propto z^{1/2}$ (or $\Delta z/z \propto z^{-1/2}$), which is often found experimentally,[4,5,72] is approximated in the region of intermediate depth.

A knowledge of the experimental depth resolution gives a quality figure of

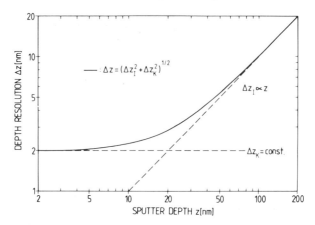

Figure 4.11 Typical dependence of the total depth resolution on depth z as predicted by equations (4.23) for a superposition of constant and depth-dependent contributions (see Table 4.1). The example is for an instrumental factor $\Delta z = 0.1z$ and an atomic mixing effect $\Delta z_k = 2$ nm[14]

the measured profile. A deconvolution procedure can be applied to reveal the shape of the original profile, as discussed in Section 4.6.

4.5 Special Profiling Techniques

Besides the general sputter profiling techniques there are some useful alternative methods which require special equipment: variation of the electron emission angle and energy in XPS and AES, crater edge profiling in SAM and ball cratering in combination with SAM.

4.5.1 Variation of the electron emission angle and the electron energy

These methods are non-destructive and are used to study compositional variations in the outermost atom layers of the solid. We shall first consider varying the electron emission angle. The principle of profiling by variation of the effective escape depth by variation of the electron emission angle is obvious from the general expressions equations (4.6) and (4.8), where φ is the angle between the normal to the sample and the direction of emission of the Auger electrons or photo-electrons. If λ_i is a continuous variable, equation (4.6) means that the intensity is given by a Laplace function $\mathscr{L}$ of the real concentration $c_i(z)$.[4] This is an integral transformation with the conditions 0 for $z < 0$ and $\exp[-(1/\lambda_i)z]$ for $z > 0$. Thus equation (4.6) can be expressed as

$$I\left(\frac{1}{\lambda_i}\right) = \text{constant } \mathscr{L}[c_i(z)] \tag{4.23a}$$

$c_i(z)$ is then determined by the inverse Laplace transformation $\mathscr{L}^{-1}$ according to

$$c_i(z) = \text{constant } \mathscr{L}^{-1}\left[I\left(\frac{1}{\lambda_i}\right)\right] \tag{4.23b}$$

The general equations (4.23a,b) are applicable for both AES (neglecting r_B) and XPS. The problem is that equation (4.6) has to be solved for an unknown integral $c_i(z)$. For some special cases, the solution can be found in mathematical tables.

For a concentric hemispherical analyser (CHA) with extraction lens, as most frequently used in XPS, the experimental determination of $I(1/\lambda_i)$ is straightforward by tilting the sample, as discussed by Fadley[10] and more recently by Ebel,[11,98] although problems may arise with surface roughness and at the angle $\varphi > 80°$, due to total reflection of the emitted electrons.[10,11,98]

Because of the limitations in sensitivity (cf. Section 4.3.2), the depth range covered is restricted to $< 3 \lambda^0$, i.e. < 5 nm. Furthermore, the depth resolution is limited by the experimental error of the intensity measurement. The most important application of the method of emission angle variation is the detection of the thickness of thin layers, as contamination[4,47], implantation[42] or segregation layers. The simple case of a homogeneous layer of thickness d of an element A on a substrate B is given by the normalized intensities (according to the solution of equations 4.6 and 4.8)[3,4]:

$$\frac{I_A}{I_A^0} = 1 - \exp\left(-\frac{d}{\lambda_{A,A}^0 \cos \varphi}\right) \tag{4.24a}$$

$$\frac{I_B}{I_B^0} = \exp\left(-\frac{d}{\lambda_{B,A}^0 \cos \varphi}\right) \tag{4.24b}$$

where $\lambda_{A,A}^0$ is the inelastic mean-free path of A electrons in A and $\lambda_{B,A}^0$ that of B electrons in A. For the special case of $\lambda_{A,A}^0 \approx \lambda_{B,A}^0$ the ratio $(I_B/I_B^0)/(I_A/I_A^0)$ is shown in Figure 4.12 as a function of the emission angle φ for different values of d/λ^0.[47]

Using a CMA for the electron energy analysis, complications with respect to the analyser geometry arise which restrict the useful angle variation and influence the signal intensity by shadowing effects.[4] A possibility to circumvent these difficulties is to use the CMA with a rotable drum device with a slit which cuts out a segment (e.g. 4° or 12°) of a defined azimuth θ from the acceptance cone.[4] If the fixed angle of the sample surface normal to the CMA axis is α, the relation between the azimuth angle and emission angle is given by:[4,47]

$$\cos \varphi = \sin \alpha \sin \varphi_A \cos \vartheta + \cos \alpha \cos \varphi_A \tag{4.25}$$

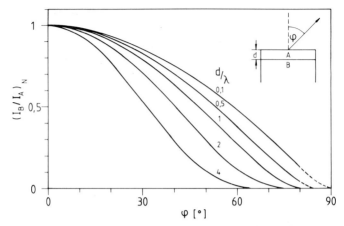

Figure 4.12   The ratio of equations (4.24a,b), $(I_B/I_B^0)/(I_A/I_A^0) = (I_B/I_A)_N$, plotted as a function of the emission angle φ for different values of d/λ (assuming $\lambda_{A,A}^0 = \lambda_{B,A}^0 = \lambda$)

where the CMA analysing angle $\varphi_A = 42.3°$ and $\alpha \leq \varphi_A$. For $\alpha = 30°$, it follows from equation (4.25) that $0.3 \leq \cos \varphi \leq 0.97$ for $0 \leq \vartheta \leq 180°$. Figure 4.13 shows an example taken with this geometry ($\alpha = 30°$, double-pass CMA), where the thickness of a contaminating layer on Nb_2O_5 consisting of carbon and oxygen hydrides was determined using the C $1s$ and O $1s$ peaks. Here, the O $1s$ peak could be deconvoluted in two peaks for the oxide and

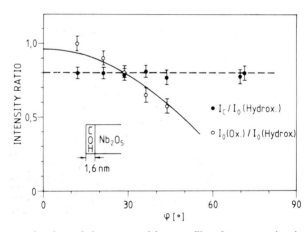

Figure 4.13   Determination of the composition profile of a contamination layer on Nb_2O_5 consisting of carbon and oxygen hydrides by the method of emission angle (φ) variation (CMA with rotable transmission slit) using the XPS intensity ratios $I(O\ 1s$ oxide$)/I(O\ 1s$ hydroxide$)$ and $I(C\ 1s)/I(O\ 1s$ hydroxide$)$ according to Figure 4.12. The constancy of the latter ratio shows an almost homogenous layer of $d/\lambda(O) = 0.71$, which gives the thickness $d = 1.6$ nm with $\lambda(O) = 2.3$ nm[47]

hydroxide (0.7 eV binding energy difference). Assuming that the oxide peak stems totally from the Nb_2O_5, the ratio of equations (4.24a,b), as shown in Figure 4.14, was used to calculate the respective $d/\lambda(O) = 0.71$. The constant ratio $I(C\ 1s)/I(O\ 1s$ hydroxide) shows that there is no detectable variation of the composition of the contamination layer with depth. Note that by use of signals of the same subshell the asymmetry term in quantitative XPS[38]

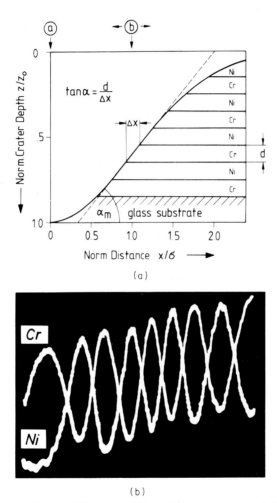

(a)

(b)

Figure 4.14   Crater edge profiling of a NiCr multilayer structure.[61] (a) Schematic of crater edge profiling for the case of a gaussian ion beam with standard deviation $\sigma \tan \alpha_m = 0.6\, z_0/x(\sigma)$. The NiCr multilayer structure profiled is schematically shown. (b) Line scans for the Cr (529 eV) and Ni (848 eV) intensities across the region indicated in (a) with single-layer thickness $d = 11.5$ nm, $d_e = 10\ \mu m$, $\tan \alpha_m = 1.9 \times 10^{-4}$

cancels. Otherwise an additional intensity variation with ϑ must be taken into account.[99]

In summary, the main advantage of the use of emission angle variation to obtain depth profiles is, particularly in XPS, that it is non-destructive and yields an absolute depth scale in terms of λ^0. However, it is restricted to very thin layers $\leq 3\lambda^0$ (about ≤ 5 nm) and requires appropriate experimental equipment (see also Chapters 3 and 9).

The second method, by varying the electron energy, also makes use of equations 4.23 and 4.24) and relies on the energy dependence of the escape depths $\lambda^0 \cos \varphi$. For instance, if we compare the relative intensities of low energy, I_b, and high energy, I_B, peaks from the substrate with the homogenous overlayer of thickness d with the relative intensities of the clean substrate I_b^0 and I_B^0, respectively, it is easy to show that

$$d = \frac{\lambda_{bA}^0 \lambda_{BA}^0}{\lambda_{BA}^0 - \lambda_{bA}^0} \cos \varphi \, \ln \left(\frac{I_b^0 / I_B^0}{I_b / I_B} \right). \qquad (4.26)$$

4.5.2 Crater edge profiling with SAM (scanning Auger microscopy)

In the technique of crater edge profiling[61,100] the crater effect of a static Gaussian ion beam on depth resolution[62] (see Section 4.4.2) is used in a positive manner to obtain a depth profile. It is related to the conventional angle lapping technique. As depicted in Figure 4.14(a), the slope of the crater can be used to transform a displacement of the well-focused electron beam (x coordinate parallel to the sample surface) into a depth scale (z coordinate) by a relation similar to equation (4.17). An example for determination of the profile of a NiCr multilayer sandwich sample is shown in Figure 4.14(b).[61] The instrumental depth resolution as determined by the electron beam diameter d_e is[61]

$$\Delta z_e \approx d_e \tan \alpha \qquad (4.27)$$

where $\tan \alpha$ is the slope of the crater. The accuracy of the profile depends on the accuracy of the shape of the crater, which is proportional to the intensity distribution within the ion beam (to be measured by a Faraday cup). However, with respect to factors dependent on the escape depth, sample composition and topography, as well as with respect to bombardment-induced effects, the limitations in depth resolution are expected to be similar to general profiling.[62] The main advantage of this technique is that a depth profile can be obtained at ease after sputtering so that information loss is prevented and the signal-to-noise figure can be improved.

4.5.3 Profiling by ball cratering and SAM

This technique[101] is a special case of the well-established angle lapping technique. Like the latter, it is useful for profiles of several micrometres depth

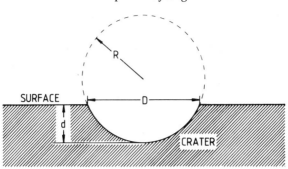

Figure 4.15 Schematic illustration of the geometry of ball cratering. (After Walls *et al.*[101])

which consume a tremendous sputtering time in general equipment. It over-comes many of the difficulties of angle lapping by using a rotating steel ball coated with fine diamond paste. The geometry of the crater is illustrated schematically in Figure 4.15.[101] Since the radius R of the ball is known (typi-cally 1–3 cm), the depth d of the crater is given by $D^2/8R$, where D is the crater diameter. The lateral position of the electron beam on the spherical crater can be related to depth by the geometry shown in Figure 4.15. Assum-ing a perfectly smooth crater the depth z is related to the distance x by which the electron beam is moved from the crater edge[101] by

$$z = d - R + \left[R^2 - \left(\frac{D}{2} - x\right)^2\right]^{1/2} \tag{4.28}$$

The inherent depth resolution is given by

$$\Delta z_b \approx \frac{d_e}{R}[2R(d - z)]^{1/2} + m \tag{4.29}$$

where d_e is the diameter of the electron beam and m is an additional term taking into account surface roughness. In practice, depth resolution is limited mainly by the surface roughness generated during wear process. Equation (4.29) shows that the relative depth resolution $\Delta z_b/z$ is improved for larger depths as shown by Walls, Hall and Sykes.[101] Lea and Seah[102] have given criteria for the use of general AES sputter profiling and mechanical lapping with respect to optimum Δz as a function of total sputtered depth.

   The advantage of ball cratering with respect to depth resolution shows up for depth profiles over large depths ($>2 \mu m$). Compared to conventional sputter profiling and to crater edge profiling it has the advantage that an absolute and linear depth scale is established with a defined surface topo-graphy, in spite of the fact that some materials (very soft or very brittle) may lead to difficulties (smearing and cracking). For smaller depths ($<2 \mu m$), sputter profiling is a more sensitive technique than mechanical abrasion and has an inherently better absolute depth resolution.

4.6 Deconvolution Procedures for Sputtering Profiles

Assuming that the first step in quantification of a sputtering profile, i.e. the determination of the true (linear) depth scale (Section 4.3.1) and the concentration scale (Section 4.3.2), has been made, we still have to consider all the effects which may have distorted the true profile (Section 4.4). The task is then the solution of equation (4.14):[49,53]

$$I(z) = \int_{-\infty}^{+\infty} c(z')g(z - z')\, dz \tag{4.30}$$

where $c(z')$ is the true profile, $I(z)$ the depth and intensity scale corrected concentration profile and $g(z-z')$ is the resolution function. The latter is the result of all the contributions to sputtering profile distortion in a specific experiment, as discussed in Section 4.4. If a single contribution is known it can be treated separately. This is particularly useful for the escape depth effect (together with the back-scattering effect in AES), given by

$$g_{\lambda, r_B}(z - z') = r_B(z)\exp\left[- \frac{(z - z')}{\lambda_i} \right] \tag{4.31}$$

Inserting equation (4.31) into equation (4.30) shows the similarity with equation (4.6), for which the explicit solution is given by equation (4.9) for $r_B(z) = $ constant. The back-scattering factor can also be taken into account by an appropriate exponential function.[3,39] A first-order approximation of quantification using a rectangular approximation of equation (4.6) by the product $r_B(X_A/X_B)\lambda_i(X_A/X_B)$ for binary systems A/B has been given by Morabito and others.[6,37] In this case, a composition-dependent calibration factor is applied to the measured profile. It should be noted that the corrections for back-scattering and escape depth yields a profile which is still broadened by the other factors discussed in Section 4.4.

Even if the different resolution functions originating from these factors are asymmetric (as, for example, from atomic mixing[79-83], their superposition can be approximated with sufficient accuracy by a Gaussian resolution function:

$$g(z - z') = \left(\frac{\pi \, \Delta z^2}{2} \right)^{-1/2} \exp\left[- 2\left(\frac{z - z'}{\Delta z} \right)^2 \right] \tag{4.32}$$

where $\Delta z = 2\sigma$ is the depth resolution defined in Section 4.4.1.1. A dependence $\Delta z = f(z)$ (see Table 4.1) can be inserted into equation (4.32).

For an arbitrary resolution function—as for a Gaussian—equation (4.30) cannot be solved explicitly. In this case, two mathematical approaches have been developed:[103] the Fourier transform method and the iterative method of Van Cittert.[104] Wertheim[105] has discussed the equivalence of both methods. The method of successive iterations was first applied to depth profiling by Ho and Lewis.[48] According to this approach, the solution after the nth iteration, $c^n(z)$ is given by:

$$c^n(z) = c^{n-1}(z) + I(z) - \int_{-\infty}^{+\infty} c^{n-1}(z')g(z - z')\, dz' \quad (4.33)$$

with $c^0(z) \equiv I(z)$ ($=$ the concentration profile as measured and calibrated). It is expected that $\lim_{n\to\infty} c^n(z) \equiv c(z)$ if the iteration converges. Equation (4.33) is easy to solve by computer numerical methods.

The applicability of the Van Cittert method can be tested analogously to a proposal by Madden and Houston[106] if the true profile $c(z)$ is assumed as an error function. Thus, a 'measured' profile is obtained by folding $c(z)$ by a Gaussian function $g(z - z')$[14,49,53]

$$c(z) = 0.5\left\{1 - \text{erf}\left[2^{1/2}\left(\frac{z - z_0}{\Delta z_0}\right)\right]\right\} \quad (4.34)$$

$$g(z - z') = \left(\frac{\pi \Delta z_g}{2}\right)^{1/2} \exp\left[-\frac{2(z - z')^2}{\Delta z_g^2}\right] \quad (4.35)$$

where z_0 defines the interface (50 per cent value), Δz_0 is the 2σ value of the true profile and Δz_g is the depth resolution which defines $g(z - z')$. Inserting equations (4.34) and (4.35) into equation (4.30) we obtain a 'measured', normalized profile $I(z)$:

$$I(z) = 0.5\left\{1 - \text{erf}\left[2^{1/2}\left(\frac{z - z_0}{\Delta z}\right)\right]\right\} \quad (4.36)$$

with

$$\Delta z = [\Delta z_0^2 + \Delta z_g^2]^{1/2} \quad (4.37)$$

The convergence can be monitored by comparing the width Δz_0 of the true profile with the width of the deconvoluted profile after the nth iteration, Δz_0^n. In Figure 4.16 this convergence factor is shown as a function of the parameter $\kappa = \Delta z_0/\Delta z_g$ for $0.05 \le \kappa \le 1.5$ and for 0, 2 and 10 iterations. The influence of κ on the convergence is obvious. Figure 4.16 shows that only cases with $\kappa \ge 1$ are suitable for the true profile recovery with a reasonable number of iterations. It can be shown that for $\kappa < 1$ oscillations arise[48,106] and the true profile cannot be regenerated.

Applying the above criterion ($\kappa \ge 1$) to the measured Δz in equation (4.37) we obtain

$$\frac{\Delta z}{\Delta z_g} \ge 2^{1/2} \quad (4.38)$$

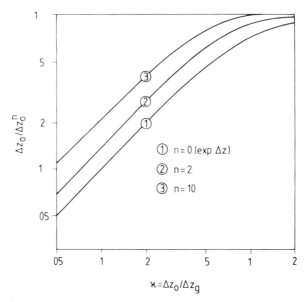

Figure 4.16   Convergence of the Van Cittert deconvolution given by the ratio Δz_0 (true profile) to $\Delta z_0^{(n)}$ (deconvoluted profile after n iterations) as a function of $\kappa = \Delta z_0/\Delta z_g$ with Δz_g defined by the resolution function[49,53]

This means that the generation of the true profile $c(z)$ is only possible if the induced broadening Δz_g is less than 70 per cent of the measured Δz. The condition given by equation (4.38) can be easily extended to the slope dI/dz of a normalized measured profile $I(z)$ if we consider the definition of the inverse maximum slope in an error function profile $(0 < I < 1)$, where $dz/dI = 1.25\,\Delta z$. Then equation (4.38) becomes

$$\Delta z_g \left(\frac{dI}{dz}\right) \le 0.566 \qquad\qquad (4.39)$$

Analysing the deviation of the 'measured' width of the profile, Δz, from the true width, Δz_0, as defined by equation (4.37) with respect to equations (4.38) and (4.39) we can distinguish between three regions (I, II, III) which are depicted in Figure 4.17.[14] If the maximum slope of the measured profile is less than 10 per cent of the Δz_g of the resolution function, the deviation between measured and true profile with respect to resolution is less than 1 per cent and therefore no deconvolution at all is required (Figure 4.17, region I). A deconvolution after the iterative method as described above can be applied in region II where the maximum slope is less than 57 per cent of Δz_g. For a steeper slope the Van Cittert method does not converge, which means that there is some ambiguity in the generation of the true profile (region III).

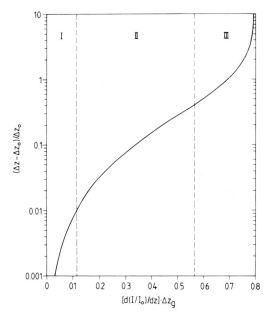

Figure 4.17   The relative deviation $(\Delta z - \Delta z_0)/\Delta z$ between the depth resolutions of the measured (Δz) and the true profile (Δz_0) as a function of the product of the maximum slope $d(I/I_0)/dz$ of the measured profile and the Δz_g of the resolution function for error function profiles.[14] The three regions indicate the cases where: I, no deconvolution is necessary; II, the Van Cittert deconvolution scheme can be applied; III, deconvolution is ambiguous (true profile assumption and comparison with the measured profile)

Particularly in this case, the method of iterative true profile assumption can be applied successfully.[3,49,53]

This is in fact a convolution method based on the assumption of a true profile and calculation of the 'measured' profile $I(z)$ by equation (4.30), introducing a known resolution function.[7,49,53] A direct comparison with the normalized experimental data shows the validity of the assumption. If deviations show up, the assumed true profile is modified and this procedure can be repeated until a sufficient agreement is obtained[49] ('trial and error' method).

In general it is convenient to approximate the true profile by thin successive layers of constant concentration $c(z_j)$ and thickness $d = z_{j+1} - z_j$ and to calculate the corresponding measured, normalized intensity $I_j(z)$ of each single layer which is given by[3,7,51]

$$I_j(z) = \frac{c(z_j)}{2} \left\{ \mathrm{erf}\left[2^{1/2}\left(\frac{z - z_j}{\Delta z_g}\right)\right] - \mathrm{erf}\left[2^{1/2}\left(\frac{z - z_j - d}{\Delta z_g}\right)\right] \right\} \qquad (4.40)$$

A calculation using equation (4.40) has to be made for each layer $c(z_j)$ at z_j for

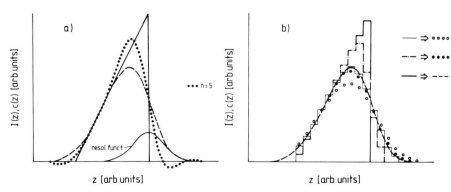

Figure 4.18   Comparison of the true profile $c(z)$ (——) calculation starting with the measured profile $I(z)$ (– – –) and the indicated resolution function.[49] (a) Van Cittert method, $n = 5$ iterations; (b) iterative true profile assumption

about $-1.5\,\Delta z_g \leq z - z_j \leq 1.5\,\Delta z_g$ ($3\,\sigma$ iterations with 99.7 per cent accuracy). The choice of d depends on the steepness of the profile and the desired precision with a reasonable lower limit of one monolayer thickness. After summing up the $I_j(z)$ for any z a profile is obtained which can directly be compared with the normalized measured profile:[3,49]

$$I(z) = \sum_j I_j(z) \tag{4.41}$$

To illustrate the two deconvolution procedures discussed above, Figure 4.18 shows as an example the case of a triangular true profile.[49] Figure 4.18(a) shows the application of the Van Cittert method and Figure 18(b) the method of iterative true profile assumption. Due to the pointed edge and the steep gradient in the profile, the generated 'true' profile in both cases shows some ambiguity with respect to the accuracy of the measurement (compare region III, Figure 4.17).

A correction of the broadening effect of the escape depth alone can also be performed by the iterative true profile assumption method. In this case, equation (4.40) has to be replaced by an equation similar to equation (4.8):[7,51]

$$I_j(z) = c(z_j)\left[\exp\left(\frac{z - z_j}{\lambda_i}\right) - \exp\left(\frac{z - z_j - d}{\lambda_i}\right)\right] \tag{4.42}$$

where λ_i is taken to be composition dependent (see Section 4.2.2) in the next better approximation.[7] The summing up of all the single-layer contributions (equation 4.41) is the same as before. Since λ_i is the effective escape depth depending on the experimental geometry, equation (4.42) offers an approach to the non-destructive depth profiling technique of the emission angle variation (see Section 4.5.1) for an arbitrary profile. Setting $\lambda_i = \lambda_i^0 \cos \varphi$, then $I(\varphi) = \sum_j I_j(\varphi)$ for any given angle φ only because the z-dependence is now

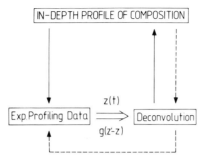

Figure 4.19   Schematic of the evaluation of the true, original profile of composition[49,53] via deconvolution of the calibrated experimental data (measured profile) (⟶) and/or by the iterative true profile assumption method − − −⟶)

replaced by the λ-dependence. The angle-dependent normalized intensity generated in this way can be directly compared to the measured angle-dependent profile (compare Figures 4.12 and 4.13).

The general scheme of both deconvolution and convolution procedures is shown in Figure 4.19. The recommended methods (I, II, III) should be chosen according to the criteria given in Figure 4.17.

4.7   Summary and Conclusions

Any quantitative evaluation of depth profiling data obtained by combination of electron spectroscopy with ion sputtering requires a quantification of the signal intensity and the depth scale together with a knowledge of the depth resolution function, provided a composition change during the analysis is negligible or known. The speed of analysis, electron background considerations and above all the small analysis spot greatly favours high spatial resolution AES (SAM) over XPS. Whereas the decisive parameters for quantitative AES and XPS (electron escape depth and electron back-scattering in AES) are comparatively well defined, a prediction of the large number of parameters (depending on instrumental and ion beam/sample interaction characteristics) appears highly speculative at present. A recommended test of the depth scale estimation and of the instrumental set-up is the use of anodized layers of Ta_2O_5/Ta with defined interface location which will be available as standards from BCR (Bureau Committee du Reference) in the near future. *In situ* measurement methods of the eroded depth (e.g. optical interferometry, X-ray analysis) seem to be the only way to get a true precise calibration of the depth scale. A test for the ion beam induced alterations of the sample is possible by varying the sputtering conditions (e.g. ion species, ion energy and incidence angle). Alternative methods to conventional sputtering, like crater edge profiling, ball cratering and effective escape depth variation, are useful approaches in special cases.

After the first step in quantification (concentration and depth scales, a finite difference to the true, original profile remains, which is determined by the experimental depth resolution or, more precisely, the depth resolution function. A knowledge of this function allows the deconvolution of the normalized measured profile to obtain the true profile by numerical methods. Alternatively, a convolution of an assumed true profile and an iterative comparison with the measured profile allows the determination of the true composition profile. In any case, the key parameter for the reliability and precision of the latter is the experimentally resolved depth which depends on a careful optimization of the profiling conditions.

Acknowledgement

Helpful discussions with Dr M. P. Seah, NPL Teddington, Dr J. M. Sanz and Dr U. Roll are gratefully acknowledged.

References

1. A. Benninghoven, *Thin Solid Films*, **31**, 89 (1976).
2. R. E. Honig, *Thin Solid Films*, **39**, 3 (1976).
3. S. Hofmann, *Surf. Interface Anal.*, **2**, 148 (1980).
4. S. Hofmann, *Analusis*, **9**, 181 (1981).
5. S. Hofmann, in (Ed. G. Svehla), *Wilson and Wilson's Comprehensive Analytical Chemistry*, Vol. IX, pp. 89–172, Elsevier, Amsterdam (1979).
6. P. M. Hall and J. M. Morabito, *Crit. Rev. Sol. State Mat. Sci.*, **8**, 53 (1978).
7. S. Hofmann, *Le Vide, No. Spec.*, March **1979**, 259.
8. I. V. Mitchell, *Phys. Bull.*, **30**, 23 (1979).
9. P. H. Holloway and J. M. McGuire, *Thin Solid Films*, **53**, 3 (1978).
10. C. S. Fadley, *Progr. Sol. State Chem.*, **11**, 265 (1976).
11. M. F. Ebel, *J. Electron Spectr. Rel. Phen.*, **14**, 287 (1978).
12. K. Maier, R. Kirchheim and G. Tölg, *Mikrochim. Acta Suppl.*, **8**, 125 (1979).
13. J. W. Coburn, *J. Vac. Sci. Technol.*, **13**, 1037 (1976).
14. S. Hofmann, in *Proc. Third Int. Conf. on SIMS (Budapest, 1981)* (Eds A. Benninghoven, J. Giber, J. Lazlo, M. Riedel and H. W. Werner), p. 186, Springer Verlag, Berlin (1982).
15. H. J. Mathieu and D. Landolt, *J. Microsc. Spectr. Electron*, **3**, 113 (1978).
16. M. P. Seah and C. Lea, *Thin Solid Films*, **81**, 257 (1981).
17. D. E. Sykes, D. D. Hall, R. E. Thurstans and J. M. Walls, *Appl. Surf. Sci.*, **5**, 103 (1980).
18. H. W. Werner, *Mikrochim. Acta Suppl.*, **8**, 25 (1979).
19. R. Behrisch (Ed.), *Sputtering by Ion Bombardment*, Vol. I, Topics in Appl. Phys., Springer Verlag, Heidelberg (1981).
20. M. P. Seah, *Thin Solid Films*, **81**, 279 (1981).
21. H. Oechsner, *Appl. Phys.*, **8**, 185 (1975).
22. H. Schoof and H. Oechsner, in *Proc. Fourth Int. Conf. Sol. Surf. and Third ECOSS, Cannes (1980)* (Eds F. Abélès and M. Croset), p. 1291, Paris (1980).
23. A. J. Bevelo, *Surf. Interface Anal.*, **3**, 240 (1981).
24. L. Young, *Anodic Oxide Films*, Academic Press, London (1961).

25. J. M. Morabito and R. K. Lewis, *Anal. Chem.*, **45**, 869 (1973).
26. P. Laty, D. Seethanen and F. Degreve, *Surf. Sci.*, **85**, 533 (1979).
27. J. W. Coburn and E. Kay, *Crit. Rev. Sol. State Sci.*, **4**, 561 (1974).
28. S. Hofmann and A. Zalar, *Thin Solid Films*, **60**, 201 (1979).
29. J. Kirschner and H. W. Etzkorn, *Appl. Surf. Sci.*, **3**, 251 (1979).
30. J. H. Thomas and S. P. Sharma, *J. Vac. Sci. Technol.*, **14**, 1168 (1977).
31. J. Kempf, in *Proc. Second Int. Conf. on SIMS (II)* (Eds A. Benninghoven, C. A. Evans, R. A. Powell, R. Shimizu and N. A. Storms), p. 97, Springer Verlag, Berlin (1979).
32. J. Kempf, *Surf. Interface Anal.*, **4**, 116 (1982).
33. M. P. Scah, *Surf. Sci.*, **32**, 703 (1972).
34. P. W. Palmberg, *Anal. Chem.*, **45**, 549 (1973).
35. M. P. Seah and W. Dench, *Surf. Interface Anal.*, **1**, 2 (1979).
36. A. Jablonski, *Surf. Interface Anal.*, **1**, 122 (1979).
37. P. M. Hall, J. M. Morabito and D. K. Conley, *Surf. Sci.*, 62, 1 (1977).
38. M. P. Seah, *Analusis*, **9**, 171 (1981).
39. S. Hofman, *Mikrochim. Acta Suppl.*, **8**, 71 (1979).
40. F. Pons, J. LeHéricy and J. P. Langeron, *Surf. Sci.*, **69**, 565 (1977).
41. J. LeHéricy, *Le Vide, No. Spec.*, March **1979**, 307.
42. M. Pijolat and G. Hollinger, *Analusis*, **10**, 8 (1982).
43. H. Iwasaki and G. Nakamura, *Surf. Sci.*, **57**, 779 (1976).
44. S. Hofmann, *Talanta*, **26**, 665 (1979).
45. J. H. Thomas III and S. Hofmann, *Surf. Interface Anal.* 4, 156 (1982).
46. C. D. Wagner, *Anal. Chem.*, **44**, 1050 (1972).
47. S. Hofmann and J. M. Sanz, Fresenins Z. *Anal. Chem.*, **314**, 215 (1983).
48. P. S. Ho and J. E. Lewis, *Surf. Sci.*, **55**, 335 (1976).
49. S. Hofmann and J. M. Sanz, in *Proc. Eighth Int. Vac. Congr., Cannes (1980)* (Eds F. Abélès and M. Croset), p. 90, Paris (1980).
50. S. Hofmann, *Appl. Phys.*, **9**, 59 (1976).
51. S. Hofmann, in *Proc. Seventh Int. Vac. Congr. and Third Int. Conf. Sol. Surf.* (Eds A. Dobrozembsky *et al.*), Vol. III, p. 2613, Berger, Vienna (1977).
52. H. J. Mathieu and D. Landolt, *Le Vide, No. Spec.*, March **1979**, 273.
53. S. Hofmann and J. M. Sanz, to be published in *Topics in Current Physics* (Ed. H. Oechsner), Springer, Heidelberg (1982).
54. S. Hofmann, J. Erlewein and A. Zalar, *Thin Solid Films*, **43**, 275 (1977).
55. R. Ludeke, L. Esaki and L. L. Chang, *Appl. Phys. Lett.*, **29**, 417 (1974).
56. H. J. Mathieu, D. E. McClure and D. Landolt, *Thin Solid Films*, **38**, 281 (1976).
57. K. Röll and C. Hammer, *Thin Solid Films*, **57**, 209 (1979).
58. R. J. Blattner, S. Nadel, C. A. Evans, A. J. Braundmeier and C. W. Magee, *Surf. Interface Anal.*, **1**, 32 (1979).
59. P. H. Holloway and R. S. Battacharya, *Surf. Interface Anal.*, **3**, 118 (1981).
60. P. H. Holloway and H. J. Stein, *J. Electrochem. Soc.*, **123**, 723 (1979).
61. A. Zalar and S. Hofmann, *Surf. Interface Anal.*, **2**, 183 (1980).
62. J. B. Malherbe, J. M. Sanz and S. Hofmann, *Surf. Interface Anal.*, **3**, 235 (1981).
63. L. Bradley, Y. M. Bosworth, D. Briggs, V. A. Gibson, R. J. Oldman, A. C. Evans and J. Franks, *Appl. Spectr.*, **32**, 175 (1978).
64. W. Hauffe, in *Proc. Third Int. Conf. on SIMS (III), (Budapest, 1981)* (Eds A. Benninghoven, J. Giber, J. Lorló, M. Riedel and H. W. Werner), p. 206, Springer Verlag, Berlin (1982).
65. K. Tsunoyama, T. Suzuki, Y. Okahasbi and H. Kishikada, *Surf. Interface Anal.*, **2**, 212 (1980).
66. S. S. Makh, R. Smith and J. M. Walls, *Surf. Interface Anal.*, **2**, 115 (1980).

67. R. Smith, T. P. Valkering and J. M. Walls, *Phil. Mag.*, **A44**, 879 (1981).
68. L. D. Chadderton, *Rad. Eff.*, **33**, 129 (1977).
69. G. K. Wehner and D. J. Haijcek, *J. Appl. Phys.*, **42**, 1145 (1971).
70. J. L. Vossen, *J. Appl. Phys.* **47**, 544 (1976).
71. P. Sigmund, in *Sputtering by Ion Bombardment* (Ed. R. Behrisch), Vol. I, Topics in Appl. Phys., Springer Verlag, Heidelberg (1981).
72. S. Hofmann, *Appl. Phys.*, **13**, 205 (1977).
73. W. O. Hofer and H. Liebl, *Appl. Phys.*, **8**, 359 (1975).
74. C. R. Helms, N. M. Johnson, S. A. Schwartz and W. E. Spicer, in *The Physics of SiO₂ and Its Interfaces* (Ed. S. T. Pantelides), p. 366, Pergamon, New York (1978).
75. M. P. Seah, J. M. Sanz and S. Hofmann, *Thin Solid Films*, **81**, 239 (1981).
76. J. Erlewein and S. Hofmann, *Thin Solid Films*, **69**, L39 (1980).
77. H. W. Werner, *Surf. Interface Anal.*, **4**, 1 (1982).
78. H. H. Andersen, *Appl. Phys.*, **18**, 131 (1979).
79. S. Taikang, R. Shimizu, T. Okutani, *Jap. J. Appl. Phys.*, **18**, 1987 (1979).
80. U. Littmark and W. O. Hofer, *Nucl. Instr. a. Meth.*, **168**, 329 (1980).
81. P. Sigmund and A. Gras-Marti, *Nucl. Instr. a. Meth.*, **182**, 25 (1981).
82. H. W. Etzkorn, U. Littmark and J. Kirschner, in *Proc. Symp. on Sputtering* (Eds P. Varga, G. Betz and F. P. Viehböck), p. 542, Perchtolsdorf (1980).
83. Z. L. Liau, B. Y. Tsaur and J. W. Mayer, *J. Vac. Sci. Technol.*, **16**, 121 (1979).
84. H. Shimizu, M. Ono and K. Nakayama, *Surf. Sci.*, **36**, 817 (1973).
85. H. J. Mathieu and D. Landolt, *Surf. Sci.*, **53**, 228 (1975).
86. P. S. Ho, J. E. Lewis, H. S. Wildman and J. K. Howard, *Surf. Sci.*, **57**, 393 (1976).
87. H. J. Mathieu and D. Landolt, *Appl. Surf. Sci.*, **10**, 100 (1982).
88. R. Kelly, *Nucl. Instr. a. Meth.*, **149**, 553 (1978).
89. S. Storp and R. Holm, *J. Electron Spectr. Rel. Phen.*, **16**, 183 (1979).
90. R. Holm and S. Storp, *Appl. Phys.*, **9**, 217 (1976).
91. H. J. Mathieu and D. Landolt, *Proc. Seventh Int. Vac. Congr. a. Third Int. Conf. Sol. Surf. (Vienna 1977)*, p. 2023, Berger (1977).
92. F. Ohuchi, M. Ogino, P. H. Holloway and C. G. Pantano, *Surf. Interface Anal.*, **2**, 85 (1980).
93. R. Kelly, in *Proc. Symp. on Sputtering* (Eds G. Varga, G. Betz and F. P. Viehböck), p. 512, Perchtolsdorf (1980).
94. S. Hofmann, *Mat. Sci. Eng.*, **42**, 55 (1980).
95. S. Thomas, *J. Appl. Phys.*, **45**, 161 (1974).
96. J. Ahn, C. R. Perleberg, D. L. Wilcox, J. W. Coburn and H. F. Winters, *J. Appl. Phys.*, **46**, 4581 (1975).
97. K. Röll, *Appl. Surf. Sci.*, **5**, 388 (1980).
98. M. F. Ebel and J. Wernisch, *Surf. Interface Anal.*, **3**, 191 (1981).
99. M. Vulli, *Surf. Interface Anal.*, **3**, 67 (1981).
100. A. Van Oostrom, *Surf. Sci.*, **89**, 615 (1979).
101. J. M. Walls, D. D. Hall and D. E. Sykes, *Surf. Interface Anal.*, **1**, 204 (1979).
102. C. Lea and M. P. Seah, *Thin Solid Films*, **81**, 67 (1981).
103. A. F. Carley and R. W. Joyner, *J. Electr. Spectr. Rel. Phen.*, **16**, 1 (1979).
104. P. H. Van Cittert, *Z. Physik*, **69**, 298 (1931).
105. G. K. Wertheim, *J. Electr. Spectr. Rel. Phen.*, **9**, 239 (1975).
106. H. H. Madden and J. E. Houston, *J. Appl. Phys.*, **27**, 3071 (1976).
107. A. Zalar, *Mikrochim. Acta.*, **I**, 435 (1980).
108. C. W. Magee, W. L. Harrington and R. E. Honig, *Rev. Sci. Instr.*, **49**, 477 (1978).

109 H. J. Mathieu and D. Landolt, *Surf. Interface Anal.* **5**, 77 (1983).

110. V. Naundorf and P. Macht. *Nucl. Instr. a. Meth.*, **168**, 40 (1980).

111. C. P. Hunt, C. T. H. Stoddart and M. P. Seah, *Surf. Interface Anal.*, **3**, 157 (1981).

112. J. B. Malherbe and S. Hofmann, *Surf. Interface Anal.*, **2**, 187 (1980).

Practical Surface Analysis
by Auger and X-ray Photoelectron Spectroscopy
Edited by D. Briggs and M. P. Seah
© 1983, John Wiley & Sons, Ltd

Chapter 5

Quantification of AES and XPS

M. P. Seah
*Division of Materials Applications,
National Physical Laboratory,
Teddington, Middlesex, UK*

5.1 Introduction

Since the very earliest applications of AES, efforts have been devoted to improve the accuracy of the quantification. However, with XPS, the main objective has, in the past, been the establishment of chemical state data, and so quantification is of more recent interest. Quantification in both techniques is treated here in one chapter since both involve electron spectroscopy and many of the new instruments have both techniques available. It is hoped that a unified methodology and symbology will aid those analysts now having to grapple with the two techniques simultaneously.

There are three basic routes to quantification, probably applicable to all forms of analysis: (a) calculation of the relevant terms from basic principles, (b) the use of published data bases and (c) the use of locally produced standards or a local data base. In practice, for the analyst, a blend of the three approaches may be most effective.

In the sections that follow the basic terms that give rise to the intensity, first in AES and then XPS, will be outlined for the simple case of homogeneous bulk alloys. This also enables the application of published or locally produced data bases, as alternative routes, to be viewed in perspective and their inherent approximations to be understood. In general, however, surface analysis is used for two main groups of sample which are not homogeneous, as discussed in Chapters 6 to 10. The first group is that of adsorbates, where surface analysis is used in its classical sense to identify and measure sub-monolayer quantities of atoms or molecules residing at the outermost atom layer of a solid, whereas the second group are those with thicker layers to be analysed with simultaneous *in situ* ion bombardment to give composition–depth profiles. These two groups will be discussed within the unified formalism at the end of the chapter.

An interesting point to start from is the simple linear expression that the intensity of the signal I_A, from element A in a solid is simply proportional to the molar fractional content, X_A, in the analysis depth. Thus:

$$X_A = \frac{I_A}{I_A^\infty} \tag{5.1}$$

where I_A^∞ is the intensity from pure A. Since, in general, I_A^∞ is not known but the ratio I_A^∞/I_B^∞ may be, where B is another constituent of the solid, the more useful form of equation (5.1) becomes[1]

[handwritten left margin: from ref. 1: $C_x = \dfrac{I_x / S_x}{\Sigma \cdot I_A / S_A}$]

[handwritten: Not true]

$$X_A = \frac{I_A/I_A^\infty}{\sum\limits_{i=A,B} I_i/I_i^\infty} \tag{5.2}$$

[handwritten top: $X_A = \dfrac{I_A}{I_A + \left(\dfrac{I_A^\infty}{I_B^\infty}\right) I_B}$ ✓]

[handwritten: This is one so it's correct. whoever is writing in this book]

where the sum is over all of the constituents of the solid. Equation (5.2) is a useful first try at quantification, but in the next two sections the terms contributing to I_A, in AES and XPS, will be discussed so that the form and extent of the necessary correction terms may be understood.

5.2 Quantification of AES for Homogeneous Binary Solids

There are two main problems in quantifying AES: establishing the correct equation to use and establishing the correct measurement to make.

5.2.1 Basic considerations

The incident electron beam, on striking the solid, penetrates with both elastic and inelastic scattering and eventually comes to rest at a depth of 1–2 μm, as shown in Figure 5.1. As discussed in Chapter 3, these energetic electrons can ionize atoms in the solid in core level X which subsequently de-excites by an electron falling from a higher level Y, the energy balance being removed by the ejection of an electron from level Z. This last electron, the so-called XYZ Auger electron, thus has a well-defined energy.

The earliest calculations of the ionization cross-section, $\sigma_X(E)$, by electrons of energy E, in AES, are those of Bishop and Rivière.[2] Following approaches originally developed for the X-ray microprobe analyser they use the Worthington and Tomlin[3] cross-section, given by

$$\sigma_{ax}(E) = 1.3 \times 10^{-13}\, bC/E_{AX}^2 \quad \text{cm}^2 \tag{5.3}$$

where $b = 0.35$ for the K shell and 0.25 for the L shell. The function C depends on the primary electron beam energy, E_p, and rises from zero at $E_p = E_{AX}$ to 0.6 at (2 to 3) E_{AX} and then declines slowly at higher energies. Taking a typical value of $C = 0.5$, the above cross-section gives an ionization efficiency of the order of 10^{-4} for the electron beam passing through an

[handwritten left margin, rotated: maximum at $E_p = (2\frac{1}{2} + \frac{1}{4}) E_{AX}$]

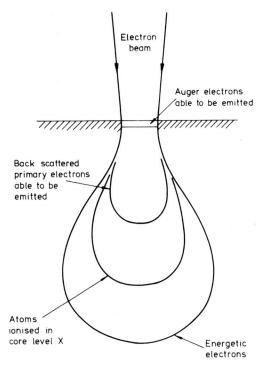

Figure 5.1   Schematic presentation of electron scattering in AES

adsorbed monolayer. Thus, for 1 per cent. of a monolayer, bearing in mind that the total secondary electron emission coefficient for many solids is around unity, the Auger electron signal will be a peak on a background which contains 10^6 times as many electrons.The measurement problems that this causes are explained in Chapters 2 and 3. Current measurements[4-6] of $\sigma_{AX}(E)$ give a better agreement with Gryzinski's[7] cross-section, the energy dependence of which is shown in Figure 5.2 together with some experimental results, although the absolute value of the cross-section is still as low as that of Worthington and Tomlin.

In the above we have considered the ionization of level X in atoms in the surface zone due to the primary electron beam of energy E_p. There is additional ionization of the level due to the back-scattered energetic electrons shown in Figure 5.2 of Chapter 3. If the true back-scattered electron spectrum is $n(E)$ per unit incident electron, the total ionization of level X is given by:

$$\sigma_{AX}(E_p) + \int_{E_{AX}}^{E_p} \sigma_{AX}(E)n(E)\, dE \qquad (5.4)$$

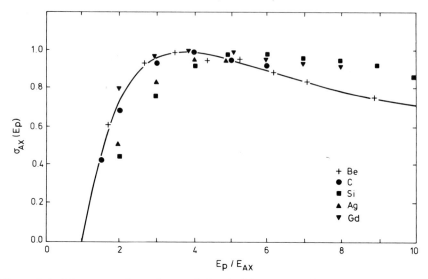

Figure 5.2   The energy dependence of Gryzinski's cross-section for ionizing level X, together with experimental measurements for Be,[4] C, Si, Ag and Gd[5]

which is, in turn, often written

$$\sigma_{AX}(E_p)[1 + r_M(E_{AX}, E_p, \alpha)] \tag{5.5}$$

where the back-scattering term r_M is dependent on the matrix M in which the A atoms are embedded and the angle α to the surface normal of the incident electron beam. Evaluations of the back-scattering term, r_M, have been made by several authors by Monte Carlo calculations[2,6,8] and empirically by Reuter.[9] The calculations are roughly in agreement for deep core levels but Reuter's relations errs on the low side for core levels below 500 eV. Shimizu and Ichimura's[6] calculations are most complete and so are included here. Figure 5.3, compiled from their work, shows how r_M increases with atomic number, Z, and reduces with the depth of the core level E_{AX} for a 5 keV electron beam incident at $\alpha = 30°$ to the surface normal. Figure 5.4(a) shows how r_M varies with Z and Figure 5.4(b) the variation with α. A final term giving additional ionization of the core level X arises from Coster–Kronig transitions. These are very rapid but occur only between shells of the same principal quantum number. Thus, for instance, a hole in the L_1 shell is filled by an electron from L_2 or L_3 and similarly L_1 can be filled from L_3. Thus, the more weakly bound levels have added ionization and give stronger Auger electron peaks.

We are now in a position to consider the total contribution to the Auger electron signal. The ionized core level X decays with a probability of Auger electron emission through the XYZ transition, γ_{XYZ}. The created Auger electron then has a probability e^{-1} of travelling a distance characterized by the

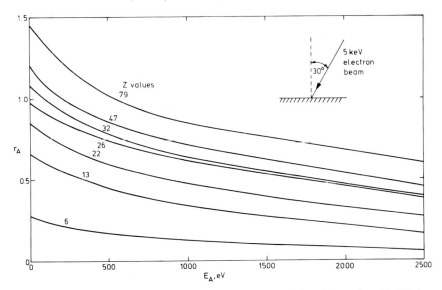

Figure 5.3 The back-scattering term, r_M, for a range of elements using 5 keV electrons at 30° from the surface normal. From the data of Shimizu and Ichimura[6]

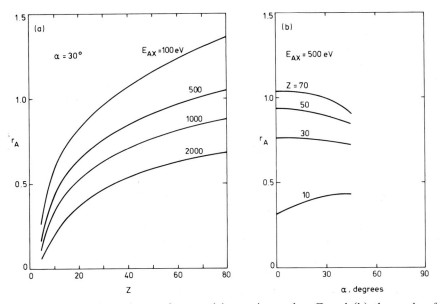

Figure 5.4 The dependence of r_M on (a) atomic number Z and (b) the angle of incidence of the electron beam α. (From the data of Shimizu and Ichimura[6])

inelastic mean free path in the matrix, $\lambda_M(E_{AXYZ})$, before being inelastically scattered and no longer appearing in the Auger electron peak. Thus the flux of Auger electrons decays as $\exp(-l/\lambda)$ as a function of the distance l from the point of origin. The characteristic depth from which Auger electrons can be emitted is simply $\lambda_M(E_{AXYZ}) \cos \theta$, where θ is the angle of emission to the surface normal. This combined term is often called the escape depth. Values of λ are in the range 2–10 atom layers, as shown in the compilation of experimental data shown in Figure 5.5.[10] It is this important phenomenon that makes both AES and XPS surface-sensitive techniques. The emitted Auger electron then is detected by an electron spectrometer with transmission efficiency, $T(E_{AXYZ})$, and an electron detector of efficiency, $D(E_{AXYZ})$. Thus the Auger electron current, I_{AXYZ}, may be written:

$$I_A = I_0 \sigma_A(E_p)[1 + r_M(E_A, \alpha)]T(E_A)D(E_A) \int_0^x N_A(z)\exp[-z/\lambda_M(E_A)\cos \theta] \, dz$$

$$(5.6)$$

where for clarity the XYZ subscript has been omitted, I_0 is the primary electron beam current and $N_A(z)$ is the A atom distribution with depth z into

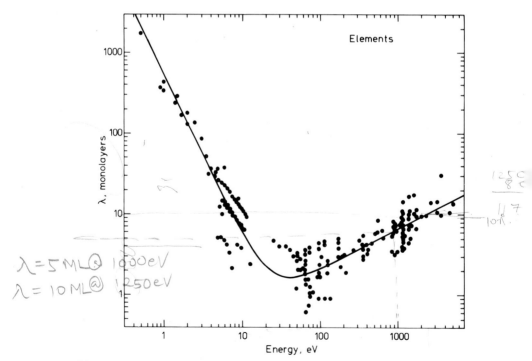

Figure 5.5 The dependence of λ on electron energy. (After Seah and Dench[10])

the sample surface. If the electron spectrometer has a large angular aperture, the current should be integrated over the appropriate solid angle.

At the present time no analyst uses equation (5.6) directly for quantification. For homogeneous binary systems, AB, the integral may be evaluated to be simply $N_A \lambda_M(E_A) \cos \theta$. We can reduce the number of unknowns further by considering the ratio of intensities I_A/I_B and that ratio, in turn, to the ratio for pure element standards, I_A^∞/I_B^∞, recorded on the same instrument. Thus

$$\frac{I_A/I_A^\infty}{I_B/I_B^\infty} = \frac{[1 + r_{AB}(E_A)]N_A \lambda_{AB}(E_A)[1 + r_B(E_B)]N_B^\infty \lambda_B(E_B)R_B^\infty}{[1 + r_{AB}(E_B)]N_B \lambda_{AB}(E_B)[1 + r_A(E_A)]N_A^\infty \lambda_A(E_A)R_A^\infty} \quad (5.7)$$

where R_A^∞ and R_B^∞ are the roughnesses of the pure element standards and N_A^∞ and N_B^∞ their atom densities. The ratio R_A^∞/R_B^∞ can be ignored if the ratio I_A^∞/I_B^∞ is established for many samples or from an *in situ* fracture surface of a homogeneous alloy of AB. Equation (5.7) may be simplified further by noting that

$$N_A^\infty = a_A^{-3} \quad \text{and} \quad N_A = a_{AB}^{-3} X_A \quad (5.8)$$

such that

$$\frac{N_A N_B^\infty}{N_B N_A^\infty} = \frac{X_A}{X_B}\left(\frac{a_A}{a_B}\right)^3$$

where X_A is the required molar fractional composition of the solid and a_M is the atom size of A derived from $1000 \rho_M N a_M^3 = A_M$ where, in turn, ρ_M is the density (in kilograms per cubic metre), N is Avogadro's number and A_M the mean atomic weight of the matrix atoms (of course, if the matrix is A this is just the atomic weight of A). From the work of Seah and Dench[10] the inelastic mean free path is given approximately by

$$\lambda_M = 0.41 a_M^{1.5} E_M^{0.5} \quad (5.9)$$

where λ_M and a_M are in nanometres and E_M is in electronvolts. Thus, combining equations (5.8) and (5.9) in equation (5.7) gives

$$\frac{X_A}{X_B} = F_{AB}^A \frac{I_A/I_A^\infty}{I_B/I_B^\infty} \quad (5.10)$$

where the Auger electron matrix factor F_{AB}^A is given by

$$F_{AB}^A(X_A \to 0) = \left[\frac{1 + r_A(E_A)}{1 + r_B(E_A)}\right]\left(\frac{a_B}{a_A}\right)^{1.5} \quad (5.11)$$

$$F_{AB}^A(X_A \to 1) = \left[\frac{1 + r_A(E_B)}{1 + r_B(E_B)}\right]\left(\frac{a_B}{a_A}\right)^{1.5} \quad (5.12)$$

From Figure 5.3 it can be seen that the ratio r_A/r_B is not very energy dependent and so F_{AB}^A varies little across the composition range from A to B and may usefully be considered as a constant for the AB system.

The above approach was originated by Hall and Morabito[11] who calculated the matrix terms for the two extremes of equations (5.11) and (5.12) for 4860 binary systems. Instead of using the Monte Carlo back-scattering calculations[6,8] they used Reuter's empirical[9] relation but, as an added alternative for Seah and Dench's[10] empirical inelastic mean free path assessments, also included Penn's[12] theoretical predictions. In this way they showed that in 86 per cent. of the binary systems:

$$F_{AB}^A(X_A \to 0) = F_{AB}^A(X_A \to 1) \tag{5.13}$$

to within 5 per cent. Hall and Morabito then noted that the distribution of the matrix factors centred about unity but that the scatter showed a width with a standard deviation factor of 1.5. If we ignore the matrix factors equation (5.10) reduces to equation (5.2). The magnitude of the error so produced is shown in Figure 5.6 which illustrates the relation between the true X_A from equation (5.10) and the reported X_A from equation (5.2) for various values of F_{AB}^A. It is surprising that many workers still use equation (5.2) when the evaluation of the matrix factors is apparently straightforward.

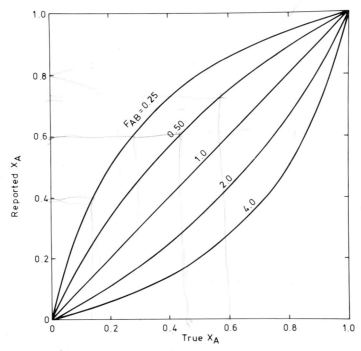

Figure 5.6   A comparison between the true composition and that deduced ignoring the matrix term

As an example, the matrix factor for carbon in iron would be

$$F_{CFe}^{A} = \left[\frac{1 + r_{C}(E_{C})}{1 + r_{Fe}(E_{C})}\right]\left(\frac{a_{Fe}}{a_{C}}\right)^{1.5}$$

$$= \frac{1.23}{1.81}\left(\frac{0.22752}{0.20696}\right)^{1.5}$$

$$= 0.783$$

and similarly for aluminium in nickel $F_{AlNi}^{A} = 0.70$ for the high-energy aluminium peak and 0.68 for the low-energy one. These matrix factors are tabled by Hall and Morabito as 0.997, 0.999 and 0.589, respectively. The differences between these two groups come not from the back-scattering terms but from the differences between the λ values from Seah and Dench[10] and from Penn.[12] It should be remembered that the experimental data shows a scatter factor of 1.3 about Seah and Dench's compilation and 1.4 about Penn's, and high accuracy is therefore not yet attainable in the matrix factors. Indeed, on average, Penn's λ values are 0.83 of Seah and Dench's and this ratio itself exhibits a scatter of 30 per cent.

In the above it has been assumed that the reference spectra have been recorded on the same instrument as that used for the analysis. If this is not the case the two terms $T(E_{A})$ and $D(E_{A})$ need consideration. It is fortunate that, today, most AES studies use spectrometers that are operated in a mode of constant $\Delta E/E$ where ΔE is the energy resolution of the spectrometer at energy E (see Chapter 2). This means that the transmission of most spectrometers, $T(E_{A})$, is simply proportional to E_{A} provided there are no stray magnetic fields. In this case the $T(E_{A})$ terms cancel out and may be ignored. The detector efficiency, however, is not constant between instruments. As shown by Davis *et al.*[1] the electron multiplier gain increases with E_{A} from zero to a peak around 300 eV and then declines slowly. This effect may be largely overcome by biasing the front of the multiplier to 150 V so that it sees no electrons of low energy. This reduces the errors due to $D(E_{A})$ variation to below 5 per cent. over the 2000 eV energy range.

So far we have avoided any discussion of what constitutes I_{A}. In fact, the discussion is valid for any measure of the Auger electron intensity provided that it can be used consistently. This causes a problem, as discussed below.

5.2.2 The measurement of intensity

In the past it has been customary to measure the peak-to-peak value of the signal in the derivative energy spectrum, as discussed in Chapter 3. This is perfectly valid and, within three main restrictions, these values may be used for I_{A} and I_{A}^{x}. These restrictions are that the peak shapes are the same in the analysis and reference spectra, that the analysers used for both spectra have

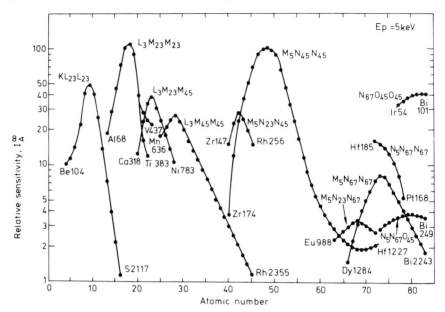

Figure 5.7 Relative sensitivity factors for derivative spectra in AES compiled from Davis *et al.*[1] The numbers at the ends of the curves indicate the energies of the peaks and the elements at these points

the same resolution and that, in both cases, the same modulation is used. For the analysis of metallic alloys with reference data taken on the same instrument this is fairly readily achieved. One can then use, for instance, for spectrometers with a resolution of 0.6 per cent. the *Handbook of Auger Electron Spectroscopy*[1] which compiles I_A^{∞} data for 3, 5 and 10 keV electron beam energies. An example of a set of relative sensitivity factors, I_A^{∞}, are given in Figure 5.7.

If now we relax two constraints, the analyser resolution and the modulation, what corrections must be invoked? We should understand these corrections since the effects of stray magnetic fields[13] and misplacement of the sample[14] affect the energy resolution of the CMA very strongly and, even though we may be using the same model analyser as in the Physical Electronics' *Handbook*,[1] we may not achieve the same resolution. Also, the precise modulation is not given in that *Handbook* although the effects of modulation are clear in McGuire's *Auger Electron Spectroscopy Reference Manual.*[15] For singlet peaks the effects of modulation voltage, V, may be described by the universal curve shown in Figure 5.8. Here it is assumed that the peak is asymmetric and that the negative wing may be described by a half width, $\frac{1}{2}W$, as shown. The positive wing will generally have a different value of W, and so only the negative wing will be considered and the intensity measured is the

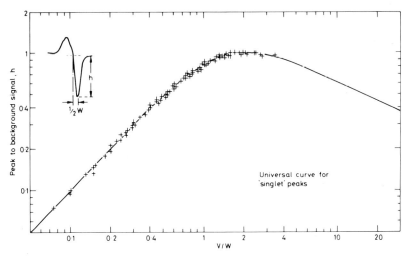

Figure 5.8 Universal curve showing the change in intensity of singlet peaks with modulation voltage, *V*, per unit Auger electron current. (After Anthony and Seah[16])

peak-to-background height, h. Thus, Figure 5.8 allows us to convert from one modulation to another for singlet peaks, providing the width W is known. For general analysis with 5 V peak-to-peak modulation and with a spectrometer of 0.5 per cent resolution, a scan from 100 to 2000 eV covers peaks that are well saturated below 1 keV to peaks that are on the linear portion in Figure 5.8 near 2000 eV.

If now we convert the data from one spectrometer to another with a different resolution, the intrinsic widths of the peaks change and their relative intensities also change. Figure 5.9 shows two spectra of the same surface using spectrometers of 0.3 and 1.2 per cent. resolution, respectively. It is clear that as the analyser resolution deteriorates the sharper peaks become attenuated more strongly than the broad peaks and that the main intensity appears to switch from copper to carbon. This change is all included in the universal curve of Figure 5.8 The copper peak at 920 eV moves from $V/W = 1.6$ to $V/W = 0.54$, losing a factor of 1.85 in hW and therefore 5.5 in h, whereas the carbon moves from $V/W = 0.71$ to 0.65 causing an intensity loss of only 1.18. Thus the relative change of the copper and carbon is 4.6, comparing very closely with the measured 4.3 from Figure 5.9.

A further word or two may clarify the use of Figure 5.8. If we consider any particular peak in the $n(E)$ or $En(E)$ spectra and double its width W at unit Auger electron current, it is clear that h will reduce a factor of 4, i.e. hW^2 is constant for small modulations V and hence hW is proportional to V/W, as is seen for low modulations in Figure 5.8. Consider now the effect of Figure 5.8 on the relative intensities of two Auger electron peaks at 500 and 1000 eV in

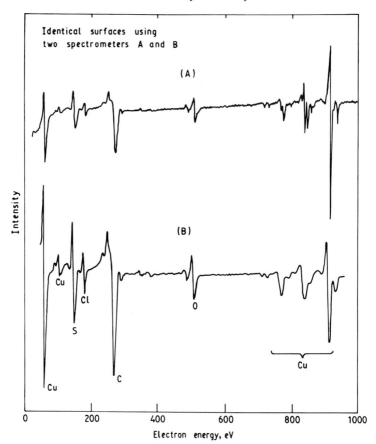

Figure 5.9 Auger electron spectrum from contaminated copper using spectrometer A at resolution 0.3 per cent and B at 1.2 per cent. (After Anthony and Seah[16])

a CMA of resolution 0.5 per cent. when the specimen is mis-positioned by 0.4 mm. The width of intense peaks will generally be governed by this resolution. This level of mis-positioning can easily occur in studying rough or fracture surfaces or where the analyst is unaware of the implications. Sickafus and Holloway[14] show that this shift could broaden the resolution to 0.6 per cent., causing the W values for the two peaks to change from, say, 2.9 and 5.2 eV to 3.4 and 6.2 eV. For 2 V peak-to-peak modulation the relative intensities of the two peaks would remain unaltered; however, at 5 V the 500 eV peak shifts from $V/W = 1.7$ to 1.5 whereas the 1000 eV peak shifts from 0.96 to 0.81, causing a 9 per cent relative loss in intensity of the 1000 eV peak with respect to that at 500 eV.

One way of reducing some of the above problems in the future would be to use a ramped modulation such that V/W is kept at, say, 1.5 or 2 through the spectrum. In this case sample misalignment would be less important. For a 0.5 per cent. resolution analyser the modulation would then be 0.75 or 1 per cent. of the energy with a lower limit of 2 V since the intrinsic width of most peaks is 1 to 1.5 eV.

A final aspect that must be considered is the effect of peak shape changes due to the changes in chemical environment.[17-19] Figure 5.10 shows such shape changes for silicon in both the derivative and direct energy spectra Additional examples can be found in Chapter 3. These shape changes are very difficult to handle and in this case, for quantification, the high-energy peak (1606–1619 eV) would be used, at which energy the analyser resolution would reduce the peaks to a similar shape. Alternatively, appropriate sensitivity factors must be established for each peak shape.

Figure 5.10 introduces the use of the direct energy spectrum, $n(E)$ or $En(E)$. It is clear that the Auger electron signal, I_A, should be related to the

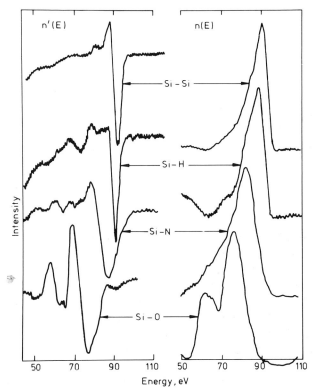

Figure 5.10 Auger electron peaks for Si showing the changes in structure and a progressive shift to lower energies with increasing electronegativity in the bond. (After Madden[17])

area under the appropriate peak in the $n(E)$ spectrum. In Figure 5.10 this appears to be straightforward and allows us to include, naturally, the lower energy peak observed in SiO_2. The problems of modulation voltage and analyser resolution now disappear. Historically the derivative spectrum is used, as discussed in Chapter 3, as the Auger electron peaks are small and often superimposed on a sloping background in the direct energy spectrum. This means that the slope is amplified along with the peak and it becomes difficult to record the peaks. With the differential, however, the background is removed and so recording and measuring peaks is very easy. Today, with the increasing use of computers to record and process spectra, the problem of background removal and the display of the Auger electron peaks in the direct spectrum can be removed and considerable efforts are under way to develop suitable background subtraction routines. Figure 5.11 gives an example of both derivative and direct spectra for a contaminated copper sample. Firstly, it should be noted that the relative sensitivity factors from the derivative spectra cannot be used in the direct spectra. This is evident since carbon and silver have similar sized peaks in the differential spectrum of Figure 5.11 but widely different areas in the direct spectrum.

The first method of constructing a background removal routine involved the use of spline polynomials.[20] It became clear early on that this method did not give a unique area to a peak but that the result depended strongly on the boundary conditions. For instance, in Figure 5.11 a smooth background drawn under the oxygen, silver and carbon peaks would give oxygen contributions over a range of 150 eV and would include approximately twice as many inelastically scattered oxygen Auger electrons as those in the peak itself. The oxygen intensity would therefore be characteristic of many times the inelastic mean free path depth, whereas the silver, summed over the peak width alone, would be characteristic of the usual depth associated with the differential spectrum. These shortcomings of background subtraction routines are discussed at length elsewhere;[21] however, it is important to be able to use the direct spectrum because measurements in high spatial resolution Auger microscopes, where the beam currents are low and counting techniques are used, may not be available through modulation techniques. An elegant analysis of the source of the background has been made by Sickafus.[22,23] He shows that the background on the low-energy side of the peak is comprised of a step due to inelastic scattering of the Auger electrons from surface atoms in the bulk, plus a rising contribution due to scattered Auger electrons originating within the bulk. This step is shown clearly in Figure 5.12 for Auger electron peaks recorded during XPS measurements of copper. This background problem is, in fact, exactly the same as that in XPS to be described in Section 5.3.2, and it seems highly likely that the solution used there would be applicable in AES; i.e. to use the background devised by Shirley,[24] which rises in proportion to the area of peak at higher energies above the background. This is described

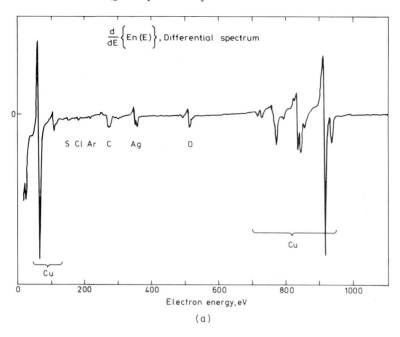

(a)

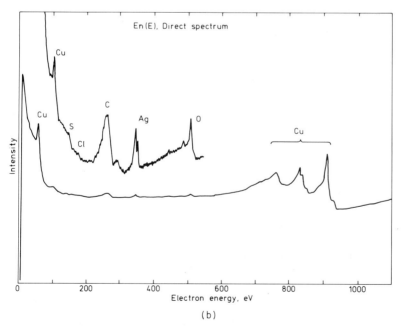

(b)

Figure 5.11   Differential and direct electron spectra from contaminated copper

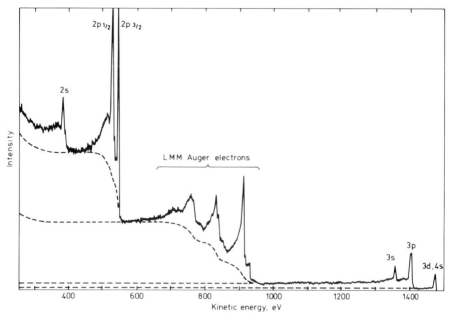

Figure 5.12 XPS spectrum from copper showing the stepped background and Auger electron peaks. (After Seah[21])

more fully in Section 5.3.2 and, at the present time, has the most promise for quantitative measurements in the direct spectrum in AES.

A slightly different solution, which does not use relative sensitivity factors but which uses libraries of standard spectra, has been proposed.[25] In this method appropriate proportions of different standard spectra are subtracted out of the original spectrum until all the peaks are accounted. This method effectively gives I_A/I_A^∞ values which may be used in the equations for quantification and has the advantage of highlighting any differences in spectra due to chemical changes or the presence of small quantities in peak overlap regions. Unfortunately, at the present time, libraries of direct spectra are not available.

In the above we have considered the relationships and measurements to quantify the AES of homogeneous materials. In the next section the analogous problems for XPS are treated. The problems in quantifying adsorbates and sputtered composition depth profiles are then presented in a unified format for both AES and XPS.

5.3 Quantification of XPS for Homogeneous Binary Solids

The analysis here parallels that in Section 5.2.

5.3.1 Basic considerations

The analogous picture to Figure 5.1 for XPS shows the characteristic X-rays penetrating the solid to a depth of many micrometers and ionizing atoms in the core level X over this depth. The ejected core electrons move through the solid and are finally ejected from the surface with some energy loss. Only those created in a zone characterized by a depth $\lambda_M(E_{AX}) \cos \theta$ from the surface escape to provide the line intensity in the XPS spectrum. As before, $\lambda_M(E_{AX})$ is the inelastic mean free path of the characteristic XPS electrons of energy E_{AX} from level X of element A in matrix M, and θ is the angle of emission of the electron from the surface normal. The E_{AX} dependence of λ is shown in Figure 5.5.

In an analogous manner to equation (5.6) we may write

$$I_A = \sigma_A(h\nu)D(E_A) \int_{\gamma=0}^{\pi} \int_{\phi=0}^{2\pi} L_A(\gamma) \int_{y=-\infty}^{\infty} \int_{x=-\infty}^{\infty} J_0(xy)T(xy\gamma\phi E_A) \int_{z=0}^{\infty} N_A(xyz)$$

$$\times \exp[-z/\lambda_M(E_A)\cos\theta] \, dz \, dx \, dy \, d\phi \, d\gamma \qquad (5.14)$$

where the subscript X for the level X has been omitted. Here $\sigma_A(h\nu)$ is the cross-section for emission of a photo-electron from the relevant inner shell per atom of A by a photon of energy $h\nu$, $D(E_A)$ is the detection efficiency for each electron transmitted by the electron spectrometer, γ, ϕ, x, y and z are given in Figure 5.13, $L_A(\gamma)$ is the angular asymmetry of the intensity of the

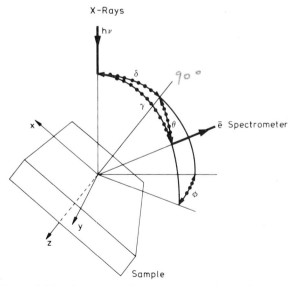

Figure 5.13   Geometry of the XPS analysis configuration

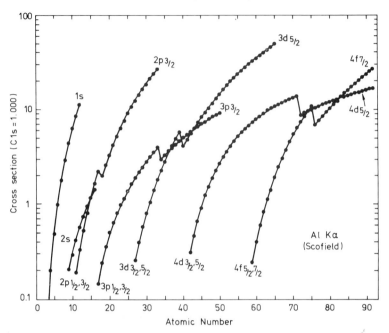

Figure 5.14   Calculated values of the cross-section $\sigma_A(h\upsilon)$ for Al $K\alpha$ radiation in
terms of the C $1s$ cross-section. (After Scofield[26])

photo-emission from each atom, $J_0(xy)$ is the X-ray characteristic line flux
intensity at point (x,y) on the sample, $T(xy\ \gamma\phi\ E_A)$ is the analyser transmission
and $N_A(xyz)$ is the atom density of the A atoms at (xyz). Unlike AES, a large
area of the sample is illuminated by the X-rays, there is no back-scattering
correction, but there is an angular anisotropy of the emission. At this stage,
although it is not needed yet, Figure 5.14 shows the calculated cross-sections
$\sigma_A(h\upsilon)^{26}$ for many useful core lines using Al $K\alpha$ radiation.

For a homogeneous material the above integral over z becomes simply
$N_A\lambda_M(E_A)\cos\theta$. If we have reference spectra of the pure elements I_A^∞ recorded
on the same instrument, equation (5.14) becomes much simplified by writing

$$\frac{I_A/I_A^\infty}{I_B/I_B^\infty} = \left[\frac{\lambda_{AB}(E_A)\lambda_B(E_B)}{\lambda_{AB}(E_B)\lambda_A(E_A)}\right]\left(\frac{R_B^\infty}{R_A^\infty}\right)\left(\frac{N_A N_B^\infty}{N_A^\infty N_B}\right) \qquad (5.15)$$

where λ_{AB} is the inelastic mean free path in the alloy, R_A^∞ and R_B^∞ are the
factors for the pure reference samples which define the intensity emitted from
the surface as a function of its roughness, and N_A^∞ and N_B^∞ are the atom
densities in the pure reference samples. Using the simplifications given be-

tween equations (5.7) and (5.8) we get

$$\frac{X_A}{X_B} = F_{AB}^X \frac{I_A/I_A^\infty}{I_B/I_B^\infty} \tag{5.16}$$

where

$$F_{AB}^X(X_A \to 0) = F_{AB}^X(X_A \to 1) = \left(\frac{a_B}{a_A}\right)^{3/2} \tag{5.17}$$

F_{AB}^X here is the exact counterpart in XPS to the matrix factors of Hall and Morabito in AES.[11] If F_{AB}^X is unity, equation (5.16) reduces to the simple equation (5.2); however, analysis of 8000 element pairs shows that the F_{AB}^X values scatter about unity with a standard deviation factor of 1.52.[27] The effect of using the full equation (5.16) in place of the simple equation (5.2) is shown in Figure 5.6, presented earlier for the AES case.

As an example of the values of F_{AB}^X we may use the previous systems of carbon in iron and aluminium in nickel. Thus

$$F_{CFe}^X = 1.153 \quad \text{and} \quad F_{AlNi}^X = 0.812$$

quite different from the values for AES.

In the above we have assumed that the reference data can be recorded on the same instrument. Because this takes a considerable effort and is open to all sorts of errors, as discussed later, it is simpler to use published sets of I_A^∞ data, of which there are many.[28-41] Later in the chapter we shall consider the relative merits of these sets but first let us consider the equivalent of equation (5.15) where the analysis is from instrument (1) and the reference data is from instrument (2). From equation (5.14), if the electron spectrometer has a small entrance aperture and if the sample is uniformly illuminated:

$$I_A = \sigma_A(h\nu)D_1(E_A)L_A(\gamma_1)J_0(1)N_A\lambda_M(E_A)\cos\theta_1 \int_{y=-\infty}^{\infty} \int_{x=-\infty}^{\infty} T_1(xyE_A)\,dx\,dy \tag{5.18}$$

The last term, the product of area analysed and the analyser transmission, is known as the analyser étendue,[42] $G_1(E_A)$. Thus the analogue of equation (5.16) is

$$\frac{I_A/I_A^\infty}{I_B/I_B^\infty} = \frac{L_A(\gamma_1)G_1(E_A)D_1(E_A)L_B(\gamma_2)G_2(E_B)D_2(E_B)}{L_B(\gamma_1)G_1(E_B)D_1(E_B)L_A(\gamma_2)G_2(E_A)D_2(E_A)} = \frac{X_A}{F_{AB}^X X_B} \tag{5.19}$$

As discussed in Chapter 2, in most XPS work the electron spectrometer is used in the constant ΔE mode. With proper design the electrons then strike the multiplier at the analyser pass energy and the detector efficiency, $D(E_A)$, is constant through the spectrum. The D terms will therefore be ignored.

The angular asymmetry factor $L(\gamma)$, which describes the intensity distribution of the photo-electron ejected by unpolarized X-rays from atoms or molecules, is given by[43]

$$L_A(\gamma) = 1 + \tfrac{1}{2}\beta_A(\tfrac{3}{2}\sin^2\gamma - 1) \tag{5.20}$$

where β_A is a constant for a given subshell of a given atom and X-ray photon. This takes no account of the diffraction effects that the electrons may undergo in being transported through a crystalline lattice. For large crystals the diffracted beams may be of some $10°$ FWHM with first-order intensity changes.[44-46] However, for finely polycrystalline or amorphous solids the diffraction effects may be ignored.

Calculated values of β are tabulated by Reilman, Msezane and Manson[43] for the important transitions of all elements using the commonly used X-ray sources (Al $K\alpha$, 1486.6 eV; Mg $K\alpha$, 1253.6 eV; and Zr $M\xi$, 151.4 eV). These calculations give a very good correlation with the data for gases,[47] but measurement problems have precluded their demonstration in solids.[48] Baschenko and Nefedov[49] argue that, for solids, the anistropy of the distribution described by equation (5.20) may be somewhat attenuated by the inclusion of elastic scattering effects which randomize the electron directions. Although this effect occurs, Baschenko and Nefedov's calculations use a

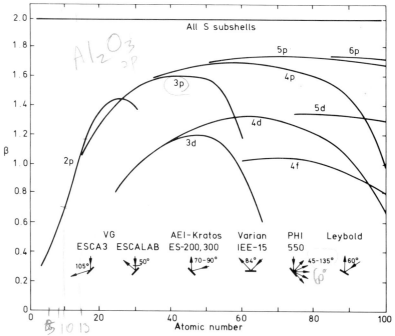

Figure 5.15   Calculated values of β. (After Reilman, Msezane and Manson.[43]) Typical commercial configurations are shown in the lower part of the diagram

Thomas–Fermi cross-section which gives over two orders of magnitude higher elastic scattering than is observed experimentally,[10] and so the effect of the randomization will probably be small. Figure 5.15 shows the β values for Al $K\alpha$ radiation. At the bottom of Figure 5.15 are shown some typical commercial arrangements but it is important to note that the geometry, even for a given manufacturer's model number, may be different from that shown since the manufacturers move the X-ray sources around to optimize the experimental configuation. Typical values of β may be 0.8 or 1.8. Figure 5.16 shows the angular dependence of $L_A(\gamma)$ for these values of β with some of the commercial geometries noted. The entrance angle about the directions noted is typically less than a few degrees except for the instruments incorporating a cylindrical mirror analyser (CMA) where the angle extends from $45°$ to $135°$, depending on the experimental set-up. Clearly, reference data taken on an instrument with one geometry will show different relative peak intensities from those recorded on an instrument with different geometry. This term explains precisely the results obtained for pure gold by Gettins and Coad,[50] where the $4p$ ($\beta = 1.63$) peak was found to be 10 per cent. stronger in the VG ESCA3 than in the PHI CMA when normalized against the $4f(\beta = 1.03$ peak. Thus, the angular asymmetry term is fairly straightforward and only the étendue, $G(E_A)$, remains.

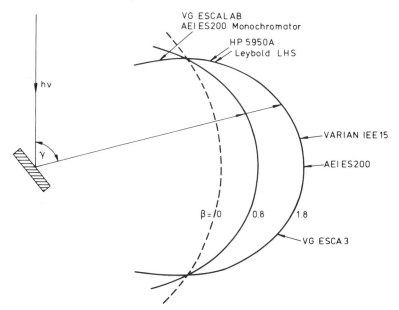

Figure 5.16 The angular intensity distributions for $\beta = 0$, 0.8 and 1.8 as a function of the relative orientation of the X-ray source and the spectrometer. General commercial configurations are shown on the right

Calculations of the étendue are reasonably straightforward in some cases but it is not always possible to evaluate the effects of off-axis electron trajectories, specimen size and roughness and of the non-uniform illumination of the sample and the aberrations of the electron optics. Because of the individual's requirements particular machines may have slightly different electron optics. Thus, it is important to have an experimental measure of $G(E_A)$. In 1978 the E-42 committee of the ASTM initiated a round robin to do this. The analysis,[51] based on the relative intensities by peak area for the XPS peaks for copper and gold, shows poor consistency with a 25 per cent. standard deviation over the energy range 300–1100 eV for one instrument, considerably worse for a second and slightly better for a third. It is not known if the scatter is an operator or an instrument effect, and all instruments systematically deviate from the expected function. A more recent brief analysis at the NPL shows that

$$G(E_A) \propto E_A^{-1} \qquad (5.21)$$

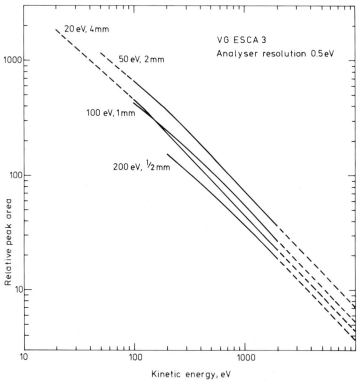

Figure 5.17   The transmitted peak area intensity as a function of kinetic energy for various slit widths and pass energies giving an analyser resolution of 0.5 eV in the VG Scientific ESCA 3. (After Seah[27])

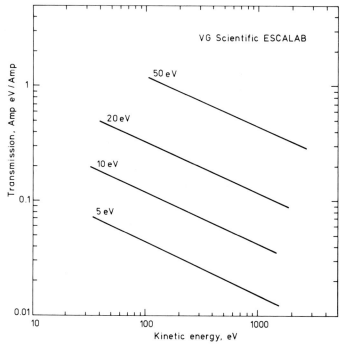

Figure 5.18   As for Figure 5.17 but for the VG Scientific ESCALAB. (Reproduced from Hughes and Phillips[53] by permission of John Wiley & Sons Ltd)

is both predicted and confirmed to within 3 per cent. over the energy range 500–1500 eV for both the VG Scientific ESCA3 and the PHI Model 550 ESCA/SAM. This simple relation is thought to be valid for the Varian IEE-15[40] and the Leybold LHS 10[52]—all when used in the constant ΔE mode. Figure 5.17 shows the calculated transmission function[27] for the VG Scientific ESCA3 for large samples and an analyser resolution of 0.5 eV. In the constant $\Delta E/E$ mode, of course, all the above spectrometers give $G(E_A) \propto E_A$.

Exceptions to the above rule of relation (5.21) are the VG Scientific ESCALAB which has a less energy-dependent function,[27,53] as shown in Figure 5.18 and the AEI ES 200/300 series spectrometers. Calculations and experiments are currently under way to establish a reference methodology for determining the combined $G(E_A)D(E_A)$ term of these and other spectrometers. As with the case for AES the measurement of I_A is not trivial, and in the next section this is considered together with reference data banks for I_A^∞.

5.3.2   The measurement of intensity and reference data banks

As discussed in Chapter 3, the XPS peaks appear on a small background and so the direct spectra are used and I_A is simply the area under the appropriate peak. For some peaks this is easy to measure; however, with others shake-up, shake-off and multiplet splitting can lead to features appearing over a wide energy range, so that the measurement of the peak area involves some decisions about the precise position of the background to use. In the literature two methods have been used to define the background. The most popular method, and the one available in most commercial data handling systems, is the straight line drawn to two suitably chosen points, such as P and Q in Figure 5.19. An alternative to this method is that of Shirley[24] in which the background intensity at a point is determined, by an interative analysis, to be proportional to the intensity of the total peak area above the background and to higher energy. This method merely assumes that each electron in the characteristic part of the peak is associated with a flat background of losses. Shirley's method gives the curved background in Figure 5.19. Again end points must be chosen as in figure 5.19 but the precise position of the end points can be relaxed. For instance, Q could shift to a lower BE by 20 eV, thus including the $K\alpha_{34}$ satellites (discussed in Chapter 3) and increasing the area for both copper and CuO using Shirley's method. However, the peak area would be reduced using the straight line method. The symmetrical problem occurs with P.

The situation becomes even worse for the straight line background when considering a mixture of two states, as shown in Figure 5.20. Figure 5.20(a)

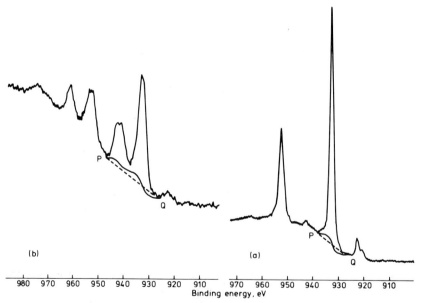

Figure 5.19   Copper photo-electron peaks from (a) metallic copper and (b) CuO powder showing background subtraction methods. (After Seah[27])

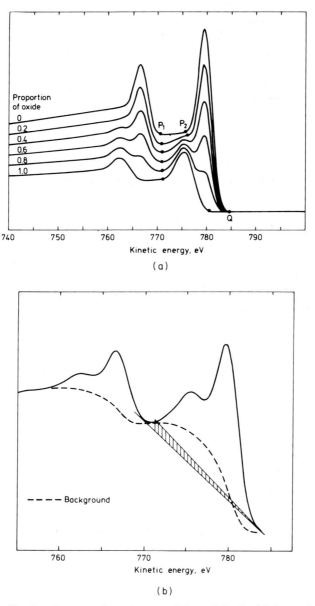

Figure 5.20 Simulated spectra for mixtures of Fe and Fe_2O_3 (a) for various proportions and (b) showing the background subtraction at 50/50 Fe/Fe_2O_3. (Reproduced from Bishop[54] by permission of John Wiley & Sons Ltd)

shows spectra for different combinations of iron and Fe_2O_3 and Figure 5.20(b), the two background subtractions for the 50/50 case. If we consider the 20 per cent. oxide curve shown in Figure 5.20(a) the point Q is easily established but there are two alternatives for P. For Shirley's background P_1 gives the best area and P_2 a slightly smaller one. However, using the straight line, Bishop[24] shows that P_1 overestimates the total area by about 33 per cent. and P_2 underestimates by 24 per cent. As shown in Figure 5.20(b) Bishop locates P close to the peaks. It could be argued that some of the tail to low kinetic energies is due to shake-up and shake-off electrons so that P should be 50 eV or so to lower kinetic energies.[55] Bishop removes this tail by matching Shirley's background slope to the tail as follows. If the signal in the ith channel, s_i, is the sum of a background and peak contributions b_i and p_i, respectively, then Shirley's method gives

$$b_i = k \sum_{j=i+1}^{N} p_j \qquad (5.22)$$

where k is a constant. Bishop weakens the contribution as the losses get further removed from the peak using

$$b_i = k \sum_{j=i+1}^{N} p_j[1 - m(j - i + 1)] \qquad (5.23)$$

where the fitting parameter is adjusted to give the correct slope at low kinetic energies. At the present time it is not clear if the fitting should be as described by Bishop, which would exclude intrinsic events of reasonable energy loss but which, however, should be included, or as suggested by Berresheim,[55] which would include extrinsic characteristic losses that should be excluded, to obtain the best results.

A further option arises in the use of Shirley's relation. In its basic form it assumes that the characteristic loss spectrum from each electron is a flat background at lower energies. This seems a reasonable approximation for metals as extrinsic events of small loss can occur by exciting electrons from just below to just above the Fermi level. However, for insulators this is no longer possible and there should be a clearly observed energy interval after the peak, before the background rises on the low-energy side, proportional to the band gap. A cursory look through the *Handbook of X-ray Photoelectron Spectroscopy*[41] in which wide-scan spectra are given for LiF, BN, polyethylene, Al_2O_3, KBr, $CaCO_3$, Sc_2O_3, SrF_2, Y_2O_3, LiI, CsOH, BaO, etc., shows that, in all cases, the background rise occurs some 10 eV or so below the highest energy peak of any group, as expected. Figure 5.21(a) extracts the peaks in SrF_2 in which this effect is observed. Both the F 1s and Sr 3d peaks clearly show the gap before the background rise, but the Sr $3p_1$ peak masks the effect for the Sr $3p_{3/2}$ peak. In these cases the straight line background will be valid unless there are additional peaks filling the gap, when the back-ground may be described by

$$b_i = k \sum_{j=i+n}^{N} p_j \qquad (5.24)$$

where n is the number of channels describing the gap.

The three different types of peak, using the same spectrometer, appropriate to these modified Shirley backgrounds are shown in Figure 5.21(b). From the above it can be seen that there is not yet a recommended universal method of defining I_A. However, the practical situation is not really all that bad since many compounds are insulators and the straight line background works well. Those that are totally metallic involve only small changes in peak shape so that any measure of the area, consistently used, will give reasonable results between metallic systems. The measure must, of course, be consistent between I_A and I_A^x. However, any method consistently used may not always generate I_A values fully consistent between both metals and insulators.

Let us now consider the reference data banks to use for I_A^x. Many sets of data are available in the literature[28-41] and it is difficult for the reader to decide which to use. They involve three basically separate approaches:

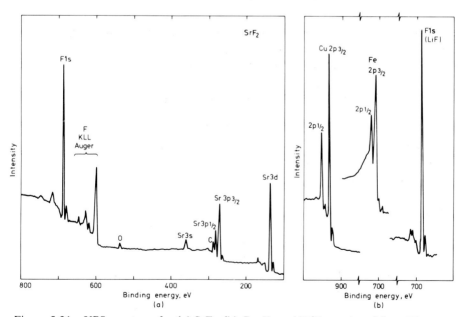

Figure 5.21 XPS spectrum for (a) SrF_2, (b) Cu, Fe and F (Reproduced from Wagner *et al.*[41] by permission of Perkin–Elmer Corporation) showing the classic Shirley step, Bishop's modification and the shifted step, respectively

measurements from standard known reference compounds in powder form,[28-32,35-38,40] measurements from elemental foils[33,35,39] and calculation.[41] The calculations are based on equation (5.14) which may be written, for a spectrometer with uniform illumination of a homogeneous target and an analyser transmission given by relation (5.21), as

$$I_A^\infty = \frac{B\sigma_A(h\nu)L_A(\gamma)N_A^\infty\lambda_M(E_A)}{E_A} \qquad (5.25)$$

where the proportionality constant B may be removed by referring the I_A values to one particular peak, says the F ls, with I_μ defined as unity. Wagner *et al.*'s[41] calculations used Scofield's calculated values[26] of $\sigma_A(h\nu)$ and Penn's[12] calculated values of $\lambda_M(E_A)$. This approach to quantification, without the use of measured I_A^∞ values from reference samples, is also used by Ebel[56,57] and by Hirokawa and Oku,[58,59] who claim a good correlation between the measured composition of alloys and their known bulk composition.

The above approach is easy to use since the I_A^∞ values may be easily calculated for all levels of all atoms from literature values of σ, λ, etc. However, the accuracy is only as good as the accuracy of these values and the accuracy with which the peak area measured in the analysis reflects all of the electrons contributing to $\sigma_A(h\nu)$. In general, measurement of the peak area over a narrow energy range, as is customary, omits that intensity which should be included but which is described as shake-up events in Chapter 3. These events occur when a conduction or valence band electron is given energy simultaneously with the initial photo-electron production so that the latter electron appears with reduced energy. The amount of intensity lost by shake-up depends on the element and the core level involved and may also depend on the chemical environment, especially for the transition metals.[35] In the $3d$ transition elements the shake-up is stronger for the $2p$ lines than the $3p$ and is associated with the unfilled d-shell since copper and zinc have the weakest effects. Calculations for the rare gases[60] show that the total shake-off probability falls from around 18 per cent. at 2 keV binding energy to 16 per cent. at 1 keV, 14 per cent. at 400 eV, 12 per cent. at 200 eV and down rapidly to 5 per cent. around 25 eV. For metals, values as high as 15–30 per cent. have been suggested.[61] Thus, calculated I_A^∞ values may not be very accurate. One can gain an estimation of their accuracy by comparing the calculated and experimental I_A^∞ values. So, let us now consider the experimental reference data sets.

A detailed analysis of these sets has been made[27] and only a few of the sets are consistent. The sets have been compared by considering their values divided by the theoretical values. All of the sets except those of Wagner,[28,35,40] Nefedov[30,31] and Evans[36-38] diverge considerably. However, these three sets agree with a scatter of only 8 per cent. as shown in Figure 5.22, for the thirty-three peaks in common. The experimental values, it should be noted,

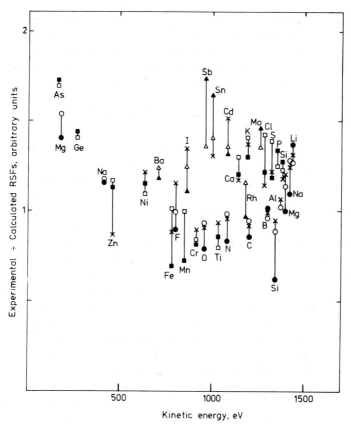

Figure 5.22 The ratio of the experimental relative sensitivity factors of Wagner (filled symbols), Nefedov (open symbols) and Evans (crosses), corrected for contamination overlayer, to that by calculation

are all for powdered samples freshly prepared just prior to insertion in the XPS instrument but all of which exhibited carbon contamination. Thus the measured intensities are all reduced by the factor $\exp[-t/\lambda_C(E_A)\cos\theta]$ where the carbon thickness is typically 0.75 nm. The values in Figure 5.22 have been corrected for this average contamination.

The systemmatic divergence of the experimental data sets from the theoretical predictions shows that the use of the common peak area measurement techniques will not give as good a quantification using calculated I_A^x values as can be obtained using one of the three experimental data sets shown in Figure 5.22. The largest data set, that of Wagner, is given in Appendix 5. A detailed analysis of relative peak intensities from many of the standard compounds by Wagner[35,62] shows that the reproducibility in quantification is of

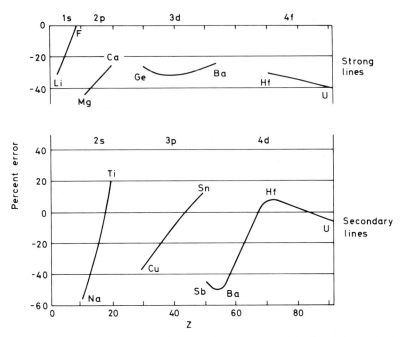

Figure 5.23 Apparent differences (per cent) between empirical and theoretical relative sensitivity factors, normalized to F 1s. (Reproduced from Wagner *et al.*[40] by permission of John Wiley & Sons Ltd)

the order of 10 per cent. and that quite large divergencies from the theoretical intensities are seen for most peaks, as shown by Figure 5.23, from Wagner *et al.*[40] This divergence need not necessarily imply an error in Schofield's cross-sections but merely that they do not reflect the experimental measurement. Recent studies by Berresheim[55] show that a better correlation can be obtained if the peak area is established with the low energy end point for Shirley's background set at the minimum in the background on the low-energy side of the peak or peak pair, even if this is 100 eV removed from the peaks. Berresheim's approach may prove impractical in general but appears to confirm the theoretical predictions.

In the above we have so far discussed how to measure the intensities in AES and XPS and how to relate this to the quantification of homogeneous binary solids. In practice there are two main uses of both techniques: the determination of adsorbate coverages or surfaces with the overlayers and the analysis of composition–depth profiles with *in situ* argon ion sputtering. These problems may now be analysed in relation to the above discussion for homogeneous materials.

5.4   Quantification of Samples with Thin Overlayers

The thin layers under consideration may be adsorbates which require analysis or a thin carbon contamination whose effect needs to be allowed for in quantifying the surface under the contamination. In general in AES work, many of the studies involve quantifying one or more adsorbates on an otherwise clean surface of one component.[63,64] Similar studies occur in XPS but a far greater problem, involving precisely the same concepts, is the analysis of a surface with a carbon or hydrocarbon contamination layer.

Following the concept discussed in Section 5.2.1, the signal of the substrate B covered by a fractional monolayer ϕ_A of A is simply given by the sum of unattenuated emission from $(1 - \phi_A)$ of the surface and an attenuated part from ϕ_A:

$$I_B = I_B^\infty \{1 - \phi_A + \phi_A \exp[-a_A/\lambda_A(E_B)\cos\theta]\} \tag{5.26}$$

and if covered by a thickness d_A of the overlayer of A:

$$I_B = I_B^\infty \exp[-d_A/\lambda_A(E_B)\cos\theta] \tag{5.27}$$

Here we assume that the correct measure for I_A is being used and that I_A^∞ is measured on the same instrument and settings. If not, the corrections already discussed must be applied. The signal from the overlayer in each of the cases is, respectively,

$$I_A = \phi_A I_A^\infty \left[\frac{1 + r_B(E_A)}{1 + r_A(E_A)}\right]\{1 - \exp[-a_A/\lambda_A(E_A)\cos\theta]\} \tag{5.28}$$

$$I_A = I_A^\infty \left[\frac{1 + r_B(E_A)}{1 + r_A(E_A)}\right]\{1 - \exp[-d_A/\lambda_A(E_A)\cos\theta]\} \tag{5.29}$$

where the back-scattering terms r_A and r_B are, of course, zero for intensities measured by XPS.

Although above we notionally take I_A^∞ as a tabulated relative sensitivity factor it must be remembered that, because of sample roughness and all the other changeable settings, it is really only possible to work with ratios I_A^∞/I_B^∞, etc. Thus from equations (5.26) and (5.28) the fractional monolayer coverage, ϕ_A, is given by

$$\frac{\phi_A\{1 - \exp[-a_A/\lambda_A(E_A)\cos\theta]\}}{1 - \phi_A\{1 - \exp[-a_A/\lambda_A(E_B)\cos\theta]\}} = \left[\frac{1 + r_A(E_A)}{1 + r_B(E_A)}\right]\frac{I_A/I_A^\infty}{I_B/I_B^\infty} \tag{5.30}$$

If the Auger or photo-electron peaks are at high energy and if ϕ_A is small, the simple relation below, analogous to equation (5.10), is obtained:

$$\phi_A = Q_{AB}\frac{I_A/I_A^\infty}{I_B/I_B^\infty} \tag{5.31}$$

where the monolayer matrix factors Q_{AB}^A for AES and Q_{AB}^X for XPS are given by

$$Q_{AB}^A = \left[\frac{\lambda_A(E_A)\cos\theta}{a_A}\right]\left[\frac{1 + r_A(E_A)}{1 + r_B(E_A)}\right] \tag{5.32}$$

and

$$Q_{AB}^X = \left[\frac{\lambda_A(E_A)\cos\theta}{a_A}\right] \tag{5.33}$$

and where the inelastic mean free paths may be calculated from equation $(5.9)^{10}$ or from the calculations of Penn[12] and the back-scattering terms from Figure 5.4 or the data of Shimizu and Ichimura.[6]

As an example we can evaluate the Q_{AB}^A factor for the metallic adsorption system of tin on iron from Figure 5.4 and equation (5.9) and tabulated data from a:

$$Q_{SnFe}^A = \frac{\lambda_{Sn}(E_{Sn})\cos\theta}{a_{Sn}}\left[\frac{1 + r_{Sn}(E_{Sn})}{1 + r_{Fe}(E_{Sn})}\right]$$

$$= 0.41(0.29986 \times 430)^{0.5}0.74\left(\frac{1.87}{1.715}\right)$$

$$= 3.76$$

where a typical value of $\cos\theta = 0.74$ has been assumed.[65] This Q_{SnFe}^A value compares very favourably with the experimental value of 3.82 that can be extracted from the data of Seah[66] using the *Handbook of Auger Electron Spectroscopy*.[1] Other experimental values of Q^A have been tabulated by the author[64] for grain boundaries where the values are twice as high since only half of the grain boundary segregant remains on the surface being analysed.

The intensity from the overlayer of thickness d can likewise be simply calculated from equations (5.27) and (5.29), but more generally one is concerned with two substrate signals under the overlayer. In XPS the I_A^∞ values given in Appendix 5 are already appropriate for samples with a 'standard' contamination level and for clean samples need to be corrected by using equation (5.27) in reverse with I_B representing the values in Appendix 5 and I_B^∞ the new sensitivity factor. For AES the sensitivity factors in ref. 1 are already appropriate to clean samples.

5.5 Quantification in Sputter–Depth Profiles

Depth profiling is discussed in detail in Chapter 4 and will only be treated briefly here. As is shown there, sputtering changes the sample composition in two ways: by atomic or cascade mixing over a depth of about 1–20 nm, depending on the sample and sputtering conditions, and by selective removal of atoms in the topmost one or two atom layers.

The sputtering yield for an atom depends on the rate of energy loss of the incident ion, the energy transfer from atom to atom and the binding energy of atoms in the outermost layers.[67] Thus, the sputtering yields of different elements in their pure state vary over an order of magnitude, as shown in Figure 5.24. Thus, if from the modified AB matrix surface the elements have sputtering yields S_A and S_B, during sputtering of the homogeneous solid, AB, the composition in the sputter modified zone of depth z_s evolves with time according to[70,71]

$$X_A(t) = [X_A(0) - X_A(\infty)]\exp\left(\frac{-t}{t_0}\right) + X_A(\infty) \qquad (5.34)$$

where $X_A(0)$ is simply X_A and $X_A(\infty)$ is the final composition in this zone. Here t_0 is the time to sputter through the depth z_s. The final measured

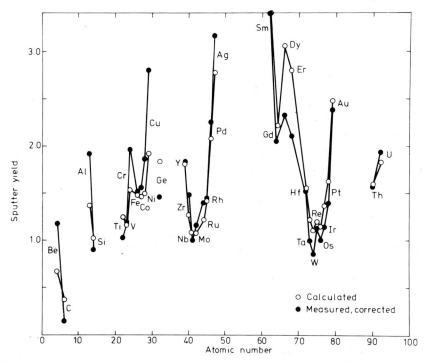

Figure 5.24   A comparison between the sputtering yields calculated[67] for pure elements (○) and the corrected experimental[68] data (●). (After Seah[69])

composition $X_A^m(\infty)$, for X_A small, is related to the sputtering yields by

$$\frac{X_A^m(\infty)}{X_B^m(\infty)} = \frac{X_A}{X_B}\left[\phi + \frac{S_B}{S_A}(1 - \phi)\right] \qquad (5.35)$$

where ϕ is $\exp(-Z_S/\lambda \cos\theta)$, A and B are assumed to have similar energy peaks and λ is thus the inelastic mean free path of either element. Using the above one may define a sputter-modified matrix factor:

$$F_{AB}^S = \left[\phi + \frac{S_B}{S_A}(1 - \phi)\right]^{-1} F_{AB} \qquad (5.36)$$

and use F_{AB}^S in place of F_{AB} in the earlier equations (5.10) and (5.16) for quantitative analysis. It is, however, unfortunate that one cannot simply calculate F_{AB}^S since S_A and S_B cannot be identified with pure element sputtering yields. This is easy to understand by considering a simple model of A and B atoms in the alkali halide structure of interpenetrating f.c.c. lattices. In their elemental state the bond energies of A and B, ε_{AA} and ε_{BB}, may be widely divergent so the pure element sputtering yields may differ by an order of magnitude. However, if A and B have the same mass, and since both atoms have bond energies ε_{AB} in the f.c.c. lattice, they may then have similar yields. The problem is much more complex than this since in general the sputtering yield is composed of collisional, thermal spike and thermal evaporation terms shown clearly in the experimental studies of Szymonski, Ouereijnder and De Vries.[72] The relative importance of these terms as well as their magnitudes are matrix dependent. The problem of compositional changes during sputtering is still an unsolved problem and the interested reader will find the reviews of Kelly[73] and Coburn[74] stimulating, as well as the work of others.[75-79] The most practical solution at the present time is to compare the sputtered profile with a sputtered standard. The use of high-energy peaks can also help by reducing the difference between F_{AB} and F_{AB}^S in equation (5.36).

References

1. L. E. Davis, N. C. MacDonald, P. W. Palmberg, G. E. Riach and R. E. Weber, *Handbook of Auger Electron Spectroscopy*, 2nd Ed., Physical Electronics Division, Perkin-Elmer Corp., Minnesota (1976).
2. H. E. Bishop and J. C. Rivière, *J. Appl. Phys.*, **40**, 1740 (1969).
3. C. R. Worthington and S. G. Tomlin, *Proc. Phys. Soc.*, **A69**, 401 (1956).
4. K. Goto, K. Ishikawa, T. Koshikawa and R. Shimizu, *Surf. Sci.*, **47**, 477 (1975).
5. D. M. Smith and T. E. Gallon, *J. Phys.*, **D7**, 151 (1974).
6. R. Shimizu and S. Ichimura, Quantitative analysis by Auger electron spectroscopy, Toyata Foundation Research Report I–006 No. 76–0175 Osaka (1981); S. Ichimura and R. Shimizu, *Surf. Sci.*, **112**, 386 (1981).
7. M. Gryzinski, *Phys. Rev.*, **A138**, 305 (1965).
8. A. Jablonski, *Surf. Interface Anal.*, **2**, 39 (1980).

9. W. Reuter, in *Proc. Sixth Int. Conf. on X-ray Optics and Microanalysis* (Eds. G. Shinoda, K. Kohra and T. Ichinokawa), p. 121, Univ. of Tokyo Press (1972).
10. M. P. Seah and W. A. Dench, *Surf. Interface Anal.*, **1**, 2 (1979).
11. P. M. Hall and J. M. Morabito, *Surf. Sci.*, **83**, 391 (1979).
12. D. R. Penn, *J. Electron Spectrosc.*, **9**, 29 (1976).
13. P. H. Holloway and D. M. Holloway, *Surf. Sci.*, **66**, 635 (1977).
14. E. N. Sickafus and D. M. Holloway, *Surf. Sci.*, **51**, 131 (1975).
15. G. E. McGuire, *Auger Electron Spectroscopy Reference Manual,* Plenum, New York (1979).
16. M. T. Anthony and M. P. Seah, *J. Electr. Spectrosc.*, in press.
17. H. H. Madden, *J. Vac. Sci. Technol.*, **18**, 677 (1981).
18. F. P. Netzer, *Appl. Surf. Sci.*, **7**, 289 (1981).
19. C. G. Pantano and T. E. Madey, *Appl. Surf. Sci.*, **7**, 115 (1981).
20. R. Hesse, U. Littmark and P. Staib, *Appl. phys.*, **11**, 233 (1976).
21. M. P. Seah, *Surf. Interface Anal.*, **1**, 86 (1979).
22. E. N. Sickafus, *Phys. Rev.*, **B16**, 1436 (1977).
23. E. N. Sickafus, *Phys. Rev.*, **B16**, 1448 (1977).
24. D. A. Shirley, *Phys. Rev.*, **B5**, 4709 (1972).
25. Y. E. Strausser, D. Franklin and P. Courtney, *Thin Solid Films*, **84**, 145 (1981).
26. J. H. Scofield, *J. Electron Spectrosc.*, **8**, 129 (1976).
27. M. P. Seah, *Surf. Interface Anal.*, **2**, 222 (1980).
28. C. D. Wagner, *Anal. Chem.*, **44**, 1050 (1972).
29. C. K. Jorgensen and H. Berthou, *Trans. Faraday Disc.*, **54**, 269 (1972).
30. V. I. Nefedov, N. P. Sergushin, I. M. Band and M. B. Trzhakovskaya, *J. Electron Spectrosc.*, **2**, 383 (1973).
31. V. I. Nefedov, N. P. Sergushin, Y. V. Salyn, I. M. Band and M. B. Trzhakovskaya, *J. Electron Spectrosc.*, **7**, 175 (1975).
32. H. Berthou and C. K. Jorgensen, *Anal. Chem.*, **47**, 482 (1975).
33. M. Janghorbani, M. Vulli and K. Starke, *Anal. Chem.*, **47**, 2200 (1975).
34. L. J. Brillson and G. P. Ceasar, *Surf. Sci.*, **58**, 457 (1976).
35. C. D. Wagner, *Anal. Chem.*, **49**, 1282 (1977).
36. J. M. Adams, S. Evans, P. I. Reid, J. M. Thomas and N. J. Walters, *Anal. Chem.*, **49**, 2001 (1977).
37. S. Evans, R. G. Pritchard and J. M. Thomas, *J. Phys.*, **C10**, 2483 (1977).
38. S. Evans, R. G. Pritchard and J. M. Thomas, *J. Electron Spectrosc.*, **14**, 341 (1978).
39. M. Vulli and K. Starke, *J. de Micros. Spectros. Electron.*, **2**, 57 (1977).
40. C. D. Wagner, L. E Davis M. V. Zeller, J. A. Taylor, R. H, Raymond and L. H Gale, *Surf. Interface Anal.*, **3**, 211 (1981).
41. C. D. Wagner, W. M. Riggs, L. E. Davis, J. F. Moulder and G. E Muilenberg, *Handbook of X-ray Photoelectron Spectroscopy,* Perkin-Elmer Corp., Minnesota (1979).
42. D. W. O. Heddle, *J. Phys.*, **E4**, 589 (1971).
43. R. F. Reilman, A. Msezane and S. T. Manson, *J. Electron Spectrosc.*, **8**, 389 (1976).
44. C. S. Fadley, *J. Electron Spectrosc.*, **5**, 725 (1974).
45. C. S. Fadley, S. Kono, L. G. Petersson, S. M. Goldberg, N. F. T. Hall, J. T. Lloyd and Z. Hussain, *Surf. Sci.*, **89**, 52 (1979).
46. S. Evans, E. Raftery and J. M. Thomas, *Surf. Sci.*, **89**, 64 (1979).
47. M. O. Krause, *Phys. Rev.*, **177**, 151 (1969).
48. M. Vulli, *Surf. Interface Anal.*, **3**, 67 (1981).
49. O. A. Baschenko and V. I. Nefedov, *J. Electron Spectrosc.*, **17**, 405 (1979).

50. M. Gettings and J. P. Coad, *Surf. Sci.*, **53**, 636 (1975).
51. C. J. Powell, N. E. Erikson and T. E. Madey, *J. Electron Spectrosc.*, **17**, 361 (1979).
52. H. G. Noller, H. D. Polaschegg and H. Schillalies, *J. Electron Spectrosc.*, **5**, 705 (1974).
53. A. E. Hughes and C. C. Phillips, *Surf. Interface Anal.*, **4**, 220 (1982).
54. H. E. Bishop, *Surf. Interface Anal.*, **3**, 272 (1981).
55. K. Berresheim, *Surf. Interface Anal.* (to be published).
56. M. F. Ebel, *Surf. Interface Anal.*, **1**, 58 (1979).
57. M. F. Ebel, *Surf. Interface Anal.*, **2**, 173 (1980).
58. K. Hirokawa and M. Oku, *Z. Anal. Chemie*, **285**, 192 (1977).
59. K. Hirokawa, M. Oku and F. Honda, *Z. Anal. Chemie*, **286**, 41 (1977).
60. T. A. Carlson, *Photoelectron and Auger Spectroscopy,* Plenum, New York (1978).
61. C. J. Powell, in *Quantitative Surface Analysis of Materials* (Ed. N. S. McIntyre), ASTM STP 643, p. 5, American Society for Testing and Materials, Philadelphia (1978).
62. C. D. Wagner, in *Quantitative Surface Analysis of Materials* (Ed. N. S. McIntyre), ASTM STP 643, p. 31, American Society for Testing and Materials, Philadelphia (1978).
63. M. P. Seah, *J. Catal.,* **57**, 450 (1979).
64. M. P. Seah, *J. Vac. Sci. Technol.* **17**, 16 (1980).
65. M. P. Seah, *Surf. Sci.*, **32**, 703 (1972).
66. M. P. Seah, *Surf. Sci.*, **40**, 595 (1973).
67. P. Sigmund, *Phys. Rev.*, **184**, 383 (1969); and proof corrections in *Phys. Rev.*, **187**, 768 (1969).
68. N. Laegreid and G. K. Wehner, *J. Appl. Phys.*, **32**, 365 (1961).
69. M. P. Seah, *Thin Solid Films*, **81**, 279 (1981).
70. H. Shimizu, M. Ono and K. Kakayama, *Surf. Sci.*, **36**, 817 (1973).
71. P. S. Ho, J. E. Lewis and J. K. Howard, *J. Vac. Sci. Technol.*, **14**, 322 (1977).
72. M. Szymonski, H. Overeijnder and A. E. De Vries, *Rad. Effects*, **36**, 189 (1978).
73. R. Kelly, *Surf. Sci.*, **100**, 85 (1980).
74. J. W. Coburn, *Thin Solid Films*, **64**, 371 (1979).
75. R. Kelly and N. Q. Lam, *Rad. Effects*, **19**, 39 (1973).
76. Z. L. Liau, B. Y. Tsaur and J. W Mayer, *J. Vac. Sci. Technol.*, **16**, 121 (1979).
77. H. J. Mathieu and D. Landolt, *Appl. Surf. Sci.*, **3**, 348 (1979).
78. N. J. Chou and M. W. Shafer, *Surf. Sci.*, **92**, 601 (1980).
79. T. Narusawa, T. Satake and S. Komiya, *J. Vac. Sci. Technol.*, **13**, 514 (1976).

Practical Surface Analysis
by Auger and X-ray Photoelectron Spectroscopy
Edited by D. Briggs and M. P. Seah
© 1983, John Wiley & Sons, Ltd

Chapter 6

Applications of AES in Microelectronics

R. R. Olson, P. W. Palmberg, C. T. Hovland and T. E. Brady
Perkin-Elmer, Physical Electronics Division,
6509 Flying Cloud Drive, Eden Prarie, Minnesota 55344, USA

6.1 Introduction

The direction of development in silicon digital and analogue microelectronics has been towards increased circuit density, driven by desires for higher levels of system integration, higher speed, lower power dissipation and higher yield. This miniaturization is being achieved through changes in circuit architecture, changes in circuit elements and their structure, use of new materials and through reductions in the lateral and vertical dimensions of structures. Concurrent with the drive for miniaturization in very large-scale integration (VLSI) is the development of new devices and structures in both silicon and in compound semi-conductors to produce transducers and all types of electro-optical devices.

The role of analytical instrumentation is becoming increasingly critical as all aspects of the technology associated with increased circuit density are pushed towards their physical limitations. Given the very small dimensions of the structures involved, one desires an analytical technique with a very high spatial and depth resolution. Scanning Auger electron spectroscopy is such a technique. Because the mean free path of Auger electrons is only a 5–20 Å, the Auger signal samples only the first few atomic layers of the specimen surface. Modern, commercially available Auger microprobes are capable of focusing the primary electron beam into a spot smaller than 50 nm in diameter. Such an instrument has an excellent scanning electron microscopy (SEM) capability and provides elemental analysis, which allows one to clearly image and identify submicrometre features which can be correlated with surface topography in the SEM micrograph.

Auger electron spectroscopy (AES) adds analytical capability to the SEM in a quite different way than energy dispersive X-ray analysis (EDX). The surface sensitivity of AES samples a layer which is orders of magnitude thinner than the approximately 1 μm depth analysed in EDX (see Chapter 1).

Thin contamination layers, surface composition, thin film structures and thin interfacial layer composition are growing ever more important in their influence on device performance; as the lateral and vertical dimensions of circuit elements shrink, the surface-to-volume ratio increases and surfaces and interfaces become increasingly important. Effects such as surface leakage currents and contact resistances, which were secondary effects in larger structures, can dominate state-of-the-art device performance characteristics. A surface-sensitive analytic technique such as AES is needed to determine the influence of surface and interfacial chemistry on the operating characteristic of the device. AES also differs from EDX in that it is sensitive to low atomic number elements. All the elements above helium exhibit unique, easily identifiable Auger spectra, with nearly uniform sensitivity throughout the Periodic Table.

The lateral resolution of AES is also much better than EDX, which is on the order of a micrometre, due to electron scattering effects within the solid. Lateral resolution in scanning Auger microscopy is largely determined by the diameter of the incident beam; chemical spatial resolution of less than 100 nm is routinely achieved.

The objective of this chapter is to discuss the areas of processing technology where AES is utilized and to present a few illustrative examples. A bibliography is included at the end of the chapter for those readers interested in further microelectronics applications. Excellent review articles on applications of AES in microelectronics include an informative article by Thomas[1] and a good overview paper by Holloway[2] which includes an extensive reference list.

6.2 Silicon Microelectronics Technology

6.2.1 Raw material evaluation

In silicon integrated circuit technology, scanning Auger microscopy (SAM) can be used as a powerful tool in characterizing nearly every processing step. Requirements on the condition of the raw material, the silicon wafers, are becoming ever more stringent for acceptable yields on high-density device structures with large area dies. The surface of the wafer must be free of contamination such as polishing residues, etching residues, inclusions, particulates or films, which can nucleate faults in subsequent epitaxial growth and form defects in subsequent oxidation or other process steps involving thermal cycling.

Surface impurities may segregate to the critical oxide–semi-conductor interface during oxidation. Surface contamination can also have an adverse effect on the adhesion of deposited layers. AES can be used to evaluate the effectiveness of various cleaning techniques and identify residues on the wafer surface from etching, solvents and deionized water. Wafers which optically appear to have a surface haze can be analysed to determine whether the cause is topography or contamination.

The SAM technique is ideally suited for identifying particulates and inclusions on the wafer surface, since one can place the focused electron beam on the particulate and record its Auger spectrum to determine its chemical composition. By combining SAM with *in situ* ion beam sputter etching one can perform thin analysis (TFA) on a very small depth scale (in nanometres) to measure the thickness of very thin oxides or contamination layers.

In cases where a higher sensitivity to homogeneously distributed surface impurities is needed, the 'static' SIMS (secondary ion mass spectroscopy) technique provides a complementary analysis method. SIMS systems are available as attachments to most SAM systems. SIMS generally is not as quantitative as AES, but has much greater sensitivity (for a large analysis area) for most elements. In combination with AES, the higher sensitivity of SIMS can be used to extend surface analysis to surface concentrations in the parts per million range.

The thin film analysis (TFA) technique is useful on film structures up to a few micrometres in thickness by using ion beams capable of sputter etching materials at rates of tens of nanometres per minute. For structures where film thicknesses are greater than a few micrometres, mechanical sectioning techniques may be more successful since the sputter etch rate in polycrystalline materials varies with crystal orientation and causes roughening. Other effects such as whisker growth and cone formation can also limit the depth resolution achievable after etching through thick structures, as discussed in Chapter 4.

Depth–composition profiles obtained with the TFA technique are very useful, and often show that adhesion, etchability and contact problems thought to be associated with the bulk composition of thin films are actually near-surface or interface composition problems.

Other solid raw materials which are used in microelectronics processing, such as sputter deposition targets and evaporation source material, can also be characterized by SAM analysis. In polycrystalline metals, the impurities which are in the material usually reside on the grain boundaries. By breaking a sample in the UHV SAM test chamber, one can produce an intergranular fracture, and these fresh grain surfaces can be analysed using SAM. Often, such surfaces show high concentration of impurities which in the bulk analysis lie in the parts per million or parts per billion concentrations. An example of *in situ* fracture of polycrystalline silicon and correlation of SAM data with electron beam induced current measurements has been done by Kazmerski *et al.*[3]

6.2.2 Oxidation

Oxidation processing steps can also be evaluated with the AES and TFA techniques. TFA can be used to measure oxide thickness and thus determine differences in oxide growth rates due to orientation or doping effects. Segregation of impurities to the oxide surface or oxide–semi conductor interface can also be studied. Also, many researchers have employed AES to study the nature of the Si/SiO_2 interface, and a wealth of literature exists.[4-9]

The energy position and shape of the silicon Auger peaks are dependent on whether the parent atom is bound to other silicon atoms, nitrogen atoms or oxygen atoms (see Chapters 3 and 5). This chemical effect has been used to advantage by those studying the interface region; however, the spectra do not give direct information about interface states or surface states. These chemical state effects have also been used to investigate silicides. Many other elements such as aluminium, magnesium, phosphorus, boron, sulphur, carbon, etc., also exhibit Auger peak shape changes and energy shifts.

The present trend in MOS is towards thinner gate oxides, with greater demands for high breakdown fields and greater oxide uniformity and integrity. SAM can be used to analyse localized oxide defects in the very small gate region or in nitride films.

AES can also characterize the oxidation of other materials on the wafer–polycrystalline silicon, silicon nitride, interconnect metal alloys and silicides. This characterization leads to improved yield in processing and faster problem solutions in failure analysis.

6.2.3 Photo-lithography, patterning and stripping

In the photo-lithography process, AES can be very helpful even though thick layers of organic photo-resist are not well suited for AES analysis; electron beam damage effects and charging of the surface pose problems for the analyst. Such surfaces are more suited to characterization by XPS. Inorganic photo-resists may prove to be more amenable to AES analysis. Photo-resist adhesion problems can be addressed by AES analysis before the photo-resist application and to check the completeness of the photo-resist stripping after the IC has been patterned. Surface residues following a stripping operation, such as an oxygen plasma treatment, can be easily detected.

Thomas[1] shows a very interesting example where photo-resist has been used as a mask for a high-dose arsenic ion implant. Following the implantation the wafer was ashed in an oxygen plasma; however, a complete removal of the resist was not possible. SAM analysis of the residual photo-resist showed the residue to have a very high arsenic concentration. This characterization suggested a change in the process to achieve a complete etch removal.

6.2.4 Wet chemical or plasma etching

In wet or plasma etching, SAM can be used to characterize etch rates and the etch uniformity and to check on the effects of defects and contamination on the completeness of an etch. Often there exists an interfacial layer of different composition which alters etching characteristics. An incomplete etch can result in high contact resistances or complete lack of contact to a desired area. SAM can be used to identify such etch resistant residues, which can also lead to subsequent corrosion and adherence problems. Etching can also leave surfaces with a modified surface chemistry which may produce undesirable effects.

In plasma or ion beam sputter etching the high resolution SAM can be used to check for redeposition of materials on side walls of steps and to study topography changes due to etching. When polycrystalline silicon is used in gate electrodes or interconnects, its doping uniformity (when heavily doped) and its oxidation properties can be characterized.

6.2.5 Doping by diffusion or ion implantation

The minimum detectable concentration of impurities using AES is on the order of one part in a thousand of a surface monolayer. This translates into doping levels of 10^{19} per square centimetre. The detection limit depends on the element, which Auger transition is used, the matrix, the primary electron beam voltage and current, the recording time and the spectrometer transmission and resolution. The dependence of the signal-to-noise ratio in the recorded spectrum is

$$\frac{S}{N} \propto \left(\frac{1}{1 + 1/S/B}\right)^{1/2} T_s^{1/2}(I_p Y)^{1/2}$$

where S/B = ratio of the Auger peak above the background to the background (depends on Auger transition energy matrix, primary beam voltage, analyser resolution)
T_s = transmission of the spectrometer
I_p = primary electron beam current
Y = recording time per point

The achievable signal-to-noise ratio limits how small a concentration one can detect. When optimizing the beam current and the recording time by using a computer to signal average spectra, it is possible to extend element sensitivity somewhat beyond the 10^{19} per cubic centimetre range, but not enough to detect dopants in most device structures. AES is useful, however, for characterizing the first step of the normal two-step diffusion process. Prior to a higher temperature diffusion step, a lower temperature is normally used during deposition of the dopant onto the surface. SAM can be used to characterize the thickness, uniformity and concentration of the pre-deposit. Also, in high-dose shallow ion implants, TFA can be used to determine the uniformity and the depth profile of the implant.

Unwanted surface impurities can also be introduced during ion implantation or thermal diffusion processing. A poor vacuum in the ion implanter may leave a hydrocarbon residue, or sputtering from apertures or fixtures in the implanter may leave deposits on the wafer surface. During thermal diffusion processing, SAM can be used to detect both surface segregation and nucleation of new phases that may occur.

6.2.6 Metallization systems

The largest and best established application of AES is in thin film analysis (TFA) where an ion beam is used to sputter etch into the sample while the composition of the immediate surface in the centre of the etched area is measured using AES. By measuring the Auger signals as a function of the sputter etching time, one can generate a plot of the composition of the sample as a function of depth. This information is invaluable for looking at thin film composition, thickness and uniformity. The composition distribution through the thickness of the film in alloy metalizations can be studied, revealing segregation and annealing effects in multicomponent films, impurities in thin films and the composition, structure and interface chemistry of metal semi-conductor junctions in both ohmic contacts and Schottky barrier diodes. The electrical properties, such as the barrier height in such junctions, can be related to the depth versus composition analysis.[10-12]

TFA is also used to (a) monitor silicide metalization or contacts to deter-

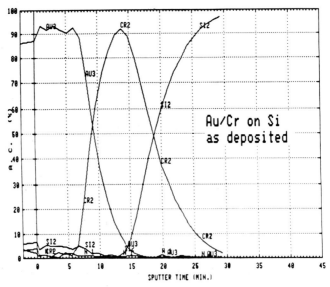

Figure 6.1 TFA depth profile of a gold film on chromium on silicon, as deposited

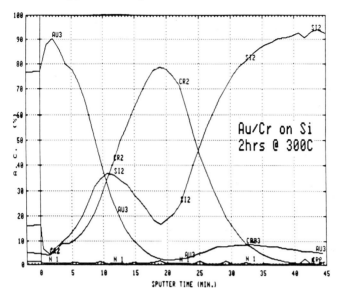

Figure 6.2 TFA depth profile of the thin film structure in Figure 6.1, after heat treatment. The diffusion barrier has failed

mine whether the refractory metal has fully reacted with the silicon, (b) evaluate film stoichiometry in co-deposited silicides, (c) check for aluminium spiking through shallow emitter junctions, (d) test the compatibility of layered materials in complex multilayer, multicomponent material, multilevel metalization structures and (e) check interlayer diffusion and the effectiveness of diffusion barriers. Figure 6.1 shows a TFA depth profile of a sputter deposited thin film structure consisting of a gold film on chromium on silicon.[13] The chromium layer was intended to be a diffusion barrier to prevent the gold and silicon from interdiffusing. Figure 6.2 shows a depth profile of the structure after a heat treatment of 2 hours at 300 °C. The TFA analysis shows that the barrier has failed. To produce a new structure, the partial pressure of nitrogen in the sputter deposition system was intentionally increased during the chromium deposition, resulting in the structure shown in Figure 6.3. This chromium layer with added nitrogen is a more effective diffusion barrier. The TFA depth profile of this structure following heat treatment is shown in Figure 6.4. A whole series of materials such as titanium, tungsten, molybdenum, chromium, platinum, nickel, etc., are now being used in metallization systems and analytical techniques will be necessary for their successful implementation. AES is also used to analyse contacts between layers in multilevel systems, such as metal to polysilicon ohmic contacts, and to evaluate the role of interfacial chemistry in adhesion problems between layers.

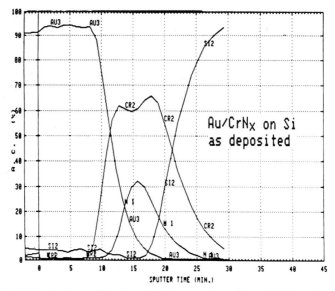

Figure 6.3 TFA depth profile of a gold film on chromium with intentionally added nitrogen on silicon, as deposited

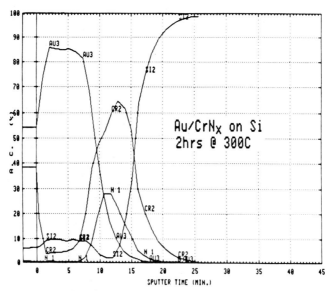

Figure 6.4 TFA depth profile of the thin film structure in Figure 6.3, after heat treatment

As mentioned earlier, the area of analysis in AES is determined mainly by the diameter of the primary electron beam. To understand a material's properties one must understand the microstructure of the material. This had led to the development of small beam scanning Auger electron microprobes. Auger imaging shows the lateral distribution of elements over a surface.

To create an Auger image of elemental distribution, the focused electron beam is stepped across the sample surface and the intensity of an Auger peak above the background of inelastically scattered electrons is measured at each point in a rectangular matrix of points. The subtraction of this background level is essential since the Auger peak-to-background ratio in the energy spectrum of the total secondary electrons emitted from the sample is characteristically small, and this background level is very strongly modulated by the sample topography. The Auger signal measured in this way is then used to modulate the photo CRT from black to white, so that the whiter areas map the areas of higher concentration of the element.

SAM has been used to examine segregation effects in an aluminium, 4 weight per cent copper metallization system. Large area Auger depth–composition studies[14–16] of sputter deposited unheated films indicated enhanced copper segregation at the $Al_{(96)}Cu_{(4)}/SiO_2$ interface, with some evidence of

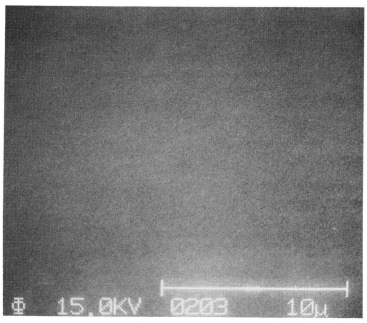

Figure 6.5   Secondary electron image of the surface of the Al–Cu deposit before heat treatment

Al–SiO$_2$ interaction. Heating these films to temperatures at or above 500 °C was found to decrease the apparent copper segregation and give a more uniform copper profile. In a more recent study, similar films were analysed with a high resolution scanning Auger microprobe (SAM) to investigate possible material irregularities which could have been missed in a large area analysis. Al$_{(96)}$Cu$_{(4)}$ films on SiO$_2$ were analysed before and after heating to 500 °C.

A SEM image, Figure 6.5, and SAM images (not shown) of the unheated specimen surface indicated a featureless and uniformly distributed aluminium–copper film. A depth profile, Figure 6.6, shows uniform copper distribution through the film with some copper segregation at the interface approximately 8500 Å below the surface. SEM and SAM images at the interface, Figures 6.7 and 6.8 respectively, are mostly featureless and laterally uniform.

After heating to 500 °C, a depth profile (averaged over a 400 μm^2 area by rastering the electron beam), shown in Figure 6.9, indicated copper segregation at the surface, uniform copper distribution through the film and no copper segregation at the interface. With these data only, it could be concluded that the heating has improved film quality, since the apparent interfacial copper segregation has been eliminated. The SEM image from the heated film surface, shown in Figure 6.10, reveals considerable inhomogeneity, including, however, some features about 1 μm in diameter. Corresponding

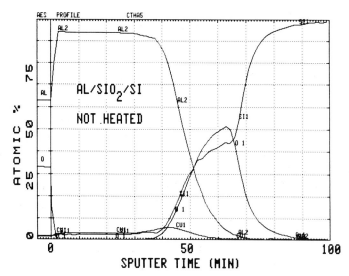

Figure 6.6   Large analysis area depth profile of Al–Cu deposit before heat treatment, showing copper enrichment at the Al–Cu/SiO$_2$ interface

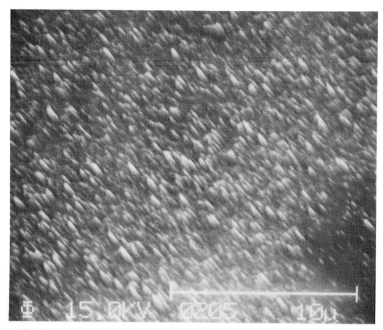

Figure 6.7   Secondary electron image of the Al–Cu film before heat treatment, after etching the sample by argon ion bombardment to a depth of approximately 11 000 Å. Note that the small cone structures point in the direction of the incident ion beam

SAM images for copper and aluminium in Figure 6.11 indicate that copper segregation has indeed occurred in the aluminium film. Auger point analysis in high copper areas showed a copper-to-aluminium atomic ratio of about 1/2, much larger than the bulk ratio.[17]

To determine the three-dimensional nature of the segregation, the film was ion-etched and various areas along the sputter crater were examined. A SEM image and corresponding SAM images for copper, aluminium and silicon obtained at a depth of approximately 7500 Å into the film are shown in Figures 6.12 and 6.13. The silicon image completes the picture, identifying the black areas in the aluminium map not high in copper, as SiO_2. Point analysis again showed the high copper areas to have a copper-to-aluminium atomic ratio of about 1/2. The aluminium–copper phase diagram suggested the formation of a theta phase of the alloy, segregated within a more pure aluminium. Similar series of data (not shown) were collected from films heated to 150, 200, 300 and 400 °C. Even after annealing at 150 °C for only 30 minutes, some evidence of copper segregation was noted.

To summarize, this study shows that a large-area depth–composition analysis of a heated aluminium–copper film suggests only a small change in

(a)

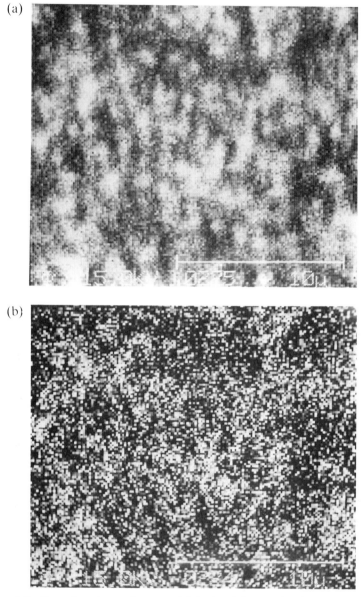

(b)

Figure 6.8   Scanning Auger images of (a) aluminium and (b) copper of the area shown in Figure 6.7. The elements are homogeneously distributed

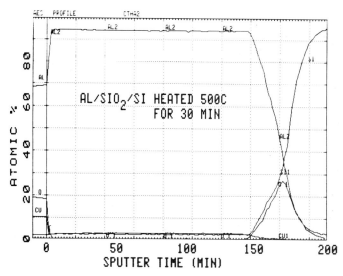

Figure 6.9   Large analysis area depth profile of the Al–Cu deposit after heat treatment. Note the apparently uniform copper depth distribution

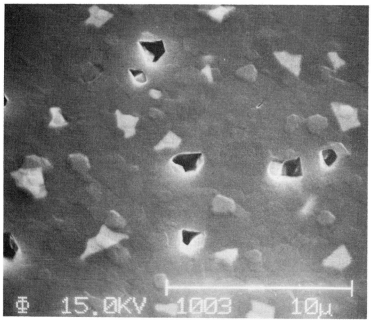

Figure 6.10   Secondary electron image of the Al–Cu deposit surface after heat treatment. Note the voids and new phase segregation

(a)

(b)

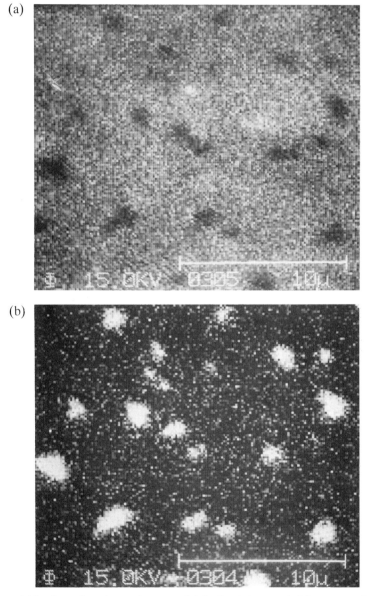

Figure 6.11 Scanning Auger images of (a) aluminium and (b) copper of the area
shown in Figure 6.10. Note that the new phase is copper-rich

Figure 6.12 Secondary electron image of the Al–Cu film, after heat treatment, at the bottom of the sputter etched crater. Note the different sputter etch rates on the various phase grains

copper distribution after heating. However, high spatial resolution SAM results show a much more complicated segregation pattern, demonstrating the need for Auger imaging. Indeed, heating causes the formation of an Al_2Cu theta phase which extends from the SiO_2 interface to the surface. The data suggests that a more complete, detailed study in which both annealing temperature and time are systematically varied would lead to a better understanding of the system kinetics. Furthermore, such analysis applied to processed devices will reveal the thermal history by the extent of copper segregation in the aluminium. Similar studies can be performed on aluminium–silicon and aluminium–copper–silicon systems, as well as on other metallization systems.

6.2.7 CVD/plasma deposition

In chemical vapour deposition (CVD) and plasma deposition techniques, TFA is also used widely in the analysis of the deposited films. The composition versus depth information is invaluable in analysing thin films of complex composition such as SiO_xN_y films where the oxygen-to-nitrogen ratio changes through the film. This information complements such methods as optical

(a)

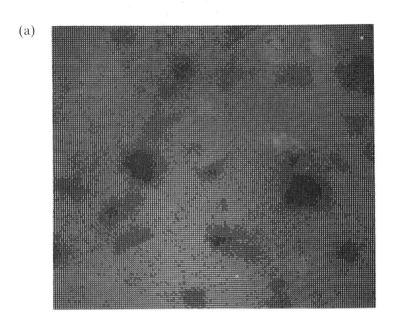

(b)

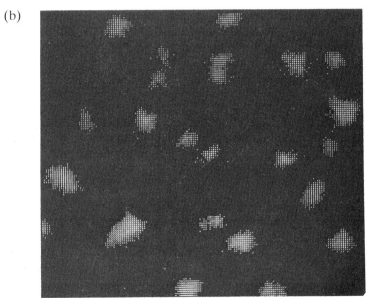

(c)

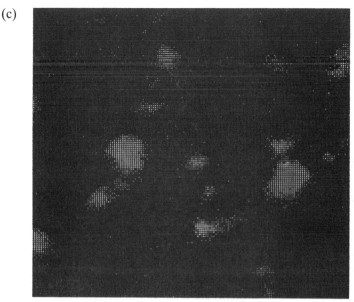

Figure 6.13   Scanning Auger images of (a) aluminium, (b) copper and (c) silicon of
the area shown in Figure 6.12

elipsometry, where the measurement is an integral over a much larger area and through the depth of the film, and one must assume an average optical index of refraction. The example which follows is typical of AES applications in this area of integrated circuit (IC) technology.

Phosphosilicate glass (PSG) is frequently used in the manufacturing of semi-conductor devices as an insulating or encapsulating layer because it has properties that can improve reliability.[18,19] It has also been shown that PSG layers with high phosphorus content can accelerate corrosion and cause other reliability problems.[20-23] For this reason the P_2O_5 concentration in PSG layers must be monitored closely.

The techniques commonly used to measure P_2O_5 content—wet chemistry methods, electron probe, X-ray fluorescence and neutron activation—are used to provide a bulk concentration for P_2O_5 and are frequently used for process monitoring.[24] These techniques, however, are unable to provide detailed information about the distribution of phosphorus within a passivation layer. To determine how much P_2O_5 is present at a particular location or interface, it is necessary to use scanning Auger electron spectroscopy with sampling area or probe size compatible with the geometry of the device and which also has an analysis volume that is small enough to look at surfaces and interfaces.

The techniques for applying AES to phosphosilicate glass, methods for quantifying the data, potential problems and some applications of the technique will be described. In this study the derivative of the electron energy distribution was used to display and interpret the Auger electron spectra. The samples were phosphosilicate glass layers containing from 2 to 20 weight per cent P_2O_5. The bulk compositions were determined using the wet chemical methods or X-ray fluorescence.

While performing Auger analysis on insulating samples, such as phosphosilicate glasses, sample charging can occur. The effect of sample charging on an analysis is dependent on how much surface charge has built up. A small surface charge can cause the spectral lines to shift in energy or shift to several energies, broadening the peak, making reliable peak height measurements difficult. A large surface charge can make it impossible to obtain any spectra at all. Of special interest to this study is the fact that the phosphorus Auger line occurs at a low energy (120 eV) and is, therfore, easily distorted by small surface charges. Also, the phosphorus is mobile in the glass and large charges on the surface may cause movement of the phosphorus within the glass. To avoid sample charging it has been found essential to mount the sample at a grazing angle with respect to the electron beam, which increases the secondary electron emission from the sample. In this study the samples were mounted at a 30° angle. It was found that if the samples were mounted at 60° or 90° (normal to the electron beam) severe sample charging occurred, making it difficult to obtain useful Auger electron spectra.

When analysing microelectronic devices or other thick (>1 μm) insulating structures, mounting the sample at a grazing angle (30°, for example) may not always have enough of an effect to completely reduce the surface charge. In these cases, some additional sample preparation may be necessary. Several techniques have been successfully used to reduce sample charging by providing better electrical contact to the sample. Flat samples can be mounted under a thin metal plate (mask) which has a small (~3 mm) hole in it. The sample is then held in place by the mask which provides improved electrical contact near the analysis area and minimizes the exposed area of insulating material. Rough samples can be wrapped in a thin metal foil formed to the shape of the sample. Indium and aluminium foil masks have been used successfully on packaged devices and other irregularly shaped objects to reduce sample charging. It has been found in this study that by using these techniques most PSG samples can be analysed with AES.

Another area of concern for this study is the effect the electron beam has on the sample. It has been shown that if the current density of the electron beam is too high, sample damage can occur. In SiO_2 the Si—O bonds can be broken leaving some elemental silicon present on the surface. However, it has been shown that by using lower current density this problem can be avoided.[25,26] Also, in the case of phosphosilicate glasses it has been shown that the P_2O_5 composition and distribution can be altered when using an electron beam with a high current density. It has been observed that if the current density is high enough, the phosphous will be attracted to the region irradiated by the electron beam causing a pile-up of phosphous near the surface and the phosphous on the surface can be desorbed. This process can continue until all the P_2O_5 within the diffusion path length of the irradiated region has been drawn to the surface, ionized and desorbed.[27,28] These effects of ionization, desorption and movement of P_2O_5 towards the surface can cause a great deal of confusion in the data taken at high current levels. By using low current density electron beams these effects can be minimized. It was found in this study and has been reported by others[27,28] that the current density should be less than 10^{-4} A/cm^2.

In this study 5 keV Ar$^+$ ions were used to sputter etch the samples. The ion beam, in all cases reported, reduced the P_2O_5, causing a visible chemical shift in the Auger spectrum,[28] even when lower energy ion beams were used. The reduction effect, however, seems to be constant. It did not cause a problem with attempts to quantify the amount of P_2O_5 in the glass.

Standard spectra from SiO_2 and phosphosilicate films with known P_2O_5 concentrations were obtained and compared to silver which has a defined relative sensitivity factor of 1. This method of quantifying Auger electron spectra is described in the *Handbook of Auger Electron Spectroscopy*.[17] Standard spectra were obtained at several electron beam voltages, with a current density of approximately 10^{-4} A/cm^2, on sputtered samples and the resulting

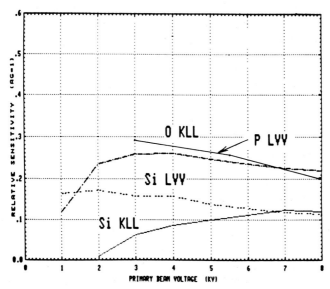

Figure 6.14   Relative sensitivity factors for phosphorus, silicon and oxygen in PSG

Bulk composition			AES
Wt % P_2O_5	Wt % P	Atom % P	Atom % P
9.2	4.0	2.4	2.6
10.2	4.4	2.6	2.8
14.3	6.2	3.8	3.6
17.5	7.6	4.6	4.8
21.5	9.4	5.8	5.8

Figure 6.15   A comparison of the bulk composition of PSG samples to the composition as determined by AES

sensitivity factors are plotted in Figure 6.14. Several samples of phosphosilicate glass with known P_2O_5 content, as determined by bulk methods (X-ray fluorescence and wet chemical techniques), were analysed by AES, using the sensitivity factors to quantify the spectra. It was found that the quantitative information obtained with AES agreed to within 10 per cent of the values obtained by bulk methods (see Figure 6.15). An important observation in quantifying PSG films is that films containing high concentrations of P_2O_5 are not stable in moist environments. Under normal ambient conditions the PSG film will become depleted of phosphorus with time. In this study PSG films with more than 15 weight per cent P_2O_5 were found to be unstable over long periods of time. Figure 6.16 is a phosphorus profile from a sample that originally contained approximately 6.0 atom per cent (22 weight per cent P_2O_5) phosphorus. After being stored for several months it became depleted of phosphorus near the surface.

A device was de-capped and analysed for phosphorus as a function of depth. The resultant compositional depth profile (Figure 6.17) shows that three distinct layers are present. The first is a passivation layer containing approximately 1.2 atom per cent phosphorus. Below the passivation there are two layers: first a layer containing approximately 3.0 atom per cent. phosphorus and below it a layer of silicon dioxide with no phosphorus. This same procedure can be used to study the distribution of phosphorus as a function of depth in a device.

From this study, one may conclude that:

(a) AES can be used to determine the phosphorus content of phosphosilicate glass films and also provide information as to the distribution of phosphorus as a function of depth.
(b) Electron beam damage can be minimized by controlling the current density of the electron beam.

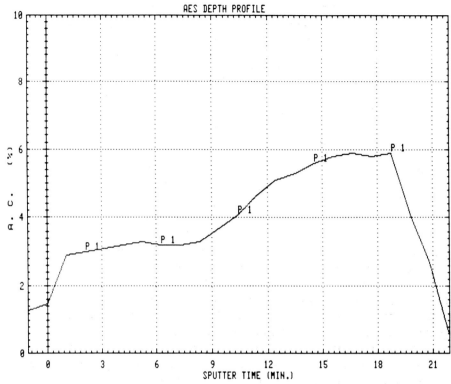

Figure 6.16   TFA depth profile showing only the phosphorus signal in a PSG film. Note the phosphorus depletion at the surface

(c) Ion beam damage will occur but it does not present a serious problem for making quantitative measurements.
(d) Sample charging can be avoided in most cases with proper sample preparation.
(e) Finished devices can be analysed to provide information which can be used to improve reliability.

Checking plating integrity on lead frames and controlling corrosion due to surface contamination on lead frames and packages is yet another area where AES analysis is being applied.

6.2.8   Packaging and bonding

The final part of integrated circuit manufacturing is the packaging of the completed die. In both wire and tape bonding schemes it is the immediate

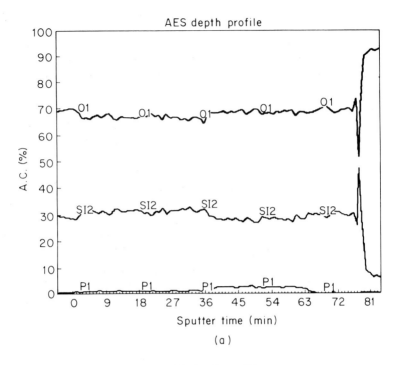

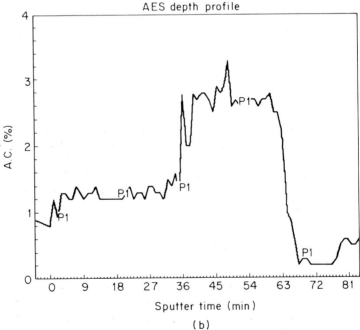

Figure 6.17 (a) TFA depth profile of a PSG film. (b) An expansion of the low concentration range, showing the phosphorus depth distribution

surface chemistry of the pad and mating lead which determines the bondability, strength and electrical integrity of the bond. Very thin oxides, or alkali metal contamination layers which can inhibit bonding can only be detected by a surface-sensitive method.

The attaching of the die to the header is another processing step where thin film composition, diffusion and surface chemistry are critical. Hoge and Thomas[29] have shown examples where the success of die-attach achieving high mechanical strength and low thermal resistance using a AuSi eutectic pre-form depended on the handling of the pre-form and its surface chemistry. Very thin silicon dioxide layers were shown to inhibit the die-attach. Brady and Hovland[30] have shown the use of SAM in studying the diffusion of nickel through gold films in a copper–nickel–gold thin film system which was used in attaching dies to a ceramic substrate in a hybrid circuit. Nickel from the intermediate nickel layer which was intended as a diffusion barrier between the gold and copper was found to diffuse along the gold grain boundaries to the surface, where it reacted with the atmosphere to form a very thin nickel oxide layer. This 5 nm thick layer inhibited the bonding of the gold-backed die to the hybrid substrate.

6.3   Other Microelectronics Technologies

Modern surface analysis techniques are also being used in microelectronic technologies other than silicon where materials problems are as, or even more, difficult. In GaAs, AlGaAs and all the other compound semiconductors, as well as in Josephson junction technologies where very thin films are used, the application of high resolution scanning Auger microscopy will play a critical role in the development and implementation of those technologies.

In crystal growth techniques, such as molecular beam epitaxy (MBE), *in situ* Auger electron spectroscopy is routinely used to check substrate cleanliness, monitor the thermal desorption of carbon and oxygen from GaAs substrates, check film composition and understand dopant incorporation kinetics. Diffusion between thin layers and lateral diffusion can be monitored using TFA and SAM, respectively.

6.4   Conclusion

In microelectronics the constant push for higher performance and higher levels of system integration will continue. In the past, most of the advances in circuit density have been through changes in circuit and device architecture, rather than any great change in absolute line widths on ICs. This situation is now changing and the trend is to scale down structures requiring thinner oxides and smaller line widths. This puts more burden on the process architect who must use new techniques and new materials. If he is to see and understand what is going on, the process developer needs modern analytical tools, whose spatial resolution is appropriate for the scale of the structures.

Surface analytical techniques should by no means be limited to only process development. The process engineer who is trying to maintain a production line needs help in 'fighting fires'. If the yield of working devices on a process drops to zero, it is a crisis situation, and the process engineer needs immediate access to high-throughput, fast-response analytical equipment.

In the past few years, quality and reliability have become by-words and real issues in microelectronics. The integrated circuit industry is realizing that quality cannot be tested into a product. Quality comes through constant long-term effort in many fronts and is commensurate with process yield optimization. Quality requires feedback from all levels into product design, process design and process execution. Modern surface analytical techniques such as SAM will have an ever-expanding role in providing the information for feedback for product reliability optimization and process yield optimization.

AES is not yet generally used as an in-process monitor technique, except in a few particular cases. As the technique matures and lower-cost, more highly automated equipment becomes available, this situation will slowly change.

Acknowledgement

This chapter is dedicated to the memory of Thomas Brady. Thanks are due to J. F. Moulder, J. M. Burkstrand, D. F. Paul, G. Lemons of Signetics, R. McDonald of Intel and R. S. Nowicki of Exxon Enterprises.

References

1. Simon Thomas, *Proceedings of Test and Measurement Expo* (1982).
2. P. H. Holloway, Characterization of electron devices and materials by surface-sensitive analytical techniques, *Appl. Surf. Sci.*, **4**, 410–444 (1980).
3. L. L. Kazmerski, O. Jamjoum, P. J. Ireland and R. L. Whitney, A study of the initial oxidation of polycrystalline Si using surface analysis techniques, *J. Vac. Sci. Technol.*, **18**, 960–964 (1981).
4. S. A. Schwarz, C. R. Helms, W. E. Spicer and N. J. Taylor, High resolution Auger sputter profiling study of the effect of phosphorus pileup on the $Si–SiO_2$ interface morphology, *J. Vac. Sci. Technol.*, **15**, 227–230 (1978).
5. A. C. Adams, T. E. Smith and C. C. Chang, The growth and characterization of very thin silicon dioxide films, *J. Electrochem. Soc.*, **127**, 1787–1794 (1980).
6. H. S. Wildman, R. F. Bartholomew, W. A. Pliskin and M. Revitz, Variations in the stoichiometry of thin oxides on silicon as seen in the Si *LVV* Auger spectrum, *J. Vac. Sci. Technol.*, **18**, 955–959 (1981).
7. C. R. Helms, N. M. Johnson, S. A. Schwarz and W. E. Spicer, Studies of the effect of oxidation time and temperature on the $Si–SiO_2$ interface using Auger sputter profiling, *J. Appl. Phys.*, **50**, 7007–7014 (1979).

8. C. R. Helms, Y. E. Strausser and W. E. Spicer, Observation of an intermediate chemical state of silicon in the Si/SiO₂ interface by Auger sputter profiling, *Appl. Phys Lett.*, **33**, 767–769 (1978).

9. S. S. Olevskil and V. P. Repko, *Use of Auger Spectroscopy to Investigate SiO₂ Films*, pp. 1192–1194, Scientific Research Institute of the Horological Industry, September 1979.

10. H. R. Grinolds and G. K. Robinson, Pd/Ge contacts to n-type GaAs, *Solid State Electronics*, **23**, 973–985 (1980).

11. H. R. Grinolds and G. Y. Robinson, Study of Al/Pd₂Si contacts on Si, *J. Vac. Sci. Technol.*, **14**, 75–78 (1977).

12. L. P. Erickson, A. Waseem and G. Y. Robinson, Characterization of ohmic contacts to InP, *Thin Solid Films*, **64**, 421–426 (1979).

13. R. S. Nowicki and J. F. Moulder, Studies of CrN–Au as an improved die attach metallization system, *Thin Solid Films*, **83**, 209–216 (1981).

14. S. Mader and S. Herd, *Thin Solid Films*, **10**, 377 (1972).

15. A. J. Learn, *Thin Solid Films*, **20**, 261 (1974).

16. D. R. Denison and L. D. Hartsough, *J. Vac. Sci. Technol.*, **17**, 1326 (1980).

17. L. E. Davis, N. C. MacDonald, P. W. Palmberg, G. E. Riach and R. E. Weber, in *Handbook of Auger Electron Spectroscopy*, Physical Electronics Division, Pekin-Elmer Corp., Eden Prairie, Minnesota (1976).

18. D. R. Kerr, J. S. Logan, P. J. Burkhardt and W. A. Pliskin, *IBM J.*, **18**, 376 (1964).

19. M. M. Schlacter, E. S. Schlegel, R. S. Keen, R. A. Lathlaen and G. L. Schnable, *IEEE Trans. on Electron Devices*, **ED 17–12**, 1077 (1970).

20. W. M. Paulson and R. W. Kirk, *Proc. Twelfth Rel. Phys. Symp.*, 172 (1974).

21. R. B. Comizzoli, *RCA Rev.*, **37**, 483 (1976).

22. K. Maeda, J. Sato and Y. Ban, *Jpn. J. Appl. Phys.*, **16**, 729 (1977).

23. N. Nagasima, H. Suzuki, K. Tanaka and S. Nishida, *J. Electrochem. Soc.*, **121**, 434 (1974).

24. A. C. Adams and S. P. Murarka, *J. Electrochem. Soc.*, **126**, 334 (1979).

25. S. Thomas, *J. Appl. Phys.*, **45**, 161 (1974).

26. J. S. Johannessen, W. E. Spicer and Y. E. Strausser, *J. Appl. Phys.*, **47**, 3028 (1976).

27. T. Inoue, *Jpn. J. Appl. Phys.*, **16**, 851 (1977).

28. G. Queirolo and G. U. Pignatel, *J. Electrochem. Soc.*, **127**, 2438 (1980).

29. C. E. Hoge and S. Thomas, Some considerations of the gold–silicon die bond based on surface chemical analysis, *Eighteenth Annual Proc. Rel. Phys.* (1980).

30. T. E. Brady and C. T. Hovland, Scanning Auger microprobe study of gold–nickel–copper diffusion in thin films, *J. Vac. Sci. Techol.*, **18**, 339 (1981).

Bibliography

Survey articles

P. H. Holloway, Application of surface analysis for electronic devices, *Appl. Surface Analysis ASTM STP*, **1980**, 5–23 (1980).

R. K. Lowry and A. W. Hogrefe, Applications of Auger and photo-electron spectroscopy in characterizing IC materials, *Solid State Techn.*, **January 1980**, 71–75 (1980).

G. E. McGuire, Choosing between ESCA and Auger for surface analysis, NBS Special Publication 400–23, pp. 175–182 (March 1976).

G. E. McGuire and P. H. Holloway, Use of X-ray photoelectron and Auger electron spectroscopies to evaluate microelectronic processing, *Scanning Electron Microscopy,* **I**, 173–202 (1979).

N. K. Wagner, Scanning Auger analysis of III–V semiconductor heteroepitaxial interfaces, pp. 165–169.

Miscellaneous

D. J. Brinker, E. Y. Wang, W. H. Wadlin and R. N. Legge, Auger study of the interface in evaporated tin oxide–GaAs MIS solar cells, *J. Electrochem. Soc.,* **128**(9), 1968–1971 (1981).

A. Christou, W. T. Anderson and K. J. Sleger, Enviromental effects on GaAs MESFETs, International Electron Devices Meeting, Washington, D.C., pp. 389–393 (1978).

G. D. Davis, T. S. Sun, S. P. Buchner and N. E. Byer, Anodic oxide composition and Hg depletion at the oxide–semiconductor interface of $Hg_{1-x}Cd_xTe$, *J. Vac. Sci. Technol.,* **19**(3), 473–476 (1981).

J. P. Duchemin, J. P. Hirtz, M. Razeghi, M. Bonnet and S. D. Hersee, GaInAs and GaInAsP materials grown by low pressure MOCVD for microwave and optoelectronic applications, *J. Crystal Growth,* **55**, 64–73 (1981).

H. Elabd and A. J. Steckl, Structural and compositional properties of the PbS–Si heterojunction, *J. Appl. Phys.,* **51**(1), 726–737 (1980).

H. Elabd and A. J. Steckl, Auger analysis of the PbS–Si heterojunction, *J. Electronic Materials,* **9**, 525–549 (1980).

S. J. Ingrey, F. R. Shepherd and W. D. Westwood, Thermal diffusion of chromium in polycrystalline CdSe films, *J. Vac. Sci. Technol.,* **17**(1), 481–484 (1980).

H. Kuwano, S. Miyake and T. Kasai, Dry cleaning of Si surface contamination by reactive sputter etching, *Japn. Appl. Phys.,* **21**, 529–533 (1982).

N. Lieske and R. Hezel, Chemical composition and electronic states of MNOS structures studies by Auger electron spectroscopy and electron energy-loss spectroscopy, Inst. Phys. Conf. Series No. 50, Chapter 3, pp. 206–215 (1980).

M. Matsuura, M. Ishida, A. Suzuki and K. Hara, Laser oxidation of GaAs, *Jpn. J. Appl. Phys.* **20**, L726–L728 (1981).

J. Pernas, M. Erman, J. B. Theetan, F. Simondet, A. Gheorghiu, M. L. Theye and L. Nevot, Characterization of thin films of amorphous GaP using optical and electron spectroscopy, *Thin Solid Films,* **82**, 377–391 (1981).

Shin-ichi Takahashi, Influence of supersaturated melt for InP growth on InP–InGaAsP interface of double-heterostructure laser wafers, *J. Appl. Phys.,* **52**(10, 6104–6108 (1981).

Depth profiles of ion implants

K. Nagahama, K. Nishitani and M. Ishii, Thermally stable metallization to InSb, *Japn. J. Appl. Phys.,* **20**, 1171–1172 (1981).

Y. S. Park. J. T. Grant and T. W. Haas, The determination of sulfur–ion-implantation profiles in GaAs using Auger electron spectroscopy, *J. Appl. Phys,* **50**, 809–812 (1979).

Y. S. Park, W. M. Theis and J. T. Grant, Characterization of ion implants in GaAs by AES and GDOS, *Appl. Surf. Sci.* **4**, 445–455 (1980).

G. U. Pignatel and G. Queirolo, Further insight on boron diffusion in silicon obtained with Auger electron spectroscopy, *Thin Solid Films,* **67**, 233–237 (1980).

S. A. Schwarz, C. R. Helms, W. E. Spicer and N. J. Taylor, Abstract: Auger-sputter profiling study of phosphorus pileup at the Si–SiO₂ interface, *J. Vac. Sci. Technol.*, **15**, 1519 (1979).

Ohmic contacts

W. T. Anderson, Jr., A. Christou and J. E. Davey, Smooth and continuous ohmic contacts to GaAs using epitaxial Ge films, *J. Appl. Phys.*, **49**, 2998–3000 (1976).
A. Aydinli and R. J. Mattauch, Au/Ni/SnNi/n-GaAs interface: ohmic contact formation, *J. Electrochem. Soc.*, **128**, 2635–2638 (1981).
P. A. Barnes and R. S. Williams, Alloyed tin–gold ohmic contacts to n-type indium phosphiode, *Solid State Electronics*, **24**, 907–913 (1981).
M. Watanabe, M. Kishimoto, Y. Hiratsuka, S. Mitani and T. Mori, Study of contact failures caused by organic contamination on Ag–Si contacts, *IEEE Trans. on Components, Hybrids, and Manufacturing Technology*, **CHMT-5**, No. 1, 90–94 (1982).

Metallization systems

J. E. Baker, R. J. Blattner, S. Nadel, C. A. Evans, Jr., and R. Nowicki, Thermal annealing study of Au/Ti–W metallization silicon, *Thin Solid Films*, **69**, 53–62 (1980).
C. A. Chang and N. J. Chou, Ambient effects on the out-diffusion of GaAs through thin fold films, *J. Vac. Sci. Technol.*, **17**, 1358–1359 (1980).
C. C. Chang and G. Quintana, Diffusion kinetics of Au through Pt films about 2000 and 6000 Å thick studied with Auger spectroscopy, *Thin Solid Films*, **31**, 265–273 (1976).
A. Christou, L. Jarvis, W. H. Weisenberger and J. K. Hirvonen, SEM, Auger spectroscopy and ion backscattering techniques applied to analyses of Au/refractory metallizations, *J. Electronic Materials*, **4**, 329–345 (1975).
J. W. Dini and H. R. Johnson, *Optimization of Gold Plating for Hybrid Microcircuits*, pp. 53–57, Sandia National Laboratories (January 1980).
P. B. Ghate, J. C. Blair, C. R. Fuller and G. E. McGuire, Applications of Ti: W barrier metallization for integrated circuits, *Thin Solid Films*, **53**, 117–128 (1978).
P. J. Grunthaner, F. J. Grunthaner, D. M. Scott, M. A. Nicolet and J. W. Mayer, Oxygen impurity effects at metal/silicide interfaces: formation of silicon oxide and suboxides in the Ni/Si system, *J. Vac. Sci. Technol.*, **19**, 641–648 (1981).
K. J. Guo, J. D. Wiley, J. H. Perepezko, J. E. Nordman, D. B. Aaron, E. A. Dobisz, D. E. Maeisen and R. E. Thomas, Amorphous metal diffusion barriers, *Proc. Second Conf. on High Temperature Electronics and Instrumentation*, Houston, 7–8 Dec. (1981).
P. M. Hall, N. T. Panousis and P. R. Menzel, Stength of gold-plated copper leads on thin film circuits under accelerated aging, *IEEE Trans. on Parts, Hybrids and Packaging*, **PHP-11**, No. 3, 202–205 (1975).
W. S. Lee, D. K. Skinner and J. G. Swanson, Structural evidence for a low temperature interaction of aluminum and InP, *Thin Solid Films*, **70**, L17–L19 (1980).
C. P. Lee, J. L. Tandon and P. J. Stocker, Alloying behaviour of Au–Ge/Pt ohmic contacts to GaAs by pulsed electron beam and furnace heating, *Electronics Letters*, **16**, 849–850 (1980).
G. Luzzi and L. Papagno, Vacuum-deposited thin films studied by Auger spectroscopy, *J. Vac. Sci. Technol.*, **16**, 1004–1006 (1979).

G. E. McGuire, J. V. Jones and H. J. Dowell, The Auger analysis of contaminants that influence the thermocompression bonding of gold, *Thin Solid Films*, **45**, 59–68 (1977).

T. J. Magee and J. Peng, Si epitaxial regrowth and grain structure of Al metallization on (100) Si, *J. Appl. Phys.*, **49**, 4284–4286 (1978).

H. Morkoc, A. Y. Cho, C. M. Stanchak and T. J. Drummond, Properties of Au/W/GaAs Schottky barriers, *Thin Solid Films*, **69**, 295–299 (1980).

R. S. Nowicki, J. M. Harris, M. A. Nicolet and I. V. Mitchell, Studies of the Ti–W/Au metallization on aluminum, *Thin Solid Films*, **53**, 195–205 (1978).

M. Ogawa, Alloying behaviour of Ni/Au–Ge films on GaAs, *J. Appl. Phys.*, **51**(1), 406–412 (1980).

K. Oura, S. Okada, Y. Kishikawa and T. Hanawa, Surface structure of epitaxial, Pd_2Si thin films, *Appl. Phys. Lett.*, **40**, 15 138–140 (1982).

T. A. Shankoff, C. C. Chang and S. E. Haszko, Controlling the interfacial oxide layer of Ti–Al contacts with the CrO_3–H_3PO_4, *J. Electrochem. Soc.*, **125**, 467–471 (1978).

R. J. Thompson, D. R. Cropper and B. W. Whitaker, Bondability problems associated with the Ti–Pt–Au metallization of hybrid microwave thin film circuits, *IEEE Trans. on Components, Hybrids, and Manufacturing Techn.* **CHMT-4**, No. 4, 439–445 (1981).

H. S. Wildman and G. C. Schwartz, Auger profiling and electrical resistivity studies of the interface between evaporated Al–Cu layers, *J. Vac. Sci. Technol.*, **20**, 396–399 (1982).

J. D. Wiley, J. H. Perepezko and J. E. Nordman, *High Temperature Metallization System for Solar Cells and Geothermal Probes*, Sandia National Laboratories, December 1980.

J. D. Wiley, J. H. Perepezko, J. E. Nordman and Guo Kang-Jin, Amorphous metallization for high-temperature semiconductor device applications, *Proc. First High Temperature Electronics Conf.* Tucson, March 1981, pp. 35–38 (1981).

Schottky barriers

A. Amith and P. Mark, Schottky barriers on ordered and disordered GaAs (110), *J. Vac. Sci. Technol.*, **15**, 1344–1352 (1978).

A. Aydinli, R. J. Mattauch, GaAs Schottky barrier interface by AES, *Proc. of the Southeastern Conf.*, pp. 14–17 (1980).

C. C. Chang, S. P. Murarka, V. Kumar and C. Quintana, Interdiffusions in thin-film Au on Pt on GaAs (100) studied with Auger spectroscopy, *J. Appl. Phys.*, **46**, 4237–4243 (1975).

C. A. Crider and J. M. Poate, Growth rates for Pt_2Si and PtSi formation under UHV and controlled impurity atmospheres, *Appl. Phys. Lett*, **36**, 417–419 (1980).

C. M. Garner, C. Y. Su, W. A. Saperstein, K. G. Jew, C. S. Lee, G. L. Pearson and W. E. Spicer, Effect of GaAs or Ga_xAl_{1-x} as oxide composition on Schottky-barrier behavior, *J. Appl. Phys.*, **50**, 3376–3382 (1979).

H. R. Grinolds and G. Y. Robinson, Low-temperature sintering of Pd/Ge films on GaAs, *Appl. Phys. Lett.*, **34**, 1, 575–577 (1979).

E. Hokelek and G. Y. Robinson, Aluminum/nickel silicide contacts on silicon, *Thin Solid Films*, **53**, 135–140 (1978).

B. W. Lee, L. Jou, P. Mark, J. L. Yeh and E. So. Surface composition and characteristics of oxide-free $Ga_{1-x}Al_xAs$ (110) Schottky barriers, *J. Vac. Sci. Technol.*, **16**, 514–516 (1979).

G. E. McGuire and W. R. Wisseman, Diffusion studies of Au through electroplated Pt films by Auger electron spectroscopy, *J. Vac. Sci. Technol.*, **15**, 1701–1705 (1978).

A. McKinley, A. W. Parke and R. H. Williams, Silver overlays on (110) indium phosphide: film growth and Schottky barrier formation, *J. Phys. C. Solid St. Phys.*, **13**, 6723–6736 (1980).

M. Ogawa, Alloying reaction in thin nickel films deposited on GaAs, *Thin Solid Films*, **70**, 181–189 (1980).

K. Oura, S. Okada and T. Hanawa, Thermally induced accumulation of silicon on palladium silicide surfaces as studied by Auger electron spectroscopy, *Appl. Phys. Lett.*, **35**, 705–706 (1979).

G. Y. Robinson, A study of metal-semiconductor contacts on indium phosphide, Rome Air Development Centre, RADC-TR-81-169, July (1981).

P. F. Ruths, S. Ashok, S. J. Fonash and J. M. Ruths, A study of Pd/Si MIS Schottky barrier diode hydrogen detector, *IEEE Transactions on Electron Devices*, **Ed-28**, 1003–1009 (1981).

P. E. Schmid, P. S. Ho, H. Foll and G. W. Rubloff, Electronic states and atomic structure at the Pd_2Si–Si interface, *J. Vac. Sci. Technol.*, **18**, 937–943 (1981).

D. K. Skinner, New method of preparing (100) InP surfaces for Schottky barrier and ohmic contact formation, *J. Electronic Materials*, **9**, 67–78 (1980).

N. Szydio and J. Oliver, Behaviour of Au/InP Schottky diodes under heat treatment, *J. Appl. Phys.*, **50**, 1445–1449 (1979).

L. L. Tongson, B. E. Knox, T. E. Sullivan and S. J. Fonash, Comparative study of chemical and polarization characteristics of Pd/Si and $Pd/SiO_x/Si$ Schottky-barrier-type devices, *J. Appl. Phys.*, **50**, 1535–1537 (1979).

Passification and encapsulation

T. Inada, H. Miwa, S. Kato and E. Kobayashi, Annealing of Se-implanted GaAs with an oxygen-free CVD Si_3N_4 encapsulant, *J. Appl. Phys.*, **49**, 4571–4573 (1978).

T. Inada, T. Ohkubo and S. Sawada, Chemical vapor deposition of silicon nitride, *J. Electrochem. Soc.*, **125**, 1525–1529 (1978).

T. Ito, T. Nazaki and H. Ishikawa, Direct thermal nitridation of silicon dioxide films in anhydrous ammonia gas, *J. Electrochem. Soc.*, **127**, 2053–2057 (1980).

P. A. Leigh, *Investigation of CVD Silicon Nitride Encapsulation for Gallium Arsenide*, British Telecom Research Laboratories, 29 July 1981.

K. V. Vaidyanathan, M. J. Helix, D. J. Wolford, B. G. Sreetman, R. J. Blattner and C. A. Evans, Jr., Study of encapsulants for annealing GaAs, *J. Electrochem. Soc.*, **124**, 1781–1784 (1977).

X. Wang, A. Reyes-Mena and D. Lichtman, Interface composition studies of thermally oxidized GaAs using Auger depth profiling, *J. Electrochem. Soc.*, **129**, (1982).

T. Yoshimi, H. Sakai and K. Tanaka, Analysis of hydrogen content in plasma silicon nitride film, *J. Electrochem. Soc.*, **127**, 1853–1854 (1980).

MBE

F. Alexandre, C. Raisin, M. I. Abdalla, A. Brenac and J. M. Masson, Influence of growth conditions on tin incorporation in GaAs grown by molecular beam epitaxy, *J. Appl. Phys.*, **51**, 4296–4304 (1980).

J. R. Arthur and J. J. LePore, Quantitative analysis of $Al_xGa_{1-x}As$ by Auger electron spectroscopy, *J. Vac. Sci. Technol.*, **14**, 979–984 (1977).

R. D. Dupuis, L. A. Moudy and P. D. Dapkus, Preparation and properties of $Ga_{1-x}Al_xAs$–GaAs heterojunctions grown by metallorganic chemical vapour deposition, Inst. Phys. Conf. Ser. No. 45, Chapter 1, pp. 1–9 (1979).

C. M. Garner, C. Ẏ. Su, Y. D. Shen, C. S. Lee, G. L. Pearson and W. E. Spicer, Interface studies of $Al_xGa_{1-x}As$–GaAs heterojunctions, *J. Appl. Phys.*, **59**, 3383–3389 (1979).

S. Ichimura, M. Aratama and R. Shimizu, An interpretation of quantitative analysis of Arthur and LePore for $Al_xGa_{1-x}As$ by using AES. *J. Vac Sci. Technol.*, **18**, 34 (1981).

J. S. Johannessen, J. B. Clegg, C. T. Foxon and B. A. Joyce, Interface composition profiles of MBE grown GaP films on GaAs substrates, *Physica Scripta*, 3 **1980**, 440–443 (1980).

K. Ploog, Surface studies during molecular beam epitaxy of gallium arsenide, *J. Vac. Sci. Technol.*, **16**, 838–846 (1979).

K. Ploog and A. Fischer, Surface segregation of Sn during MBE of n-type GaAs established by SIMS and AES, *J. Vac. Sci. Technol.*, **15**, 255–259 (1978).

C. E. C. Wood and B. A. Joyce, Tin-doping effects in GaAs films grown by molecular beam epitaxy, *J. Appl. Phys.*, **49**, 4854–4861 (1978).

SiO_2 and SiO_2–Si interface

T. J. Chuang, Electron spectroscopy study of silicon surfaces exposed to XeF_2 and the chemisorption of SiF_4 on silicon, *J. Appl. Phys.*, **51**, 2614–2619 (1980).

Y. E. Strausser, C. R. Helms, S. Schwarz and W. E. Spicer, Factors affecting depth resolution and chemistry in Auger electron spectroscopy depth profiles of MOS and MNOS structures, *Thin Solid Films*, **53**, 37 (1978).

Practical Surface Analysis
by Auger and X-ray Photoelectron Spectroscopy
Edited by D. Briggs and M. P. Seah
© 1983, John Wiley & Sons, Ltd

Chapter 7

AES in Metallurgy

M. P. Seah

*Division of Materials Applications, National Physical Laboratory,
Teddington, Middlesex, UK*

7.1 Introduction

Metallurgy is not a new discipline but has evolved at a modest pace over the last few thousand years by a process of trial and error, in keeping with the requirements of the relevant civilizations. Today, sophisticated metallurgical products are found in all aspects of modern life, both domestic and industrial. The properties, in terms of corrosion resistance, integrity at high and low temperatures, wear resistance and weight reduction for a given application, have all improved steadily over the years in response to a growing competition in the market place. As the materials have become more complex, and the specifications more stringent, the advances have only occurred through a proper understanding of the material's behaviour within its multiparameter environment. In particular, much recent motive has come from the high cost of energy where, for instance, a development in turbine blade material could lead to a more efficient aero engine which would give considerable advantages to the engine maker—the aircraft manufacturers fitting his engines and the airline running these aircraft. A second thrust has come from the consideration of source material supply where the questions of possible alternatives for elements with a limited or uncertain source, and the extent to which materials can be recycled or reclaimed, become more and more pressing.

A considerable contribution to this advancement on all the frontiers of metallurgy has been made by surface analysis and, in particular, by using AES. These areas involve corrosion,[1-3] oxidation,[4,5] carburizing[6] and nitriding,[7] soldering,[8] sintering[9] and powder metallurgy,[10] machining,[11] wear[12] and lubrication,[13] the development of hardmetals[14] and heavy metals, as well as a wide range of general interfacial fracture problems. Some of these problems are discussed later in Chapter 10 and many also involve the use of sputter depth profiling, as discussed in Chapter 4. The interpretation of those involving depth profiling may be viewed, to some extent, as an extension of electron microprobe X-ray analysis work, but at higher depth resolution. However, the interpretation of the interfacial fracture problems is a separate discipline which only became possible with the development of AES.

In this chapter we concentrate on the interfacial problems, and in particular fracture, as being model examples of how the presence of sub-monolayer concentrations of elements, following classical adsorption theories, adsorb at internal interfaces, such as grain boundaries and phase interfaces, and cause catastrophic failures in engineering materials. To provide some perspective, it is useful to look at the well-known example of a catastrophic intergranular fracture in the rotor of the Hinkley Point Power Station turbine generator. During a routine overspeed test in 1969 this rotor disintegrated, destroying much of the turbine with it.[15] The rotor initiating this failure was reassembled, as shown in Figure 7.1. It originally had turbine blades around the periphery

Figure 7.1   The reassembled $3Cr\frac{1}{2}Mo$ rotor that initiated the turbine failure at Hinkley Point. (Reprinted by permission of the Council of the Institution of Mechanical Engineers from D. Kalderon, *Proc. Inst. Mech. Eng.*, **186**, 341 (1972))

Figure 7.2   Scanning electron micrograph of the fracture surface of the failed rotor (micrograph edge 200 μm). (After Seah[16])

and was shrunk onto a shaft passing through its core. The scanning electron micrograph of the fracture surface is shown in Figure 7.2.[16] This shows that the fracture had followed the prior austenite grain boundary path very closely with little deformation or energy expending processes. Such a fracture is said to be brittle and, to the engineer, is almost equivalent to his steel rotor being replaced by glass! This behaviour is not a characteristic of the particular alloy steel and its heat treatment, since some of the other rotors were quite safe. The behaviour results from the segregation of impurities in the steel to the grain boundary sites, the level of these impurities varying from cast to cast and, unless especial care is taken, exceeding the value at which catastrophe becomes inevitable.

To understand this problem in more detail, an atomic view of the grain boundary is necessary. The Bragg–Nye bubble raft shown in Figure 7.3 clearly illustrates the degree of misfit at the boundaries of two ordered rafts representing crystals. It is easy to see from this figure that atoms slightly larger than the matrix atoms would have an associated strain energy when at

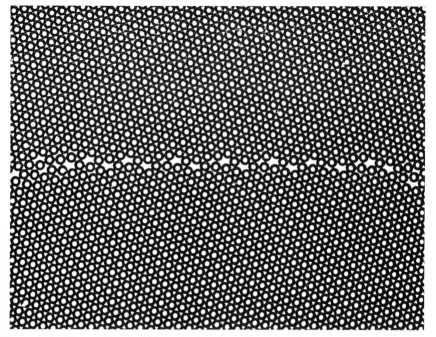

Figure 7.3 Bragg–Nye bubble raft showing a boundary between two close-packed
rafts representing ordered crystals

the matrix sites (i.e. in solution) but that the atom could replace certain
matrix atoms at the grain boundary and, at the same time, reduce the
pre-existing strain there. This shift of solute atoms from the grain interior,
concentrating at the grain boundary, free surface or other interface, is called
equilibrium segregation and the extent to which it occurs is govered thermo-
dynamically through the appropriate free energy of segregation. A more complete visualization of the process is shown in Figure 7.4 where a solid with a
free surface and internal defects is held in an isothermal enclosure at such
a temperature that the diffusion processes approach equilibrium. In this case
the chemical potential of the solute species is constant throughout the enclosure and the solute species atoms occur both in solution and in the vapour state
as well as partitioning to the internal defects, as shown in Figure 7.4.

The partitioning of the solute atoms between the grain boundary and the
lattice was first predicted by McLean in 1957,[18] although the bright intercrys-
talline fractures caused by arsenic, phosphorous and sulphur in steels were
noted as early as 1894.[19] McLean set out a programme of work at NPL to
measure solute segregations in copper and iron using the predicted dependence of the surface[20] and grain boundary[21] energies on segregation. This
work was very successful but was very slow and tedious and was limited to

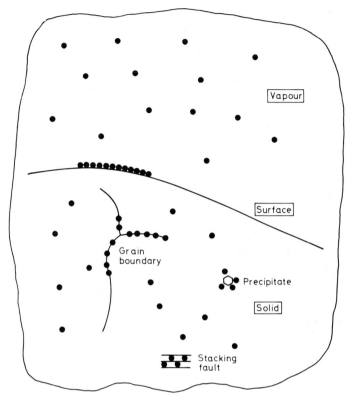

Figure 7.4   Adsorption sites in crystalline solids. (After Hondros and Seah[17])

binary alloy systems. With the use of AES to identify directly those elements which have segregated, these restrictions vanish and experiments are possible at the lower temperatures more relevant to the materials problems. Also very important is the ability of AES to diagnose the segregant in instances of commercial failures. The only drawback to the use of surface analysis in the study of grain boundary segregation is the need to fracture the samples along the grain boundary in the instrument. This is no problem for temper brittle materials but some instances of intergranular failure only occur under particular regimes of cyclic stress[22] or temperature[23, 24] that are not easy to simulate in the AES instrument. If the intergranular failure does not occur *in situ*, a measurement of grain boundary segregation is not possible.

The range of elements for which grain boundary segregation measurements have been made in iron is shown in Figure 7.5. It is quite likely that many of the other elements would segregate from solid solution in iron under appropriate conditions in binary or ternary alloys. In surface segregation experiments on steels, boron and the transition metals have all been observed as

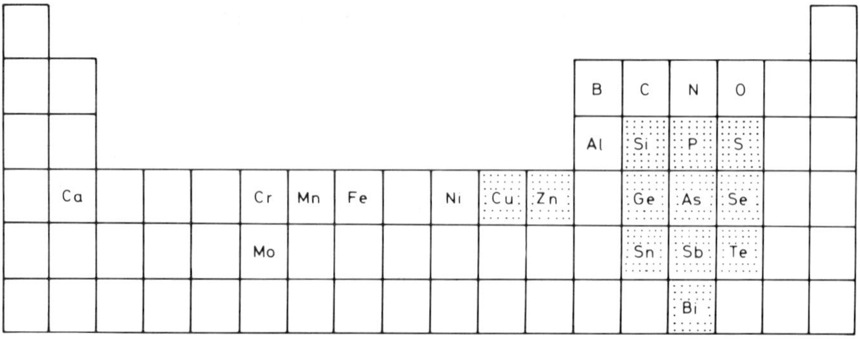

Practical Surface Analysis

													B	C	N	O		
												Al	Si	P	S		
Ca				Cr	Mn	Fe		Ni	Cu	Zn		Ge	As	Se		
				Mo								Sn	Sb	Te		
													Bi			

Figure 7.5 Elements segregated to grain boundaries in iron, as measured by AES. Those known to cause embrittlement are shown on a dotted ground

well as the alkali metals. Similar results occur for copper and nickel and it is likely that, in commercial purity materials, segregation is not the exception but the rule.

In the sections that follow, the characterization of grain boundary, interphase and surface segregations, by AES, will first be presented. This is followed by a section describing the theory of segregation—the energetic terms promoting the segregation and the kinetic terms defining the rate at which equilibrium is approached. The final sections deal with the effects of segregation on basic materials parameters and hence their overall effect on metallurgical properties and the prognosis for remedial actions.

7.2 Characterization of Segregation

In metallurgy there is considerable interest in both grain boundary and surface segregations. The surface segregation measurements are easier and so have been used[25] in attempts to obtain data for the grain boundary. Although this is possible, in principle, there is not yet sufficient known of the precise numerical values of the appropriate terms to allow any meaningful transference between the two, except in the simplest binary systems.[26] Another approach used by some workers to simplify the grain boundary fracture problem is to fracture the samples outside the UHV system.[27] This leads to very dubious and unquantifiable results and should only be used to provide an indication of possible segregants.[28]

The quantification of AES for measuring segregants is described in Chapter 5 and only a brief reminder will be given here. It will be remembered that

$$\phi_A = Q_{AB} \frac{I_A/I_A^{\infty}}{I_B/I_B^{\infty}} \qquad (7.1)$$

from equation (5.31). This may be further simplified approximately to

$$\phi_A = mK_{AB}\frac{I_A}{I_B} \qquad (7.2)$$

where ϕ_A is the level of surface or grain boundary segregant, I_A and I_B are the peak-to-peak intensities designated in the *Handbook of Auger Election Spectroscopy*,[29] $m = 0.5$ for surfaces and 1 for grain boundaries, and

$$\frac{K_{AB}}{2} = \left[\frac{\lambda_A(E_A)\cos\theta}{a_A}\right]\left[\frac{1 + r_A(E_A)}{1 + r_B(E_A)}\right]\frac{I_B^\infty}{I_A^\infty} \qquad (7.3)$$

where all of the terms are as defined in Chapter 5. For convenience, K_{AB} values for segregants in iron, referred to the iron peak at 703 eV, are presented in Table 7.1, calculated from equation (7.3) using the inelastic mean free path data $\lambda(E)$ of Seah and Dench,[30] the back-scattering terms $r(E)$ of Ichimura and Shimizu[31] and the relative sensitivity factors I_A^∞ of the *Handbook of AES*.[29] Remember that this quantification assumes that the segregant atoms pack at the same density as they would in the pure elemental state. The accuracy of the inelastic mean free path data is around 30 per cent standard deviation and so these K values can be no more accurate than that figure.

Table 7.1 K_{AB} values for grain boundary segregants in iron using a 5 keV electron beam at 30° to the surface normal

Element	eV	K_{AB}	Element	eV	K_{AB}	Element	eV	K_{AB}
Li	43	3.2	Co	775	7.2	Sn	430	1.8
Be	104	3.8	Ni	848	6.9	Sb	454	2.6
B	179	4.4	Cu	920	7.9	Te	483	3.7
C	272	4.6	Zn	994	10.3	I	511	5.6
N	379	4.4	Ga	1070	13.9	Cs	563	14.0
O	503	2.8	Ge	1147	16.9	Ba	73	4.1
F	647	3.0	As	1228	21.2	La	78	1.9
Na	990	7.6	Se	1315	28.6	Ce	82	2.8
Mg	45	3.2	Br	1396	39.2	Hf	185	7.1
Al	68	2.4	Rb	1565	69.1	Ta	179	7.2
Si	92	1.9	Sr	1649	76.9	W	169	7.3
P	120	1.5	Y	1746	88.8	Re	176	8.7
S	152	1.0	Zr	147	5.8	Ir	171	12.7
Cl	181	0.9	Nb	167	4.3	Pt	64	1.9
K	252	1.3	Mo	186	3.5	Au	69	1.9
Ca	291	3.0	Ru	273	2.3	Hg	76	1.9
Sc	340	4.9	Rh	302	1.8	Tl	84	2.1
Ti	418	3.7	Pd	330	1.1	Pb	94	2.3
V	473	3.4	Ag	351	1.4	Bi	101	2.5
Cr	529	4.6	Cd	376	1.4			
Mn	542	7.5	In	404	1.6			

7.2.1 Measurements on grain boundaries

7.2.1.1 *Fracturing the sample*

Grain boundary segregation measurements are either made for the diagnosis of industrial failures or as part of a research programme on laboratory alloys. Whereas the former are studied in the 'as received' condition the latter require a treatment cycle. Here the samples are all heat treated to the appropriate temperature, with or without stress, for the appropriate time outside the UHV system. Depending on the requirements, the material may be in a massive form and heated in air, or machined to their final size and heated in argon or vacuum. The samples are then quenched to or below room temperature and, if necessary, machined to size. Machining after the air-exposed heat treatment removes any loose scale and also a layer of material which will be oxygen saturated along the grain boundaries. If the segregation is to be measured at a given temperature the quench must be sufficiently fast to limit the movement of the segregating species. Of course, once at room temperature most samples can be stored indefinitely. For many systems the bulk diffusivity increases tenfold for every 50 °C temperature rise. Thus, even with the small pre-machined samples being quenched into iced brine (the metallurgists fastest quench), segregation measurements can only be made for temperatures up to 900 °C in iron. This limits studies in iron to the α phase.

The prepared samples are then inserted in the AES system and UHV is achieved by bake-out unless an airlock is available (see Chapter 2). The 200 °C bake-out causes no redistribution of substitutionally diffusing solutes but will remove any hydrogen embrittlement and will allow a full redistribution of the interstitial solutes. Most samples are then fractured by impact on a liquid nitrogen cooled stage. The high strain rate of impact and the low temperature makes all body-centred cubic (b.c.c.) materials brittle and some face-centred cubics (f.c.c.) also. Suitable areas of the fracture surface are then imaged using the focused electron beam to raster the surface, after the manner of the SEM. Auger electron maps with a spatial resolution currently around 1 μm (except for the more expensive instruments which can achieve 0.1 μm) allow interesting areas to be selected so that point quantitative analyses can be made of the element segregation levels. Care has to be exercised in the fracture process since some steels change from ductile failure above room temperature to intergranular fracture in the range 0 to −40 °C, and then to cleavage failure. Such samples must be fractured in the correct temperature range and not merely as cold as possible. For weldment problems where the sample should fail in a ductile manner, apart from the critical regions of interest, a tensile fracture stage[32] can be used. For high temperature problems hot tensile stages have been designed[33] and, to assist failure in steels, hydrogen ambients[34] or the cathodic charging of the sample[35] can be used. In the latter case the precise conditions of charging depend on the sample. Generally, charging may be achieved drawing a current density of 20 mA/cm^2 between platinum electrodes, at room temperature, for two hours in 0.5M H$_2$SO$_4$ solution with a hydrogen recombination poison such as As$_2$O$_3$ added at 0.03 gm/l. In many cases these samples may be stored at liquid nitrogen temperatures but have a very short life at room temperature and should not be put through a bake-out schedule.

Some general words of caution should be added to the above procedures. After fracture, the surfaces of metals such as iron and tungsten are very

Figure 7.6   SEM micrograph of intergranular and intra-crystalline cleavage in α-iron
(micrograph edge 0.75 mm)

reactive and, unless the vacuum conditions are particularly good, the peaks due to carbon and oxygen will be seen to increase linearly with time. This can make precision measurements of these important elements very difficult. Again, chlorine occurs as a result of finger contamination and nitrogen can occur through air leaks. If an imaging facility is available with sufficient resolution, the intra-crystalline cleavage and intergranular fracture regions can be clearly distinguished, as shown in Figure 7.6. Any contamination will show up on both types of region whereas segregants can only appear on the grain boundaries. The contaminants do not come solely from the instrument but can come from the material itself by surface diffusion. In many metals the alkali metals and elements such as lead and bismuth are very unsoluble and form very small, effectively pure, precipitates. After fracture, material from the precipitates, exposed to the surface, can diffuse over the whole surface covering the cleavage and intergranular regions alike, making accurate segregation measurements difficult.[36] This effect is only removed by retaining the sample at liquid nitrogen temperatures during analysis. The problem of surface diffusion also dominates the work on hot tensile stages. Here the fracture is slow and the sample may be at high temperatures for an hour before cooling so that all measurements must be treated with great caution.

7.2.1.2 *Characterizing the segregated region*

Figure 7.7 shows the AES spectrum from the grain boundaries of the rotor that caused the Hinkley Point failure, discussed above. The rotor was temper brittle and, according to the quantification shown in Table 7.1, contained 45 per cent of a monolayer of phosphorus at the grain boundaries. This sort of spectrum raises several immediate questions: (a) how far does the phosphorus concentration extend from the grain boundary, (b) how representative is one spectrum of the boundaries in the steel, (c) is half of the segregant left on each side during fracture and (d) what other elements present are also segregated, as distinct from being in precipitates such as carbides, and, if they are segregated, are they active in the particular failure process under study?

The answer to the first question covering the localization of the segregant about the fracture plane may be established by sputtering the fracture away using an *in situ* ion beam. The results for many systems, both metallic and compound, are shown in Figure 7.8. The exponential decay, with a characteristic of 0.3 to 0.6 nm, agrees with the prediction for the sputtering of the atoms sited on the outermost atom layer of the surface.[37] Thus, it is clear from Figure 7.8 that the segregant atoms occupy a zone only one atom wide at the grain boundary and that it is their bonds that have ruptured in the fracture process. Analysis of both halves of the fracture surface show that, in general, the fracture path divides the segregant equally onto the two faces.[38] This is only true on average and, microscopically, one would not expect it to be valid

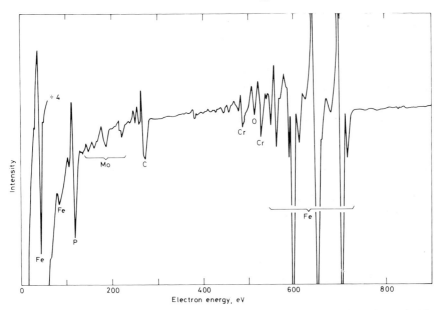

Figure 7.7   Auger electron spectrum from the grain boundary surface shown in Figure 7.2, exhibiting 0.45 monolayers of phosphorus segregated at the grain boundary. (After Seah[28])

in all cases. Indeed, in the fracture of interphase interfaces the segregants tend to remain with the more reactive surface, as observed by Johnson and Smartt for graphite nodules in iron.[39]

Care must be exercised in interpreting the sputter profiles for elements which may be in precipitate form at the grain boundary as, for instance, the chromium and molybdenum signals in the spectrum of Figure 7.7. Some of the chromium and molybdenum may be segregated and some may be in precipitates of $Cr_{23}C_6$ and Mo_2C. It is important to know the levels segregated and the sputter profile, shown in Figure 7.9, may be thought to give this information. The residual signals after long sputtering times have been attributed to the thick carbides whereas the initial decays for the chromium and molybdenum, following that for phosphorus, indicate the level of segregation. Quantitative analysis of the spectra[40] indicate grain boundary coverages of precipitates, $Cr_{23}C_6$ and Mo_2C, of 7 and 5 per cent by area, respectively, with 17 percent of a monolayer of chromium and 12 per cent of a monolayer of molybdenum segregated. However, this analysis makes no allowance for the differential sputtering discussed in Chapter 4. It has been shown[41, 42] that the surface composition of a stoichiometric material will change exponentially from the original composition to a new altered level as a result of preferential sputtering. The change occurs during the sputtering of the first few atom

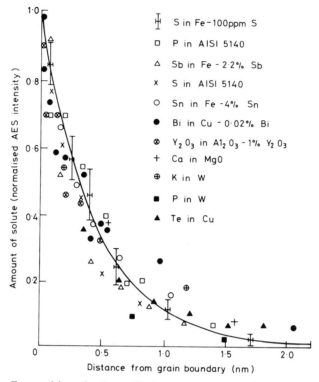

Figure 7.8 Composition–depth profile for segregants at grain boundaries using AES and argon ion etching. The continuous curve shows the expected result for atoms exactly in the grain boundary plane

layers and is expected to occur for carbides as well as all other materials.[43] All of the models of sputtering indicate that the metal will be depleted in carbides so that some or all of the decays seen for chromium and molybdenum, in Figure 7.9, may result from preferential sputtering. A clear separation of the preferential sputtering and segregation terms is not possible at the present time and requires reference profiles from UHV cleaved stoichiometric carbides.

Analysis of many grain boundary faces shows that the segregation is generally fairly uniform with one boundary behaving much like another,[32,44] as shown in the sharp histograms of Figure 7.10 for segregants in iron. Included in this diagram is the histogram for the rotor steel of the Hinkley Point failure discussed in the Introduction. The uniformity between boundaries is largely to be expected since the greatest number of boundaries, and especially those that fail will be random high-index, high-energy boundaries. Instances do

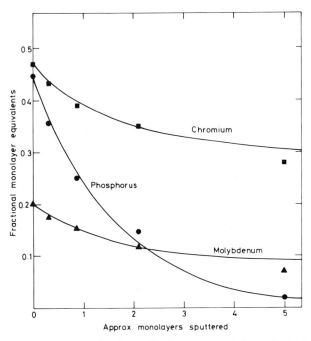

Figure 7.9   Average sputter–depth profile from ten grain boundaries in the $3Cr\frac{1}{2}Mo$ rotor steel. (After Seah[40])

occur, however, when wide variations in the segregation levels are observed and, since the onset of fracture is clearly related to the boundaries with the highest segregation level, this level must be estimated by fitting the data to a Weibull distribution.[46] The bismuth segregation in Figure 7.10 fits such a distribution.

One reason for expecting different levels of segregation at different boundaries arises from the different degrees of disorder there. Watanabe, Kitamura and Karashima,[47] in a very careful study on the iron–tin system, show that a twentyfold increase in segregation occurs as the angle of a tilt boundary increases up to 60° but that no minima occur at coincidence boundaries. On the other hand, in the iron–phosphorus system, Suzuki, Abiko and Kimura[48] correlate the intensity of segregation with the index of the fracture surface rather than the overall misorientation of the boundary. Thus, not only is the segregation highest for the most disordered boundaries but, after fracture, the segregant is retained on the higher index surface as shown in Figure 7.11(a). Figure 7.11(b) shows the anisotropy of the segregation in the polar diagram of the fracture faces and it is clear that the low index faces have considerably lower segregation.

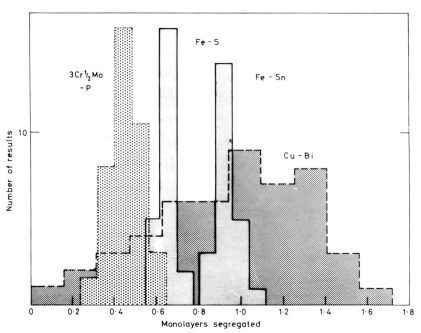

Figure 7.10   Histograms of segregation to grain boundaries for sulphur and tin in iron at 600 °C (reproduced from Seah and Hondros[32] by permission of the American Institute of Physics), for bismuth in copper at 700 °C (after Powell and Woodruff[45]) and for phosphorus in 3Cr½Mo steel at 450 °C (after Hondros and Seah[17])

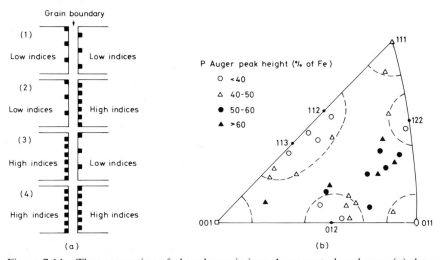

Figure 7.11   The segregation of phosphorus in iron, 1 per cent phosphorus, (a) showing the segregation atoms (dark squares) in relation to the fracture process and (b) the level of segregation in relation to the relative orientation of each grain to the grain boundary. (Reproduced from Suziki Abiko and Kimura[48] by permission of Pergamon Press Ltd)

7.2.2 Measurements on free surfaces

7.2.2.1 *Heating the sample*

Measurements of surface segregation involve both the alloying elements and the impurities. Whilst both are important to the metallurgist, workers in the alloy catalyst field have concentrated on the former. Most of these measurements are made using AES. Small samples of material are usually mounted on a hot stage in the apparatus, cleaned on the front by ion etching and then analysed as a result of some heat treatment schedule *in situ* in the UHV. In work at NPL, special lightweight stages have been devised which heat small samples to 1000 °C whilst consuming only a few watts of power. This reduces outgassing in the vacuum system and enables a good vacuum environment to be maintained. This is important since the balance of the segregation of other constituents can be changed if the surfaces become contaminated with oxygen or carbon. For instance, the surface of nickel–gold alloys, which are usually nickel depleted, can become nickel rich in the presence of oxygen or hydrogen. A second advantage is that magnetic fields associated with the use of high power can upset the operation of some spectrometers so that measurements must then be taken with the power switched off. With the low power sources the samples cool in a few seconds so that quenches from 850 °C are sufficiently fast to retain the segregation of that temperature.

A further problem that exists with surface segregation concerns the evaporation of volatile species. True equilibrium measurements require no net evaporation from the surface. Tables of elemental vapour pressures as a function of temperature are given by Honig and Kramer[49] and it is seen that many of the interesting segregants are also very volatile. This limits the maximum temperature for meaningful measurements generally to below 850 °C. At higher temperatures the segregant is evaporated and the bulk content of the sample falls so that eventually no more segregant is available. Evaporation can be reduced by partially enclosing the surface or working at the bottom of a crevice.[50] However, this is only likely to increase the operating temperature range by about 50 °C. A check on the evaporating species is easily made by collecting emitted atoms on a nearby cold substrate which may subsequently be analysed by AES.

7.2.2.2 *Characterizing the segregated region*

Sputter depth profiles, as described above, show that surface segregants are located on the outer atom plane. Compared with grain boundary segregation, much more work has, however, been completed in binary alloy systems and on well-characterized, low-index single-crystal faces. Much of the data of binary systems is included in a recent analysis by Seah[51] and involves a wide range of substrates including silver, gold, copper, iron, nickel, palladium,

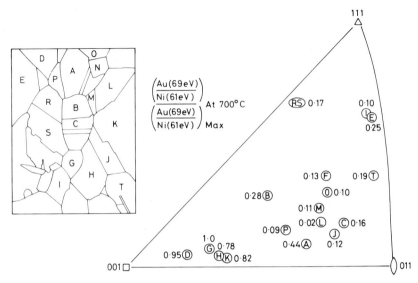

Figure 7.12   Orientation dependence of gold segregation on nickel grains at 700 °C and the relative disposition of the grains on the surface. (Reproduced from Johnson *et al.*[52] by permission of the American Institute of Physics)

platinum and ruthenium, with a similar range of segregants. Histograms of the scatter of the level of segregation with crystal face have not been published since a uniform result over the surface has been indicated in most studies. However, in a detailed study of gold segregation in nickel, Johnson *et al.*[52] analysed the segregation and surface orientation of eighteen grains, as shown in Figure 7.12. The letters correspond to the grains shown in the inset and the numbers are approximately proportional to the level of segregation of gold for the nickel grains in the polar diagram. The distribution of surface segregation shown in Figure 7.12 is broadly as expected. The driving force for segregation is the lowering of the surface energy.[20] Thus the segregants will be most concentrated on the faces where the surface energy of the pure nickel can be most reduced by gold atoms satisfying and minimizing the local dangling bond energy requirements. Such a condition appears to occur on the (014) plane where the segregant level maximizes.

The above sections describe the general characterization of segregation that is possible with Auger electron spectroscopy. In the next section the theory of segregation will be described and illustrated by measurement with AES.

7.3   Theory of Segregation

The theories of segregation have developed rapidly in the last decade in response to the growing body of AES measurements available for interpretation. The adsorption theories for the solid–solid interface and the solid–vacuum surface have been established as direct analogues or developments of theories well known in the field of gas adsorption on the free surfaces of solids. In this section the theories will be presented as a series with successively more complex adsorption conditions. These describe the final equilibrium state but, in practical situations, events are limited by diffusion and so the kinetics of the segregation process are then presented.

7.3.1 The Langmuir–McLean theory for surface and grain boundary segregation in binary systems

In the earliest theory specifically for grain boundaries, McLean[18] proposed a model with the solute atoms populating grain boundary and lattice sites with an energy difference, ΔG, the free energy of segregation. Using simple statistical mechanics he found that the system energy was minimized for a fractional monolayer of segregant, X_b, at the grain boundary, where

$$\frac{X_b}{X_b^0 - X_b} = \frac{X_c}{1 - X_c} \exp\left(-\frac{\Delta G}{RT}\right) \qquad (7.4)$$

Here X_b^0 is the fraction of the grain boundary monolayer available for segregated atoms at saturation and X_c is the bulk solute molar fraction. Segregation measurements obeying the form of equation (7.4) have been reported for several systems at free surfaces[17] but the theory is best illustrated by the data of Kumar and Eyre[53] for the grain boundary segregation of oxygen in molybdenum, shown in Figure 7.13.

7.3.2 The free energy of grain boundary segregation in binary systems, ΔG_{gb}

Values of ΔG_{gb} were estimated by McLean[18] from the elastic strain energy released by the segregation of solute atoms. This gave values correct to a factor of two; however, a higher accuracy in estimating ΔG_{gb} was provided by Seah and Hondros.[32] Using the gas adsorption theory of Brunauer, Deming, Deming and Teller, known as the truncated BET theory, they write[54] the solid-state analogue as

$$\frac{X_b}{X_b^0 - X_b} = \frac{X_c}{X_c^0} \exp\left(\frac{-\Delta G'}{RT}\right) \qquad (7.5)$$

where X_c^0, the solid solubility, is the important parameter for which considerable data exists in metallurgical handbooks. In the dilute limit, if X_b^0 is one monolayer, we may write:

$$\beta_{gb} = \frac{X_b}{X_c} = \frac{K}{X_c^0} \qquad (7.6)$$

where $K = \exp(-\Delta G'/RT)$ and β_{gb} is the grain boundary enrichment ratio.

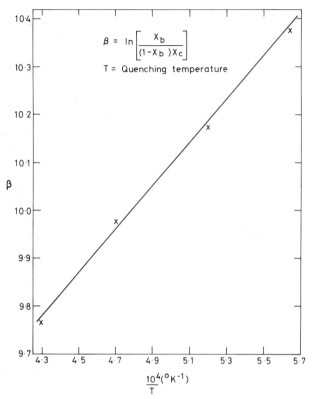

Figure 7.13 Langmuir–McLean plot for the grain boundary segregation of oxygen in molybdenum (reproduced from Kumar and Eyre[53] by permission of The Royal Society). The bulk oxygen level of 6–9 p.p.m. is well within solid solubility and the free energy of segregation, ΔG, is $(37414 + 65.2T)$J/mol

Equation (7.6) provides an excellent description of the mass of experimental data, from AES and other techniques, which show that β_{gb} is dependent on X_c^0 for solubilities which range from 100 p.p.m. to 100 per cent. The compilation of this data is shown in Figure 7.14 from which it is clear that

$$\beta_{gb} = \frac{4.3 \overset{\times}{\underset{\div}{}} 2.5}{X_c^0} \qquad (7.7)$$

The chart on the right-hand side of Figure 7.14 shows the values of β_{gb}, obtained in binary and higher order alloys, for which X_c^0 is not known. This shows the important effect that the segregation of impurities such as phosphorus, tin and antimony appears to be an order of magnitude higher in alloy steels than in pure iron. This point will be reconsidered later.

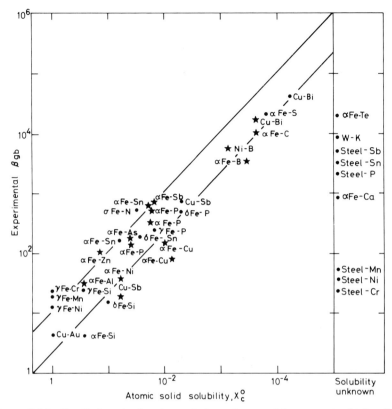

Figure 7.14 Predictive plot for the grain boundary enrichment ratio. (After Seah[40])

Thus the truncated BET theory has a powerful predictive value for binary systems and it is seen that the free energy of grain boundary segregation, ΔG_{gb}, over a range from 0 to -100 kJ/mol, is equal to $(\Delta G_{sol} - 9 \pm 5)$kJ/mol. Many other adsorption theories have been developed for binary systems[54] but for general predictive value the truncated BET seems best.

7.3.3 The free energy of surface segregation in binary systems

Calculations of the free energy of surface segregation have variously been attempted through bond energy terms [55–57] and strain release terms.[56–58] The bond energy terms give $\Delta G = \gamma_A^S - \gamma_B^S$ where γ_B^S is the surface energy (in kilojoules per mole) of the pure matrix atoms B and γ_A^S that of the pure segregant A. For solid metals the surface energies are related to the melting points so that ΔG may be estimated from the difference in the melting points of the segregant and matrix atoms, T_A^M and T_B^M.

Using this approach Seah[51] combines the surface energy and strain release terms through a regression analysis with the experimental data for many systems to derive a prediction for the surface segregation enrichment ratio, β_A^S:

$$\ln \beta_A^S = \frac{24(T_B^M - T_A^M) + 1.86\,\Omega + M4.64 \times 10^7 a_B (a_A - a_B)^2}{RT} \quad (7.8)$$

where Ω is the molar regular solution parameter for the AB binary system and a_A and a_B are the atom sizes of A and B atoms, calculated as described in Chapter 5 in relation to equation (5.8). The parameter M is unity for $a_A > a_B$ and zero for $a_A < a_B$. The first term represents the effect of dangling bond energies and is the dominant term for segregant systems in which the atom sizes differ by less than 10 per cent., the second term is a small contribution due to the chemical interaction of A and B and the last term is that due to the

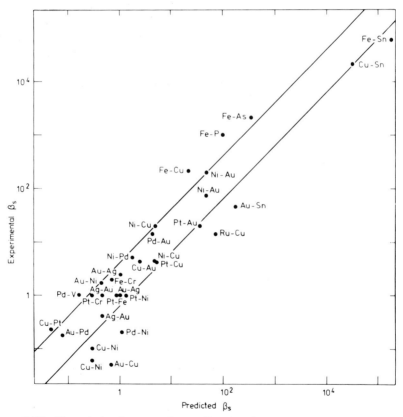

Figure 7.15   The relation between the predicted surface segregation enrichment ratio and that observed experimentally. (After Seah[51])

strain energy release. The strain release term only contributes for segregant atoms larger than the matrix but not for the reverse case since the strain is then always small.[59] Equation (7.8) agrees well with experiment and allows prediction of ΔG_s to a standard deviation of 9 kJ/mol over the range of surface segregation free energies from -80 to 20 kJ/mol. Figure 7.15 shows the correlation of this prediction with the experimental values.

7.3.4 Site competition in simple ternary systems

Many measurements have been reported in ternary and more complex systems but few have fully characterized the phenomenon of site competition. The first analysis by Seah and Lea[50] showed that, for segregation levels below one monolayer, a simple linear site competition could occur. A strongly segregating species could displace a weakly segregating species on, approximately, an atom-for-atom basis. Such a competition was first demonstrated for tin and sulphur on an iron substrate and more recently for both antimony[61] and oxygen[62] with sulphur. In all of the above cases if the element displaced by sulphur is strongly segregated to the extent of close to, or more than, one monolayer in the absence of sulphur, the initial displacement goes more strongly than on a one-for-one atom basis until, at approximately 25 per cent. of a monolayer of sulphur, the behaviour switches to that described above. This is illustrated in Figure 7.16.

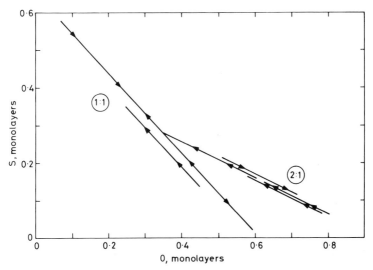

Figure 7.16   Site competition between oxygen and sulphur on the surface of iron, illustrating the regimes of 1 : 1 and 2 : 1 oxygen-to-sulphur replacement ratios. (Reproduced from Thomas *et al*.[62] by permission of the American Institute of Physics)

Site competition as described above is not expected at grain boundaries since, even at a full monolayer of segregation, further solute atoms can segregate and reduce the strain and matrix atom dangling bond energies simply because there are matrix atoms on both sides of the interface. These extra bonding sites, compared with the free surface, make competition less likely and, for the tin–sulphur and oxygen–sulphur systems, no grain boundary site competition is observed at all.[32, 63] Site competition at grain boundaries in iron has also been observed between nitrogen and sulphur,[64] but not under all conditions.[63] More important is the reported displacement of phosphorus by carbon.[65] This last observation is very important for the fracture problem and will be discussed more fully later.

7.3.5 More complex binary systems

The above theories assume that the segregated atoms are non-interacting. If an interaction energy ω is allowed between adjacent adsorbate atoms, such that they attract (ω negative) or repel (ω positive) each other, the solid-state analogue of the Fowler adsorption theory[66] may be derived:

$$\frac{X_b}{X_b^0 - X_b} = \frac{X_c}{1 - X_c} \exp\left[\frac{-\Delta G - Z_1\omega(X_b/X_b^0)}{RT}\right] \qquad (7.9)$$

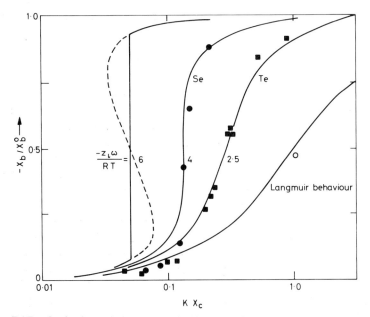

Figure 7.17 Grain boundary Fowler isotherms with the experimental results of Pichard *et al.*[67] for Se and Te in iron. (After Hondros and Seah[54])

If ω is zero this theory reduces to that of Langmuir and McLean. However, as ω becomes more and more negative the segregation shows progressively sharper rises as the temperature falls until eventually the rise in segregation is discontinuous at a certain temperature. This transition in behaviour is shown in Figure 7.17, together with the AES measurements for selenium and tellurium in iron.[67] For $Z_1\omega$ below $-4RT$ the curves become S-shaped, as shown by the curve at $Z_1\omega = -6RT$. For this final curve the segregation does not follow the S-shaped solution to equation (7.9) but moves discontinuously from A to B. The way this arises can be seen from the plots of G as a function of X_b in which the high coverage minimum eventually has a lower absolute energy than that at the low coverage.[68]

7.3.6 Segregation in ternary systems

In 1975 Guttmann[69] extended the Fowler theory to allow for interactions between two co-segregating species in multicomponent systems. This development, vital to explain the segregation behaviour resulting in intergranular failures in engineering materials, gives

$$\frac{X_{bi}}{X_{bi}^0} = \frac{X_{ci}\exp{-\Delta G_i/RT}}{1 + \sum_{j=1}^{2} X_{cj}[\exp(-\Delta G_j/RT) - 1]} \tag{7.10}$$

where X_{b1}^0 and X_{b2}^0 are the molar fractional monolayers segregated by the impurity and alloying elements of bulk contents X_{c1} and X_{c2}, respectively. Additionally,

$$\begin{aligned}\Delta G_1 &= \Delta G_1^0 + \alpha_{12}' X_{b2}\\ \Delta G_2 &= \Delta G_2^0 + \alpha_{12}' X_{b1}\end{aligned} \tag{7.11}$$

where ΔG_1^0 and ΔG_2^0 are the free energies of segregation of the impurity and alloying elements separately in the matrix. The interaction coefficient α_{12}' refers to the changes in nearest neighbour bond energies in forming the alloy–impurity bonds and are obtained from measurements of the effects of the alloy elements on the solubilities of the impurities.[70] If α_{12}' is negative, segregation of the alloying element enhances segregation of the impurity. In a similar manner to the Fowler result, shown in Figure 7.17, as α_{12}' becomes more and more negative, the curves exhibit a more pronounced S-shape and eventually exhibit a region in which the segregation shows a discontinuous increase from a low level to a high one. The general effect of Guttmann's coupling can be seen in Figure 7.14. Alloy element additions with a negative value of α_{12}' cause the impurity element solubility to fall. Simultaneously, through equations (7.10) and (7.11) the segregation level rises. Thus, the addition of the alloy element causes the point for the original binary system in Figure 7.14 to move diagonally upwards to the right but to stay within the general correlation.[71]

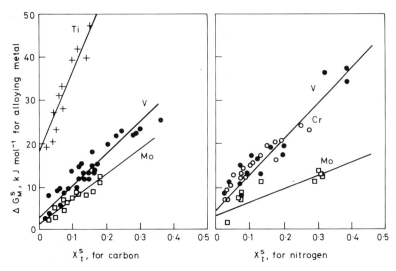

Figure 7.18   The increase in the free energy of surface segregation for alloy elements in steel as a function of the carbon and nitrogen segregation. (Reproduced from Dumoulin and Guttman[72] by permission of Elsevier Sequoia SA)

The general behaviour of this model is now well accepted and allows an accurate description of some of the behaviour observed in embrittlement problems. The increase in the free energy of segregation of one species by a second, described in equation (7.11), has been validated in careful experiments on surface segregation in steel by Dumoulin and Guttmann,[72] as shown in their data of Figure 7.18.

Guttmann's original theory was modified by Seah[73] to remove the site competition which had not been observed in practice. In equation (7.10) the sum over the j terms is simply replaced by the ith term. This simplifies the calculations and gives a better description of the experimental results. A range of more complex theories is detailed by Guttmann and McLean[71] and by Guttmann.[74]

7.3.7   The kinetics of segregation

In practical situations where segregation is important, the segregant atoms often have insufficient time to reach their full equilibrium level as defined by the adsorption theories. In this section, therefore, the kinetics of the segregation are analysed.

Most models of the kinetics follow McLean's[18] approach. Solute atoms are assumed to segregate to a grain boundary from two infinite half crystals of uniform solute content or to a surface from one infinite half crystal. Diffusion in the crystals is described by Fick's laws and the ratio of the solute in the grain boundary to that in the adjacent atom layer of the bulk is given by the constant enrichment ratio, β, of equation (7.6). The kinetics of the segregation are thus described by

$$\frac{X_b(t) - X_b(0)}{X_b(\infty) - X_b(0)} = 1 - \exp\left(\frac{FDt}{\beta^2 f^2}\right)\text{erfc}\left(\frac{FDt}{\beta^2 f^2}\right)^{1/2} \tag{7.12}$$

where $F = 4$ for grain boundaries and 1 for the free surface, $X_b(t)$ is the boundary content at time t, D is the solute bulk diffusivity and f is related to the atom sizes of the solute and matrix, b and a, respectively, by $f = a^3 b^{-2}$. For short times equation (7.12) approximates to

$$\frac{X_b(t) - X_b(0)}{X_b(\infty) - X_b(0)} = \frac{2}{\beta}\frac{b^2}{a^3}\sqrt{\left(\frac{FDt}{\pi}\right)} \tag{7.13}$$

Equations (7.12) and (7.13) are, in fact, limited extremes of a general problem. In practice, β is only constant for dilute systems with low segregation levels. As segregation proceeds β generally falls as a result of saturation. If β starts high and falls rapidly as the segregation saturates, equation (7.13) is valid up to saturation.[75, 76] A detailed analysis for the saturation occurring in the Langmuir–McLean adsorption theory has been presented by Rowlands and Woodruff.[77] Their analysis, re-interpreted in Figure 7.19, shows how the time dependence of the segregation changes from equation (7.12) to equation (7.13) as the final equilibrium segregation level $X_b(\infty)$ approaches the saturation level X_b^0.

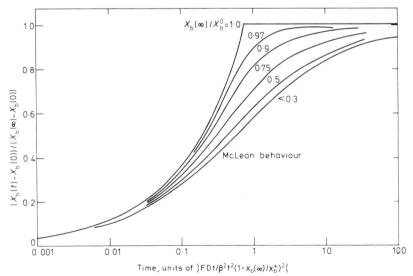

Figure 7.19 The kinetics of segregation in binary systems for varying degrees of saturation in the final equilibrium level. (After Rowlands and Woodruff[77])

The more complex problem of the kinetics of segregation in ternary systems has been studied by Tyson.[78] Tyson uses McLean's approach but incorporates Guttmann's theory to define the grain boundary concentration. For a ternary system with a dilute impurity and a non-dilute alloy element he finds that equation (7.13) is often a good description if the final segregation level is near saturation and equation (7.12) similarly if the final level is low. However, for certain critical cases, in which the two interacting species have similar values of $D\beta^{-2}$, the segregation curves exhibit an intermediate plateau.

The kinetics of de-segregation have, in the past, been described as much faster than the kinetics of segregation. This is not true and the relationships are as given in equations (7.12) and (7.13) with $X_b(o)$ and $X_b(\infty)$ transposed.

The above discussion is valid for wholly enclosed surfaces or for grain boundaries. For surface segregation experiments in a vacuum system using AES, evaporation of the segregant is possible. Lea and Seah[75] evaluated the effect of surface evaporation concomitant with surface segregation, as shown in Figure 7.20. The results show initially no divergence from the McLean result of equation (7.12). However, as the segregation builds up, the evaporation rate increases and the bulk material begins to be depleted of solute. The segregation goes through a maximum and eventually irreversibly falls to a low value as the substrate purifies. The numbers against the curves, E, are proportional to the evaporation rate of the segregant. As an illustration, the surface segregation of tin in Fe–0.22 per cent Sn at 700 °C represents a situation of

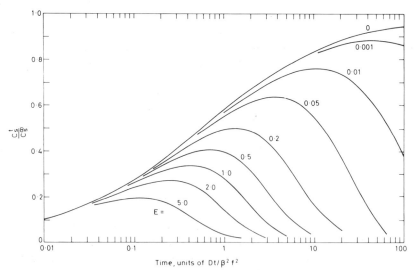

Figure 7.20   The time dependence of surface segregation with various evaporation rates described by a reduced parameter E. (After Lea and Seah[75])

weak segregation with $E = 0.007$ and the segregation peak occurs after 15 minutes.[75]

7.4 Materials Effects of Segregation

The segregation of solute species to surfaces and grain boundaries in a solid produces a zone with a discrete composition and discrete properties, vital to the overall performance of the material in an analogous manner to the importance of the properties of cement in brick buildings. For many of the engineering properties of materials, the effects of segregation on the interface energy, cohesion and diffusion are the important functions. Other properties are also important for specific types of failure but first we shall outline the effects of segregation on the basic materials properties.

7.4.1 Segregation and basic materials parameters

The effect of segregation on surface and interfacial energies is now well established.[20, 21] Combining the Gibbs adsorption theory[21] with equations (7.4) or (7.5) yields[49, 79]

$$\gamma_b = \gamma_b^0 + \Gamma_b^0 RT \ln(1 - X_b) \tag{7.14}$$

where γ_b and γ_b^0 are the interface or surface energies in the segregated and pure states, respectively, and Γ_b^0 is the mole per square metre of segregant in one atom layer. This relation, amply supported by experiment, shows that the interface energy is depressed by segregants and that those segregants that are highly surface-active depress the energy most.

The effect of segregants on grain boundary diffusion has been established much more recently. Combining the theory of Borisov, Golikov and Scherbedinsky[80] and equation (7.14), Bernardini *et al.*[81] find for the self-diffusivity, D_{gb},

$$D_{gb} = D_{gb}^0 \left[1 + \left(b_v - \frac{2\beta a^3}{b^3} \right) X_c \right] \tag{7.15}$$

where D_{gb} is the diffusivity for the segregated boundary, D_{gb}^0 is for the clean boundary and b_v is a small term denoting the effect of the solute on the bulk diffusivity D_v, given by

$$D_v = D_v^0 (1 + b_v X_c) \tag{7.16}$$

Equation (7.15) shows that the removal of the disorder by the segregation of solutes reduces the rapid transit paths normally associated with grain boundaries. The experiments show that order-of-magnitude reductions are possible for the grain boundary self-diffusivity and that similar but stronger effects occur for the solute diffusivity there. In the past grain boundary diffusion measurements have always been made using radiotracers; however, as Bernardini, Lea and Hondros[82] point out, they may now be determined for all systems using AES measurements of solute transport through thin foils.

The effects of segregants on cohesion has been the subject of many papers and much controversy, which it is not appropriate to discuss here. The basic divergence of opinion now centres round the assertion, on the one hand, that the segregant occupies a hole in the boundary and in so doing locks up metallic bonding electrons, consequently weakening the pre-existing metal–metal bonds[83] and, on the other hand, that the segregant replaces a host atom so replacing a metal–metal bond by a weaker metal–segregant bond.[84] In the regular solution approximation this latter approach agrees with the thermodynamic theory[85] and leads to the prediction that it is the atomic

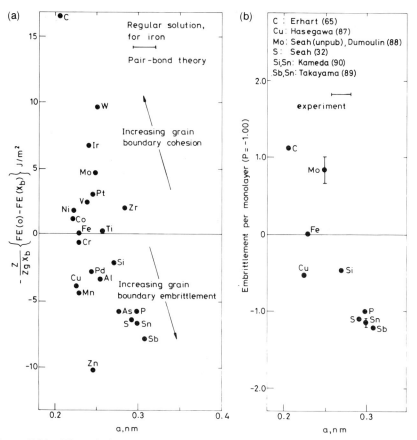

Figure 7.21   The relative embrittling/ductility effects of equivalent levels of segregants in iron: (a) the theoretical predictions (after Seah[73]) and (b) the measurements relative to phosphorus (after Seah[91])

bond energy per unit area that governs the propensity of segregants to weaken or strengthen grain boundaries. These terms may all be calculated from the data of Hultgren *et al.*[86] Figure 7.21(a) shows the relative effect of different segregants in iron calculated in this way. The ordinate value may be treated as a relative scale but is approximately 800 times the effect on the ideal work of fracture per per cent. of monolayer segregated. From Figure 7.21(a) it is clear that phosphorus, tin, antimony and arsenic will embrittle steel whereas carbon and molybdenum will reduce this embrittlement. The avail-

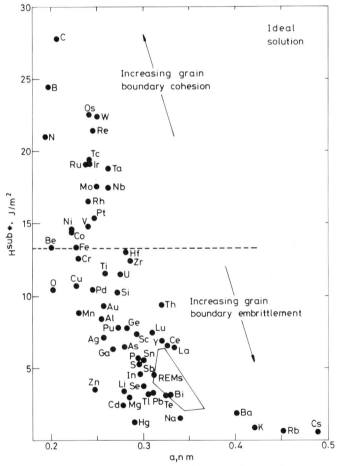

Figure 7.22   The general embrittlement/ductility plot for matrix and segregant elements in the ideal solution approximation (after Seah[73]). As an example, if iron is taken as the matrix, all points above the dotted line are ductilizing and those below are embrittling, to an extent dependent on their displacement from the line

able experimental results[32,64,87–90] for the relative effects of segregants, normalized to phosphorus, have been analysed by Seah,[91] as shown in Figure 7.21(b). These results support the original calculations very well except for carbon which appears to give the correct remedial effect but less strongly than originally proposed.

The calculations for segregants in iron may be generalized to all matrices in the ideal solution approximation. The results, shown in Figure 7.22, give the bond energy density on the ordinate axis. Those elements appearing below any other will, if segregated, reduce the grain boundary cohesion of that second element, roughly in proportion to the vertical separation of the elements in the figure. Thus the observed embrittlement of copper by bismuth,[92] tungsten by potassium[93] and nickel by sulphur[94] are all in agreement with this work and, in particular, the improvement in ductility of platinum–rhodium–tungsten alloys by boron[95] is anticipated in this diagram.

7.4.2 Temper brittleness

The analysis of the interface cohesion above is central to the problem of low temperature intergranular failure. The greatest incidence of this, in practice, occurs when low alloy steels are heated in the temperature range 450–550 °C. The name 'temper brittleness' merely arises from this embrittlement occurring in the steel's tempering range. Analysis of many commercial failures shows that phosphorus from typical commercial levels of 80 to 160 p.p.m. causes the failures in CrMo steels and that a combination of phosphorus and tin are involved in NiCrMo steels.[96]

If steels are tested by impact over the temperature range −100 to +100 °C it will be found that, at high temperatures, they are ductile and, at low temperatures, brittle. The temperature of the transition should be as low as possible but will rise roughly linearly with the segregation if intergranular failures are involved. Thus the measurements of transition temperatures for a SAE 3140 steel containing 150 p.p.m. of phosphorus, shown in Figure 7.23, should reflect the phosphorus segregation level. The predictions of the segregation of phosphorus in this steel using Guttmann's ternary segregation theory with McLean's kinetics show a perfect agreement of the curves for constant segregation with the curves for constant transition temperature, and hence allow extension of the data to periods of up to 30 years, as shown in Figure 7.24.[73] Some care must be taken in making such extensions, in practice, since over a long time the steel's hardness will generally reduce, improving the ductility. Also the partitioning of the alloy elements, between their incorporation in the carbides and being in solution in the matrix, will change, causing decreases or increases in segregation and embrittlement above that calculated by the simple approach.[88, 98]

In support of Guttmann's theory is his own wide range of work reviewed

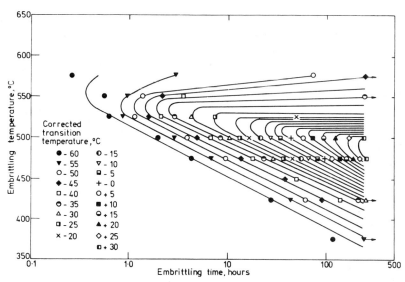

Figure 7.23   The time–temperature diagram for the embrittlement of SAE 3140 steel with 150 p.p.m. of phosphorus. (After Carr *et al.*[97])

recently[74] and the work of McMahon and his co-workers who show that nickel and chromium increase the segregation of phosphorus,[99] antimony[100] and tin.[101] However, recent work by Erhart and Grabke[65] and by Briant[102] all indicate that a totally different view may be more appropriate. In a study of a series of alloys of iron with carbon, chromium and phosphorus separately and in combination, Erhart and Grabke show that the addition of chromium to an iron–phosphorus alloy does not alter the segregation of phosphorus but that the addition of carbon suppresses the phosphorus segregation about fivefold. This is due to direct site competition where the sum of the fractional mono- layers of segregated phosphorus and carbon is constant. The effect of chromium addition to the iron–carbon–phosphorus alloy is to lock up the

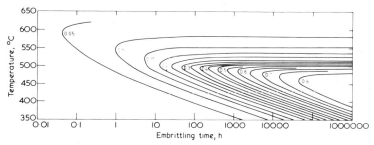

Figure 7.24   The segregation diagram appropriate to Figure 7.23. (After Seah[73])

carbon in carbide precipitates so that the carbon no longer segregates and the phosphorus segregation returns to the higher value close to that for the iron–phosphorus alloy. Thus the addition of chromium increases the phosphorus segregation, not through direct coupling, but through the carbon–phosphorus site competition. Segregation levels, similar to that above for the iron–chromium–carbon–phosphorus alloy have been observed by Briant[102] in a NiCr steel. Additionally, Briant finds that increases in the chromium, manganese and nickel levels have no effect on the phosphorus segregation. Clearly more work must be done in this area to clarify this controversy in order to understand the problem of temper brittleness. To some extent the problem is solved in a practical sense either by using better steel-making routes and purer ores to remove the phosphorus and tin or to use reactive additivies, as shown by Seah, Spencer and Hondros,[103] which fully neutralize the effects of the impurities.

7.4.3 Stress corrosion cracking

Intergranular failures in stress corrosion cracking are not as simple as the direct cohesion failures of temper brittleness. Here segregations can occur which do not affect the material's impact strength or ductility but which do cause failures in service, in certain electrolytes. For instance, Lea and Hondros[104, 105] show that certain segregants reduce the time to failure of strained mild steels in ammonium nitrate whilst no effect is seen in paraffin. The relative effect of the segregants is found to be proportional to the difference in the tabulated equilibrium electro-potential of each segregant and that of iron. The mechanism giving rise to this effect is thought to be anodic dissolution at the crack tip promoted by an electrochemical cell formed between the atoms of the segregant and matrix, the one promoting dissolution of the other depending on the electropotential relative to that of iron. In order of their potency, calcium, phosphorus, sulphur, aluminium, copper, nickel, tin, antimony, arsenic and zinc are found to be deleterious.

7.4.4 Stress relief cracking and creep embrittlement

These failures occur in creep-resistant steels under stress at their working temperatures. Creep embrittlement occurs at low stress during service and stress relief cracking occurs during the annealing cycle necessary to remove the stresses formed in the heat-affected zone of weldments. The intergranular failures are often associated with grain boundary cavities which have nucleated and grown, during the time under stress at temperature, until the net material section is so reduced that failure ensues under modest loads.

Following the theory of Skelton[106] which shows that the rupture strain, ε_r, due to cavitation is given by

$$\varepsilon_r \propto \left(\frac{D_{gb}}{N} \right)^{1/5}$$

where N is the cavity nucleation rate, Seah[23] proposed that the impurities could be responsible for embrittlement by two separate routes: (a) through surface segregation to the embryo cavity surface, so reducing the surface energy and increasing N, and (b) through grain boundary segregation, so reducing D_{gb} as shown in equation (7.15). Both routes decrease the rupture strain and hence increase the embrittlement. The nucleation problem has been studied in more detail from this basis recently[107] and additional evidence is being established for the role of phosphorus and tin surface and grain boundary segregation in the grain boundary cavitation failure of low alloy steels.[108, 109]

7.5 Conclusion

As shown in Figure 1.2 of Chapter 1, AES has a major impact in many of the value-added technologies. In metallurgy, long-standing problems are being solved and new materials and processes are being developed. In this chapter a classical adsorption problem is analysed and its impact in a whole range of materials failure problems defined. The use of scanning AES has led to the fruition of this field and to an understanding of the precise atomistics[110] of the fracture processes involved—whether they involve the segregant effect on bond strengths, electrochemistry, energetics of interfaces or material transport. This understanding has led metal producers and users to respond to the problem with effective remedial procedures, for instance by the use of additives,[103, 111] as discussed above, or by modification of the alloy element content.

References

1. D. A. Stout, G. Gavelli, J. B. Lumsden and R. W. Staehle, *Applied Surface Analysis, ASTM STP*, **699**, 42 (1980).
2. G. R. Conner, *Applied Surface Analysis, ASTM STP*, **699**, 54 (1980).
3. J. B. Mathieu, H. J. Mathieu and D. Landolt, *J. Electrochem. Soc.*, **125**, 1039 (1978).
4. C. Lea, *Met. Sci.*, **13**, 301 (1979).
5. E. Bullock, C. Lea and M. McLean, *Met. Sci.*, **13**, 373 (1979).
6. H. J. Grabke, W. Paulitschke, G. Tauber and H. Viefhaus, *Surf. Sci.*, **63**, 377 (1977).
7. H. J. Grabke, *Mater. Sci. Eng.*, **42**, 91 (1980).
8. C. Lea, *Proc. Eighth Int. Vac. Congr.*, Vol. II, *Vacuum Technology and Vacuum Metallurgy*, p. 468, Cannes (1980); *Le Vide les Couches Minces*, **1201**, Suppl., 468 (1980).
9. D. V. Edmonds and P. N. Jones, *Met. Trans.*, **10A**, 289 (1979).

10. R. E. Waters, J. A. Charles and C. Lea, *Met. Technol.*, **8**, 194 (1981).
11. C. T. H. Stoddart, C. Lea, W. A. Dench, P. Green and H. R. Pettit, *Met. Technol.*, **6**, 176 (1979).
12. H. J. Mathieu, D. Landolt and R. Schumacher, *Wear*, **66**, 87 (1981).
13. R. Schumacher, E. Gegner, A. Schmidt, H. J. Mathieu and D. Landolt, *Tribology*, **13**, 311 (1980).
14. C. Lea and B. Roebuck, *Met. Sci.*, **15**, 262 (1981).
15. D. Kalderon, *Proc. Inst. Mech. Eng.*, **186**, 341 (1972).
16. M. P. Seah, *Surf. Sci.*, **80**, 8 (1979).
17. E. D. Hondros and M. P. Seah, *Int. Met. Revs.*, **22**, 262 (1977).
18. D. McLean, *Grain Boundaries in Metals*, Oxford University Press, London (1957).
19. J. O. Arnold, *J. Iron Steel Inst.*, **45**, 107 (1894).
20. E. D. Hondros and D. McLean, Monograph 28, Society of Chemical Industry, London (1968).
21. E. D. Hondros, in *Proc. Melbourne Conf. on Interfaces*, (Ed. R. C. Gifkins), Butterworths (1969).
22. C. J. Beevers, *Met. Sci.*, **14**, 418 (1980).
23. M. P. Seah, *Phil. Trans. Roy. Soc.*, **A295**, 265 (1980).
24. M. G. Nicholas and C. F. Old, *J. Mater. Sci.*, **14**, 1 (1979).
25. J. Woodward and G. T. Burstein, *Met. Sci.*, **14**, 529 (1980).
26. C. Lea and M. P. Seah, *Scripta Metall.*, **9**, 583 (1975).
27. H. G. Suzuki and M. Ono, *Trans. Iron and Steel Institute of Japan*, **12**, 251 (1972).
28. M. P. Seah, *Surf. Sci.*, **53**, 168 (1975).
29. L. E. Davis, N. C. MacDonald, P. W. Palmberg, G. E. Riach and R. E. Weber, *Handbook of Auger Electron Spectroscopy*, 2nd ed., Physical Electronics Industries Inc., Minnesota (1976).
30. M. P. Seah and W. A. Dench, *Surface Interface Anal.*, **1**, 2 (1979).
31. S. Ichimura and R. Shimizu, *Surf. Sci.*, **112**, 386 (1981).
32. M. P. Seah and E. D. Hondros, *Proc. Roy. Soc. London*, **A335**, 191 (1973).
33. J. M. Walsh, K. P. Gumz and N. P. Anderson, *Quantitative Surface Analysis of Materials, ASTM STP*, **643**, 72 (1978).
34. R. P. Wei and G. W. Simmons, *Scripta Metall.*, **10**, 153 (1976).
35. M. Smialowski, *Hydrogen in Steel*, Pergamon, New York (1962).
36. M. Menyhard, *Surf. Interface Anal.*, **1**, 175 (1979).
37. M. P. Seah, J. M. Sanz and S. Hofmann, *Thin Solid Films*, **81**, 239 (1981).
38. B. D. Powell and H. Mykura, *Acta Metall.*, **21**, 1151 (1973).
39. W. C. Johnson and H. B. Smartt, *Met. Technol.*, **6**, 41 (1979).
40. M. P. Seah, *J. Vac. Sci. Technol.*, **17**, 16 (1980).
41. H. Shimizu, M. Ono and K. Nakayama, *Surf. Sci.*, **36**, 817 (1973).
42. P. S. Ho, J. E. Lewis and J. K. Howard, *J. Vac. Sci. Technol.*, **14**, 322 (1977).
43. R. Kelly, *Surf. Sci.*, **100**, 85 (1980).
44. B. C. Edwards, B. L. Eyre and G. Gage, *Acta Metall.*, **28**, 335 (1980).
45. B. D. Powell and D. P. Woodruff, *Philos. Mag.*, **34**, 169 (1976).
46. J. Kameda and C. J. McMahon, *Met. Trans.*, **11A**, 91 (1980).
47. T. Watanabe, S. Kitamura and S. Karashima, *Acta Metall.*, **28**, 455 (1980).
48. S. Suzuki, K. Abiko and H. Kimura, *Scripta Metall.*, **15**, 1139 (1981).
49. R. E. Honig and D. A. Kramer, *RCA Review*, **30**, 285 (1969).
50. M. P. Seah and C. Lea, *Philos. Mag.*, **31**, 627 (1975).
51. M. P. Seah, *J. Catal.*, **57**, 450 (1979).
52. W. C. Johnson, N. G. Charka, R. Ku, J. L. Bomback and P. P. Wynblatt, *J. Vac. Sci. Technol.*, **15**, 467 (1978).

53. A. Kumar and B. L. Eyre, *Proc. Roy. Soc. London,* **A370**, 431 (1980).
54. E. D. Hondros and M. P. Seah, *Met. Trans.,* **8A**, 1363 (1977).
55. F. L. Williams and D. Nason, *Surf. Sci.,* **45**, 377 (1974).
56. P. Wynblatt and R. C. Ku, in *Proc. ASM Materials Science Seminar 'Interfacial Segregation'* (Eds. W. C. Johnson and J. M. Blakely), p. 115, ASM, Metals Park (1979).
57. J. J. Burton and E. S. Machlin, *Phys. Rev. Lett.,* **37**, 1433 (1976).
58. A. R. Miedema, *Zeits. f. Metallkunde*, **69**, 455 (1978).
59. N. H. Tsai, G. M. Pound and F. F. Abraham, *J. Catal.,* **50**, 200 (1977).
60. C. Lea and M. P. Seah, *Surf. Sci.,* **53**, 272 (1975).
61. G. T. Burstein and J. Q. Clayton, *Scripta Metall.,* **13**, 1099 (1979).
62. M. T. Thomas, D. R. Baer, R. H. Jones and S. M. Bruemmer, *J. Vac. Sci. Technol.,* **17**, 25 (1980).
63. R. H. Jones, S. M. Bruemmer, M. T. Thomas and D. R. Baer, *Met. Trans.,* **12A**, 1621 (1981).
64. G. Tauber and H. J. Grabke, *Ber. Bunsenges. Phys. Chem.,* **82**, 298 (1978).
65. H. Erhart and H. J. Grabke, *Metal Sci.,* **15**, 401 (1981).
66. R. H. Fowler and E. A. Guggenheim, *Statistical Thermodynamics,* Cambridge University Press (1939).
67. C. Pichard, M. Guttmann, J. Rieu and C. Goux, *J. de Phys.,* **36**, C4-151 (1975).
68. M. P. Seah, *J. Phys. F: Metal Phys.,* **10**, 1043 (1980).
69. M. Guttmann, *Surf. Sci.,* **53**, 213 (1975).
70. M. Guttmann, *Met. Sci.,* **10**, 337 (1976).
71. M. Guttmann and D. McLean, in *Proc. ASM Materials Science Seminar 'Interfacial Segregation'* (Eds W. C. Johnson and J. M. Blakely), p. 261, ASM, Metals Park (1979).
72. Ph. Dumoulin and M. Guttmann, *Mat. Sci. Eng.,* **42**, 249 (1980).
73. M. P. Seah, *Acta Metall.,* **25**, 345 (1977).
74. M. Guttmann, *Phil. Trans. Roy. Soc.,* **A295**, 169 (1980).
75. C. Lea and M. P. Seah, *Philos. Mag.,* **35**, 213 (1977).
76. S. Hofmann and J. Erlewein, *Surf. Sci.,* **77**, 591 (1978).
77. G. Rowlands and D. P. Woodruff, *Philos. Mag.,* **40**, 459 (1979).
78. W. R. Tyson, *Acta Metall.,* **26**, 1471 (1978).
79. M. P. Seah, *Proc. Roy. Soc.,* **A349**, 535 (1976).
80. V. T. Borisov, V. M. Golikov and G. V. Scherbedinskiy, *Physics Metals Metallogr.,* **17**, 80 (1964).
81. J. Bernardini, P. Gas, E. D. Hondros and M. P. Seah, *Proc. R. Soc. London,* **A379**, 159 (1982).
82. J. Bernardini, C. Lea and E. D. Hondros, *Scripta Metall.,* **15**, 649 (1981).
83. C. L. Briant and R. P. Messmer, *Philos. Mag.,* **B42**, 569 (1980).
84. M. P. Seah, *Acta Metall.,* **28**, 955 (1980).
85. J. P. Hirth and J. R. Rice, *Met. Trans.,* **11A**, 1501 (1980).
86. R. Hultgren, P. A. Desai, D. T. Hawkins, M. Gleiser and K. K. Kelly, *Selected Values of the Thermodynamic Properties of Elements,* ASM, Metals Park (1973); and *Selected Values of the Thermodynamic Properties of Binary Alloys,* ASM, Metals Park (1973).
87. M. Hasegawa, N. Nakajima, N. Kusunoki and K. Suzuki, *Trans JIM,* **16**, 641 (1975).
88. Ph. Dumoulin, M. Guttmann, M. Foucoult, M. Palmier, M. Wayman and M. Biscondi, *Met. Sci.,* **14**, 1 (1980).
89. S. Takayama, T. Ogura, Shin-Cheng Fu and C. J. McMahon, *Met. Trans.,* **11A**, 1513 (1980).
90. J. Kameda and C. J. McMahon, *Met. Trans.,* **12A**, 31 (1981).

91. M. P. Seah, *Materials Science Club Bulletin,* **64**, 2 (1981).
92. E. D. Hondros and D. McLean, *Philos. Mag.,* **29**, 771 (1974).
93. R. P. Simpson, G. J. Dooley and T. W. Haas, *Met. Trans.,* **5**, 585 (1974).
94. W. C. Johnson, J. E. Doherty, B. H. Kear and A. F. Giamei, *Scripta Metall.,* **8**, 971 (1974).
95. C. L. White, J. R. Keiser and D. N. Braski, *Met. Trans.,* **12A**, 1485 (1981).
96. E. D. Hondros, M. P. Seah and C. Lea, *Met. Mater.,* January **1976**, 26 (1976).
97. F. L. Carr, M. Goldman, L. D. Jaffee and D. C. Buffum, *Trans. AIME,* **197**, 998 (1953).
98. B. L. Edwards, B. L. Eyre and G. Gage, *Acta Metall.,* **28**, 335 (1980).
99. R. A. Mulford, C. J. McMahon, D. P. Pope and H. C. Feng, *Met. Trans.,* **7A**, 1183 (1976).
100. R. A. Mulford, C. J. McMahon, D. P. Pope and H. C. Feng, *Met. Trans.,* **7A**, 1269 (1976).
101. A. K. Cianelli, H. C. Feng, A. H. Ucisik and C. J. McMahon, *Met. Trans.,* **8A**, 1059 (1977).
102. C. L. Briant, *Scripta Metall.,* **15**, 1013 (1981).
103. M. P. Seah, P. J. Spencer and E. D. Hondros, *Met. Sci.,* **13**, 307 (1979).
104. C. Lea and E. D. Hondros, *Proc. Roy. Soc. London,* **A377**, 477 (1981).
105. E. D. Hondros and C. Lea, *Nature,* **289**, 663 (1981).
106. R. P. Skelton, *Met. Sci.,* **9**, 192 (1975).
107. D. McLean, *Metals Forum,* **4**, 44 (1981).
108. C. A. Hippsley, J. F. Knott and B. C. Edwards, *Acta Metall.,* **30**, 641 (1982).
109. W. Hartweck and H. J. Grabke, *Scripta Metall.,* **15**, 653 (1981).
110. M. P. Seah and E. D. Hondros, *Proc. NATO Advanced Research Institute 'Atomistics of Fracture'*, Calcatoggio, Corscia, May 1981, NATO Conf. Ser., VI, Materials Science v. 5, p. 855, Plenum, New York.
111. C. J. Middleton, *Met. Sci.,* **14**, 107 (1981)

Practical Surface Analysis
by Auger and X-ray Photoelectron Spectroscopy
Edited by D. Briggs and M. P. Seah
© 1983, John Wiley & Sons, Ltd.

Chapter 8

Applications of Electron Spectroscopy to Heterogeneous Catalysis

T. L. Barr

*Corporate Research Center,
UOP Inc. Ten UOP Plaza, Des Plaines,
Illinois 60016, USA*

8.1 Introduction

The principal theme of this chapter is concerned with the involvement of electron spectroscopies (primarily X-ray photo-electron spectroscopy (XPS) or ESCA) in the field of heterogeneous catalysis. The chapter is intended to represent a broad coverage of that involvement. Although an extensive (and occasionally detailed) survey of the existing literature is presented, the goal is to cover most of the important developments rather than provide a complete review.

Catalysis was one of the areas of application of ESCA specifically encouraged by Siegbahn in his pioneering expositions,[1] and based partially upon that encouragement, a number of XPS-catalyst studies, both basic and applied, were reported during the first few years of commercial ESCA, 1969–1974. Many of these attempts were not very fruitful, however, and, in fact, in some cases immaturity produced confusion and even errors. Generally speaking, however, sufficient success was achieved, particularly in the basic areas, to promote a second, more productive phase (1975–present). It is this author's belief that this second phase has been completed and we are now on the verge of a vastly more productive period. This chapter therefore concentrates on developments in electron spectroscopic techniques and application during the latter period. In recent years numerous different catalyst fields have been examined. For brevity only three are described herein; however, this should be sufficient to provide some feeling for the scope and magnitude of the present-day influence of electron spectroscopy in catalysis.

Electron spectroscopic studies in catalysis are complimented to a certain extent by their proximity to the large 'basic' field of surface studies of controlled reactions on single-crystal metals and alloys. While the latter area can be viewed as an essential 'underpinning' for surface studies in heterogeneous catalysis, the exact points of contact between the two fields are, as we shall see, still somewhat limited. In this context, it is very revealing to examine three reviews written around 1975 by Yates,[2] Yates and Madey,[3] and Fischer[4] that stress the importance of the interrelationships between studies in fundamental surface science and basic and applied catalysis. These articles were united in the premise that the aforementioned interaction should be a major factor in surface analysis in the years following 1975. All of these reviews, therefore, contain detailed predictions concerning the (then) future. It is interesting to compare their excellent predictions against the accomplishments summarized at the end of this chapter.

The multitechnique approach seems to play a major role in the aforementioned basic studies. Often, in fact, XPS and AES play relatively minor roles in basic studies. On the other hand, it would appear that although ion techniques and structural analysis methods are assuming growing importance in applied catalysis, the primary tools are still the electron spectroscopies, particularly XPS.

Numerous excellent reviews of basic chemisorption studies have recently appeared; therefore, we will concentrate in this chapter on the (less well covered) 'applied' areas. Although an extensive review of the work of the former area is not appropriate here, some idea of the interaction between the applied and basic field may be obtained from the description of selected examples of the latter, particularly in the section devoted to platinum metal catalysis.

Before beginning this presentation, a few statements are also in order about terminology. First, the term *fundamental* will be employed in this chapter to denote those studies designed largely to improve or extend the general capability of electron spectroscopy. Most of these topics have been covered in some detail elsewhere in this text. For that reason, they will be described herein only insofar as their attributes are unique to catalysis. Otherwise the reader is referred to the appropriate section elsewhere in this book. The phrase '*basic* catalyst studies', on the other hand, will be used to designate that research designed to simulate catalyst properties by examining materials purposefully constructed to reflect only the properties of interest. Most of the efforts of researchers such as Somorjai, Madix and White (and, in fact, the entire field of chemisorption) fall into this category. This chapter considers only the highlights of these studies. Readers interested in more detailed descriptions should consult the many recent reviews. The primary concern of this chapter is in the *applied* area that features the application of electron spectroscopy to commercial and simulated commercial catalyst systems. These systems are composed of metals, oxides, sulphides and other species, labelled as *dopants*, generally placed onto a variety of wide-band gapped oxides, designated as *supports*.

All of the materials under consideration in this chapter were examined in their solid state. Although some of the catalysts were unsupported (see below), they are all considered to be *heterogeneous catalysts* (at least for the purposes of this chapter). This is an important point because it implies that most (if not all) of the processes of interest, at least, initiate on the surfaces of these catalyst materials, and therefore if techniques can be devised to examine these surfaces before, during and following the reactions in question, then useful information can be obtained. Not considered in this statement, or indeed in this chapter, is the very important (and as yet unanswered) question: what constitutes a catalyst surface? This is particularly uncertain in the case of commercial catalysts, where the materials involved are often configured such that much of their (apparent) active surface is isolated inside pores or compressed into spheres or extrudates. In point of fact, a large part of the active surface of a catalyst, while still being accessible to reactants (under pressure), may be largely inaccessible to traditional surface analysis tools, e.g. an electron spectrometer. In fact, the variable and often uncertain marriage of the material's surface to the analysis tool remains a major concern.

All of this means that the *applied* surface scientist is required to be very adroit in 'presenting' the material so that the areas of interest are as 'visible' as possible to the spectrometer. No matter how adroit she or he may be, however, some uncertainty exists in this area. Of even more importance is the related assumption, commonly made in electron spectroscopy (particularly ESCA), that a spectrum represents a statistical survey of that material, i.e. the spectrum can be used to portray a reasonable *average* of the state of the surface of that catalyst. No detailed or even semi-detailed studies are reported in this chapter to defend this supposition. (In fact, disturbingly few studies of this problem have been attempted.) Readers will have to form their own conclusions regarding this controversial point; however, one should keep in mind that hundreds of studies have been reported in which the supposition was assumed to be valid and in most of these cases, ESCA information was found that directly correlated with independent, chemical and physical changes.

8.1.1 Synopsis of pre-1975 electron spectroscopy–catalyst studies

The status of catalyst-related research employing electron spectroscopy during the first half of the 1970s was also described in review articles by McCarrol[5] and Tompkins.[6]

Tompkins devoted his attention mainly to a review of the status at that time of 'surface physics' or what we have labelled as basic studies. Carbon monoxide adsorption and reaction on single crystals and valence band spectra from both UPS and XPS were his paramount concerns. McCarroll, on the other hand, pointed out in the beginning of his review that 'heterogeneous catalysis had, in practical terms, little benefited' from the developments 'in the (then) last decade' in surface physics. He also noted that 'the practical value to the world . . . of a fully developed scientific basis to catalysis would be enormous'.[5]

McCarroll described pertinent advances in a variety of basic and applied areas, and also argued for the feasibility of the application of surface physics to problems in industrial catalysis, despite the apparent differences in the fastidious nature of the surfaces in the two areas. He explained how LEED and RHEED studies of single-crystal surfaces modified by overlayers of adsorbents can be pertinent to applied problems if the gases employed are similar to those that deposit on the metals in commercial catalysts. For example, the 'reorientation' of the Ni(111) face in the presence of sulphur was found to be promoted by an overlayer of Ni(100) (2×2)–S.[7] McCarroll also pointed out that, as a result of this reorientation the catalyst may present to the petroleum molecules a new surface with substantially modified activity and selectivity.

After a brief discussion of dehydrocyclization on platinum (similar to that studied by Somorjai, see below), McCarroll described some early applications of electron spectroscopy to catalysis. The XPS studies of Carberry, Kuczynski and Martinez[8] demonstrated that γ-irradiation induced the migration of alkali metals (e.g. cesium) to the surface of supported silver catalysts, thus enhancing the ability of this catalyst to oxidize ethylene.

Preliminary AES studies by Van Santen and Sachtler[9] and others, of the plantinum alloys that were suspected to be the precursors of the bimetallic-reforming catalysts for dehydrocyclization were also described. The relevance of this work was doubted by McCarroll. Subsequent developments, however, seem to have justified these initial efforts (see below).

A preliminary XPS study of the Co–Mo–S–Al_2O_3 system by Stevens and Edmonds[10] was also mentioned by McCarroll. He pointed out that this represents 'a more correct use of XPS' unaware, of course, of the numerous controversies that would soon arise in the growing literature on this system (see below). The observation of Stevens and Edmonds[10] that Mo^{IV} in the form of MoS_2 plays a key role in the performance of this catalyst system proved to be rather prophetic.

The AES studies of Bhasin[11] were also described. In these studies, judgement as to the expected performance of a Cu/SiO_2–Al_2O_3 catalyst in the presence of lead and magnesium was based upon the surface to bulk ratios of lead, magnesium, aluminium and silicon. In view of suspected beam damage effects, some of these conclusions may be called to question, but the results still point to the potential utility of AES.

A number of applied XPS-catalyst studies appeared in the early 1970s that were not reviewed by McCarroll. Most of these seem to have been influenced by the pioneering paper of Delgass, Hughes and Fadley.[12] These papers included a study of rhodium adsorbed on carbon by Brinen and Melera.[13] The potential of XPS as a tool for the type of relative quantitative study that is so important in making catalyst judgements is indicated in this work, particularly if caution is employed to assure that the characteristics of the support (pore size, etc.) are properly taken into consideration (see the subsequent section on quantitation). Brinen also utilized XPS to examine catalyst poisoning.[14] In another early study Matienzo *et al.*[15] provided vital XPS binding energies for the subsequent studies of nickel catalysts.

One of the XPS studies of supports was by Lindsay *et al.*[16] They described binding energy shifts between different aluminium oxide systems. Extending the quantitative arguments of Delgass, Hughes and Fadley,[12] and Brinen and Melana,[13] Temperé, Delafosse and Contour[17] employed XPS to examine zeolites with differing silicon-to-aluminium ratios. (Zeolites are materials with potential as supports and catalysts.) Evidence of extensive surface dealumination was found as the bulk silicon-to-aluminium ratio increased. Changes in the oxidation state of a cerium cation were also detected by monitoring the changes in the Ce $3d$ binding energies.

Additional studies of the $Co-Mo-S-Al_2O_3$ catalyst system were presented by Armour *et al.*,[18] who found that sulphiding reduced the binding energies of molybdenum and cobalt. The same system was studied by Cannesson and Grange[19] who used XPS data to postulate, for the first time, the presence of Mo^{5+} as well as MoS_2.

Although the XPS-catalysts studies mentioned above and the others published between 1970 and 1975 were individually quite informative, collectively they confirmed the arguments of Yates and Madey[2,3] (and reiterated by McCarroll[5]) that the situation in 1975 was one of, as yet, unrealized promise.

8.2 Fundamentals of Electron Spectroscopy—As Applied to Catalysis

8.2.1 Introduction

Most XPS studies begin with qualitative general scans and then rapidly expand to include the most intense core-level spectra. The latter often provide a wealth of quantitative and chemical information and many XPS analyses never branch beyond this stage. Very often, however, subtle changes in the composition, structure or chemistry of a system are the features of principal interest, and these changes may not 'register' in the 'common' parts of the core spectra. For this reason, other, less used and generally less well-understood, features of XPS have been tested and exploited in studies that have recently included catalytically related materials. Among the most promising of these features are XPS-induced loss, Auger and valence band spectra. All of these have been described elsewhere in this text; therefore, they are examined here only insofar as they represent novel ventures into the employment of XPS to study catalysts or catalytically related materials. In addition (and for the same reason), short statements about recent work in EELS, synchrotron radiation and pre-adsorption are included.

Before discussing these 'finer' points, however, it should prove instructive (and somewhat humbling) to consider the results of two recent surveys[20,21] that addressed the question of accuracy and precision for both XPS and AES quantitation and binding energies.

8.2.2 Comparative energy and intensity surveys

A very important catalyst-related survey study was published in 1977 by Madey, Wagner and Joshi.[20] It included contributions from a wide spectrum of industrial and academic XPS and AES laboratories (including that of the author). The study featured the compilation and comparison of line position and intensity measurements from 'standard' silica gel, alumina and NaA materials. Substantial dispersion was found in all these data (see Figure 8.1), not only when their absolute values were compared but also when several rather ingenious relative schemes (designed to eliminate random factors) were employed. The size of both the individual and collective dispersions indicated that many questions related to sample handling, charging (and its removal), quantitation and even energy scale calibration were still unanswered, particularly when *non-conductive* samples were being examined.

As a result of the lack of consistency in the above results, it was decided to 'return to basics' and examine three pure, polycrystalline, *conductive* metal foils of nickel, copper and gold. The dispersions registered in this study[21] were substantially reduced over those reported in the first study; however, the variances in both quantitation (intensity) and binding energy were still much larger than many had anticipated. In many ways, the second study was even more discouraging than the first, since these metals have commonly been employed to calibrate most electron spectrometers. Viewed objectively, these studies have been of tremendous value to electron spectroscopists (particularly those dealing with non-conductive materials), because they have inspired a period of introspection during which the concepts of calibration and gauging have been seriously examined and practiced. Most systems are functioning with greater precision (and accuracy) because of these studies. One should still be wary, however, in accepting interpretations based on singular measurements, particularly if the crux of that interpretation hangs on small differences in line energy and/or intensity.

8.2.3 Novel techniques and topics

8.2.3.1 *XPS-induced loss spectra*

Loss spectra have been both a curiosity and nuisance in ESCA since its founding by Siegbahn more than two decades ago. Initially, the appearance of

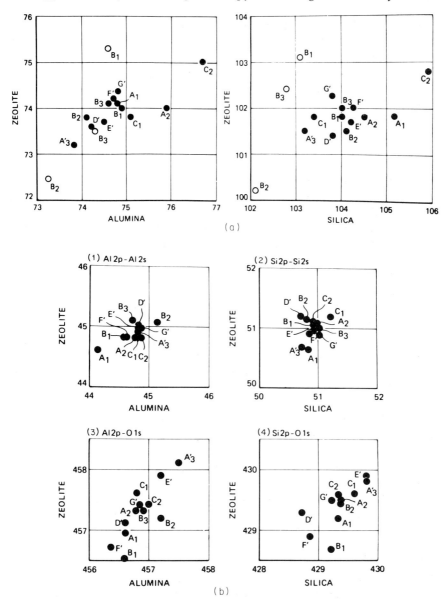

Figure 8.1 (a) Al $2p$ binding energy for zeolite versus Al $2p$ binding energy for alumina. Open circles: electron flooding; solid circles, C $1s$ reference. Si $2p$ binding energy for silica versus Si $2p$ binding energy for zeolite. Open circles: electron flooding; solid circles, C $1s$ reference. (b) XPS line energy difference plots: (1) Al $2p$–Al $2s$ for zeolite versus Al $2p$–Al $2s$ for alumina. (2) Si $2p$–Si $2s$ for zeolite versus Si $2p$–Si $2s$ for silica. (3) Al $2p$–O $1s$ for zeolite versus Al $2p$–O $1s$ for alumina. (4) Si $2p$–O $1s$ for zeolite versus Si $2p$–O $1s$ for silica. (Reproduced from Madey *et al.*[20] by permission of Elsevier Scientific Publishing Co., Amsterdam)

these variously sized 'satellites' on the high binding energy side of the core, Auger and valence peaks were not understood. However, diligent (and sometimes contradictory) experimental and theoretical efforts gradually discerned that these secondary peaks result from a variety of diverse inelastic events that depend as much upon the speed of the ejected photo-electron as the possibility for secondary transitions. In any case, as discussed in Chapter 3, a complicated mixture of shake-up (and-off), multiplet, plasmon, band-to-band and other transitions contribute to these satellite peaks.

Detailed discussions of the physical origins of the mutliplet and shake-up phenomena have been published by Fadley[22] and others.[23] Of all of the loss effects mentioned above, these two are probably the most understood, and are certainly the most exploited. They tend to produce prominent features in the XPS spectra of many transition metal systems. Shake-up features are being utilized in the analysis of catalyst systems, particularly where transition metal dopants are employed. A typical example was reported by Ng and Hercules,[24] during the study of a nickel–tungsten–alumina catalyst. In that study, the appearance (and disappearance) of the multiplet and shake-up satellites of nickel, as its oxidation state changes with treatment of the catalysts, was a central feature in the analyses. However, this type of analysis is not as straightforward as it may seem. Thus, its utilization during attempts to describe the Co–Mo–S–Al$_2$O$_3$ catalyst system at various stages of its treatment has led to some interesting controversies,[25,26] since it appears that in some circumstances the calcogenides of certain transition metals may not exhibit any satellite spectra (or chemical shifts), even when the corresponding oxides of the same oxidation state exhibit dramatic satellites and shifts.[27]

The use of XPS satellite spectra as a means for chemical analysis has recently been expanded by the present author to include the relatively weak loss lines exhibited by a number of 'ceramic' (support-type) oxides, e.g. Al$_2$O$_3$, SiO$_2$,[28] and other materials that might loosely be described as wide-band gapped semi-conductors.

Beginning with a series of indium compounds,[29,30] it was demonstrated that:[31] (a) finite (albeit small), reproducible loss lines can be generated for most systems with a non-zero band gap (see Figure 8.2); (b) for a series of compounds with common cation (e.g. indium) or anion (e.g. antimony) these loss lines are shifted from one another by an amount that generally exceeds that of the corresponding (Siegbahn) chemical shift by as much as several electronvolts (see Table 8.1); (c) despite their small size, these loss lines are often detectable for mixed systems; (d) quantum mechanical analyses of the causes of these loss lines reveal that they are primarily caused by intrinsic (hole) and extrinsic (photo-electron) induced excitation of the surrounding valence band plasma with some (often significant) perturbations due to band-to-band and other non-collective effects; and (e) despite their origin in the final (hole) state of the system, it has been demonstrated theoretically and

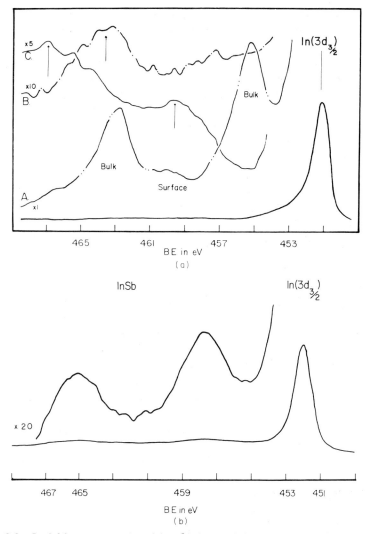

Figure 8.2 In 3*d* loss spectra for: (a) In⁰(A), InN(B) and In₂O₃(C) and (b) In Sb. (After Barr *et al.*[31])

experimentally[32] that the splitting of these loss lines from their principal (no-loss) peaks is independent of the Fermi edge (charging) of the material and is, to first order, also independent of relaxation effects. These loss splittings are, in fact, dependent upon rather simple parameters of the emitting systems (e.g. the valence-state electron density, etc.). Therefore, *the differences between the splitting of sets of loss lines (loss shifts) are often chemical shifts* that may be

Table 8.1   Identified loss lines (relative to their corresponding
principal XPS peaks) for selected In systems, in electronvolts

	In^0	InSb	InN	In_2O_3
$In(3d_{5/2})_1$	443.9	444.1	444.3	444.85
$In(3d_{3/2})_2$	451.3	451.5	451.7	452.3
S_1	—	—	11.1	13.2
B_1	11.5	13.3	15.5	18.5
S_2	8.2	—	11.4	13.5
B_2	11.8	13.3	15.8	18.8
$B_1 \times S_1$	20.1	—	26.5	31.9
$2 \times B_1$	23.1	26.4	30.9	37.2
$2 \times B_2$	23.2	26.2	31.2	37.5
$3 \times B_1$	—	40.4	46.3	—
$3 \times B_2$	35.2	—	—	—

S – probable surface plasmon, B – probable bulk plasmon.

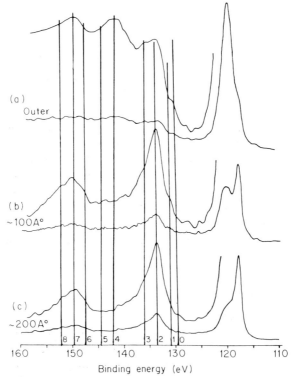

Figure 8.3   Representative XPS-induced loss spectra for Al 2s for (a) Al_2O_3 on Al^{10},
(b) mixed Al_2O_3–Al^0 system and (c) Al^0 with less than one monolayer cover of Al_2O_3.
(After Barr[28])

Table 8.2   Identification of probable loss lines in Al_2O_3–Al^0 system

	Source of photo-electron	Site of loss transition	Type of loss transition	Separation from Al lines in Al_2O_3	Separation from Al lines in Al^0
P_1	Al2p	—	—	0.0 (75.5)	0.0 (72.9)
P_2	Al2s	—	—	0.0 (120.4)	0.0 (118.0)
0	Al^0	Al^0 surface	S	9.1	11.5
1	Al_2O_3	Al^0 surface	S	11.1	13.5
2	Al^0	Al^0	B	13.5	15.9
3	Al_2O_3	Al^0	B	16.0	18.4
4	Al^0	Al_2O_3	B	21.9	24.3
5	Al_2O_3	Al_2O_3	B	24.2	26.5
6	Al_2O_3–Al^0	Al^0/interface	B × S/S × S	27.1	29.5
7	Al^0	Al^0	2 × B	29.5	31.9
8	Al_2O_3	Al^0	2 × B	31.9	34.3

S — surface emission
B — bulk emission

Loss splittings (Primarily plasmon)

	bulk	surface*
Al^0	15.8	11.3
Al_2O_3	24.4	17.4

$$* \, W_S = \frac{W_B}{\sqrt{2}}$$

employed to assist in differentiating between systems in much the same way as the original (Siegbahn) chemical shift. This has been done, for example, to describe physical changes noted in the growth of Al_2O_3, following the exposure of polycrystalline Al^0 to room temperature oxygen.[32] Similar features were successfully monitored during a study of the SiO_2–Si^0 interface.[28] The key loss lines exhibited in these cases were of extrinsic (electron only) origin, indicative of the passage of the photo-electron from a sublayer through a thin overlayer. The loss transitions that occur in this case permit XPS determination of the layer morphology (e.g. Al_2O_3 on Al^0, or vice versa; see Figure 8.3 and Table 8.2) in a manner not previously possible.[28,32] Such determinations may be very helpful in subsequent combined chemical–morphological studies of complicated layer systems.

8.2.3.2   X-ray induced Auger spectra

As mentioned above, XPS-induced Auger spectra may also be useful in the study of catalyst systems. In the past limited, but important, practical uses have been made of these final-state relaxation effects (see the discussion in Chapter 3). The fact that these Auger lines suffer dramatic changes with changes in chemistry that are often much more distinct that the changes experienced by the corresponding photo-electron peaks has been employed in the analysis of inorganic materials[33] and corrosion byproducts.[34,35,36] Corresponding XPS analyses of the changes experienced by species similar to the dopants (in use) on a catalyst have yet to be exploited. Potential applications of the *Auger parameter* (see also Appendix 4) to study catalyst supports, zeolites and related materials have recently been described in detail by Wagner and colleagues.[37] This technique may provide a powerful tool for analysis of typical catalyst systems because the parameter is (to first orders) independent of the charging shift (and thus is, in this sense, similar to the *loss shift*). Since Auger peaks are generally quite substantial for the materials commonly employed as catalyst supports, the Auger parameter may prove to be a sensitive barometer of such hard to read effects as metals–support interaction, support sintering, etc.

8.2.3.3 *XPS (and UPS) generated valence band spectra and EELS*

The valence band region is often loosely defined as approximately the first 20 eV of binding energy. In this region, the principal contributions to the PES arise from the energy levels that have suffered extensive molecular (orbital) or solid (band) involvement. (Some researchers may for convenience expand this region to cover the first 50 eV.)

Very little use has been made of the XPS-generated valence band spectra to study catalysts or related systems. There have been a number of studies, on the other hand, employing UPS and synchroton radiation sources, to examine some of the basic problems in catalysis. As an example, consider the studies of Gland, Sexton and Fisher[38] in which UPS was employed in conjunction with EELS and thermal desorption to examine the interaction of O_2 with the Pt(111) surface. Three temperature-dependent states of interaction were observed: (a) molecular oxygen (below 120 K); (b) atomic oxygen (150–500 K) and (c) subsurface oxide (1000–1200 K). UPS data (valence band spectra) were integral parts of the final analyses of these results, providing evidence of the filling and depletion of molecular orbitals, but the vibrational information of EELS also was necessary to truly establish state formation.

A related study by Kishi and Roberts[39] combined UPS with XPS to study the temperature programmed adsorption of NO by iron surfaces. Both He^{II} and He^I valence band spectra were reported and correlated with the simultaneous generation of (XPS) O $1s$ and N $1s$ core level peaks. In this manner, not only adsorbate characteristics but also desorption and dissociation were monitored. Molecular NO apparently adsorbs at low temperatures (~85 K), whereas these units were completely desorbed near room temperature. An uncertainty about the bonding of the NO species to iron remained, suggesting that simultaneous EELS studies (such as Gland, Sexton and Fisher[38]) would probably be useful. The importance of including XPS in these studies was demonstrated by the results. Thus NO dissociation was detected, even at 85 K, by monitoring the newly emerging N $1s$ peak at ~397 eV, indicative of the formation of Fe–N species. A variety of other studies employing some UPS have been reported. Several are described in the next section.

EELS has yet to be employed to study a commercial or simulated catalyst because of its present restriction to single-crystal substrates. It has been utilized, however, in a number of studies similar to the above described work of Gland, Sexton and Fisher. For example, Thomas and Weinberg[40] have employed EELS and thermal techniques to examine the adsorption of NO on Ru(001). Both bridged and linear forms of adsorbed NO are indicated, as is the order of dissociation. Further useful results, particularly on non-planar surfaces, may have been obtained employing other types of electron spectroscopy.

The use of XPS-generated valence band spectra to study catalytically related materials was demonstrated by the present author in an examination of the properties of zeolites and clays.[28,41] Distinctive variations in band structure were documented (see Figure 8.4) and the selective degradation of the

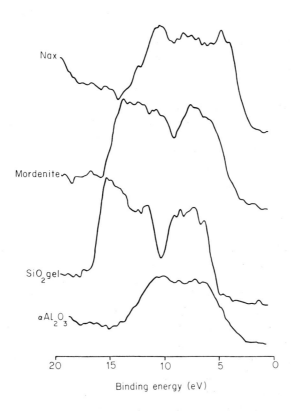

Figure 8.4 Representative valence band spectra (0–20 eV). (After Barr[41])

Table 8.3   (a) Selected band widths* in electronvolts (±0.1 eV)

Component	Band width
α-Al$_2$O$_3$	9.1
NaA	9.1
CaA	9.2
NaX	9.8
CaX	9.8
NaY	10.4
CaY	10.1
SiO$_2$	10.4

*Based upon widths at half band height.

(b) Reduction in band widths for selected materials, in electron volts (±0.1 eV), following sputter etching

Component	Band width
SiO$_2$	0.1
NaY	0.3
CaY	0.3
Bentonite	0.3
Kaolinite	0.7
Catapal	0.7
γ-Al$_2$O$_3$	0.8
NaA	0.7

zeolites was monitored through these spectra (see Table 8.3), thus permitting differentiating of these materials.

8.2.3.4   Photo-electron spectra generated from synchrotron radiation sources

Synchrotron radiation sources for photo-electron spectroscopy provide a number of advantages that may have utility in catalyst studies. For example, Miller *et al.*[42] have used the expansive variable energy range to locate the 2p and 2s levels of chemisorbed oxygen on Pt6 (111) × (100) crystal. Enhanced coverage with increased temperature is shown despite the lack of 'true' oxide formation. These arguments are also useful in studies of the adsorption of water and carbon monoxide as precursors to their oxidation (see below).

8.2.3.5   Pre-adsorption

Pre-adsorption, particularly of ambient, reactive gases, may markedly modify the results attributed above to *metals*. Platinum, for example, is assumed to suffer relatively easy reduction to P^0 (in flowing H_2) and this may promote catalytic processes on the surface of a dispersed phase of this element. This may, in fact, not be the case since it appears that all surfaces, even in rather pristine environments, adsorb some O_2, and even when no oxide growth is suspected there may still be oxygen involvement. In fact, McCabe and Schmidt[43] have demonstrated that oxygen-covered planes of Pt(110) are distinctly more reactive to $H_2 + CO$ than Pt^0. These studies utilized AES to assist in confirming the oxide layers and densities.

8.2.4 Chemisorption and basic catalysis studies

As pointed out in the introduction, an appreciation of the development of this area is very useful in any attempt to understand applied catalytic surfaces. A number of extensive reviews have already been published dealing with the application of electron spectroscopy to chemisorption and basic catalysis. These reviews are far too extensive to warrant repeating them in this chapter. The interested reader is advised to examine the articles by Brundle and Todd,[44] Gomer,[45] Robers and McKee,[46] Somorjai[47] (and references therein) in order to gain an appreciation of these developments. Several recent basic studies that this author feels to be of particular interest to applied catalysis are:

(a) The investigations by Madix[48-51] of a common reactant, formic acid, adsorbed onto various different metal surfaces permitting the detection of intermediates and subsequent suggestions of mechanisms. The concept of a 'pressure gap' as proposed by others was also challenged by Madix.
(b) Roberts and associates[52-54] studied the ability of various transition metals to adsorb and dissociate N–O. XPS results were key ingredients in detect-

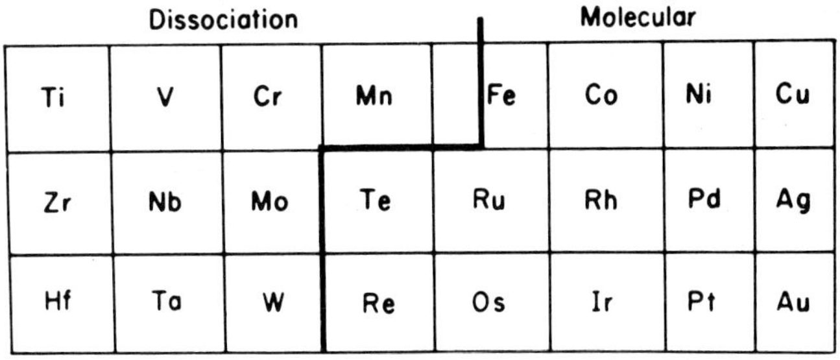

Dissociation				Molecular			
Ti	V	Cr	Mn	Fe	Co	Ni	Cu
Zr	Nb	Mo	Te	Ru	Rh	Pd	Ag
Hf	Ta	W	Re	Os	Ir	Pt	Au

Figure 8.5 Carbon monoxide reactivity pattern at 290 K; dissociative chemisorption occurs to the left of the heavy line. With iron, dissociation is comparatively slow at this temperature, and so is a borderline case. (Reproduced from Roberts[54] by permission of Academic Press Inc.)

ing patterns that lead to predictions of periodic behaviour based upon the electronic properties of transition metals (see Figure 8.5).

(c) White and associates[55–63] recently employed Auger as one of the tools in their studies of the reaction(s) of $CO + O_2$ on platinum and palladium.

(d) Burns and colleagues[64–66] have also conducted similar AES studies and developed mechanisms for $SO_2 + CO$ on platinum.

(e) Ertl *et al.*[67] have provided a detailed analysis of the mechanism for the production of NH_3 from N_2 and H_2 on iron. Langmuir–Henschel kinetics were proposed for the rate determining process based upon adsorbed atomic $(N + H)$ species.

8.3 Electron Spectroscopy and Applied Catalysis

8.3.1 Recent reviews related to this subject

An interesting review of applied heterogeneous catalysis and its interface with surface science was published in late 1975 by Boudart.[68] This contribution was similar in scope to that of Fischer[4] (see above) and in many ways the former superseded the latter.

Several informative reviews dealing specifically with the application of electron spectroscopy to catalysis have also recently been published. Two of the more illuminating are those of Brinen[25] and Briggs.[69] Both articles deal primarily with applications to applied catalysis and they are particularly interesting because of their unique points of view and the examples chosen.

All of Brinen's examples[25] were of supported catalysts. The nature of the resulting support is a feature of paramount concern, as was demonstrated in the aforementioned study of rhodium (a relatively inactive dopant) on carbon, where carbon was apparently chosen because many of its physical properties could be varied in a controlled manner.[13] The Brinen review[25] also contains an extensive discussion of the $Co–Mo–S–Al_2O_3$ catalyst, the most extensively reported catalyst system in the XPS literature. This system seems to exhibit easy to measure, but hard to interpret, changes in quantitation, layering and oxidation state (see Section 8.3.5.3 below).

Briggs,[69] on the other hand, began his study by describing what he refers to as the three 'essential' aspects of XPS studies of catalysts: (a) deactivation, (b) quantitation and (c) catalyst modelling (i.e. derivation of structure/property relationships). Three examples are provided in each of these areas, but only one was of a supported catalyst. Of all the points considered, Briggs finds dealing with catalyst modelling to be the most important, but as he points out in this example it is, at present, the least understood.

Another review by Kelley[70] was not exclusively intended for catalysis, but it contained vital catalytic implications because it dealt extensively with surface-to-bulk ratios and surface segregation. Many applied catalysts are bi- and trimetallic systems where one of the metals (or its oxide) commonly acts as the primary catalytic agent (for hydration, dehydrogenation, oxidation, etc.), whereas the other metals (or oxides, sulphides, etc.) function primarily as *promoters* or *attenuators* or in some other manner *modify* the activity of the former metals. Many researches suspect that these multimetallic systems exist in the functioning catalyst in alloy[71] or cluster forms on their supports. Kelley's article addresses, in a collective sense, both empirical and experimental methods for determining (and continuously monitoring) the state of mixing of the various species (on the surface). TEM and ion spectroscopies, as well as UPS and AES methods, were described. In a later extension of these arguments Kelley[72] also described the application of XPS to these problems.

8.3.2 Problems and techniques in the application of electron spectroscopy to catalysis

8.3.2.1 *Introduction*

The two major areas of uncertainty in the use of electron spectroscopy to study supported catalyst systems are the techniques employed to determine chemistry and quantitation. Interestingly, these features seem to have been treated with reasonable success in individual laboratories, but universally accepted procedures, particularly for detecting binding energy shifts indicative of changes in chemistry, do not yet exist (see the survey results discussed above in Section 8.2.2). The primary reasons for the lack of consensus in these areas would seem to be the inability to fix the binding energy zero (charge shifting problem) and the inhomogeneity that characterizes many catalysts and often destroys the previously described statistical nature of electron spectroscopy. For brevity, the following discussion of these points concentrates upon their effect on XPS results, although many of the arguments apply equally well to AES.

8.3.2.2 *The charge shift problem and energy referencing*

The surface chemistry of both the support and the dopants in a catalyst have generally been studied with XPS. As mentioned above in the discussions of the general features of electron spectroscopy, there are two principal problems in applying this type of analysis to catalysts: (a) the dopants are often so dilute that it is very difficult to detect any chemical change (this feature is discussed in more detail in the subsequent sections dealing with dopant chemistry) and (b) the binding energies of the generally non-conductive support materials are difficult to fix because of the charge shifting problem (see Appendix 2 and Figure 8.6 a and b). As described elsewhere in this text, the

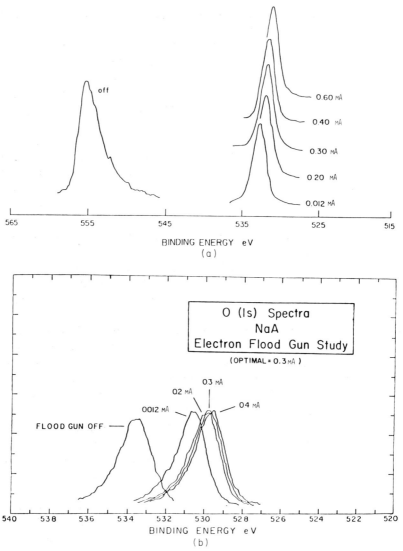

Figure 8.6 Examples of electron flood gun treatment of XPS charging shifts of O 1s lines. (a) SiO_2—note the substantial size of the flood gun off shift. (b) NaA—note small size of charge shift

morphology, as well as the composition of systems, plays a major role in the size of the resulting charge shift. In addition, the separate complication of differential charging may arise. The latter problem is difficult to handle, particularly if one employs the common practice of fixing the binding energy scale against a specific value for one of the atoms in the support, and then referencing all dopant binding energies against that value. None of these difficulties have been solved or eliminated in the general sense, but the following description may, at least, provide useful insight and suggestions.

The general problem of the charging shift has recently been reviewed and carefully defined by Lewis and Kelly.[73] Wagner[74] and Windawi and Wagner[75] have also described charging, and have demonstrated how the homogeneity of materials affects this problem. Windawi[76] has also described some aspects of how a lack of homogeneity may lead to differential charging.[14]

These ideas have recently been extended in a presentation by this author,[77] wherein the causes of differential charging were defined and models constructed based upon a variety of morphological units composed of two species of distinct charging characteristics. In this work, it was shown that even when a dopant is contiguously mixed (dispersed) into a support, that dopant *may* experience some form of differential charging. Also demonstrated in this work[77] and that of Lewis and Kelly[73] was the fact that the charging shift is only part of the referencing problem. The other part results from the 'floating' (lack of coupling) of the Fermi edge of many materials off the Fermi edge of the spectrometer due to a lack of thermal equilibrium. This feature occurs for non-conductive samples and for conductive materials insulated from the spectrometer. If this occurs, the XPS binding energies registered against the zero (Fermi edge) of the spectrometer are not valid. Thus binding energies realized for catalyst dopants are not only shifted through charging (inherent, for example, to the 'chemistry' of the support oxide), they are also shifted to a greater or lesser degree due to this 'floating Fermi edge'. All of these effects depend upon the morphological conditions of the system. Some specific examples and models are indicated in Figures 8.6 and 8.7.

These problems have resulted in the examination and implementation of a variety of procedures for removing the charging shifts and determining valid binding energies. Unfortunately, none of these procedures are universal panaceas. The most commonly employed techniques seem to be:

(a) As mentioned above, the binding energies of certain peaks for key elements in the support, e.g. Al $2p$ in Al_2O_3, have been assumed to have a fixed value, no matter what the status of the system. Sometimes this approximation is acceptable; however, there are also times when it may negate the determination of such important features as metals–support interactions. In general, it has been described by the present author as an oversimplification that is often incorrect.[28]

(b) To circumvent the above-mentioned difficulty, many authors have suggested that the C 1s peak produced by adsorbed, adventitious carbon could be assigned a fixed value, e.g. 285.0 eV, and all other binding energies could be referenced to it. The assets and problems with this procedure have been recently reviewed by Swift,[78] who advised a great

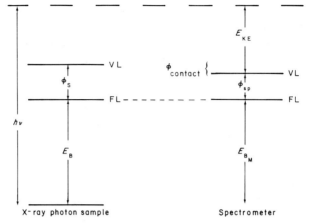

$$E_{B_M} = h\nu - \phi_S - (\phi_{sp} - \phi_S) - E_{KE}$$

$$E_{B_M} = h\nu - \phi_{sp} - E_{KE}$$

$$E_B = E_{B_M} \quad \text{Due to thermal equilibrium} \\ \text{(Fermi level coupling)}$$

$$\therefore E_B = h\nu - \phi_{sp} - E_{KE}$$

Spectrometer factors   Measured

(a)

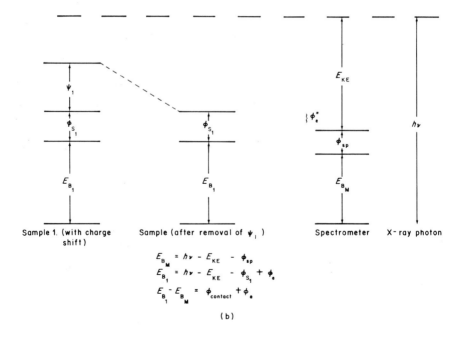

$$E_{B_M} = h\nu - E_{KE} - \phi_{sp}$$

$$E_{B_1} = h\nu - E_{KE} - \phi_{S_1} + \phi_e$$

$$E_{B_1} - E_{B_M} = \phi_{contact} + \phi_e$$

(b)

deal of caution in employing this technique. This procedure has also been analysed in this laboratory,[38,77] where it has been concluded that many practioners have failed to grasp the significance of the two, independent, but interrelated problems mentioned above, i.e. *the charging shift* and the *lack of Fermi edge coupling between the spectrometer and the sample.* Because of the complexity of the interrelationship of these effects with respect to (i) the support, (ii) the dopants and (iii) the adventitious impurities (such as carbon), it is suggested that it may be necessary to try to *first remove the charging shift separately* (perhaps with a weak current of electrons, see below) before tackling the Fermi edge problem and fixing the binding energy scale as, for example, against a value for the C 1s line. It must be noted, however, that *even for the latter procedure to work, all identified species must be contiguous to the carbon employed as a reference.* Otherwise differential charging and inaccurate binding energy referencing will result.

(c) Removal of the charging shift through application of an electron flood gun is another procedure often employed to attempt to produce valid binding energies for materials such as supported catalysts (see Figure 8.6). However, as pointed out by Windawi[76] and amplified by Barr,[77] this procedure will not work for discontiguous systems that produce differential charging, even if it does succeed for most contiguous systems. Further, as described by Lewis and Kelly[73] and amplified above, removal of the charging shift of a contiguous system does not in itself promote Fermi-level coupling. This means that either determination of the sample work function, forced Fermi edge pinning[79] or binding energy referencing (as described above) are still necessary to achieve valid binding energies.

(d) A fourth procedure of circumventing the charging problem and establishing useful (and perhaps valid) binding energies, for materials like supported metal catalysts, involves the use of parametric features of the total spectrum that are independent of both the charging shift and the floating Fermi edge problem. Several such parameters exist, and each seems to have individual, but perhaps limited, merit. One method is the *Auger parameter* (described in more detail elsewhere in this book). This technique permits the registration of a characteristic value for each system (based on the splitting of the principal photo-electron and Auger peaks)

Figure 8.7   Energy level diagrams for the XPS experiment. (a) Conductive sample in Fermi level contact with spectrometer. Original Siegbahn experiment. (b) Semiconductor or insulator sample 1, showing charging ψ and its removal. Note lack of Fermi level coupling. ϕ_{tot} is the energy needed to remove ψ_1 (the outer potential or charging shift) and bring the electron from the sample vacuum level to the spectrometer vacuum level. Note that, for simplicity, this energy is assumed not to affect ϕ_{S_1} or E_{B_1}

that exhibits many of the features of a chemical shift, but is independent (to first order for contiguous materials) of the charging shift and floating Fermi edge. The procedure was introduced by Wagner[37] and has recently been extended and expanded upon by Castle and West.[80] In a paper that is particularly important to those interested in catalyst supports, Wagner *et al.* have extended the procedure to cover a number of oxygenated compounds of aluminium and silicon.[81]

Another parametric procedure of potential utility is the generation of XPS-induced loss spectra as described earlier in this chapter.[29,30,31] Successful application of this technique to SiO_2,[28] Al_2O_3[32] and zeolites[28] has been described elsewhere. In these studies it was demonstrated that the loss shifts could not only assist in chemical but also morphological determinations. To first order, these loss shifts depend only on the electronic properties of the emitting material and are also independent of the status of the Fermi edges.

During considerations of the rectification of the problems associated with the charging shift, it is often forgotten that like many problem areas some very useful side-effects may arise. For example, it has been demonstrated[28] that through knowledge of the presence, size and response of a charge shift (and/or the disparity in Fermi edges) to an electron flood gun, one may be able to provide a quasi-microscopic description of the surface (and sometimes sub-surface). This is achieved by empirically relating the experimental results to the appropriate model.[77]

8.3.2.3 *Quantitative analysis of catalyst surfaces*

A number of papers have been published dealing specifically with the subject of the use of XPS to determine the quantitation of catalyst surfaces. The results indicate, however, that additional fundamental work is needed.

The basic concepts of XPS quantitation have been discussed in Chapter 5. Catalyst surfaces present a number of particular problems because of the inhomogeneities that often occur (e.g. multiple phases, laterial composition variations, crystallite growth and particulate clustering, etc.). An early assessment of the homogeneous case and some of these inhomogeneities was presented by Brinen.[25] In addition, Scharpen[82] provided an excellent account of how the lack of dispersion of platinum on SiO_2 may complicate the interpretation of quantitative results. Since then several models have been developed specifically designed to describe the dispersion problem, i.e. when one phase (e.g. a dopant) is poorly dispersed into another (e.g. a support), as well as the effects of support porosity. Space limitations do not permit a description of these models; the interested reader is therefore directed to the original dispersion studies of Angevine, Vartuli and Delgass,[83] Kerkhof and Moulijn,[84] and Windawi and Wagner[75] for details. Fung[85] has added factors to these models designed to account for specific shape variations in the undispersed phase. The recent ideas of Nefedov and others[86,87] related to the effect of elastic scattering on quantitation should also be considered.

8.3.3 Platinum metal catalysis

8.3.3.1 Studies of 'molecular state' catalysis by Somorjai

Before describing the application of electron spectroscopy to applied platinum metal catalysis, it should prove instructive to examine the efforts of Somorjai and colleagues. This work was carried out primarily on platinum metal single crystals and was designed to delineate some of the basic aspects of many reforming-type reactions. Of all the basic examinations designed to simulate applied catalysis, these provide perhaps the closest match. Most of these studies utilized various aspects of surface science as their principal *modus operandi,* with LEED (and recently EELS) playing the primary role. Conventional electron spectroscopy has also been utilized, particularly AES. (XPS has recently seen increasing use.)[88] In these investigations, Somorjai and associates have attempted to relate certain aspects of metals catalysis to the reactive behaviour of a variety of key organic gases on variously construed faces of (primarily) platinum single crystals. No supports, in the classic sense, were employed; thus, the crystal faces of platinum must not only function as catalyst reactants but also provide the adsorbent sites. The five principal variables monitored in most of these studies were: (a) the type of organic reactants and products; (b) the surface faces (with special cuts to produce terraces, steps and kinks); (c) the temperature (reactions were monitored between 75 and 550 K); (d) the pressure P of the gases, where $10^{-4} < P < 100$ atmospheres (the upper limit represents the use of a special high-pressure apparatus); and (e) the various extents and locations (on the single-crystal faces) of retained carbon byproducts of the reactants listed in (a). The time of attachment of these carbonaceous layers was also monitored.

Somorjai and colleagues have discovered that (under the conditions employed) the most common reactions of alkanes and olefins (dehydration, cyclization, isomerization, hydrogenolysis, etc.) seem to prefer the sites (on Pt metal single crystals) of high Miller indices (low symmetry).[89] Thus, relatively low temperature chemisorptions and reactions that do not occur, for example, on the (100) faces will occur with relative ease on the (111) face, particularly if steps and kinks are present. In fact, Somorjai and associates have discovered that the above-mentioned bond breaking and rearrangement processes occur selectively at these sites. This means that different single-crystal faces of platinum provide the two key ingredients of a good catalyst, i.e. (a) the energetics needed to *activate* a reaction and (b) the *selectivity* needed to promote one reaction over several competing processes.[8]

The patterns exhibited by the above-mentioned surface faces were determined by LEED. The qualitative and quantitative characteristics of the adsorbed species were adjudged by AES, and the reactants and products were evaluated by transport techniques.[90] As mentioned above, high pressure studies were made possible through the use of a unique UHV system with a built-in high-pressure cell.[90]

In particular, the Pt(111) face was found to be much more reactive for dehydrocyclization than the Pt(100) face.[91] In addition, it was postulated that the hydrogenolysis reaction occurred preferentially on *kink* sites of the Pt(111) face, whereas dehydrocyclization seems to prefer the *step* sites of that face.[92] These features were summarized by Somorjai and Zaera in a recent review article.[93] In that publication, representations were presented of the various sites for these reactions (Figure 8.8) and the relative activities and selectivities for each (see Figure 8.9).[93] Note that the number of platinum atoms located on the terraces (above the steps and kinks) also plays a part.

It was also determined that these reactions are catalysed in the presence of thin, ordered layers of carbonaceous byproducts.[94] A thick, amorphous layer

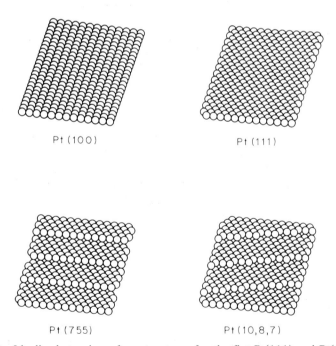

Pt (100) Pt (111)

Pt (755) Pt (10,8,7)

Figure 8.8   Idealized atomic surface structures for the flat Pt(111) and Pt(100), the stepped Pt(755) and the kinked Pt(10,8,7) surfaces. (Reprinted with pemission from Somorjai and Zaera[93]. Copyright 1982 American Chemical Society)

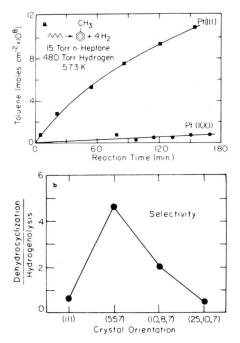

Figure 8.9 (a) Toluene accumulation curve measured as a function of reaction time for *n*-heptane conversion over Pt(111) and Pt(100) surfaces. (b) Dependence of dehydrocyclization selectivities at 15 torr of *n*-heptane, 480 torr of hydrogen, and 573 K, as a function of surface structure. (Reprinted with permission from Somorjai and Zaera[93]. Copyright 1982 American Chemical Society)

of this 'coke', on the other hand, seemed to poison (block) the catalyst sites, particularly when the carbon has lost its mobility. Similar observations have been made by other surface scientists studying related reactions.[95] Employing rather elaborate[14]C studies,[96] Somorjai and colleagues have been able to extend this thesis to include the proposed states of degradation (sequential dehydrogenation) of these carbonaceous deposits.[97] From this argument, models have been constructed of catalytically active, mobile C_2H_3 units, that with further heating or use become immobile as C_2H on top of the former species. Finally adsorbed, graphite clumps begin to grow and the rates of all reactions seem to decrease exponentially with the concentration of this species.[93] Figure 8.10 (a), (b) and (c) is a pictorial rendition of these carbon deposit effects.

It has also been shown that selective insertion of certain metals (e.g. tin) may prevent hydrogenolysis and thereby enhance dehydrocyclization. This may be accomplished through the blockage of kink sites by the tin and its extraction of electrons from the corners, thus activating steps.[98] These fea-

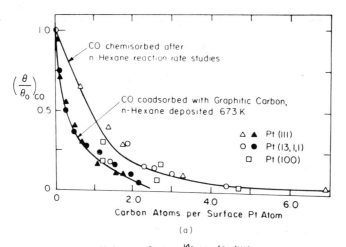

(a)

Hydrogen Content $^{14}C_2H_4/Pt$ (III)

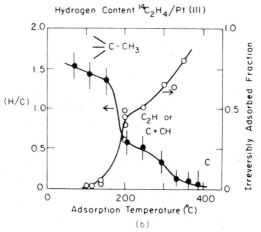

(b)

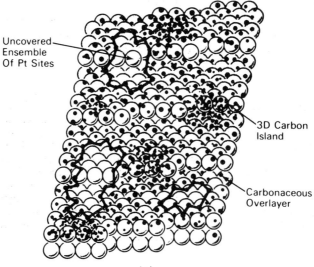

(c)

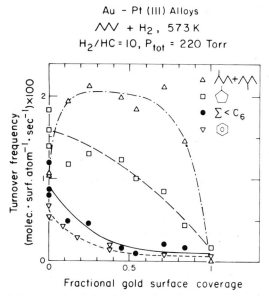

Figure 8.11   Activity and selectivity changes for *n*-hexane conversion as a function of gold coverage in the gold–Pt(111) alloy system. (Reprinted with permission from Somorjai and Zaera[93]. Copyright 1982 American Chemical Society)

tures may relate to those that are attributable to the bimetallic catalysts which are needed to produce higher octane gasoline. Recently Sachtler *et al.*[99] have attempted to further bridge the gap between these single-crystal catalysts and the bimetallic area. Layering structures of gold on platinum single crystals (and vice versa) were examined. In addition to some reconstruction, it was noted that gold on Pt(100) noticeably enhanced the rate of the dehydrogenation of cyclohexane to benzene. Further gold on Pt(111) seemed to enhance isomerization of *n*-hexane (at 575 K), whereas dehydrocyclization and other reactions were retarded by this gold layer[100] (see Figure 8.11).

Figure 8.10   (a) Fractional concentrations of uncovered platinum surface sites determined by CO adsorption–desorption as a function of surface carbon coverage on the (100), (111) and (13,1,1) platinum surfaces. A comparison is made between the CO uptake determined following *n*-hexane reaction studies and CO uptake determined when CO was coadsorbed with 'graphite' surface carbon. (b) Composition and reactivity of ethylene-^{14}C chcmisorbed on Pt(111) at temperatures between 20 and 370 °C; the irreversibly adsorbed fraction determined by radiotracer analysis displays an excellent correlation with the average hydrogen content (H/C) of the strongly adsorbed species. (c) Model for the working surface composition of platinum reforming catalysts. (Reprinted with permission from Somorjai and Zaera[93]. Copyright 1982 American Chemical Society)

The effects of several additional reactants including oxygen, have recently been described. Pre-oxidation was shown (by AES) to alter the rates of the aforementioned reactions, enhancing the dehydrogenation, apparently because the strongly bonded oxygen changes the electronic structure of the Pt(111), in or around the kinks.[101]

Somorjai and Zaira[93] have also recently expanded these types of studies to include another platinum metal, rhodium. They particularly designed their experiments to try to delineate effects realized through changes in the oxidation state of that metal. In this case, EELS was employed along with AES to study the reduction of NO by CO over Rh(331) (a reaction of interest in air pollution chemistry).[102] The NO dissociates, apparently producing an

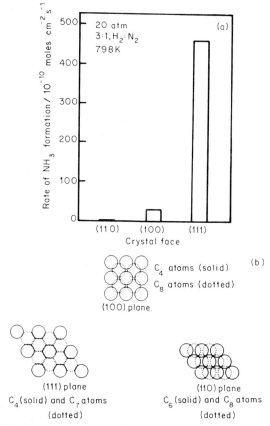

Figure 8.12   (a) Surface structure sensitivity of iron-catalysed ammonia synthesis. (b) Idealized atomic structure for the low-index planes of iron: Fe(100), Fe(111) and Fe(110). Reprinted with permission from Somorjai and Zaera[93]. Copyright 1982 American Chemical Society)

adsorbed oxygen atom intermediate. In addition, Watson and Somorjai[103] have employed XPS as well as AES to study the hydrogenation of CO over rhodium catalysts (reactions of interest in Syn Gas technology). XPS indicated that Rh^0 produced mainly alkanes and olefins, whereas to get aldehydes and other oxidized products, Rh_2O_3 was needed. However, anhydrous Rh_2O_3 easily reduced to the metal retarding the oxidation reactions. Hydrated Rh_2O_3, on the other hand, was only partially reduced during the process, producing patches of the oxide and metal. In a recent extension of this study it was discovered that $LaRhO_3$ stabilized the reactant, producing almost exclusively oxidized products. Rh^{+1} seemed to be a probable intermediate.[104] Efforts employing EELS to extend these studies to include rhodium supported on Al_2O_3 have recently been initiated.[105]

The development of the combined low-high pressure chamber, described above, has also permitted studies of other important chemical reactions over non-platinum metal crystals. For example, AES and the above-mentioned chamber were employed to examine the use of iron as a catalyst to generate NH_3 from N_2 and H_2 at 798 K and 20 atmospheres.[106] The reaction was found to dramatically favour the Fe(111) face over the Fe(100) and Fe(110) faces (see Figure 8.12).

Somorjai, in his recent review, speculated on how the above information may lead to commercial versions of 'high technology molecular catalysts'. This vision may be somewhat premature, but future practical utility of these ideas seems certain. In the meantime, one cannot help but be fascinated by the wealth of information, particularly the selective promotion of reactions on low symmetry sites, and the staged promotion followed by poisonous behaviour of the developing coke layers.

8.3.3.2 *Introduction to modern platinum metal catalysis*

The use of platinum as a catalyst has existed for many years, but it was not until the discovery of the reforming process by Haensel of UOP[107] that the special catalytic status of platinum became apparent. The system gradually evolved into one in which platinum in the form of a complex was placed on a high surface area support, often γ-Al_2O_3.[108] This system was subsequently reduced, apparently generating microscopic Pt^0 particles, extensively dispersed over the alumina. The Pt^0 units were so tiny that they could not be detected with conventional X-ray diffraction.[109] This system was often augmented with halide ions to promote the acidity of the alumina, thus making a *bifunctional* catalyst, suitable for the selective promotion of the dehydrocyclization reactions needed to enhance the octane number of high-performance gasolines.[108,109] This system, in slightly modified form, also found utility in the rapidly growing petrochemical industry, particularly as a dehydrogenation catalyst to generate olefins.[108] Thus, by the mid-1960s platinum metal catalysis for high-octane gasoline production (reforming), petrochemicals and even modified oxidation purposes was a major separate branch of catalysis. The reforming system was not without its shortcomings, however, as it tended to lose activity and selectivity if called upon to refine low quality crudes at low pressures and high temperatures.[109] Therefore, even before the science of the platinum-only catalyst was understood, catalyst scientists and engineers were experimenting with additional dopants to improve the system. Out of these efforts, in the later 1960s came bimetallic catalysis.

Many of the most important applied catalysts are multimetallic systems whose functionality has been attributed by some to alloying.[71] This is particularly true for the Group VIII–1B system, where the former is the catalytic species and the latter functions as a modifier. It was postulated that the selective behaviour of these catalysts depends as much upon the number of 'contiguous surface sites' as their composition.[110] Thus, for example, bimetallic systems are postulated to behave in one way, if they may be loosely configured as –A–B–A–B–A– (where A represents the Group VIII species and B the Group 1B component) and in another way if they may be configured as –A–B–A–B–A–A–A–B–A.[111] The former, totally dispersed system, seems to favour the fissure or establishment of —C—H bonds, while the latter (A clustered environment) tends to promote severance and/or rearrangement of the —C—C bond.[71,111] Stated in another way, it is apparent that contiguous, relatively large clusters of a Group VIII element (e.g. platinum) will promote reactions that *may* lead to coking-type processes that rapidly deactivate the catalyst, whereas catalysts composed primarily of microcrystallites of platinum (separated from each other by the alloying (?) of Group 1B elements or such elements as rhenium) seem to experience largely dehydrogenation.[71,108] The support for these bimetallic 'alloys' (generally alumina) may also provide acidic sites that may separately promote selective cyclization (aromatization).[108] This acidity may be enhanced by halide addition and proper thermal treatment.[108,112] Thus, these systems may be referred to as bimetallic–bifunctional catalysts.[71] The Group 1B or VIIB (and sometimes Group IVA) species included are often referred to as modifiers, since they promote the longevity and selectivity of the catalyst, while inhibiting (attenuating) runaway side reactions that deactivate the system.[111]

It should be noted that several researchers have provided compelling studies that dispute the alloy concept in bimetallic platinum metal catalysis.[113] It is their general contention that the modifiers retain positive oxidation state following reduction of the catalyst and that these species interact with the platinum in a way that prevents clustering of the latter into debilitating large crystallites.[113]

The published literature on the *characterization* of these catalysts are extremely sparse.[110] The relative amounts of the metal dopants are generally very small and the identification of the acid sites on supports has long baffled catalyst scientists.[113] A wide spectrum of questions has been asked about the composition, macro and micro structure, and chemical and metallurgical characteristics, before, during and after use.[111] Electron spectroscopists, for example, have been asked to contribute information about such important features as: (a) the degree (if any) of the alloying;[111] (b) the type and position of the acid sites;[113] (c) the oxidation state(s) of the metals involved;[110] (d) identification of cluster growth;[113] and (e) the chemical and physical characteristics of the carbonaceous deposits.[71,93] Very few of these questions have seemingly been answered with any success. The complicated nature of the commercial catalysts, plus their multiplicity of purpose and sparse ingredients, have led to various simulations involving non-reactive supports (e.g. SiO_2[114]) and unrealistic increases in dopants. The results, so far, have proved useful, but satisfactory detailed explanations of 'real' platinum metals catalysis awaits well-directed published studies on commercial systems.[115]

The electron spectroscopic studies of platinum metal catalysis to be described are divided into five categories depending upon the principal feature being examined. Note that the concept of the platinum metal has been extended to include not only the light and heavy platinum metal triads but also the heavy coinage metals (silver and gold) and (for reforming purposes) rhenium. Copper and nickel are not included.

8.3.3.3   *XPS spectra of the key elements and their oxides*

There have only been a few XPS studies of platinum metals that are useful for catalytic studies. Winograd and colleagues[116] have produced several interesting studies of platinum, palladium and silver and their oxides. A detailed study of the elemental metals, palladium, platinum and rhodium, their oxides and hydroxides was also reported by the present author.[117] In addition, Cimino *et al.*[118] have recently presented a detailed study of rhenium and its multiplicity of oxides as a preliminary study for a subsequent investigation of supported rhenium catalysts. Probably the best study of gold was contained in the aforementioned ASTM sponsored study edited by Powell, Erickson and Madey.[21] Many of these studies are summarized along with a collection of other binding energy results recently published by the Physical Electronics Division of Perkin-Elmer.[119]

All of these studies are of importance to catalysis because one of the chief concerns in the latter area is the possible back-and-forth changes experienced by the elemental metals and oxides during the oxidation and reduction steps commonly utilized during catalyst set-up. In addition, the effects of various degrees of air exposure must be registered. The above-mentioned study from our laboratory has demonstrated that many metals, including, to varying degrees, most of those under consideration, oxidize in O_2 or air to form oxide layers of increasing oxidation state; these, in air, terminate (passivate) in thin layers of hydrated oxides or hydroxides.[117] A crude summary of the apparent 'natural passivation' layers formed in room temperature air for the metals in question has been presented.[117] Note that platinum hardly passivates at all and that the rate (and perhaps degree) of this process is *retarded, but not removed,* in moderate vacuums (e.g. 10^{-3} torr). These observations are therefore of importance in XPS studies of catalysts, since most are exposed to air either during their (poorly confined) use or transport to the characterization chamber. Thus, with the exception of gold (and possibly platinum), one should expect some degree of surface oxidation of all metals in a catalyst due simply to this exposure. It should be noted, however, that results obtained on metal foils such as those described[117] do not necessarily transfer exactly to metals dispersed onto large surface area supports. For these reasons, care must be taken to properly interpret the true set-up and/or use the status of a catalyst, particularly when trying to adjudge the oxidation state changes of the metals involved. Contrary to the objections raised by some critics, however, these problems do not make surface analysis of these catalysts impossible, as it is often possible to avoid, ignore, remove or read out the degree of this (air-induced) oxidation. It is important to remember that the set-up, use and extraction of commercial catalysts are *not* conducted under conditions that will prevent some degree of the above-mentioned air passivation. Therefore, processes conducted in the ultra-high vacuum confines of the conventional surface analysis chamber may not duplicate the corresponding status of commercial catalysts. The significance of a 'pressure gap', as discussed (negated) in the work of Madix[49] and implied in the studies of Somorjai and Zaera,[93] is therefore a significant factor in this discussion.

The most important binding energies reported in the above-mentioned studies have been listed.[119] The spectra reported for the rhenium compounds by Cimino *et al.*[118] may be modified somewhat by the above-mentioned air exposure problem. In addition, at least two 'types' of the PtO species were documented in the studies of Winograd and colleagues.[116] This laboratory reported only one $Pt^{II}O$, but evidence for two more 'states' has been found. The one with a smaller chemical shift, PtO_a, possibly indicates incomplete oxidation of Pt^0. This may have significance with respect to the 'electron withdraw' effect in platinum metal catalysis, as described by Clarke and Creaner[115] and Burch.[113]

8.3.3.4 XPS studies of the dopants in reforming catalysts

Several studies of simulated commercial platinum metal reforming catalysts using XPS have succeeded in providing useful information despite limitations due to the very restricted 'visibility' of the sparsely populated dopants and the natural inhibition provided by the oxide supports. Escard *et al.*,[120] for example, have employed XPS, thermogravimetric analysis and XRD to investigate the status of the surface of hexachloroplatinic acid (H_2PtCl_6), under the thermal conditions employed to impregnate a platinum metal catalyst in both air and inert atmospheres (H_2PtCl_6 is a common 'Pt' source in commercial catalysts).[113] Using comparative spectral analyses, the surface species of the H_2PtCl_6 were found to suffer extensive thermal decompositon. Binding energy analysis established that this decomposition apparently produced a mixture primarily composed of $Pt^{II}Cl_2$ and Pt^0 'interacting' with adsorbed oxygen. The XRD analysis indicated that this decomposition was primarily, but not entirely, restricted to the surface. It appears, therefore, that the decomposition may have been induced in part by the atmospheric conditions and partially by the analysis beam, and although the effect is depth-limited, it should influence the resulting catalyst set-up. Czáran, Finster and Schnabel[121] have examined the same problem but have tried to more closely simulate actual catalyst preparations by employing charcoal and aluminosilicate supports. They also observed the aforementioned decomposition (even on the two supports!) to Pt^0 and Pt^{II} species, but they found that the extent of this process varied depending upon the status and type of support. Apparently the amount of dispersion greatly influences the degree of decomposition. They also noted that $Pt(NH_3)_4Cl_2$, another common platinum source for platinum metal reforming, did not decompose following exchange onto the aluminosilicate support.

Bozon-Verduraz *et al.*[122] also employed XPS and ultra-violet–visible spectroscopy to study the products formed following both impregnation and subsequent oxidation and reduction of $PdCl_4^{2-}$ onto SiO_2, TiO_2 and Al_2O_3. In this case, the authors felt obliged to use unrealistically large concentrations of the platinum metal dopant in order to facilitate XPS analysis. The principal concern was to detect metals–support interactions. In this case the $PdCl_4^{2-}$ complex appeared to be retained in the oxidized sample. Following reduction there was evidence of a quasi-complex formed with the support (?) containing Cl^-. They postulated this involvement because of noticeable broadening of the Pd 3*d* lines for the reduced catalyst. The support seemed to play an obvious, but as yet undefined, role in this complex formation. Differences were detected in the amounts of complex formed on the various supports, with Al_2O_3 exhibiting the largest amounts of the three. The system with the most 'complex' in the reduced form ($Pd–Al_2O_3$) was also found to be the most attenuated in its ability to promote oxidation of ethylene.

In a similar vein, XPS, electron microscopy and electron diffraction were employed by Griffiths and Evans[123] to study the changes of silver particles supported on graphite during attempts to oxidize the latter species. These results were reported to show enhanced reactivity of the carbon, resulting from the production of a silver–oxygen–carbon interface. The authors denied the significance, however, of a shift in the Ag 3*d* spectra following reduction that appeared to be substantial enough (to this author) to signify oxidation of the silver. In this case, as in many other related studies, charge-shifting may have played an undetermined role.[28]

XPS was also utilized by Batista-Lcal, Lester and Lucchesi[124] to examine the changes realized by $AuCl_3$ and other gold-containing systems when they were thermally treated, both alone and while supported on Al_2O_3. Heating in air to 350 °C caused a complete reduction of the Au^{III} to Au^0, whereas heating to 110 °C reduced most of the Au^{III} to Au^I. On the other hand, $AuCl_3$ supported on Al_2O_3 at room temperature exhibited no reduction. The authors suggested that the experiments must be conducted with care, as some X-ray-induced reduction was detected.

Schipiro *et al.*[125] employed XPS to study the states produced by various Re^{VII} species doped onto SiO_2 and Al_2O_3 supports. The reducibility of certain rhenium compounds on γ-Al_2O_3 was found to be exceeded by that for the same compounds on SiO_2. *This was assumed by Shipiro et al.*[125] *to be due to the much larger metals–support interaction of Re^{VII} in the former (Al_2O_3) case.* Reduction of these systems in H_2 at 300 °C produced an intermediate rhenium oxidation state, perhaps Re^{IV}. Heating both of the reduced Al_2O_3 and SiO_2 supported systems in air to 400 °C seemed to reoxidize the rhenium to Re^{VII}. *All of these features demonstrated the utility of XPS to delineate changes in electronic properties for metals dispersed over 'active' (Al_2O_3) and*

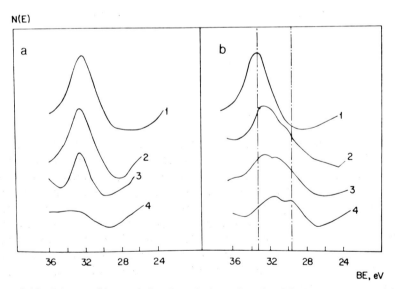

Figure 8.13 Line positions of the Ge 3*d* photo-line for (a) Ge–Al_2O_3 and (b) Pt, Ge–Al_2O_3 (reference: C 1*s* = 284.6 eV). (1) Calcined, 525 °C; (2) hydrogen-treated, 450 °C; (3) hydrogen-treated, 550 °C; (4) hydrogen-treated, 650 °C. (Reproduced from Bouwman and Biloen[126] by permission of Academic Press Inc.)

'inactive' (SiO₂) supports. Apparently this is a key factor in catalyst behaviour.

Very few studies have been reported in which electron spectroscopy was employed to study multimetallic platinum metal catalysts, designed to simulate present-day commercial reforming systems. Perhaps the most interesting and encompassing have originated in the Shell-Amsterdam Research Laboratories, in the groups of Sachtler, Biloen and others. Both the Pt–Ge–Al₂O₃ and the Pt–Re–S–SiO₂ systems have been investigated, but in each case the authors were forced by the 'visibility problem' to use substantially more metal dopants than commonly employed in the corresponding commercial systems. *XPS studies of the reduced Pt–Ge–Al₂O₃ system*[114,126] *produced Pt 4f spectra that were interpreted as resulting from electron-deficient Pt⁰.* The oxidation state of germanium seemed, in this case, to be split between Ge²⁺ and Ge⁰ when the system was reduced in H₂ at 625 °C, and between Ge⁴⁺ and Ge²⁺ for reductions at 550 °C in the presence of platinum. These studies were the first detailed XPS examinations claiming evidence of bimetallic alloy formation (see Figure 8.13 and Table 8.4). Evidence was also presented apparently showing both metals diffusing into the SiO₂ support. XPS, IR and chemisorption techniques indicated that sulphur promotes the separation of the planinum atoms in the Pt–Re–S–SiO₂ alloy (?) catalyst.[127] The authors depicted this concept as follows:

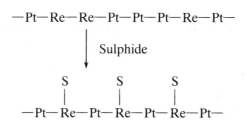

This transformation seemed to be indicated in the changes experienced in the Re 4f XPS spectra following addition of sulphide. The authors also noted that the production and maintainance of the 'dispersed' platinum metal atoms seemed to ensure catalytic selectivity and stability for the reforming process. The substitution of SiO₂ for the more conventional Al₂O₃ support was done to avoid the acidity of the latter and also the interference of the Pt 4f lines by the Al 2p lines. This alteration dramatically changes the catalytic behaviour of this system from that typical of the standard commercial (Al₂O₃-supported) system because of the substantial contribution of the acidity of alumina in the function of the latter.

In the non-reforming area, Angevine and Delgass[128] employed XPS to study a Ru–Mo catalyst on SiO₂ or Al₂O₃. This is a catalyst employed in methanation reactions. In an examination of the precursor, Ru–SiO₂ and

Table 8.4   Pt $4d_{5/2}$ line positions for some reference systems*. (Reproduced from Bouwman and Biloen[126] by permission of Academic Press Inc.)

System	BE, eV	ΔBE†	ΔBE‡
Metal	314.1	0	
PtO_2	318.0	3.9	
H_2PtCl_6	317.5	3.4	
Na_2PtCl_4	316.7	2.6	
$PtGe_{0.72}$ alloy	315.5	1.4	
'Pt^0'/γ-Al_2O_3	313.5	−0.6	0
Pt/Al_2O_3			
Calcined, 525 °C	317.0	2.9	3.5
Hydrogen-treated, 650 °C	314.6	0.5	1.1
Pt, Ge/Al_2O_3			
Calcined, 525 °C	316.7	2.6	3.2
Hydrogen-treated, 650 °C	315.1	1.0	1.6

*C $1s$ = 284.6 eV.
†Reference: platinum metal.
‡Reference: 'Pt^0'/γ–Al_2O_3, which is obtained by measurement of the $(Au–Pt)_{bulk}4d_{5/2}$ distance and measurement of the Au $4d_{5/2}$ line position of 0.1 monolayer gold on top of γ–Al_2O_3.

Ru–Al_2O_3 systems, it was discovered that the former permitted (hydrogen-treated, 450 °C) reduction of ruthenium completely to Ru^0, whereas in the latter (Al_2O_3) case 45 per cent of the ruthenium remained oxidized. Angevine and Delgass[128] attributed this effect to the strong metals–support interaction between ruthenium and Al_2O_3 that does not exist for SiO_2. On the other hand, for the Mo–SiO_2 system, the Mo^{VI} was only reduced under these same conditions to Mo^{IV} whereas in Ru–Mo–Al_2O_3 the molybdenum was reduced to Mo^0. *The authors interpret this to mean that the bimetallic effect (between ruthenium and molybdenum) supersedes the metals–support interaction.*

8.3.3.5   Studies of clustering and other morphological variations in platinum metal catalysis

XPS, UPS and AES have been applied in conjunction with XRD, electron microscopy, LEED and thermal spectroscopies to delineate the micro and macro morphological features of both the dopants and the supports in commercial and simulated platinum metal catalysts. These studies are significant because it has been discovered that the behaviour of these catalyst systems depend crucially upon the degree of both the sintering of the support and the clustering of the dopants. In particular, it has been discovered that use of a platinum metal catalyst under conditions of extreme physical (e.g. thermal) strain may promote growth of platinum metal crystallites. As these crystallites grow, both the degree and type of subsequent catalytic reactions change dramatically. For the purposes of promoting long-lived dehydrocyclization reactions (a primary producer of high-octane gasoline in reforming), it is necessary (as mentioned above) to impede the growth of these crystallites. Therefore, any combination of the above-mentioned characterization techniques that can accurately assess this type of problem is of potential importance to the catalyst industry. Until recently electron spectroscopy has not been a major contributor in this area, but this appears no longer to be the case.

An excellent description of the crystallite growth problem and several attempts to characterize it (although not with electron spectroscopy) was presented recently in the work of Dautzenberg and Walters.[129] In that study XRD, TEM and H_2 (BET) chemisorption were employed to study the effects of heat treatment on the state of dispersion of platinum on Al_2O_3. Once again, unrealistically large amounts of platinum (2 weight per cent) were utilized to enhance visibility. With XRD these authors were able to detect the growth of platinum metal crystallites for clusters larger than 30 Å. A similar range of particle sizes were characterized by TEM. It was also discovered that as long as the hydrogen-to-platinum ratio remained near 1, the platinum crystallites were small. Growth of these crystallites (to an average size of 100–150 Å diameter) was accompanied by a linear decrease in the H/Pt ratio to ~0.5. (Numerous independent measurements have been made of H/Pt and in most of these, values of 1 are associated with dispersed catalytically (active) platinum metal, and 0.5 with clustered catalytically (unactive) platinum metal. Giordano *et al.*[130] have recently extended these H/Pt studies to include chloride addition in both oxidative and reductive atmospheres. They have also provided a detailed critical review of the existing models of hydrogen adsorption on platinum; see Figure 8.14.) Growth of platinum metal crystallites seems to occur through lateral migration of Pt^0 induced by heating in air in excess of 500 °C. Dauztenberg and Walters found[129] that heat treatment in H_2, to as high as 650 °C, produced a reduction in the H/Pt ratio, but no apparent platinum metals agglomeration; thus BET measurements alone may not necessarily reveal the size of the platinum metal crystallites.

Because of the apparent restrictions in the detectable size range of XRD, TEM and also BET measurements, a number of researchers have turned to electron spectroscopy as a means for studying this problem. Success is, as of yet, only marginal. Several related questions were of obvious interest in these studies including: (a) the ability to realize meaningful surface-to-bulk quantitation; (b) the determination of support pore size; (c) the detection of metals segregation; and (d) examples of surface diffusion.

As mentioned earlier in this chapter, Brinen and Schmitt[131] employed rhoduim on carbon as a test system to study the distribution of a metal dopant, in particular from the point of view of surface-to-bulk quantitation. These authors carefully documented the many deceptive quantitative changes

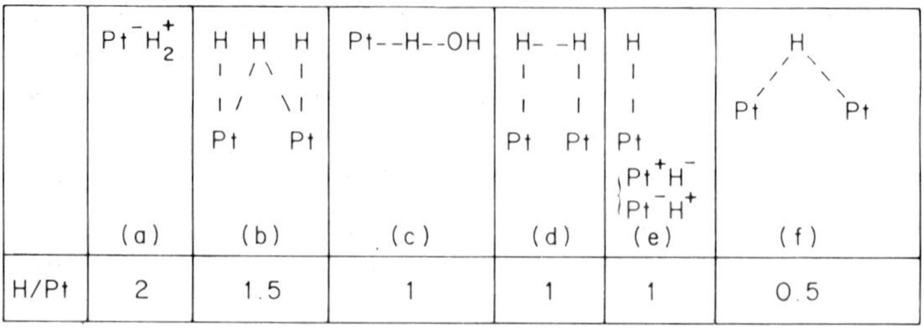

	Pt⁻H₂⁺	H H H I /\ I I / \I Pt Pt	Pt--H--OH	H- -H I I I I Pt Pt	H I I Pt \Pt⁺H⁻ /Pt⁻H⁺	H / \ Pt' `Pt
	(a)	(b)	(c)	(d)	(e)	(f)
H/Pt	2	1.5	1	1	1	0.5

Figure 8.14 Various H–Pt species and stoichiometries. (a) Adsorption of the H_2 molecules gives rise to a weak bond. (b) Simultaneous adsorption such as (e) and (f) (see below). (c) Hydrogen is strongly chemisorbed, due to the presence of vicinal hydroxyl groups; this situation is more pronounced on less dehydroxylated surfaces. Bonding involves Pt^{4+} according to:

$$O^{2-} Pt^{4+} \square^{2-} (OH)_{Al_2O_3} + \tfrac{1}{2} H_2 \longrightarrow O^{2-} Pt^{3+} H\, OH^-$$

(d) Some authors consider the bonding and dissociation of molecular hydrogen on Pt or other metallic surfaces having d electrons as due to a super-exchange or indirect exchange, a well-known phenomenon from the magnetic interaction in insulators. The bonding involves a system comprising two vicinal Pt atoms and the H_2 molecule, i.e. interaction between the unpaired d electrons of the Pt (spin-'up' or -'down') and the paired electrons of molecular hydrogen (spin-'up' or -'down'). The model implies a system with four electrons and three particles. Such hydrogen chemisorbed species are easily removed under vacuum at room temperature while they could be desorbed already at $-20\ °C$ in temperature programmed desorption. (e) Hydrogen is positioned over the surface Pt atoms through a weak covalent bonding (reversible chemisorption) with a weakly ionic character. (f) Hydrogen is interstitially located within the surface Pt atoms giving rise to irreversible chemisorption. The bonds are not located necessarily over two specific Pt atoms. (Reprinted from Giordano et al.[130], p. 585, by courtesy of Marcel Dekker, Inc.)

that occurred due to the poor diffusion of the rhodium and the resulting photo-electron escape depth.

Stoddart, Moss and Pope[132] utilized AES to describe the metals distribution in a Pd–Ni alloy on SiO_2 catalyst by exploiting the small rasterable beam of the AES. Because excellent agreement was obtained between AES and the X-ray determined bulk concentration, no electron beam damage was suspected. Several important effects of diffusion were suggested by the AES results, e.g. air exposure seemed to draw nickel out of the bulk alloy to the surface, to form NiO.

Similar XPS and SEM studies have been conduced of platinum metal oxidation catalysts. For example, Contour et al.[133] examined the distribution of metals in a Pt–Rh ammonia oxidation system. The SEM registered evidence of separate crystallite growth by both platinum and rhodium upon combustion. XPS quantitation indicated that during oxidation Rh_2O_3 was segregated to the surface, perhaps as a result of PtO↑. The authors noted that these effects may be particular to ammoxidation.

Takasu *et al.*[134] have employed a combination of XPS and UPS in conjunction with electron microscopy to delineate the degree of aggregation of different dispositions of palladium on SiO_2. These authors achieved excellent microscopic resolution permitting the registration of metal particle sizes down to ~10 Å. They were able to note changes in both UPS and XPS that seemed to correlate with particle size. The matrix effect of Kim and Winograd[135] was suggested as a possible source for the resulting (rather small) changes in XPS binding energies and line widths. The authors also noted that these effects were consistent with polarization changes suggested in the calculations of Cini.[136] One must not exclude the possibility, however, that some of the effects detected by Takasu *et al.*[134] may have been due to the differential charging that may arise due to the size and distribution inhomogeneity of the particles (see above). It is unfortunate that these authors chose to begin their excellent study with the statement that 'surface-sensitive techniques such as photoelectron spectroscopy' are 'unsuitable for investigation' of 'supported metal-catalysts'. Dozens of industrial electron spectroscopy laboratories have shown this statement to be incorrect. Had the authors replaced the word 'unsuitable' with 'difficult' it would have been valid.

Tauszik and others[137] have attempted to characterize the metals dispersion on a thermally treated Pd–Cl–Al_2O_3 system with both AES and XPS. In these studies the authors employed metals loadings that were typical of commercial catalysts to try to avoid the adverse dispersion problems caused by excessive loading. Both AES and XPS agreed that the amount of surface palladium remained roughly the same throughout calcination and reduction, while the amount of surface Cl^- was greatly reduced. XPS results indicated that in H_2 the reduction step was only moderately successful with respect to the surface palladium whereas when hydrazone was employed more extensive surface reduction of palladium occurred.

Mason and Baetzold[138] published an interesting article that pertained indirectly to the ability of XPS to assist in the determination of crystallite cluster growth of platinum and related metals. Studying silver metal crystallite growth on a carbon support, these authors were able to delineate some changes in the XPS binding energies of silver that seemed to correlate directly with the particle size of the crystals.

Fung has created quantitative models[85] (see above) that may increase the applicability of XPS in this area by combining the concepts of Brinen and Schmitt[131] and Contour *et al.*,[133] with the particle size effects suggested by Mason and Baetzold[138] and the particle location studies of Takasu *et al.*[134] In these models relative XPS peak intensities were directly related to specific sizes, shapes and distributions of the peak generator. Fung demonstrated how the model may be employed by calculating the degree of sintering of platinum on SiO_2 following reduction. Extension of these potentially useful models into many other applied areas are being considered.

8.3.3.6   Studies of various poisons in platinum catalysts

One of the primary tasks originally assigned to electron spectroscopy in the study of catalysts was the detection and classification of poisons. In the case of platinum metal catalysis the subject is very complex because certain species act as promoters under some circumstances and poisons under many others. Carbon, sulphur and, in fact, some of the elements employed as alloy promoters (e.g. Sn^0) seem at times to promote catalytic reactions, but may under certain other conditions (usually when the species in question is in 'excess') inhibit the catalytic functionality of platinum[113] Sulphur is an interesting example, being utilized as a promoter in conjunction with rhenium[127] and yet sulphur is also known to selectively suppress the dehydrogenation of cyclohexane to benzene over a platinum bifunctional catalyst.[139] Surface analyses designed specifically to examine sulphur on platinum catalysts are therefore to be expected.

Originally it was hoped that AES could be utilized in these studies, because of its excellent surface sensitivity, small rasterable spot size and easy depth profiling capability. It was thus hoped that Auger could be employed to determine the surface locations, distributions with depth and characteristic adjacencies of the poisons. Largely because of beam damage problems, this promise has yet to be realized, but a few interesting studies do exist. Bhasin[140] has, for example, employed AES to try to assess the status of a variety of poisonous(?) impurities on a commercial Pd–Al_2O_3 hydrogenation catalyst. The catalyst was examined after use and extensive deactivation. A number of impurities were detected, including sulphur and chlorine. However, the amount of surface iron detected indicated that it was the probable deleterious poison (if any!) for this system. AES was also utilized, along with LEED, by Szymerska and Lipski[141] to study the segregation of sulphur to the surface of Pd(100). The study demonstrated that H_2 reduction induced sulphur to migrate to the surface of the palladium, from the bulk, thus apparently poisoning the ability of palladium to dissolve hydrogen. In addition, AES, XPS, UPS and EELS were employed by Matsumoto *et al.*[142] to study the interaction of sulphur with palladium surfaces. Polycrystalline palladium foils were employed, and once again sulphur was detected migrating with ease from the bulk to the surface. In this particular study, the XPS results revealed evidence of a possible electron transfer from palladium to sulphur whereas the EELS results seemed (for this non-single-crystal situation) to be inconclusive.

8.3.3.7   The chemical and physical changes of platinum metal cations after their exchange into zeolites

XPS has also been employed to examine the platinum metal cations on zeolites. The principal points of interest seemed to be: (a) the degree of exchange of the platinum metal cations; (b) the changes in oxidation state experienced by those cations during set-up and use of the zeolite; (c) the extent of the interaction between these cations and the zeolite; and (d) the particular reactions experienced by these cations.

For example, Vedrine *et al.*[143] have employed XPS to monitor the exchanges of palladium and platinum ions into a type Y zeolite. Impregnation of palladium was found to produce positive line shifts in the Pd $3d$ spectra, perhaps indicating successful exchange of the Pd^{2+} ion for Na^+ ions. Subsequent reduction of this system produced Pd^0, but *there was also some electron-deficient palladium suggested by the resulting Pd 3d spectra, perhaps indicative of Pd^I formation.* The impregnated platinum also appeared to be successfully exchanged and also exhibited some evidence of electron deficiency following reduction. It appears that the electron-deficient metals were sacrificing electrons to the acidic nature(?) of the zeolites, yet no noticeable change occurred in the XPS spectra of the latter. The authors also suggested a charge-transfer complex between the platinum metal and the zeolite acidic unit as another means for explaining these results. They also noted that this electron deficiency should enhance the electron-donating capability of the 'partially reduced' cations.

Okamoto *et al.*[144] also utilized XPS to study Y zeolites following exchange of the Na^+ ions with Rh^{III}. Reduction of this system in vacuum produced varying mixtures of Rh^I and Rh^0. These results were realized through comparative binding energy analysis. The Rh^I was suggested as the active site for ethylene adsorption and subsequent hydrogenation, whereas acetylene was assumed to be attached only to the Rh^0. The charge shifting problem may have compromised some of these qualitative and quantitative conclusions. Calculations were made of the surface Cl/Rh ratios employing techniques similar to those advanced by Brinen.[25] These results suggested that some of the rhodium may have migrated to the bulk during heating. Perhaps implementation of the more demanding ideas of Kerkhof and Moulijn[84] might modify this supposition. Simultaneous ESR analysis also indicated that addition of HCl may convert most Rh^0 to RhH and RhCl, apparently facilitating ethylene dimerization.

Finally, an interesting XPS study, also involving rhodium-doped zeolites, was recently reported by Anderson and Scurrell.[145] This investigation was somewhat unique in that it assumed that the zeolite was primarily functioning as a support. The catalysts were intended for carbonylation reactions of alcohols. The system was constructed by exchanging a chloroamine complex of rhodium onto an X zeolite and oxidizing the resulting 'mixture' to drive off the ammonia ligands. Heating the 'mixture' in N_2 to 600 °C or H_2 to 340 °C reduced the rhodium, but the final oxidation state of rhodium was uncertain. Binding energy measurements indicated that some Rh^0 was probably involved (see Figure 8.15). 'Standard' XPS quantitation indicated some preferential

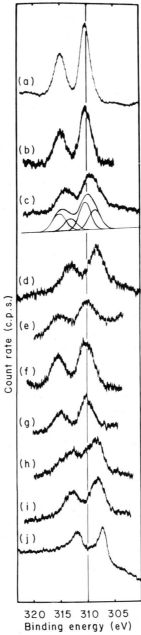

Figure 8.15 Rh 3d electron spectra for some Rh-containing samples (for each spectrum, first temperature is that at which analysis was performed). (a) $[Rh(NH_3)_5Cl]Cl_2$, 10 °C; [(b)–(j), $Rh(NH_3)_5$ ClX]: (b) −80 °C; (c) −80 °C, after 1 h of analysis; (d) 400 °C, vacuum (400 °C, 0.5 h); (e) −60 °C, vacuum (400 °C, 0.5 h); (f) 25 °C, O_2

concentration of Rh(Cl)$_3$X at the surface, with the binding energy data suggesting possible Cl$^-$ bridges. The binding energy references employed for these results may, however, be suspect. The authors also suggested that the reduction of Rh^{III} to Rh^I may occur in the presence of carbonyls. Substitution of a less acidic support (e.g. SiO$_2$) for the zeolite dramatically reduced the binding energy shifts and the activity for carbonylation, suggesting a substantial rhodium-support interaction (exchange?) in the presence of zeolite.

8.3.4 Zeolites—catalytic cracking and other processes

8.3.4.1 Introduction to zeolites and catalytic cracking

Of all of the materials described in this chapter, the crystalline aluminosilicates called zeolites or molecular sieves probably have the most diverse use history (and potential). This fact results both from their unique, regular structural features and their ability to experience dramatic changes in their adsorptive and reactive properties, though the change of the cations need to balance the fundamental negative charge carried by the aluminosilicate (structural) unit. This negative charge arises, of course, because aluminium must pick up an additional electron in order to 'duplicate' silicon in the aluminosilicate lattice. Because of the crystallographic regularity of these lattice patterns, the aluminium occupies quite specific sites and in order to establish local neutrality the cations must be in close proximity to these sites, generally located in varying sized cavities in the 'super' lattice.[146] Simple cations (e.g. H$^+$, Na$^+$, Ba^{2+}, etc.) differ substantially in size and charge; therefore cation exchange may be an extremely selective event, as certain sites are often too confining to accommodate large and/or multicharged cations.[146] It is, therefore, possible under certain conditions to control the distribution of cations in some zeolites and, through this, control some of the properties of the zeolite. In addition, the repetitive, regular structural patterns of the basic units of the zeolites create super lattice structures, with variably sized channels, windows, cavities and pores (see Figure 8.16). For these reasons, particular zeolites with particular cations may be employed for selective adsorbencies and reactions.[146] Thus, zeolites provide the capability to 'tailor' certain aspects of heterogeneous catalysis[108] by controlling not only what species are able to adsorb (and desorb) from various sites but also the time of occupancy of these sites.

---

or N$_2$ (400 °C, 0.5 h, 1 atm); (g) 25 °C, CO (100 °C, 17 h, 1 atm) or CO + CH$_3$I (100 °C, 15 h, 225 + 225 torr) after O$_2$ (450 °C, 4–5 h, 1 atm); (h) 25 °C, CO + H$_2$O (SVP at 25 °C) (100 °C, 20 h, 1 atm) after O$_2$ (450 °C, 2 h, 1 atm); (j) 25 °C, H$_2$ (340 °C, 1.5 h, 1 atm) after O$_2$ (400 °C, 0.5 h, 1 atm); (k) Rh–C (5 per cent), 25 °C, vacuum (400 °C, 2 h). (Reproduced from Anderson and Scurrell[145] by permission of Academic Press Inc.)

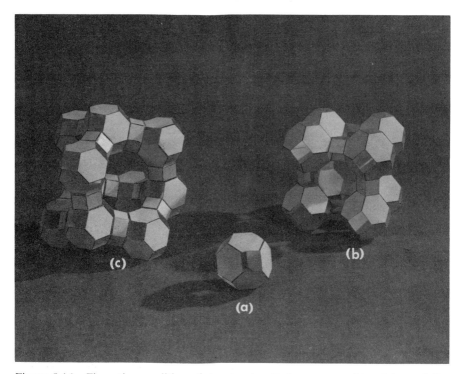

Figure 8.16   Figurative rendition of structural units for some zeolites: (a) a sodalite cube, (b) A type structure and (c) faujasite structure. Note that each vertex represents a Si or Al, connected by off-axis oxygen. The cation sites are not shown

Because of the dramatic variations in different zeolites, an entirely new area of 'shape selectively' was established.[147]

Of all of the catalytic use areas for zeolites, catalytic cracking has been, by far, the most important.[108] Continuous control of catalytic cracking has plagued scientists and engineers for years, due to its propensity to rapidly coke the catalyst being employed and thereby destroy the latter's selectivity and eventually its activity.[108] Present-day petroleum crudes are also often high in sulphur and as the temperature of the catcracking units are increased to enhance activity, sulphur and carbon disposition become substantial problems. The process of cracking involves the selective fissure of C—C bonds, with subsequent hydrogenation to form useful (primarily gasoline) molecules out of heavier crude feedstocks. Originally (until the late 1930s) thermal cracking was employed, but this rather inefficient process has been almost entirely replaced by catalytic processes.[108] Acid sites are needed on the catalyst to promote the above-mentioned reactions. Initially clays were employed as the catalyst, but gradually these were replaced by synthetic amorphous silica alumina systems.[108,148]

In the mid-1960s synthetic faujasites (see Figure 8.16) were introduced as catalysts and today about 90 per cent of all catalytic cracking is carried out on zeolites.[108] The most successful cracking catalysts seem to be composed of the Y-type faujasites (see Figure 8.16), with the original sodium cation replaced by a mixture of protons and rare earth[149] or alkaline earth[108] cations. These cations seem to dramatically enhance the (selective) acidity of the Y system, permitting it substantial activity that can be maintained for long periods.[108,149] Zeolites are (as described in other sections of this chapter) also employed in many other catalytic processes, including hydrogenation, oxidation and polymerization (e.g. the syngas methanation process and the dimerization of ethylene),[108] as well as numerous uses in the adsorbent industry. The recent introduction of 'shape selectivity'[147] and the type ZSM-5 zeolites[150] by Mobil have permitted development of uses in such areas as selective reforming and alkylation (e.g. the removal of low-octane hydrocarbons as LPG and the dehydropolymerization of methanol to gasoline).[108]

Faujasites are formed from a basic sodalite unit and have crystalline structures as depicted in Figure 8.16. The Si/Al ratio R for faujasites ranges from $\sim1 < R < \sim3$ and the corresponding number of cations vary accordingly. (Note that it is possible to 'decationize' the faujasite system by splitting some of the aluminosilicate bonds and removing the cations.[146] In this case both the positive and negative charges are carried in the lattice. This system apparently has excellent catalytic potential.[108]) The Y-type faujasite, generally synthesized as NaY, has a Si/Al ratio of ~2.5.

Among the questions commonly asked of surface scientists about zeolites are to adjudge and quantify: (a) the degree of success achieved in the above-mentioned cation exchanges (as opposed to having the new cations simply 'dispersed' onto the other surface); (b) the changes experienced by these cations following use of the zeolite; and (c) the integrity of the surface of the zeolite itself during this use. The latter is particularly important because many zeolites are prone to adverse surface effects such as dealumination during thermal treatment and use.[146] In addition, junk from incomplete crystallization or internal collapse of a zeolite may gather on the outer surface or in the pores preventing proper reactant and product flow. This fact may be detected using XPS by attempting to identify the difference between surface zeolite and surface alumina or silica, etc.[28,41]

The ZSM-5 zeolites mentioned above also exhibit a well-controlled structure that is markedly different from that of the faujasites (see Figure 8.16). The Si/Al ratio R for the ZSM-5 system seems to vary from $\sim15 < R < 100+$. This zeolite is often prepared with organic amine cations.[146]

8.3.4.2 Studies of the aluminosilicate structural unit

Although a number of the interesting features of zeolites seem at first glance to be designed for AES, surface studies employing this technique have been largely unproductive due probably to electron beam damage effects. XPS studies of zeolites began with the pioneering study of Delgass, Hughes and Fadley.[12] These were soon followed by studies by Minachev *et al.*[151] in which results were obtained, indicative of the degree of cation exchange. This early promise seemed somewhat dampened by the above-mentioned results of Temperé, Delafosse and Contour,[17] in which evidence was provided by XPS of a progressive growth of the surface Si/Al ratio, R_s, that far exceeded increases induced in the bulk, R_b, although both were comparable for A systems (Si/Al $\equiv$ 1) (see Table 8.5a). This work apparently indicated the presence of extensive surface clusters rich in silicon for systems with large (Si/Al) R_b ratios, suggesting either surface dealumination of the material or possibly

Table 8.5 (a)   Bulk and superficial silicon and aluminium composition of various synthetic zeolites. (Reproduced from Temperé *et al.*[17] by permission of North-Holland Publishing Co.)

	Linde molecular sieve			Norton zeolon
	NaA	NaX	NaY	NaZ
$(Si/Al)_b$	1	1.25	2.5	5
$(Si/Al)_s$	2	2.6	6.5	11.5

(b) (After Barr[41])

	Ratio of photo-electron intensity $(Si/Al)^*$ computed from bulk concentrations	Ratio of $(Si/Al)(2p)^*$ from peak heights
NaX	1.8	1.9
NaY	3.7	4.0
NaA	1.5	1.55

*It is difficult to attach any measure of the relative precision of these results. In view of the method and the approximations discussed above, they are probably no better than ±10 per cent.

(c) (Reproduced from Knecht and Stork[152] by permission of Springer-Verlag)

Sample	Ratio in the bulk	Ratio in the surface
Zeolite A No. 1	Na/Al/Si = 0.96 : 0.99 : 1	Na/Al/Si = 0.99 : 0.65 : 1
Zeolite A No. 2	Na/Al/Si = 0.90 : 0.90 : 1	Na/Al/Si = 0.95 : 0.67 : 1

even selective damage done during the analysis. Some researchers have questioned various aspects of this quantitation and the extent of the noted progressive change (see Table 8.5b). In any case, it became apparent that surface and bulk quantitation for zeolites might display marked disagreement. These problems were exacerbated by the XPS results of Kneckt and Stork[152] who reported quantitative studies that indicated *surface dealumination for several A-type zeolites,* including copper exchanged forms (see Table 8.5c).

Further uncertainty was provided by the researchers in this laboratory,[153] when in 1978 it was reported that reasonable quantitative precision could be expected for XPS studies of all zeolites constructed from the basic sodalite cage structure. The surface ratio R_s was shown to grow faster than R_b, but the ratio R_s/R_b for Y zeolites was found to be ~1.2 (see Table 8.5b) compared to the ~2.7 found by Temperé, Delafosse and Contour.[17] If the extent of surface dealumination is a variable, rather tenuous phenomena, depending upon a variety of factors such as the state of hydration, then these discrepancies are not unexpected.

Progressive shifts in the binding energies of all of the elements in zeolites were also detected by this author in an XPS study of various systems.[28,41,153] These shifts were accompanied by rather dramatic reductions in the line widths of the appropriate oxygen, aluminium, silicon and metals XPS peaks for these zeolites, compared to the equivalent lines for the 'precursor' SiO_2 and Al_2O_3 systems.

All of our results have been further analysed and synthesized by the author in several recent publications.[28,41] A variable amount of surface dealumination was still found (during XPS analysis) that did seem to increase with R_b, but, as stated above, in a less pronounced manner than reported by Temperé, Delafosse and Contour.[17] The reason for this dealumination is as yet unknown, but its detection is unique and important to subsequent delineation of the status of the outer pores in these zeolite systems. The binding energy shifts described above were shown to follow reproducible patterns that permitted the positive identification of particular zeolites even if they were present in relatively small amounts. Specific binding energies and line widths were also found for several types of silicas and aluminas, thus permitting the specific identification of the constituents in the mixture following dealumination and other 'mixing' processes. Some of these results are presented in Table 8.6 and Figure 8.17. Note that both flood gun correction and C 1s referencing have been applied in arriving at these values. A discussion of the justification for this procedure was presented earlier in this chapter. Studies were also generated of the characteristic valence bands and loss spectra for these systems.[28,41] These results were coupled with the above-mentioned binding energy and line width data to try to establish chemically sound arguments for the patterns realized. The fact that all elemental binding energies, includ-

Table 8.6 (a) Representative binding energies, in electronvolts (±0.1 eV). Charge shifts removed C 1s = 284.4 eV

	SiO₂	ZSM-5	ReY	CaY	NaY	Bentonite	CaX	NaX	Kaolinite	CaA	NaA	γ-Al₂O₃
Si 2p	103.35	102.90	102.70	102.40	102.35	102.50	102.25	101.75	102.25	101.40	100.90	—
Al 2p	—	74.28	74.45	74.15	74.00	74.60	74.35	73.70	74.10	73.50	73.20	73.60
O 1s	532.65	532.25	531.95	531.60	531.55	531.80	531.40	530.85	531.30	530.65	530.20	530.35
Na 1s	—	—	—	—	1071.00	1073.15*	—	1072.05	1071.75*	—	1071.45	—
Na 2s	—	—	—	—	63.80	64.05*	—	63.55	—	—	63.00	—
Ca 2p₃/₂	—	—	—	348.20	—	—	347.90	—	—	347.50	—	—

*Impurity. Note that there are also small to moderate quantities of other cations in the zeolites due to imcomplete exchange.

(b) Line widths for zeolite peaks listed in Table 8.4, in electronvolts (±0.15 eV)

	ReY	Ca(Na)Y	KY	NaY	CaX	NaX	Na(Ca)A	K(?)A	NaA
Si 2p	1.84	1.84	1.71	1.68	1.82	1.77	2.00	1.84	1.72
Al 2p	1.88	1.71	1.50	1.55	1.86	1.60	1.90	1.78	1.56
O 1s	2.00	2.05	2.08	2.05	2.08	1.65	1.88	2.10	1.75
Na 2s	—	—	—	1.65	—	1.67	1.67	1.9*	1.65
Na 1s	—	—	—	1.95	—	1.95	1.90	—	1.92
C 1s	2.0*	1.8*	1.50	1.50	1.94	1.52	1.66	—	1.44
Ca 2p₃/₂	—	1.92	—	—	—	—	1.90	—	—
K 2p₃/₂	—	—	1.70	—	—	—	—	—	—

*Precision reduced to ±0.3 eV.

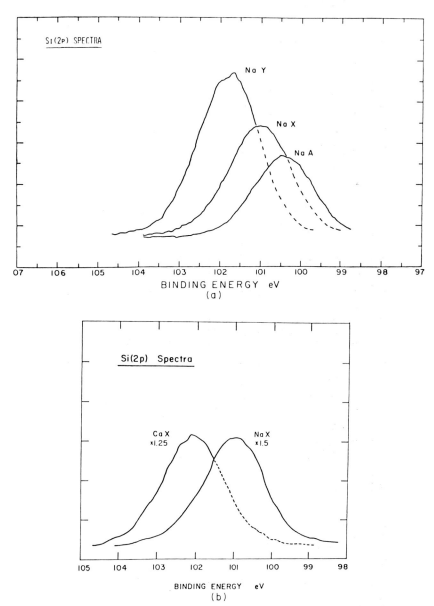

Figure 8.17 Representative binding energy shifts (in Si 2p) experienced by zeolites. (a) Resulting from change in structures, A → X → Y. (b) Resulting from change in cation Na → Ca. (After Barr[28])

ing the cations, silicon, aluminium and oxygen values, appeared to shift in the same direction was initially quite disturbing. However, it became apparent that the true 'chemical' shifts were not expressed by variations in these elemental binding energies, but rather by changes in cluster (groups of elements) patterns. Thus, Si–O and cation–Al–O groups were identified. These ideas were supported by the appearance of similar groups in the selective (residual) damage realized by zeolites during Ar^+ ion sputter etching.[28,41] Based upon these ideas, the XPS variations noted for different zeolites were related to differences in: (a) the Si/Al ratio; (b) the zeolite structural unit, e.g. $A \rightarrow X \rightarrow Y$; (c) the cations, e.g. $Na^+ \rightarrow K^+ \rightarrow Ca^{2+}$; and (d) the Dempsey charge field.[154]

8.3.4.3 Studies of exchanged cations and catalytic processes

Most XPS studies of zeolite systems have not centered around the sieves themselves or the catalytic cracking process. Generally, the aforementioned cations have been the primary XPS concern, with the results intended for hydrogen reduction and polymerization reactions. In this vein, XPS was employed by Wichterlová et al.[155] as one of the tools used to study the substitution of Al^{3+} into (and subsequent steam 'stabilization' of) a NH_4Y type Y zeolite. Al^{3+} was reported to decrease the dehydroxylation capability of the steam-stabilized Y zeolite. The XPS results were not extensively utilized, except to generate R_s values that showed some dealumination close to that realized in this laboratory.

Bravo, Duryen and Zambaulis[156] also employed XPS to assist in a study of the use of carbon monoxide to reduce $NaCu^{II}X$. Evidence for the reduction of copper was provided, in addition to indications of a suspected structural change. The latter was assumed to cause the dramatic broadening detected in the O 1s spectrum following redox treatment. This modification was attributed to 'occlusion' of active oxygen species in the zeolite lattice.

Several studies were reported in the previous section in which platinum metal cations were substituted for the sodium cations in X and Y sieves. In those cases, electron spectroscopy was employed primarily to describe changes in the platinum metals. Other researchers have also used XPS to study similar systems, but were also interested in the changes of the zeolites during such reactions as the production of methane or the water gas shift $(CO_2 + H_2)$. For example, Pederson and Lunsford[157] have examined the RuY system with XPS to not only describe the success of the Ru^{3+} for Na^+ exchange but also how the zeolite functions as a whole. XPS results indicated only partial success in the exchange with the balance of the ruthenium adsorbed onto the outer surface. (These suppositions were based upon R_s values that were similar to those found by the author, perhaps due to the employment of a common (HP) ESCA system.) Thus, in the presence of O_2

RuO_2 was discovered on the outer surface and some Ru^0 was found in the zeolite cavities when O_2 was excluded (see Figure 8.18). It appears that ruthenium is much easier to oxidize than platinum under these conditions.[158] All of these features were based upon binding energy variations that were among the first utilized to adjudge the success or failure of cation exchange in zeolites. In a subsequent study, Gustafson and Lunsford[159] have employed XPS to examine another RuY system. In this case they reported on the ruthenium in the zeolite lattice (but not necessarily exchanged) both before and following reduction (of the Ru^{III} to the metal). The latter sites were found to be catalytically active. Comparison with Ru–Al_2O_3 and Ru–SiO_2 failed to produce concrete correlations between support acidity and catalytic activity.

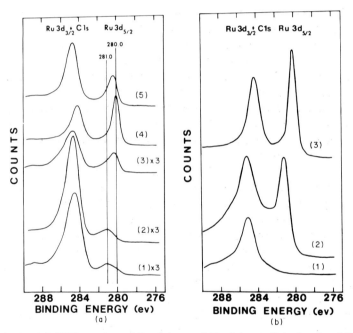

Figure 8.18   (a) XPS spectra of Ru $3d_{5/2}$ and Ru $3d_{3/2}$ + C $1s$ levels of 2 per cent RuNaY: (1) degassed in flowing He to 400 °C followed by reduction in flowing H_2 for 16–18 h at 400 °C, (2) treated as in (1) with N_2 substituted for He, (3) treated as in (1) followed by heating in flowing CO–H_2 mixture for 24 h at 280 °C and then in H_2 for 18–20 h at 300 °C, (4) treated as in (1) with air substituted for He and (5) impregnated 1 per cent Ru–NaY treated as in (1). (b) XPS spectra of Ru $3d_{5/2}$ + C $1s$ levels of 2 per cent RuNaY: (1) after degassing in flowing air in increments of 100 °C/h to 200 °C, (2) to 300 or 400 °C and (3) after degassing in air to 400 °C followed by reduction in H_2 for 16–18 h at 400 °C. (Reproduced from Pederson and Lunsford[157] by permission of Academic Press Inc.)

Fajula, Anthony and Lunsford[160] have also used XPS to assist in the analysis of a PdY zeolite intended as a syngas catalyst. They found that methanol production seemed to depend upon the presence of small crystallites of Pd^0, a feature detected by XPS. Methane production, on the other hand, appeared to be induced primarily by the acidity of the zeolite.

Platinum metal exchange into zeolites was also utilized for promotion of skeletal transformation (polymerization) reactions. As reported above, Okamoto *et al.*[144] studied RhY with XPS to try to delineate both changes in oxidation state and variations in site occupancy for this system. In a similar study, Kuznicki and Eyring[161] have discovered (using XPS) that the Rh^{III} substituted into a Y lattice may be substantially reduced to the metal employing milder conditions than utilized to reduce the other platinum metals. These authors feel that the presence of this zeolite enhances the rate of reduction. The rhodium in RhA, however, failed to reduce and furthermore this zeolite was not active in polymerization. Therefore, reduction of rhodium ions to the metal was suspected to be the process that activates catalytic polymerization. This also was the case for palladium and ruthenium.

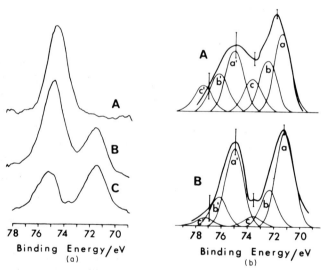

Figure 8.19   (a) Photo-electron spectrum from (Na) Y-zeolite samples in the vicinity of the Al 2*p* and Pt 4*f* lines. Spectrum A, Al 2*p* line from (Na) Y-zeolite. Spectrum B, Al 2*p* and Pt 4*f* lines from 3 weight per cent Pt(Na) Y-zeolite. Spectrum C, residual Pt 4*f* spectrum after stripping a normalized spectrum A from spectrum B. (b) The Pt $4f_{7/2,5/2}$ photo-electron profiles associated with 3 weight per cent Pt(Na) Y-zeolite (spectrum A) and 3 weight per cent (Na) Y-zeolite (spectrum B). The best-fit decomposition into three doublet components is also shown. These doublets may be formally indexed as Pt^0(a,a'), Pt^+(b,b') and Pt^{2+}(c,c'). (Reproduced from Larkins *et al.*[162] by permission of Elsevier Scientific Publishing Co., Amsterdam)

In another study Larkins *et al.*[162] investigated the exchange of platinum for lanthanum and sodium in Y zeolite systems. They reported XPS evidence for electron deficiencies detected in the platinum apparently due to different electrostatic fields at different cation sites (i.e. the Dempscy fields[154]). Particle size and incomplete exchange effects may influence these results. The electron difficiency of platinum may actively catalyse dehydrogenation and other catalytic processes leading to polymerization (see Figure 8.19).

Badrinarajanan *et al.*[163] employed XPS to investigate the status of nickel exchanged onto a ZSM-5 zeolite that has retained substantial acidity. This catalyst was intended to promote both polymerization and alkylation reactions. The Ni^{II} was found by XPS not to reduce to Ni^0 (detected by a lack of binding energy shifts) under circumstances similar to that employed above for rhodium. This lack of reduction was attributed by these authors to an extensive Ni^{II} zeolite interaction through hydroxide groups. Therefore, since this catalyst was active for polymerization, the authors discounted the need for reduction of the exchanged cations (in this case nickel) to create an active (polymerization) catalytic promoter.

8.3.5   Cobalt–molybdenum catalysts-desulphurization

8.3.5.1   Introduction to the catalyst—its uses and pecularities

As described earlier in this chapter, a number of studies of the $Co-Mo-Al_2O_3$ catalyst system (with or without sulphur) were published before 1975. This system has a variety of catalytic uses, the primary ones being hydrogenation and the removal of sulphur from feedstocks, residues and coals (by hydrodesulphurization).[108] Although the presence of sulphur is sometimes useful, generally it is a deleterious poison that unfortunately pollutes, in unacceptably large amounts, much of the world's readily available hydrocarbon fuels (e.g. Saudi Arabian crude and western US coal). Therefore, a detailed understanding and proper utilization of hydrodesulphurization (HDS) catalysis is an economic necessity.

Long before the development of modern surface science, MoO_3 on alumina was shown to catalyse the HDS process and after a few years of trial and error it was discovered that several other transition metal oxides did also. It was also demonstrated that cobalt or nickel in the presence of sulphide 'promote' the process.[108] For a variety of reasons, the Co–Mo system is generally preferred for this process, although others, particularly the Ni–W system are extensively used for desulphurization.[108] It was discovered that proper preparation of the Co–Mo system required not only strict monitoring of the amount of the dopants but also the order of their addition.[164]

From the point of view of XPS, this system seems very attractive because cobalt, molybdenum and sulphur have numerous common oxidation states that readily produce substantial binding energy shifts. In addition, the Co–Mo system generally functions as a 'dopant-rich' catalyst (unlike reforming) with 3–5 weight per cent CoO and 12–15 weight per cent MoO_3 commonly employed.[164] Substantial amounts of sulphur are also generally added to 'activate' this system.[108] On the negative side, it was soon discovered that during set-up and use of the catalyst, these substantial components were experiencing a number of complicated interactions and other changes that sometimes drastically shifted the resulting binding energies whereas at other times the binding energies hardly changed.[164] These involvements included possible strong interactions between the dopants and the support (to create such species as $CoAl_2O_4$) and complex preferential layerings of one dopant on top of the other.[164] (The latter may have compromised many of the early quantitative studies of this system using XPS.)

By the time most of these problems were recognized (and became themselves the separate objective of many XPS studies!) the Co–Mo catalyst system was already the most widely published applied catalysis system in the XPS literature. In recent years the numbers of these publications have multiplied at an increasing rate. There are several apparent reasons for this 'primary exposure': (a) the aforementioned abundance of dopants; (b) the natural tendency to 're-examine complex materials with improved instrumentation'; (c) at first glance the system seems rather easy to prepare. It can be produced, for example, by incipient wetness techniques, rather than the seemingly complex ion exchange (zeolites) and co-precipitation (platinum metal reformers) procedures needed for many other catalysts. This 'ease of preparation' is an advantage to surface scientists who are often unschooled in the art of catalyst preparation. And (d), proprietary constraints, which often restrict the publication of studies of commercial catalysts do not seem (for unknown reasons) to apply to the Co–Mo system. This is an important point for academic surface scientists who are perhaps unfamiliar with the nature of the proprietary constraint employed in industrial catalysis. It should not be presumed, therefore, that because surface studies of the Co–Mo system are the most published, that they are necessarily the most plentiful. Industrial laboratories may be examining other more proprietary systems with greater diligence. The fruits of these unpublished labours will, if successful, first appear as a part of an improved catalyst program and only later as results shared with the general scientific community. The continuing growth in numbers of surface science positions in industrial research is, of course, an indirect means of monitoring the success of these enterprises.

As mentioned above, the Co–Mo system has a number of uses, but only hydrodesulphurization (HDS) will be featured in this chapter. Its use as a hydrogenation catalyst will be considered as part of the next section, as will be the $Ni–W–Al_2O_3$ system, the chief substitute for the Co–Mo system.

8.3.5.2 *Reference XPS data needed to study the Co–Mo system*

Before seriously examining the Co–Mo XPS literature, the reader should be familiar with the XPS results obtained on the precursor materials, i.e. Co^0, Mo^0, CoO, Co_3O_4, Co_2O_3, MoO_2, Mo_2O_5, MoO_3, the hydroxides, the corresponding sulphides, aluminates and mixed Co–Mo–O systems. Binding energies and some of the spectra for the elemental metals, several of the oxides and a few hydroxides (or hydrated oxides) were presented in the general study of natural passivation by the author.[117] Note that in this study uncertainty existed as to whether the oxidized surface species produced on CoO, as a result of further air exposure, was Co_2O_3 or Co_3O_4. Results obtained by others seems to indicate that it is Co_3O_4.[165]

Brundle has conducted a detailed study of the status of molybdenum as the elemental metal and the changes of that species following adsorption of controlled doses of O_2, CO and H_2O.[166,167] The study includes UPS, as well as XPS results, and is a much more careful analysis of the *onset* of oxidation than the previous ones by this author. Cimino and DeAngelis[168] have also studied various states of molybdenum oxides, and have added the important consideration of how those binding energies and peak structures change as these oxides are subjected to support by γ–Al_2O_3 and silica. In addition, these authors have investigated the catalytically important compounds, Na_2MoO_4 and $CoMoO_4$.

One of the first studies of important cobalt compounds was conducted by Frost, McDowell and Woolsey.[169] In that study peak structures and binding energies were presented for $Co^{II}O$ and $Co^{II}S$. Cobalt and CoO were also two of the featured systems in the general XPS study of transition metal oxides by Fiermans, Hoogewijs and Vennik.[165] Kim provided a detailed XPS study of the electronic structure of CoO, featuring a careful and catalytically very useful analysis of the key satellite spectra.[170] These same satellites were also featured in a description of the XPS spectra of CoO, CoOOH, and $Co(OH)_2$ by McIntyre and Cook,[171] who pointed out and carefully documented important small differences in the satellite splittings. Chuang, Brundle and Wandett[172] have also documented these same satellite shifts, but have extended the study to include important modifications due to subsequent passivation resulting from ambient air exposure and also the chemical changes induced from Ar^+ ion bombardment. Some of the important features of these studies, particularly as they may be useful in subsequent catalyst analyses, are listed in the above references.

8.3.5.3 *XPS results for the cobalt—molybdenum catalyst system*

Many of the difficulties with the interpretations of the Co–Mo–S–Al_2O_3 system occur because the obvious solutions to problems often do not work.

For example, to avoid the possible production of $CoAl_2O_4$ (a deleterious product of the strong metals–support interaction) one might consider utilizing SiO_2 as a support. This, however, seems to promote the production of $CoMoO_4$, another detrimental species.[173] Difficulties such as these seem pervasive in XPS studies of the Co–Mo catalyst system.

In order to properly assess the states and changes experienced by this complicated system it was obvious that researchers needed to examine each component individually under a variety of use conditions. A study of this type was reported by Haber *et al.*[174] in which XPS was employed to examine the changes experienced by MoO_3, MoO_2 and $CoMoO_4$ when they were employed to oxidize aldehydes through the corresponding acids to CO_2. Surprisingly, the MoO_2 species proved to be the most active oxidizing agent, apparently due to an enhanced ability of the 'reduced surface' to adsorb and retain not only the aldehydes but also the acid intermediates; $CoMoO_4$ was found to be entirely unsuccessful at these reactions.

Declarck-Grimce *et al.*[175] studied the behaviour of the two separate oxide catalysts $CoO–Al_2O_3$ and $MoO_3–Al_2O_3$ with XPS before using the same approach to examine the combined catalyst in the presence of sulphide.[176] They found XPS evidence that H_2 reduction without S^{2-} results in $Co^{2+} \rightarrow Co^0$ in $Co–Al_2O_3$, whereas when molybdenum and sulphide are present, H_2 reduction of Co^{2+} seems to produce Co_9S_8. Too much cobalt in the original $Co–Mo–S–Al_2O_3$ system seems to generate $CoMoO_4$ from the excess. The latter was postulated to be a poison.[176]

In 1977, Okamoto *et al.*[177] began a detailed series of XPS studies of a $Co–Mo–Al_2O_3$ system utilized to perform HDS reactions on a mixture of thiophene and H_2. They described, in some detail, the rather confusing mix of explanations that existed at that time for the activated mixtures of reactants, including: (a) synergistic action between cobalt and molybdenum; (b) intercalation of MoS_2 by Co^{2+}; (c) a somewhat nebulous 'complex' of cobalt and molybdenum; and (d) the so-called surface-enrichment model. (As we shall see later, some form of nearly all of these have survived to form the more concise models favoured today). Okamoto *et al.*[177] felt justified in attributing the catalytic behaviour of the mixed *oxide* catalyst to molybdenum, whereas Co^{2+} acts as a 'promoter'. The XPS spectra for the oxide catalyst seemed to indicate that the molybdenum in the fresh (ready to use) catalyst appeared to be a mixture of Mo^{5+} and Mo^{4+}, whereas much of the Co^{2+} was reduced to the metal, Co^0, while the balance of the cobalt seemed to be converted to inactive (surface-stabilized) $CoMoO_4$. Some of this XPS evidence seemed to provide support for the suppositions of Grimblot and Bonnelle[178] that molybdenum ions occupy octahedral and cobalt ions occupy tetrahedral sites in a spinel-like arrangement in the alumina structure. The above study by Okamoto *et al.*[177] produced a critical response from Massoth,[179] who questioned the generality of the interpretation of Okamoto *et al.* because of the extremely low sulphur content in the latter's choice of state conditions. (No sulphide was added!) *In particular, Massoth attributed the presence of Co^0 to this lack of sulphide flow.* In reply to this criticism, Okamoto, Imanaka and Teranishi[180] agreed that an increase in sulphur would promote the production of cobalt sulphides, as well as MoS_2. However, they also defended their conclusions and the design of their experiments, ascribing the low sulphur content to a support effect. To check these conclusions, Okamoto *et al.*[181] employed XPS to study the effects of sulphide on Co^{2+} and Mo^{6+} under reducing conditions in a variable sulphide flow. This study demonstrated that under H_2 reduction in the presence of thiophene (only) Mo^{6+} was reduced to Mo^{4+}, with the thiophene apparently producing MoS_2, but *no* CoS was detected. When H_2S was added to the reduction stream, however, CoS was proported to be produced. Examining the problem further, Okamoto *et al.*[182] reported that the preparation (set-up) of the catalysts apparently substantially altered the surface-to-bulk Co/Mo ratio, a feature that tended to be most exacerbated by reduction, but the effect was substantially different depending upon whether H_2S was employed during reduction or not.

Okamoto *et al.*[183] also tested some of the other intermediate effects by employing XPS to examine in detail the reduction of the $Mo-Al_2O_3$ system in

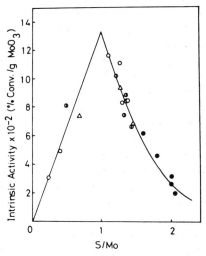

Figure 8.20 Correlation between the intrinsic activity of molybdenum for the hydrodesulphurization of thiophene at 400 °C and the sulphidation degree of molybdenum in the MoO_3/Al_2O_3 catalyst: (○)$MoO_3-Al_2O_3$ calcined at 550 °C, after the pre-reduction; (⊕) after the non-pre-treatment; (●) after the pre-sulphidation; (△)MoO_3/Al_2O_3 calcined at 700 °C, after the pre-reduction. (Reprinted with permission from Okamoto *et al.*[183] Copyright 1980 American Chemical Society)

different flowing sulphide environments. Utilizing fairly detailed quantitation similar to Brinen,[25] they discovered that the Al_2O_3 did not seem to become 'involved' during the sulphiding of the MoO_3, whereas the latter apparently exchanged oxide for sulphide before reduction. *The activity of the molybdenum to HDS also seemed to increase linearly with the creation of MoS_2,* but a *saturation point was apparently reached followed by a rapid (near linear) fall-off in HDS activity.* Thus, this system seemed to exhibit a classic volcano effect[49] (see Figure 8.20).

Detailed XPS examinations by Okamoto et al.[183] of the molybdenum species formed at different molybdenum concentrations also indicated that a *monolayer* coverage of sulphided molybdenum was possibly saturated into the tetrahedral sites of the alumina at ~12 per cent, molybdenum as postulated by Giordano et al.[184] in a non-XPS study. When this percentage of molybdenum was exceeded, Okamoto et al.[183] predicted that the excess molybdenum formed 'poisonous' deposits of MoO_3 and $Al_2(MoO_4)_3$. They further postulated that whereas presulphiding during reduction will reduce larger amounts of molybdenum to MoS and deposit some of these in octahedral sites, *only the MoS, in tetrahedral sites seemed to be active.* Based upon detailed quantitation and the observed distortion of some of the XPS peaks, *Okamoto* et al.[183] *also postulated that these tetrahedral sites were sulphur deficient with anion vacancies,* i.e. that a vacancy species such as:

was formed, where the □ represents the anion vacancy. In a related XPS study of the Mo–Al_2O_3 system, Edmonds and Mitchell[185] followed up on the earlier study[10] of the HDS system by also reporting that MoO_3 preferred to form a monolayer coverage on alumina, with saturation of that coverage at ~9 weight per cent molybdenum. *They noted also that multilayer formation of molybdenum species did not seem to occur until after the monolayer was complete.*

Another detailed series of studies of these catalyst have been generated in the research laboratory of David Hercules. Initially,[186] these studies were concerned with the effects of reduction on the Co–Mo–Al_2O_3 system, both with and without the presence of pre-sulphiding. Reduction without H_2S produced the expected decrease in Mo^{VI} and a corresponding growth in both Mo^V and Mo^{IV}. A plot of this change is presented in Fig. 8.21. Pre-sulphiding with reduction apparently produced MoS_2 and seemed to affect the cobalt, but except for noting that the cobalt in this catalyst seemed to be more rapidly

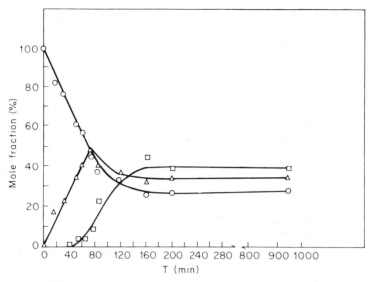

Figure 8.21   Molybdenum oxidation states as a function of reduction time. Legend: (O)Mo^{VI}; (△)Mo^V; (□)Mo^{IV}. All catalyst samples were reduced in high purity hydrogen for the specified time at 500 °C. Mole fractions were determined by deconvolution of the Mo 3p and Mo 3d envelopes using a DuPont 310 curve resolver with summer attachment. Deconvolution of 20 spectra gave the following values for the Mo 3$d_{5/2}$ binding energies: 233.0 ± 0.1, 231.9 ± 0.1, 229.9 ± 0.1. Assignment of these values to Mo^{VI}, Mo^V and Mo^{IV} was made by comparison to MoO$_3$ and MoO$_2$ where separation was found to be 3.1 ± 0.1 eV. (Reprinted with permission from Chin and Hercules[196]. Copyright 1982 American Chemical Society)

altered than CoAl$_2$O$_4$, these authors were unsure as to the resulting cobalt species.[187] Discrete phases of the sulphides were doubted.

Zingg *et al.*[188] later employed XPS to examine the MoO$_3$–Al$_2$O$_3$ catalyst system, supplementing this study with both ISS and laser Raman results. In the calcined (pre-reduced) catalyst, these researchers discovered evidence of four possible states for Mo^{VI} including a surface monolayer coverage apparently of 'active' Mo^{VI} in tetrahedral sites, thus supporting the contentions of both Edmonds[164] and Okamoto *et al.*[183] However, Zingg *et al.*[188] also found some active Mo^{VI} in octahedral sites, but all of the MoO$_3$ detected was assumed to be inactive and also in octahedral sites. Hercules and his group also discovered some inactive Al$_2$(MoO$_4$)$_3$, apparently produced through interactions between Mo^{VI} and the support in tetrahedral sites. Reduction of this system seemed to produce the above-mentioned Mo^V species (*in tetrahedral sites*) and Mo^{IV} (*in octahedral sites*). If the molybdenum was held below 15 weight per cent most of the Mo^V in the tetrahedral sites seemed to be active to thiophene HDS, perhaps in the form of a 'surface molybdate':

$$
\begin{array}{ccc}
O\!\!\!\diagdown & & \diagup\!\!\!O \\
& Mo & \\
O\diagup & & \diagdown O \\
| & & | \\
Al & & Al \\
\diagup\diagdown & & \diagup\diagdown
\end{array}
$$

After the weight per cent of molybdenum rose above 17 per cent, these 'molybdates' seemed to be inactive. At ~20 weight per cent molybdenum, the aforementioned monolayer coverage was assumed to be saturated. There appears to be substantial octahedral Mo^{IV} in the concentration range $7 <$ weight per cent molybdenum < 20, but little of this seemed to be active. Above 20 weight per cent molybdates appeared to form again in the tetrahedral sites, but this time they seemed to be mostly the inactive 'bulk' type. Note that one might speculate that the addition of sulphide may convert the surface molybdate of Zingg et al.[188] to the active sulphide (void) species reported by Okamoto et al.[183] in the tetrahedral sites. Both of these possible species are seen to 'optimize' in the same concentration ranges. It should be noted, however, that Zingg et al.[188] predict that molybdenum species in both octahedral and tetrahedral sites are needed in the oxidized catalyst to get sufficient subsequent activity. Hercules and associates have further refined these arguments, but before examining these (encompassing) publications it should prove instructive to consider some of the studies by others that influenced their present thinking.

For example, Delvaux, Grange and Delmon[189] also employed XPS to study the HDS system, but began without alumina rather than by omitting one of the metal dopants. They discovered that with the 'correct' Co/Mo ratio (?) and sufficient sulphide, Co_9S_8, MoS_2 and also a Mo^{3+} vacancy species were formed. The latter substance may be related to the sulphide vacancy species described above in the work of Okamoto et al.[183] The key to the production of these entities would appear, according to Delvaux, Grange and Delmon,[189] to be a combination of thermal variations (similar to that described by Zingg et al.[188]) the H_2S/H_2 ratio and the Co/Mo ratio. One should keep in mind, however, that these factors may vary when Al_2O_3 is present. All of this may explain why Brinen and Armstrong[190] felt that under certain conditions Co^0 may be produced rather than the aforementioned sulphide. (Note also the above-mentioned controversy between the groups of Okamoto[177,180] and Massoth[179].)

In order to test these concepts further, Delannay et al.[191] utilized XPS and electron microscopy to examine various mixtures of the entire $Co-Mo-S-Al_2O_3$ catalyst system. In this work, they noted, as before, the importance of balancing the (*surface*) concentrations—with any excess cobalt apparently coating the outer surface of the system with *inactive* Co_3O_4. Microscopy seemed to confirm that under the optimal thermal conditions, the 'proper' mixture of cobalt, sulphur (and molybdenum) resulted in the production of Co_9S_8. Delannay *et al.* also assumed that the principal molybdenum product formed following sulphiding was MoS_2; however, they made no effort to try to detect any vacancy species like that described above by Okamoto *et al.*[183] and speculated on in the early studies of the Delmon and Delannay group. They postulated that some of the excess cobalt may form inactive spinel-structured $CoAl_2O_4$. If the relative concentrations are controlled so that this does not occur, Delannay *et al.*[191] speculated that the Co_9S_8 species may be adjacent to the MoS_2, perhaps forming a complicated mixed sulphide. This suggests that a synergistic model may describe the promotional capabilities of the cobalt. Thermal treatment seemed to reduce their surface concentrations of molybdenum $[Mo]_s$.

Lycourghiotis *et al.*[192] rationalized that many of these ideas may depend, in part, upon the nature of the acid sites of the Al_2O_3. To test this, these researchers added Na^+ to the above described (pre-sulphide) system. In the first study, the authors reported the effects of Na^+ on cobalt, noting, in particular, that the alkali cation seemed to prevent cobalt dispersion. As a result, excess (deleterious) Co_3O_4 and $CoAl_2O_4$ were apparently produced. In the second paper, Lycourghiotis *et al.* noted that Na^+ also seemed to prevent the production of Mo^V following reduction of MoO_3–Al_2O_3.

Gajardo, Grange and Delmon[193] also employed XPS and diffuse reflective spectroscopy to examine various Co–Mo systems and check the concept of a stable molybdenum monolayer. They found evidence for the monolayer in the mixed system and speculated that the cobalt dispersed on the γ-Al_2O_3 (above the molybdenum) functioned as a promoter by *aiding in the molybdenum distribution* (see Figure 8.22) Addition of too much cobalt, however, seemed to create $CoAl_2O_4$ and multilayer structures that prevented monolayer molybdenum dispersion.

Many of the above-mentioned results and hypothesis were supported by the findings of other researchers. Aptekar *et al.*,[194] for instance, detected the production of the three oxidation states of molybdenum (Mo^{VI}, Mo^V and Mo^{IV}) in early XPS studies of reduced version of MoO_3–Al_2O_3 catalysts. Some of these ideas were reviewed in the aforementioned discussion by Edmonds.[164] Of particular interest in that study were the reports of the author's examination of the changes in binding energies with heat treatment and change in the per cent of molybdenum present in a MoO_3–Al_2O_3 catalyst.

Many of those results and ideas were recently reviewed by Hercules and Klein.[195] This was rapidly followed by an extension of the results obtained by the Hercules group[196] and an important consolidation of their hypothesis based upon those results. In the review article Hercules and Klein[195] pointed out that careful XPS analyses can now delineate the difference between posi-

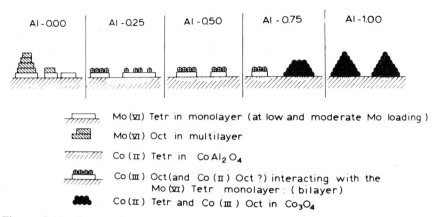

Figure 8.22 Proposed structure of CoMo–γ-Al$_2$O$_3$ catalysts. (Reproduced from
Gajardo, Grange and Delmon[193] by permission of Academic Press Inc.)

tive metals–support interactions, e.g. the observations of Okamoto *et al.*,[183]
noting the possible production of the surface vacancy complex:

and negative or deleterious metals–support interactions, i.e. the creation of
bulk-like CoAl$_2$O$_4$ or Al$_2$(MoO$_4$)$_3$. Hercules and Klein[195] also noted that XPS
studies had provided an unusual understanding of this (Co–Mo) catalyst
system, particularly through the realization of selective, sometimes subtle,
changes in surface quantitation and chemistry. Thus, for example, the well-
established point of an active monolayer of molybdenum discovered by
several XPS groups corresponded nicely to the quantitative influences of selective loadings of the dopants. In addition, it was pointed out that doping the
cobalt beyond the presently accepted limit obviously created poisonous surface layers of Co$_3$O$_4$ and possibly CoAl$_2$O$_4$. The fact that Co^{2+} seemed to
occupy first the tetrahedral and then the octahedral sites, and that these sites
varied in their catalytic importance, was another, recent discovery of XPS,
although the facts and details of these observations are still somewhat uncertain.

The XPS studies of the Mo–Al$_2$O$_3$ system were also reviewed by Hercules
and Klein[195] who pointed out that the detection of several, as of yet, ill-
defined Mo^{VI} species on the oxidized catalyst was first accomplished by XPS.
Inclusions of laser Raman studies have led to speculations, in this area, of an

active surface Mo^{VI} species and inactive $Al_2(MoO_4)_3$ depending upon loading conditions and cacination temperature. Reduction of the Mo^{VI} species seemed to lead to two different oxidation states of molybdenum depending upon the original sites of the Mo^{VI} on the alumina, i.e.

$$Mo^{VI} \rightarrow Mo^V \text{ tetrahedral sites}$$
$$Mo^{VI} \rightarrow Mo^{IV} \text{ octahedral sites}$$

The former were generally attributed to be the source of most catalytic activity.[195] This point, however, would appear to be still in dispute.

Chin and Hercules[196] have dramatically extended these arguments by providing a detailed study of a Co–Mo–Al_2O_3 catalyst system with fixed molybdenum (15 weight per cent), variable cobalt (1–9 weight per cent) and reductions with and without H_2S. Elaborate use was made of XPS, ISS and photo-acoustic spectroscope to demonstrate, among other things, that excess Co^{2+} doping (above 7 weight per cent for 15 weight per cent molybdenum) seemed to produce Co_3O_4. They note that the latter may not be detected as it is apparently converted to $CoMoO_4$, if molybdenum is also in excess (>20 weight per cent) (see Figure 8.23). The latter species were shown to be poisonous to the HDS reaction and to form naturally if the cobalt and molybdenum dopants were coprecipitated. Chin and Hercules[196] speculated that the 'positive' effects of cobalt are not seen until more than 1 weight per cent cobalt is added because this initial cobalt forms inactive $CoAl_2O_4$ in the tetrahedral sites of the alumina. The presence of this inactive species may also increase slightly between 1 and 3 weight per cent cobalt, but above the latter concentration the 'active' cobalt species becomes dominant. Reduction, according to Hercules and colleagues, does not create appreciable Co^0 unless there is an 'excess of cobalt' (>7 weight per cent for 15 weight per cent molybdenum). This fact is used to justify the presence of $CoMoO_4$ in the latter case, rather than $CoAl_2O_4$ which is difficult to reduce.

In an important companion observation, Chin and Hercules[196] discovered that ISS results seemed to indicate that the molybdenum (monolayer?) forms on top of the cobalt in the active catalyst (7 weight per cent cobalt and 15 weight per cent molybdenum), whereas the inactive coprecipitated system does not exhibit this layered structure.

Pre-sulphiding, according to Chin and Hercules,[196] definitely promotes the production of MoS_2, the possibility of a vacancy MoS_2 complex attached to the alumina at tetrahedral sites, in the manner of Okamoto *et al.*,[183] was not considered. The sulphiding of cobalt was assumed to create Co_9S_8 from some of the cobalt. This assumption was based in part upon binding energy shifts and more conclusively on quantitative deduction achieved by eliminating the proported MoS_2. (This will be commented upon below.) Sulphiding was shown not to influence the $CoAl_2O_4$; therefore a cobalt phase 'above' the

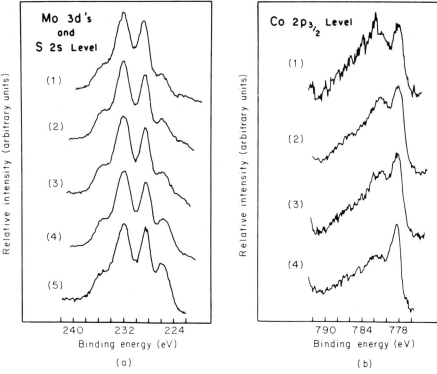

Figure 8.23   (a) ESCA spectra of Mo $3d$ doublet and S $2s$ region of sulphided catalysts. All samples were reacted with a 15 per cent H_2S–H_2 mixture for 1 h at 250 °C:
(1) MA 15, (2) CMA 315, (3) CMA 515, (4) CMA 715, (5) CMA 915. (b) ESCA
spectra of Co $2p_{3/2}$ region of sulphided CMA catalysts. Samples were reacted with a 15
per cent H_4S–H_2 mixture for 1 h at 250 °C: (1) CMA 315, (2) CMA 515, (3)
CMA 715, (4) CMA 915 wt.% of CoO and MoO_3 are respectively: MA 15 (0, 15.6),
CMA 315 (2.8, 16.5), CMA 515 (4.8, 14.2), CMA 715 (6.7, 14.2) and CMA 915
(8.9, 14.5). (Reprinted with permission from Chin and Hercules[196]. Copyright 1982
American Chemical Society)

t-$CoAl_2O_4$ was postulated. *The presence of cobalt does not seem to influence
the sulphiding of the molybdenum.* The possibility of the production of substantial Co^0 under these conditions was doubted, although the authors noted
that this feature may depend dramatically on the conditions employed for the
catalyst preparation (see above).

In order to explain these results, Chin and Hercules[196] speculate that a
cobalt (monolayer) must form on top of the t-$CoAl_2O_4$ and this layer is in
two-dimensional (proximity) contact with the active molybdenum monolayer
that lies above it. (Thus both layers form a quasi-bilayer.) The stoichiometry
of this Co–Mo bilayer (CMA) was proported to be 1:1. This was suggested to
be the HDS active layer structure for the catalyst with Co_9S_8 and MoS_2 species
formed in this structure during pre-sulphiding. Thus the molybdenum mono-

layer in the oxidized catalyst was assumed to occupy both tetrahedral (40 per cent) and octahedral (60 per cent) sites, whereas the 'active' underlayer of Co^{2+} straddles the tetrahedral sites between the underlaying Al_2O_3 and the 'overhead' molybdenum monolayer. This cobalt sublayer (Co–M) must sinergistically influence the catalytic properties of molybdenum, but does not seem to affect the latter's ability to reduce or sulphide.

Several other studies of possible importance have been conducted on problems of interest to the explanation of the functionality of the Co–Mo system. For example, Houalla and Delmon[197] examined the effect of the addition of boria to a $CoO–Al_2O_3$ catalyst. The boron was suspected to increase the acidity of the alumina, apparently resulting in the further dispersion of the cobalt and the creation of some surface $CoAl_2O_4$. In a similar vein, Vedrine *et al.*[198] employed ESR along with both UPS and XPS to examine a $MoO_3–TiO_2$ catalyst. In this study, the Mo^{VI} was first doped onto anatase. The resulting surface-to-bulk ratio displayed no segregation, i.e. $R_s/R_b \approx 1$. Following heating in air, the anatase was converted to rutile and the R_s/R_b value increased dramatically. In addition, the *bulk* molybdenum appeared to be reduced to Mo^V. This was apparently further evidence of a strong metals–support interaction by MoO_3, influenced by the thermal conditions. The direct comparison to alumina awaits further study.

8.3.5.4 Summary

Because of the complexity of the results described above and their potential importance to the functionality of this catalyst system, they are summarized at this stage:

(a) The molybdenum in the oxidized $MoO_3–Al_2O_3$ appears to exist in several Mo^{VI} species, depending upon concentration, calcination temperature and site location. As many as five species have been postulated,[241] with some of those occupying the tetrahedral sites suspected to have the greatest potential catalytic activity. The latter have been postulated to exist as a complex, *surface* molybdate[188] symbolized as:

However, the existence of this surface species has not been verified. It is generally well established that whatever this active Mo^{VI} species is, it is morphologically dispersed into monolayer coverage[184] which saturates at ~15 weight per cent molybdenum. Further addition of Mo^{VI} apparently results in the presence of a deleterious layer of MoO_3.[196] Deleterious bulk $Al_2(MoO_4)_3$ has also been postulated to be present.

(b) Reduction of the oxidized $Mo-Al_2O_3$ catalyst (without S^{2-}) creates both Mo^V and Mo^{IV} species.[194] The former are postulated to occur in the tetrahedral alumina sites, whereas the latter are suspected to arise in octahedral sites.[188]

(c) Combined reduction and pre-sulphiding of the $MoO_3-Al_2O_3$ catalyst has been shown to produce MoS_2.[195] Some evidence has been presented[183] for a sulphide vacancy complex:

but the existence of this species has not been confirmed. The existence of a monolayer of 'activate' MoS_2 seems confirmed,[195] possibly saturated in the tetrahedral sites of the alumina at ~15 weight per cent molybdenum. 'Excess' molybdenum remains inactive.[195] No Mo^0 was detected following reduction although the ability to distinquish MoS_2 from Mo^0 based solely upon XPS binding energy shifts has been questioned by the present author[27] due to a lack of a (Seigbahn) chemical shift. This effect seems characteristic of many calcogenides (see also the case for cobalt below). The supplemental use of XPS-induced loss shifts or the Auger parameter is advised if either can be detected. Some researchers have also postulated production of Mo^{III} species, but little supporting evidence exists for this uncommon ion.

(d) A number of researchers have also studied the informative $Co-Al_2O_3$ system.[175] In the oxidized form, the cobalt has been shown to interact with the alumina support, creating some $CoAl_2O_4$, apparently in tetrahedral sites. Little evidence for the existence of Co^{III} occurs, but the possibility of some Co_3O_4 remains.[195] The types of species formed were definitely shown to be dependent upon the amounts of dopants loaded and the calcination temperatures. Chin and Hercules[196] have postulated that the first 1 + weight per cent of cobalt which is loaded goes into tetrahedral sites of the alumina as inactive t-$CoAl_2O_4$. After that, continued loading produces a 'monolayer'-like cobalt effect which saturates out at concentrations that apparently may depend upon the amount of molybdenum present (whereas the reverse is apparently not true). Excess cobalt may go into octahedral sites, apparently existing as Co_3O_4.[196]

The products of reduction and pre-sulphiding of the $Co-Al_2O_3$ system are less certain than those for the $Mo-Al_2O_3$ catalyst. The production of Co_9S_8 seems indicated.[175,179,188]

(e) The creation of the mixed oxide catalyst $CoO-MoO_3-Al_2O_3$ seems to influence the eventual distribution and reductive pattern of the cobalt species more than the molybdenum. Reduction of this 'oxide' system was shown to produce mixtures of Co^{II}, Co^0, Mo^{VI}, Mo^V and Mo^{IV} species.

(f) Reduction and pre-sulphiding of the $Co-Mo-S-Al_2O_3$ catalyst produced a number of unique results, but once again the presence of cobalt seemed to have a minimimal effect on the molybdenum species. Substantial reduction of molybdenum to MoS_2 seems assured. This species was postulated to form a monolayer structure that saturates at ~15 weight per cent molybdenum.[196] The evidence, particularly that provided by ISS, strongly suggests that this monolayer covers any active cobalt species. Excess molybdenum seems to congregate on the outer surface as (poisonous) MoO_3. There is evidence for the foundation of deleterious. bulk $CoMoO_4$, but it has also been speculated, based upon the XPS evidence and the reductive patterns of this species, that it only forms from the reaction of excess Co_3O_4 and MoO_3. The 'active' cobalt on the $Co-Mo-S-Al_2O_3$ catalyst following H_2-H_2S treatment is felt to primarily exist as Co_9S_8. Evidence has been presented for the existence of substantial quantities of Co^0,[190] but others have assumed that the latter only occurs during the reduction of any excess $CoMoO_4$, formed as described above. Note that the XPS rationale for the differentiation between Co_9S_8 and Co^0 suffers from the same problem mentioned above for MoS_2 and Mo^0.[27] In any case the amount of cobalt, molybdenum and sulphur added during these processes may substantially affect these results. $CoAl_2O_4$ apparently still exists in the reduced $Co-Mo-S-Al_2O_3$ catalyst due to its resistence to reduction. Chin and Hercules[196] postulate that $CoAl_2O_4$ fills the tetrahedral sites of the alumina with the first 1+ weight per cent cobalt. Subsequent cobalt is suspected to form a monolayer substructure beneath the monolayer of MoS_2, and perhaps a quasi-bridge between the latter and the tetrahedral sites of the alumina support.[196] This cobalt is also suspected to be the species that reduces to Co_9S_8 in the H_2S flow. Although the cobalt does not itself influence the behaviour of the molybdenum during H_2-H_2S treatment, it is this bilayer MoS_2 on top of the Co_9S_8 'sandwich' that is postulated to provide the unique HDS catalytic properties of the $Co-Mo-S-Al_2O_3$ system.[196]

(g) The metals–support interactions for this catalyst have been shown to have strong negative as well as positive effects. These interactions must depend initially upon the acidity of the support as well as its physical structure.[198] The former effect has been studied by adding sodium[192] to suppress and boron[197] to enhance the acidity of the alumina. These changes have been shown to markedly affect the $CoAl_2O_4$ and CoO_3 production and also the dispersion and R_s/R_b values for the dopants. Further study is needed to clarify most of these effects.

Finally, it should be noted that the detailed set of XPS studies utilized to study this catalyst have not only employed the traditional areas of spectral analysis but also some of those newer areas described previously in this chapter. For example, Zingg et al.[188] employed the charging shift to explain some resulting peak broadening that may possibly have been misinterpreted by others. Other novel features in XPS application, however, could perhaps aid in these studies. For example, implementation of the structurally more exact quantitative procedures of Angevine, Vartuli and Delgass,[83] Kerkhof and Moulign[84] and even Fung[85] should improve the layer analyses. In addition, utilization of the Auger parameter[27] or loss shift analyses[32] may assist in sulphide detection. In closing, one must not forget the dramatic effects of even small changes in physical conditions. The results of Walton bear this out.[199]

8.4 Conclusion

The study of heterogeneous catalysis now occupies the time of large numbers of electron spectroscopists. These systems are extremely challenging because of their complicated physical and chemical status. As indicated in the earlier sections of this chapter, researchers have been forced to address such difficult issues as sample charging, *in situ* treatments, intensity and quantitative data from inhomogeneous materials, alternative chemical shifts, etc. Some of these subjects were alluded to in the aforementioned prognostications of Yates,[2] Yates and Madey[3] and Fischer.[4] Many of these areas are, however, novel features, particularly when addressed to applied catalysis.

Hopefully this chapter will have indicated the nature of the contribution of electron spectroscopy (particularly XPS) as a whole to catalysis. Several areas of substantial achievement deserving special mention include:

(a) Perhaps the most anticipated success was the application of XPS to directly detect the chemical changes (particularly in oxidation state) of the aforementioned dopants (e.g. metals, metal oxides, metal sulphides, etc.) during set-up and use of the catalysts. Numerous examples of dopant reduction, e.g. $Rh^{III} \rightarrow Rh^{I}$ (selective)—Watson and Somorjai[103]—Re^{VII}—Shipiro et al.[125]—$Pt^{II} \rightarrow Pt^{0}$ (electron deficient in the presence of germanium)—Bouwman and Biloen[126]—$Pd^{II} \rightarrow Pd^{0}$—Vedrine et al[143]—$Mo^{VI}O_3 \rightarrow Mo^{V} + Mo^{IV}$—Zingg et al.[188]—and also dopant oxidation, e.g. $Ru^{0} \rightarrow Ru^{IV}O_2$ on the Y-type zeolite—Pederson and Lunsford[157]—were described.

(b) One of the major recent achievements of the application of electron spectroscopy to catalysis has been the generation of results directly attributable to metals–support (M–S) and metals–metals (M–M) interactions. Numerous examples of the former have been published, primarily involving Al_2O_3-supported systems (e.g. Ru–Al_2O_3—Angevine and Delgass[128]—and MoO_3/Al_2O_3 $\rightarrow$ $Al_2(MoO_4)_3$—Okamoto *et al.*[182,183]). In general, it has been determined that when strong M–S or M–M interacts occur, the ability of the oxidized form of a dopant to reduce is greatly retarded (e.g. Angevine and Delgass[128] and Zingg *et al.*[188]).

(c) The importance of stratification in catalysis has been established and several aspects of this verified by electron spectroscopy. Sometimes inhomogeneities were shown to have positive catalytic effects, e.g. the formation of selective monolayers in the Co–Mo–S system—Gajardo, Grange and Delmon,[193] Chin and Hercules[196]—and in other cases inhomogeneities were shown to adversely affect catalysis, particularly when finite crystallites of platinum metals are produced, e.g. Somorjai and Zaera[93] and Mason and Bactzold.[138] The selective degradation of zeolites and subsequent surface deposition of byproducts was also monitored, e.g. Barr.[28,41,153]

(d) XPS and AES have also been successfully employed to investigate the properties and changes of catalytic promoters. Some of these were metal oxide or sulphide dopants, e.g. Re–S in platinum reforming catalysts (Biloen *et al.*[127]), while other promoters were suspected of modifying the support, e.g. F$^-$ on Al_2O_3 (Scohart *et al.*[200]) and Ce$^+$ on Al_2O_3 (Defossé *et al.*[201,202]). Equally impressive have been the numerous occasions when electron spectroscopy was employed to detect and characterize catalytic poisons, e.g. carbon on platinum (Somorjai and Zaera[93]) and Na$^+$ on Co–Mo–Al_2O_3 (Lycaurghiotis *et al.*[192]).

There is no doubt that the potential exists for direct connections between 'basic' and 'applied' catalysis. Examples of these types of studies have been produced by Somorjai and Zaera[93] and Ertl.[67] However, *until such efforts are augmented by controlled dopings of metals dispersed onto well-characterized supports (perhaps employing EELS), the direct transference of conclusions will, at best, be speculative.* It seems to this author that more effort is also needed from adroit researchers in applied catalysis to, for example, *characterize a singularly phased support onto which has been doped selected metals, microscopically dispersed through implementation of STEM observations to corroborate the status of these results. At this point controlled changes may be detected by electron spectroscopy to determine any chemical alterations, while also continuing the microscopic monitoring of the physical status.* Efforts on both sides of this present basic–applied barrier should admit some 'penetration' during the balance of the present decade.

In conclusion, application of electron spectroscopy into heterogeneous catalysis during the next eight years should result in:

(a) the useful implementation of EELS to describe supported catalysts;

(b) the realization of direct metals–adsorbate bond information in applied catalysis;
(c) quantum mechanical calculations that correlate with chemical results realized on basic and even some applied catalysts;
(d) direct correlation between structural changes (e.g. defects, dispersions, clustering) and electron spectroscopic results;
(e) quantitative procedures that accurately gauge these chemical and structural variations;
(f) techniques that describe (without overt materials damage) the lateral and depth chemistry of catalyst layers;
(g) truly interconnected multitechnique analyses;
(h) dynamic studies in applied as well as basic catalysis; and finally
(i) a number of substantive mechanistic studies in applied catalysis that not only describe the system under study but permit useful predictions of future catalyst systems of enhanced utility.

References

1. See, for example, K. Siegbahn, 'Perspectives and problems in electron spectroscopy', *J. Electron Spectrosc.*, **5**, 3 (1974).
2. J. T. Yates, Jr, *Chem. Engr. News,* August **1974,** 19 (1974).
3. J. T. Yates, Jr. and T. E. Madey, in *Science, Technology and the Modern Navy,* pp. 415–432, Office of Naval Research, Arlington, Va. (1976).
4. T. E. Fischer, *J. Vac. Sci. Technol.*, **11,** 252 (1974).
5. J. T. McCarroll, *Surf. Sci.*, **53,** 297 (1975).
6. F. C. Tompkins, *Proc. Int. Congr. Catal.*, **6,** 32 (1977).
7. J. J. McCarroll, T. Edmonds and R. C. Pitkethly, *Nature,* **223,** 1260 (1969).
8. J. J. Carberry, G. C. Kuczynski and E. Martinez, *J. Catal.*, **26,** 247 (1972).
9. R. A. VanSanten and W. M. H. Sachtler, *J. Catal.*, **33,** 202 (1974).
10. G. C. Stevens and T. Edmonds, *J. Catal.*, **37,** 544 (1975).
11. M. M. Bhasin, *J. Catal.*, **34,** 356 (1974).
12. W. N. Delgass, T. R. Hughes and C. S. Fadley, *Catal. Rev.*, **4,** 179 (1970).
13. J. S. Brinen and A. Melera, *J. Phys. Chem.*, **76,** 2525 (1972).
14. J. S. Brinen, *J. Electron Spectrosc. and Related Phenomena*, **5,** 377 (1974).
15. L. J. Matienzo, L. Yin, S. O. Grim and W. E. Swartz, Jr., *Inorg. Chem.*, **12,** 2762 (1973).
16. J. B. Lindsay, H. J. Rose, Jr., W. E. Swartz, Jr., P. H. Watts, Jr. and R. A. Rayburn, *App. Spectrosc.*, **27,** 1 (1973).
17. J. Fr. Temperé, D. Delafosse and J. P. Contour, *Chem. Phys. Lett.*, **33,** 95 (1975).
18. A. W. Armour, P. C. H. Mitchell, B. Folkesson and R. Larson, *J. Less. Common Metals,* **36,** 361 (1974).
19. P. Cannesson and P. Grange, *C.R. Acad. Sci., Paris,* **281,** 757 (1975).
20. T. E. Madey, C. D. Wagner and A. Joshi, *J. Elec. Spectrosc. and Related Phenomena,* **10,** 359 (1977).
21. C. J. Powell, N. E. Erickson and T. E. Madey, *J. Electron Spectrosc. and Related Phenomena,* **17,** 361 (1979).
22. C. S. Fadley, in *Electron Spectroscopy: Theory, Techniques and Applications*

(Eds C. R. Brundle and A. D. Baker), Vol. 2, pp. 1–156, Academic Press, New York (1978).
23. J. W. Gadzuk and M. Sunjić, *Phys. Rev.*, **B12**, 524 (1975).
24. K. T. Ng and D. M. Hercules, *J. Phys. Chem.*, **80**, 2094 (1976).
25. J. S. Brinen, in *Applied Surface Analysis* (Eds T. L. Barr and L. E. Davis), pp. 24–41, ASTM, Philadelphia (1978).
26. D. M. Hercules and J. C. Klein, in *Applied Electron Spectroscopy for Chemical Analysis* (Eds H. Windawi and F. Ho), pp. 147–189, John Wiley, New York (1982).
27. T. L. Barr, presented at the *Eleventh Annual Symposium Applied Vacuum Science and Technology*, AVS, Tampa, February 1982 (unpublished).
28. T. L. Barr, *Applications of Surf. Sci.*, **15**, 1 (1983).
29. T. L. Barr, J. E. Greene and A. H. Eltoukhy, *J. Vac. Sci. Technol.*, **16,** 517 (1979).
30. A. H. Eltoukhy, B. R. Natarajan, J. E. Greene and T. L. Barr, *Thin Solid Films*, **69,** 217 (1980).
31. T. L. Barr, J. E. Greene, A. H. Eltoukhy and B. R. Natarjan, *Thin Solid Films* (to be published).
32. T. L. Barr (to be published).
33. P. E. Larsen, *J. Electron Spectrosc. and Related Phenomena*, **4**, 214 (1974).
34. T. L. Barr and J. J. Hackenberg, *Applications of Surf. Sci.*, **10**, 523 (1982).
35. T. L. Barr and J. J. Hackenberg, *J. Amer. Chem. Soc.*, **104**, 5390 (1982).
36. T. L. Barr, *Surface and Interface Analysis*, **4**, 185 (1982).
37. See, for example, C. D. Wagner, *Anal. Chem.*, **47**, 1201 (1975); and C. D. Wagner, L. H. Gale and R. H. Raymond, *Anal. Chem.*, **51**, 466 (1979).
38. J. L. Gland, B. A. Sexton and G. Fisher, *Surf. Sci.*, **95**, 587 (1980).
39. K. Kishi and M. W. Roberts, *Proc. Roy. Soc., London*, **A352**, 289 (1976).
40. G. E. Thomas and W. H. Weinberg, *Phys. Rev. Lett.*, **41**, 1181 (1978).
41. T. L. Barr, presented at *ACS Symposium*, Chicago, January 1982 (to be published).
42. J. N. Miller, D. T. Ling, M. L. Shek, D. L. Weissman, P. M. Stefan, I. Lindau and W. E. Spicer, *Surf. Sci.*, **94**, 16 (1980).
43. R. W. McCabe and L. D. Schmidt, *Surf. Sci.*, **60**, 85 (1976).
44. C. R. Brundle and C. J. Todd (Eds.), *Adsorption at Solid Surfaces*, North Holland, Amsterdam (1975).
45. R. Gomer (Eds.), *Interactions on Metal Surfaces*, Springer-Verlag, Berlin (1975).
46. M. W. Robers and C. S. McKee, *Chemistry of the Metal-Gas Interface*, Clarendon Press, Oxford (1978).
47. G. A. Somorjai, *Chemistry in Two Dimensions: Surfaces*, Cornell U. Press, Ithaca (1981).
48. R. J. Madix, *Surf. Sci.*, **89**, 540 (1979).
49. R. J. Madix, *Adv. in Catal.*, **29**, 1 (1980).
50. J. B. Benziger and R. J. Madix, *J. of Catal.*, **65**, 36 (1980).
51. M. A. Barteau, E. I. Ko and R. J. Madix, *Surf. Sci.*, **102**, 99 (1981).
52. M. H. Matloob and M. W. Roberts, *J. Chem. Soc., Faraday Trans.*, **73** 1393 (1977).
53. D. W. Johnson, M. H. Matloob and M. W. Roberts, *J. Chem. Soc. Chem. Comm.*, **1978**, 41 (1978).
54. M. W. Roberts, *Adv. in Catal.*, **29**, 55 (1980).
55. B-H Chem., J. S. Close and J. M. White, *J. Catal.*, **46**, 253 (1977).
56. C. T. Campbell and J. M. White, *J. Catal.*, **54**, 289 (1978).

57. A. Golchet and J. M. White, *J. Catal.*, **53**, 266 (1978).
58. H. I. Lee and J. M. White, *J. Catal.*, **63**, 261 (1980).
59. J. S. Close and J. M. White, *J. Catal.*, **36**, 185 (1975).
60. T. Matsushima and J. M. White, *J. Catal.*, **40**, 334 (1975).
61. T. Matsushima, C. J. Mussett and J. M. White, *J. Catal.*, **41**, 397 (1976).
62. S. K. Shi, J. M. White and R. L. Hance, *J. Phys. Chem.*, **84**, 2441 (1980).
63. F. H. Tseng and J. M. White, *J. of Chem., Kinetics*, **12**, 417 (1980).
64. O. K. T. Wu, PhD Dissertation, U. of Illinois, Chicago (1979), O. K. T. Wu and R. P. Burns, *J. Phys. Chem.*, **84**, 1445 (1980).
65. O. K. T. Wu and R. P. Burns, *Surface and Interface Anal.*, **3**, 29 (1981).
66. P. Ho, PhD Dissertation, U. of Illinois, Chicago, 1981; P. Ho and R. P. Burns, submitted to *Applications of Surface Sci.*
67. G. Ertl, in *Chemistry and Physics of Solid Surfaces*, (Eds R. Vanselow and W. England), Vol. III, p. 11, CRC Press, Boca Raton, Florida (1982).
68. M. Boudart, in *Interactions on Metal Surfaces* (Ed. R. Gomer), pp. 275–298, Springer-Verlag, Berlin (1975).
69. D. Briggs, *Applications of Surface Sci.*, **6**, 188 (1980).
70. M. Kelley, *J. Catal.*, **57**, 113 (1979).
71. J. K. A. Clarke, *Chem. Rev.*, **75**, 291 (1975).
72. M. Kelley, presentation to the Catalyst Club, Chicago, 1981 (unpublished).
73. R. T. Lewis and M. A. Kelly, *J. Electron Spectros. and Related Phenomena*, **20**, 105 (1980).
74. C. D. Wagner, *J. Electron Spectrosc. and Related Phenomena*, **18**, 345 (1980).
75. H. Windawi and C. D. Wagner, in *Applied Electron Spectroscopy for Chemical Analysis* (Eds. H. Windawi and F. F. L. Ho), pp. 191–208, John Wiley and Sons, New York (1982).
76. H. Windawi, *J. Electron Spectrosc. and Related Phenomena*, **22**, 373 (1981).
77. T. L. Barr, presented in part at U. of Wisconsin, Milwaukee, manuscripts in preparation. See also T. L. Barr, *Chem. Phys. Lett.*, **43**, 89 (1976).
78. P. Swift, *Surface and Interface Anal.*, **4**, 47 (1081).
79. D. A. Stephensen and N. J. Binkowski, *J. Noncrystalline Solids*, **22**, 399 (1976).
80. See, for example, J. E. Castle and R. H. West, *J. Electron Spectrosc. and Related Phenomena*, **18**, 195 (1979); and R. H. West and J. E. Castle, *Surface and Interface Anal.*, **4**, 68 (1982).
81. C. D. Wagner, H. A. Six, W. T. Jansen and J. A. Taylor. *Appl. Surf. Sci.*, **9**, 263 (1981).
82. L. H. Scharpen, *J. Electron Spectrosc. and Related Phenomena*, **5**, 369 (1974).
83. P. J. Angevine, J. C. Vartuli and W. N. Delgass, *Proc. Int. Congress Catal.*, **6**, 611 (1977).
84. F. P. J. Kerkhof and J. A Moulijn, *J. Phys. Chem.*, **83**, 1612 (1979).
85. S. C. Fung, *J. Catal.*, **56**, 454 (1979).
86. O. A. Baschenko and V. I. Nefedov, *J. Electron Spectrosc, and Related Phenomena*, **13**, 405 (1979).
87. V. I. Nefedov, N. P. Serguskin, I. M. Band and M. B. Trzhaskovskya, *J. Electron Spectrosc. and Related Phenomena*, **2**, 383 (1973); V. I. Nefedov, N. P. Serguskin, Y. V. Salyn, I. M. Band and M. B. Trzhaskovskaya, *J. Electron Spectrosc. and Related Phenomena*, **7**, 175 (1975); and V. I. Nefedov and V. G. Yarzhensky, *J. Electron Spectrosc. and Related Phenomena*, **11**, 1 (1977).
88. See, for example, G. A. Somorjai, *Adv. in Catal.*, **26**, 1 (1977).
89. P. C. Stair, T. J. Kaminska, L. L. Kismodel and G. A. Somorjai, *Phys. Rev.* **B11**, 623 (1975); and B. Lang, R. W. Joyner and G. A. Somorjai, *Surf. Sci.*, **30**, 454 (1972).

90. D. W. Blakely, B. A. Sexton, E. Kozak and G. A. Somorjai, *J. Vac. Sci. Technol.*, **13**, 1091 (1976).
91. W. D. Gillespie, R. K. Herz, E. E. Peterson and G. A. Somorjai, *J. Catal.*, **70**, 147 (1981).
92. J. L. Gland, K. Baron and G. A. Somorjai, *J. Catal.*, **36**, 305 (1975); and D. W. Blakely and G. A. Somorjai, *J. Catal.*, **42**, 181 (1976).
93. G. A. Somorjai and F. Zaera, *J. Phys. Chem.*, **86**, 3070 (1982).
94. B. Lang, R. W. Joyner and G. A. Somorjai, *Proc. Roy. Soc., London*, **A331**, 335 (1975).
95. W. H. Weinberg, H. A. Deams and R. P. Merrill, *Surf. Sci.*, **41**, 312 (1974).
96. S. M. Davis, B. E. Gordon, M. Press and G. A. Somorjai, *J. Vac. Sci. Technol.*, **19**, 231 (1981).
97. M. Salmeron and G. A. Somorjai, *J. Phys. Chem.*, **86**, 341 (1982).
98. L. L. Kismodel and L. M. Falicov, *Solid State Comm.*, **16**, 1201 (1975).
99. J. W. A. Sachtler, M. A. Van Hove, J. P. Biberian and G. A. Somorjai, *Surf. Sci.*, **110**, 19, 43 (1981).
100. J. W. A. Sachtler and G. A. Somorjai (to be published).
101. C. E. Smith, J. P. Biberian and G. A. Somorjai, *J. Catal.*, **57**, 426 (1979).
102. L. H. Dubois, P. K. Hansma and G. A. Somorjai, *J. Catal.*, **65**, 318 (1980).
103. P. R. Watson and G. A. Somorjai, *J. Catal.*, **72**, 347 (1981).
104. P. R. Watson and G. A. Somorjai, *J. Catal.*, **74**. 282 (1982).
105. B. E. Koel, private communication.
106. N. D. Spencer, R. C. Schoonmaker and G. A. Somorjai, *J. Catal.*, **74**, 129 (1982).
107. V. Haensel, US Patent 2,479,109 (1949).
108. H. Heinemann, in *Catalysis: Science and Technology* (Eds J. R. Anderson and M. Boudert), pp. 1–41, Springer-Verlag, Berlin (1981).
109. M. L. Good and J. C. Hayes, presentation at the Kilpatrick Lectures at IIT, October 1982.
110. W. M. H. Sachtler and R. A. Van Santen, *Applications of Surface Sci.*, **3**, 121 (1977).
111. W. M. H. Sachtler and R. A. Van Santen, *Adv. in Catal.*, **26**, 69, (1977).
112. J. P. Bournouville and G. Martino, in *Catalyst Deactivation* (Eds B.Delmon and G. F. Froment), p. 159, Elsevier, Amsterdam (1982).
113. R. Burch, *J. Catal.*, **71**, 348, 360 (1981).
114. R. Bouwman and P. Biloen, *Proc. Seventh Intern. Vac. Cong. and Third Intern. Conf. Solid Surfaces,* Vienna, 1977, p. 1129.
115. J. K. A. Clarke and A. C. M. Creaner, *Ind. Eng. Chem. Prod. and Dev.*, **20**, 575 (1981).
116. See, for example, for platinum, K. S. Kim, N. Winograd and R. E. Davis, *J. Amer. Chem. Soc.*, **93**, 6296 (1971) and J. S. Hammond and N. Winograd, *J. Electronanal. Chem. Interfacial Electrochem.*, **78**, 55 (1977); for palladium, K. S. Kim, H. F. Grossman and N. Winograd, *Anal. Chem.*, **46**, 197 (1974); and for silver, S. W. Gaarenstroom and N. Winograd, *J. Chem. Phys.*, **67**, 3500 (1977).
117. T. L. Barr, *J. Phys. Chem.*, **82**, 1801 (1978).
118. A. Cimino, B. A. DeAngelis, D. Gazzoli and M. Valigi, *Z. Anorg. Allg. Chem.*, **460**, 86 (1980).
119. C. D. Wagner, W. N. Riggs, L. E. Davis, J. F. Moulder and G. E. Mullenberg, *Handbook of X-ray Photoelectron Spectroscopy*, Perkin-Elmer Co., Eden Prarie, Minnesota (1979).
120. J. Escard, B. Pontvianne, M. T. Chenebaux and J. Cosyns, *Bull. Soc. Chem. France*, **11**, 2399 (1975).

121. E. Czáran, J. Finster and K. H. Schnabel, *Z. Anorg. Allg. Chem.,* **443**, 175 (1978).
122. F. Bozon-Verduraz, A. Omar, J. Escard and B. Pontvianne, *J. Catal.,* **53**, 126 (1976).
123. R. J. M. Griffiths and E. L. Evans, *J. Catal.,* **36**, 413 (1975).
124. M. Batista-Lcal, J. E. Lester and C. A. Lucchesi, *J. Electron Spectrosc. and Related Phenomena,* **11**, 333 (1977).
125. E. S. Shipiro, V. I. Avaev, G. V. Antoshin, M. A. Ryashentseva and K. M. Minachev, *J. Catal.,* **55**, 402 (1978).
126. R. Bouwman and P. Biloen, *J. Catal.,* **48**, 209 (1977).
127. P. Biloen, T. N. Helle, H. Verbeek, F. M. Dautzenberg and W. M. H. Sachtler, *J. Catal.,* **63**, 112 (1980).
128. P. T. Angevine and W. N. Delgass, *Sixty-ninth Annual AIChE meeting,* Chicago, 1976, p. 55a.
129. F. M. Dautzenberg and H. B. M. Walters, *J. Catal.,* **51**, 26 (1978).
130. N. Giordano, J. C. J. Bart, R. Maggiore, C. Crisafulli, L. Sariano, G. Schernbari, S. Cavallaro and P. Antonucci, *Anali di Chemico,* **1980**, 585 (1980).
131. J. S. Brinen and J. L. Schmitt, *J. Catal.,* **45**, 274 (1976).
132. C. T. H. Stoddart, R. L. Moss and D. Pope, *Surf. Sci.,* **53**, 241 (1975).
133. J. P. Contour, G. Mouvier, M. Hoogewys and C. Ledere, *J. Catal.,* **48**, 217 (1978).
134. Y. Takasu, R. Unwin, B. Tesche and A. W. Bradshaw, *Surf. Sci.,* **77**, 219 (1978).
135. See, for example, K. S. Kim and N. Winograd, *Chem. Phys. Lett.,* **30**, 91 (1975).
136. M. Cini, *Surf. Sci.,* **62**, 148 (1979).
137. F. Garbasi and G. R. Tauszik, *J. Microsc. Spectrosc. Electron,* **2**, 245 (1977); and G. Mattagino, G. Polsonetti and G. R. Tauszik, *J. Electron Spectrosc. and Related Phenomena,* **14**, 237 (1978).
138. M. G. Mason and R. C. Baetzold, *J. Chem. Phys.,* **64**, 271 (1976).
139. V. Haensel and M. J. Sterba, *Ind. Eng. Chem. Prod. Res. and Dev.,* **15**, (1976).
140. M. M. Bhasin, *J. Catal.,* **38**, 218 (1975).
141. I. Szymerska and M. Lipski, *J. Catal.,* **41**, 197 (1976).
142. Y. Matsumoto, M. Soma, T. Onishi and K. Tamaru, *J. Chem. Soc., Faraday Trans. 1,* **76**, 1122 (1980).
143. J. C. Vedrine, M. Dufaux, C. Naccache and B. Imelik, *J. Chem. Soc., Faraday Trans. 1,* **74**, 440 (1978).
144. Y. Okamoto, N. Ishida, T. Imanaka and S. Teranishi, *J. Catal.,* **58**, 82 (1979).
145. S. L. T. Anderson and M. S. Scurrell, *J. Catal.,* **71**, 233 (1981).
146. D. W. Breck, *Zeolite Molecular Sieves,* Wiley-Interscience, New York (1974).
147. P. B. Weiz, N. J. Frilette, R. W. Maatman and E. B. Mower, *J. Catal.,* **1**, 307 (1962).
148. V. Haensel, personal communication (1982).
149. C. J. Plank, E. J. Rosinsky and W. P. Hawthorne, *I. and E.C. Prod. Res. Devel.,* **3**, 165 (1964).
150. R. J. Angauer and G. R. Landolt, U.S. Patent 3702886 (1972).
151. X. M. Minachev G V. Antoschin, E. Schapiro and T. A. Navrousov, *Izv. Akad. Nauk SSR Ser. Kim.,* **9**, 2131, 2134 (1973).
152. J. Kneckt and G. Stork, *Z. Anal. Chem.,* **283**, 105 (1977).
153. T. L. Barr, *Amer. Chem. Soc. Div. Petroleum Chem.,* **23**, 82 (1978).
154. E. Dempsey, 'Molecular Sieves', *Soc. Chem. Ind. London.* **1968**, 293 (1968).
155. B. Wichterlová, J. Novaková, L. Kubelková and P. Jiru, *Proc. Fifth Int. Conf. Zeolites* (Eds Rees and Lovat), pp. 373–381 (1980).

Applications of Electron Spectroscopy to Heterogeneous Catalysis 357

156. F. O. Bravo, J. Dwyer and D. Zambaulis, *Proc. Fifth Int. Conf. Zeolites* (Eds Rees and Lovat), pp. 749–759 (1980).
157. L. A. Pederson and J. H. Lunsford, *J.Catal.*, **61**, 39 (1980).
158. P. Gallezot, A. Alacon-Diaz, J. A. Delmon, A. J. Remouprez and B. Imelik, *J. Catal.*, **39**, 334 (1976).
159. B. L. Gustafson and J. H. Lunsford, *J. Catal.*, **74**, 393 (1982).
160. F. Fajula, R. G. Anthony and J. H. Lunsford, *J. Catal.*, **73**, 237 (1982).
161. S. M. Kuznicki and E. M. Eyring, *J. Catal.*, **65**, 227 (1980).
162. F. P. Larkins, M. E. Hughes, J. R. Anderson and K. Foger, *J. Electron Spectrosc. and Related Phenomena*, **15**, 33 (1979).
163. S. Badrinarajanan, R. I. Hodge, I. Balakrishnan, S. B. Kulkami and P. Ratmasany, *J. Catal.*, **71**, 439 (1981).
164. T. Edmonds, in *Characterization of Catalysts* (Eds J. M. Thomas and R. M. Lambert), Wiley-Interscience, John Wiley and Sons, New York (1980).
165. L. Fiermans, R. Hoogewijs and J. Vennik, *Surf. Sci.*, **47**, 1 (1975).
166. S. J. Atkinson, C. R. Brundle and M. W. Roberts, *Discussion of Faraday Soc.*, **58,** 62 (1974).
167. See, for example, C. R. Brundle *Surf. Sci.*, **48**, 99 (1975); and C. R. Brundle, lecture delivered at La Jolla Institute Workshop, 'Aspects of kinetics and dynamics of surface reactions', San Diego, 1979 (unpublished).
168. A. Cimino and B. A. DeAngelis, *J. Catal.*, **36**, 11 (1975).
169. D. C. Frost, C. A. McDowell and I. S. Woolsey, *Molec. Phys.*, **27**, 1473 (1974).
170. K. S. Kim, *Phys. Rev.*, **B11**, 2177 (1975).
171. N. S. McIntyre and M. G. Cook, *Anal. Chem.*, **47**, 2208 (1975).
172. T. J. Chuang, C. R. Brundle and K. Wandelt, *Thin Solid Films*, **53**, 19 (1978).
173. R. Prims, private communication (1980).
174. J. Haber, W. Marczewski, J. Stock and L. Ungier, *Proc. Int. Cong. Catal.*, **6**, 827 (1977).
175. R. I. Declerck-Grimce, P. Canesson, R. M. Friedman and J. J. Friplet, *J. Phys. Chem.*, **82**, 885 (1978).
176. R. I. Declerck-Grimce, P. Canessen, R. M. Friedman and J. J. Friplet, *J. Phys. Chem.*, **82**, 889 (1978).
177. Y. Okamoto, H. Nakano, T. Shimakawa, T. Imamaka and S. Teranishi, *J. Catal.*, **50**, 447 (1977).
178. J. Grimblot and J. P. Bonnelle, *J. Electron Spectrosc. and Related Phenomena*, **9**, 449 (1976).
179. F. E. Massoth, *J. Catal.*, **54**, 450 (1978).
180. Y. Okamoto, T. Imanaka and S. Teranishi, *J. Catal.*, **54**, 452 (1978).
181. Y.Okamoto, T. Shimokawa, T. Imanaka and S. Teranishi, *J. Chem. Soc., Chem. Comm.*, **47**, (1978).
182. Y. Okamoto, T. Shimokawa, T. Imanaka and S. Teranishi, *J.Catal.*, **57**, 153 (1979).
183. Y. Okamoto, H. Tomioka, Y. Katoh, T. Imanaka and S. Teranishi, *J. Phys. Chem.*, **84**, 1833 (1980).
184. N. Giordano, A. Castellan, J. C. J. Bart, A. Vaghi and F. Campadeli, *J. Catal.*, **37**, 204 (1975).
185. T. Edmonds and P. C. H. Mitchell, *J. Catal.*, **64**, 491 (1980).
186. T. A. Patterson, J. C. Carver, D.C. Leyden and D. M. Hercules, *J. Phys. Chem.*, **80**, 1700 (1976).
187. T. A. Patterson, J. C. Carver, D. E. Leyden and D. M. Hercules, *Spectrosc. Lett.*, **9**, 65 (1976).

188. D. S. Zingg, L. E. Makovsky, R. E. Tischer, F. R. Brown and D. M. Hercules, *J. Phys. Chem.*, **84**, 2898 (1980).
189. G. Delvaux, P. Grange and B. Delmon, *J. Catal.*, **56**, 991 (1979).
190. J. S. Brinen and W. D. Armstrong, *J. Catal.*, **54**, 57 (1978)
191. F. Delannay, P. Gajardo, P. Grange and B. Delmon, *J. Chem. Soc., Faraday Trans. 1*, **76**, 988 (1980).
192. A. Lycourghiotis, C. Defosse, F. Delannay, J. Lamaitre and B. Delmon, *J. Chem. Soc., Faraday Trans. I*, **76**, 1677, 2052 (1980).
193. P. Gajardo, P. Grange and B. Delmon, *J. Catal.*, **63**, 201 (1980).
194. E. L. Aptekar, M. G. Chudinov, A. M. Alekseev and O. V. Krylov, *Reaction Kinetics and Catal. Lett.*, **1**, 493 (1974).
195. D. M. Hercules and J. C. Klein, in *Applied Electron Spectroscopy for Chemical Analysis* (Eds H. Windawi and F. F. L. Ho), John Wiley and Sons, New York (1982).
196. R. L. Chin and D. M. Hercules, J. Phys. Chem., **86**, 3079 (1982).
197. M. Houalla and B. Delmon, *Appl. Catal.*, **1**, 285 (1981).
198. J. C. Vedrine, H. Praliaud, P. Meriandeau and M. Che, *Surf. Sci.*, **80**, 101 (1979).
199. R. A. Walton, *J. Catal.*, **44**, 335 (1976).
200. P. O. Scohart, S. A. Selim, J. P. Demon and P. G. Rouxhet, *J. Coll. and Interface Sci.*, **70**, 209 (1979).
201. C. Defossé, P. Canesson, P. G. Rouxhet and B. Delmon, *J. Catal.*, **51**, 269 (1978).
202. C. Defossé, P. Canesson, P. Rouxhet and B. Delmon, *J. Catal.*, **57**, 525 (1979).

Practical Surface Analysis
by Auger and X-ray Photoelectron Spectroscopy
Edited by D. Briggs and M. P. Seah
© 1983, John Wiley & Sons, Ltd.

Chapter 9

Applications of XPS in Polymer Technology

D. Briggs

Imperial Chemical Industries PLC
Petrochemicals and Plastics Division
Wilton, Middlesbrough, Cleveland, UK

9.1 Introduction

Polymers are increasingly pervasive in materials science: in plastic mouldings, sheets, fibres and films; in composites with inorganic materials; in protective coatings, sealants, adhesives; etc. The impact of XPS in this area of materials characterization has been twofold: firstly through its ability to analyse relatively intractable materials without the need for special sample preparation and secondly through its surface sensitivity.

This chapter first discusses aspects of instrumentation, sample handling and spectral information of especial relevance to the study of polymeric materials. Next the advantages and disadvantages of XPS relative to other applicable techniques are briefly discussed. The rest of the chapter is devoted to examples of the successful application of XPS in this area, concentrating on studies of polymer surface modification for improvement of adhesion, since these have shed much light onto technologically very important but empirically derived processes.

9.2 Sample Handling

No special facilities are required for polymer work in general. Contrary to the expectation of many workers who are used to studying metallic systems, polymer samples do not present serious outgassing problems and can easily be introduced into UHV systems attached to the sample probe or mount by means of double-side adhesive tape (itself polymeric!). Concern is often expressed about the possibility of silicone contamination from the release agents used in adhesive tapes. Clark[1] has reported experiments in which migration of silicones along polymer surfaces and through polymer films (not

359

under vacuum) was studied by XPS. Both kinds of migration were discerned—that along some surfaces being quite fast (e.g. several centimetres in ten days at room temperature along LDPE). There is therefore a danger of contamination from fairly long-term exposure of samples to tape, e.g. in securing samples for transportation or in the labelling of samples by attachment of self-adhesive labels which require release agents.

Contamination of polymer surfaces does not constitute a major problem provided some care is taken. In general, the sticking coefficients of residual gas molecules on polymer surfaces are rather low; nevertheless, the build-up of hydrocarbon-like material is frequently observed within the timescale of the experiment (indeed this has often been used as an energy reference in the study of polymeric materials which do not present a well-resolved 'hydrocarbon' C 1s signal; see Appendix 2). This contamination affects quantification particularly if low kinetic energy peaks (involving electrons of low escape depth) are to be measured, but perhaps more importantly it affects directly the C 1s profile. The contamination almost certainly comes from the X-ray gun window or casing in close proximity to the sample. Working with polymers routinely increases the problem because the spectrometer working pressure is usually significantly greater than base pressure over extended periods due to slow outgassing and loss of volatile low molecular weight constituents. The best preventative measure is regular system bake-out.

In the author's experience, polymer surfaces, even of those systems containing additives, are reasonably stable (in the sense that spectra are not time-dependent to a significant degree). Prolonged exposure to X-ray beams of typical flux density will often produce visual evidence of damage (discoloration). It has been reported that poly(thiocarbonyl fluoride) depolymerizes extremely rapidly and that poly(vinylidene flouride) slowly eliminates HF and crosslinks.[2] A recent study of polytetrafluorethylene (PTFE) has followed the surface changes under prolonged X-irradiation and the nature of the species lost from the surface in some detail.[3] Of the common polymers those containing chlorine, especially poly(vinyl chloride), seem to be the most sensitive to X-ray damage. Another source of damage is through thermal effects. Since the X-ray gun in some instruments can get very warm, the surfaces of good thermal insulators can then suffer appreciable temperature increases, giving rise to surface degradation, migration and/or loss of additives, etc. Cooling the sample holder may mitigate these effects in such instruments.

9.3 Instrumentation

A fundamental problem in polymer surface analysis is the occurrence of charging effects due to the insulating nature of the materials, although with conventional X-ray sources the effect is usually not greater than 5 V. As might be anticipated, the magnitude is a strong function of surface composition,[4] all other factors (especially X-ray flux and sample position) being equal. When a monochromated X-ray beam is employed the situation is much worse (largely because the window separating the X-ray source from the sample, which provides the stabilizing electrons for insulating surfaces, is now much more remote from the sample and cannot provide sufficient electron flux). In this case it is mandatory that some form of charge neutralization is used—usually a 'flood gun' to provide a low-energy electron flux over the sample surface (see Appendix 2).

In general the preferred excitation source is Mg $K\alpha$ since this leads to the smallest peak widths. A dual-anode source (usually Mg $K\alpha$ and Al $K\alpha$) is useful in cases where overlap occurs between a core level of interest and an X-ray excited Auger peak. For example, with Mg $K\alpha$ there is a strong interference between C $1s$ and Na KLL and between Na $1s$ and C 1 LMM. The usefulness of monochromated X-rays (perforce with Al $K\alpha$) in studying polymers is debatable. Compared with using Mg $K\alpha$ under the highest resolution conditions the improvement in resolution is not very significant and very careful charge compensation is required before this can be achieved. Data-collection times may be long (reduced X-ray flux) but X-ray satellites which can be a nuisance are removed (see Chapter 3). On the other hand, data processing of normal (e.g. Mg $K\alpha$) spectra can easily perform this function (see Appendix 3).

One particular instrumental facility which stands out in polymer surface analysis is the ability to charge the angle between the sample surface and the

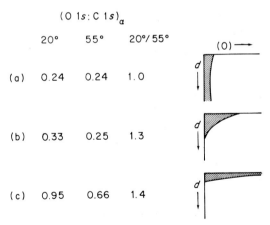

Figure 9.1 Schematic representation of oxygen–depth profiles from angular variations measurements on polymers surface oxidized in different ways: (a) 'corona' discharge treated polyethylene; (b) hot pressed polyethylene; (c) plasma oxidized polystyrene. (Reprinted with permission from Dilks[5]. Copyright 1981 American Chemical Society)

analyser entrance slit (see Chapter 2). In many cases of interest there are marked concentration profiles within the XPS sampling depth, and since argon ion etching is an inappropriate technique for polymer depth profiling (due to rapid structural and compositional alteration), non-destructive profiling by angular variation techniques assumes increased importance. A simple, but telling, example is provided in Figure 9.1.

9.4   Spectral Information

9.4.1   The information depth

The depth from which information is obtained is governed by the inelastic mean free part (IMFP) of electrons in a solid (λ). As discussed in Chapter 4, 95 per cent of the observed photo-electron signal derives from a layer 3 λs in α thick (for a flat surface) where α is the 'take-off angle' of the electrons relative to the surface.* The appropriate values of λ for polymers has been the subject of controversy, largely because two different procedures for their determination led to different results and because data from Langmuir–Blodgett (LB) films of organic materials were assumed to be comparable. Recent work on IMFPs of electrons in LB films strongly suggests this assumption to be invalid.[6] An excellent account of the issues involved in polymer IMFPs has been given[7] which concludes that λ values for polymers are similar to those for typical metals and inorganic materials.

As discussed in Chapter 4, non-destructive depth profiling can also be carried out by observation of the variation of relative intensities with KE (e.g. Clark's experiments[1] using Mg $K\alpha$ and Ti $K\alpha$ to excite polymer core levels) or by following changes in the relative intensity of two core levels widely separated in KE, e.g. O $1s$/O $2s$ or F $1s$/F $2s$. This latter measurement is particularly useful for studying the thickness of modified layers containing oxygen or fluorine on a substrate in which these elements are absent. The $1s/2s$ ratio is at a minimum for layers of thickness $\simeq 3\ \lambda_{(2s)} \sin \alpha$ (i.e. equal to the value for homogeneous materials) and increases as the thickness of the modified layer decreases. An example is given in Section 9.6.2.2.

9.4.2   Core-level binding energies

Over the last ten years Clark and others[8] have created a large body of literature on this subject through the study of pure polymers and model small molecules coupled with theoretical calculations, much of it recently reviewed by Dilks.[9] From these data some simple summaries can be made.

---

* For comparison with Chapter 5, $\alpha = 90 - \theta$, where θ is the angle with respect to the surface normal.

9.4.2.1 C 1s binding energies

(a) Carbon bound to itself and/or hydrogen only, no matter what hybridization gives C $1s$ = 285.0 eV (often used as a binding energy reference).

(b) Halogens induce shifts to a higher binding energy which can be broken down into a primary substituent effect (i.e. on the carbon atom directly attached) and a secondary substituent effect (on the neighbouring carbon atom(s)). These shifts, per substituent, are approximately:

Halogen	Primary shift, eV	Secondary shift, eV
F	2.9	0.7
Cl	1.5	0.3
Br	1.0	<0.2

(c) Oxygen induces shifts to higher binding energy by $\approx$1.5 eV per C—O bond (thus O—C—O and $>$C$=$O give similar C $1s$ binding energies, etc). The secondary effect of X in C—O—X is small ($\pm$0.4 eV) except in the case of X = NO_2 (nitrate ester) which produces an additional shift of 0.9 eV.

(d) Nitrogen functionalities have a primary substituent effect which is markedly dependent on the nature of the substituent. Thus C $1s$ shifts for —N(CH$_3$)$_2$, —NH$_2$, —NCO and —NO$_2$ are 0.2, 0.6, 1.8 and 1.8 eV, respectively. Both carbon atoms in —CH$_2$—C$\equiv$N suffer a shift of $\approx$1.4 eV. The C $1s$ shift induced by —ONO$_2$ is $\approx$2 eV.

9.4.2.2 O 1s binding energies

O $1s$ BEs from most functionalities fall within a narrow range of $\approx$2 eV around 533 eV. The extremes are seen in carboxyl and carbonate groups in which the singly bound oxygen has the higher BE.

9.4.2.3 N 1s binding energies

Many common nitrogen functionalities given N $1s$ BEs in the narrow region 399–401 eV. These include —CN, —NH$_2$, —OCONH— and —ĊONH$_2$. Quarternization, as in —NH$_3^+$, only increases the BE by $\approx$1.5 eV above that of the free amine, largely because of the effect of counterion.

Oxidized nitrogen functions have much higher N $1s$ BEs: —ONO$_2$ ($\approx$408 eV), —NO$_2$ ($\approx$407 eV) and —ONO ($\approx$405 eV).

9.4.2.4 Other core levels

The only other atoms with a variable oxidation state commonly encountered in polymers are sulphur and silicon. The primary effect of sulphur on the C $1s$

BE is very small (≈ 0.4 eV measured in a polysulphone). However, S $2p$ BEs cover a reasonable range: R—S—R (≈ 164 eV), R—SO$_2$—R (≈ 167.5 eV), R—SO$_3$H (≈ 169 eV).

There have been few reports of Si $2p$ BEs. Typical silicones having the polysiloxane structure —OSiR$_2$O— give Si $2p$ ≈ 102 eV.

It is also worth noting that halide ions can be distinguished from covalently bound halogen. In the case of fluorine the F $1s$ BE for F$^-$ is ≈ 4 eV lower than for C—F (typically 689 eV) whilst the Cl $2p$ BE for Cl$^-$ is ≈ 2 eV lower than C—Cl (typically 201 eV).

9.4.3 Shake-up satellites

As noted in Chapter 3 shake-up satellites are expected for all polymeric systems containing unsaturation. However, they have only been studied in detail for hydrocarbon polymers with pendant aromatic groups, particularly substituted polystyrenes. These studies show the shake-up satellite to be typically 6–7 eV to low KE of the primary C $1s$ peak, asymmetric and sometimes distinctly double peaked and to be up to ≈ 8 per cent of the intensity of the main peak. The structure is undoubtedly due to two transitions involving excitation from the two filled orbitals $b_{1\pi}$ and $a_{2\pi}$ to the unoccupied $b_{1\pi}^*$ orbital (i.e. $\pi \rightarrow \pi^*$).[9] These satellites are therefore indicative of aromaticity in a material. Figure 9.2 shows the use of satellite intensity to follow surface damage to polystyrene during SIMS examination.[10]

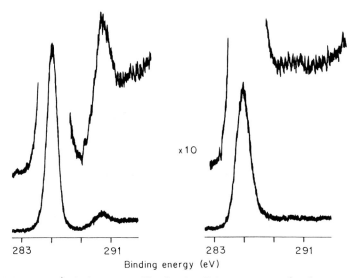

Figure 9.2 $\pi \rightarrow \pi^*$ shake-up satellite in the C $1s$ spectrum of polystyrene and its virtual elimination after a dose of 1.6×10^{14} Ar$^+$ cm^{-2} (4 keV energy). Note also the broadening of the core-level peak. (After Briggs and Wootton[10])

Attempts have been made to use $\pi \rightarrow \pi^*$ shake-up satellites for surface structure determination when core-level signals alone are inadequate for this purpose, e.g. by Clark and Dilks[11] in a study of alkane–styrene copolymers. This approach demands that the relative intensity of the satellite to the primary C 1s signal is invarient to matrix variation (change in copolymer composition, overall crystallinity, etc). That this constraint may not hold has been indicated by O'Malley, Thomas and Lee[12] in their study of polystyrene–ethylene oxide block copolymers.

The occurrence of shake-up satellites from vinylic unsaturation, at $\simeq 7$ eV from the primary C 1s peak, has been invoked in the determination of the structure of polyhexafluorobut-2-yne[13] and to rationalize discrepancies in F/C stoichiometries from plasma polymerized fluoroethylenes.[14] Atoms directly attached to unsaturated centres will also be accompanied by shake-up satellites, and this may have some analytical value.

9.4.4 Valence band spectra

The fingerprinting capacity of polymer valence band spectra has already been mentioned in Chapter 3. Figure 9.3 gives an example in which purely hy-

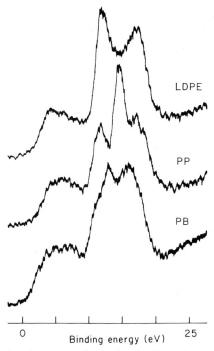

Figure 9.3 Valence band spectra from the hydrocarbon polymers: low-density polyethylene (LDPE), polypropylene (PP) and poly(but-1-ene) (PB). All these polymers give identical C 1s spectra

drocarbon polymers are distinguished. The whole subject has been carefully researched both experimentally and theoretically by Pireaux and colleagues and their results have recently been reviewed.[15] These workers have published high-resolution spectra, obtained with an X-ray monochromator, of polymers containing oxygen, fluorine and chlorine as well as hydrocarbon polymers. They have also demonstrated the sensitivity of valence band spectra to various types of isomerization (structural-, linkage- and stereo-), as well as to tacticity and geometrical conformation.

9.4.5 Functional group labelling (derivatization)

As noted in Section 9.4.2.1, there is a lack of specificity in core-level BEs, many functional groups giving rise to similar 'shifts'. This problem is even

Table 9.1 Derivatization reactions

Functional group	Reagent	Product	Ref.
$\overset{\diagup}{\underset{\diagdown}{C}}{=}\overset{\diagup}{\underset{\diagdown}{C}}$	Br_2	$\overset{\diagup}{\underset{\diagdown}{C}}{-}Br$ $\overset{\diagup}{\underset{\diagdown}{C}}{-}Br$	16, 17, 18
$\overset{\diagup}{\underset{\diagdown}{C}}{=}\overset{\diagup}{\underset{\diagdown}{C}}$	$Hg(CF_3COO)_2$ $\left.\begin{array}{c} \\ \\ \end{array}\right\}$ Cl_3CH_2OH	$\overset{\diagup}{\underset{\diagdown}{C}}{-}Hg(CF_3COO)$ $\overset{\diagup}{\underset{\diagdown}{C}}{-}OCH_2CCl_3$	19
$-CH_2OH$	$(CF_3CO)_2O$	$-CH_2OCOCF_3$	19, 20
$-CH_2OH$	$Ti(acac)_2OPr_2^i$	$-CH_2OTi(acac)OPr^i$	21
$\overset{\diagup}{\underset{\diagdown}{C}}{=}\underset{\mid}{C}{-}OH$	$ClCH_2COCl$	$C{=}\underset{\mid}{C}{-}OCOCH_2Cl$	21
$-CH_2-C\overset{\diagup\!\!\diagup O}{\diagdown}$	Br_2	$-CBr_2-C\overset{\diagup\!\!\diagup O}{\diagdown}$	21, 22
$\overset{}{\underset{\diagdown}{C}}{=}O$	$COH_5NH{-}NH_2$	$-C{=}N{-}NHC_6H_5$	21, 22, 19
$-COOH$	$NaOH$	$-COO^-Na^+$	19–23
$-COOH$	$BaCl_2$	$(-COO^-)_2Ba^{2+}$	24, 25
$-COOH$	CF_3CH_2OH $C_6H_{11}NCNC_6H_{11}$	$-COOCH_2CF_3$	19
$-COOH$	(1) KOH (2) $C_6F_5CH_2Br$	$-COOCH_2C_6F_5$	19
$-COOH$	$TlOC_2H_5$	$-COO^-Tl^+$	23
$-C{-}OOH$	SO_2	$-C{-}OSO_2OH$	21
$-NH_2$	C_6F_5CHO	$-N{=}CHC_6F_5$	19
$-NH_2$	$C_2H_5S{-}COCF_3$	$-NH{-}COCF_3$	26

more pronounced when several such groups are present in the surface, since individual peaks overlap and the whole system suffers from broadening due to substituent secondary effects. Several workers have attempted to overcome these problems by using specific derivatizing agents which react with one functional group and label it with a distinctive element. This procedure may have the additional advantage of increasing the detection sensitivity if the new element has a higher cross-section than, for example, carbon, nitrogen or oxygen. Table 9.1 lists the reactions which have so far been investigated.

The reader is referred to the original literature for reaction conditions and estimates of success. Besides being specific, the derivatizing reaction should proceed rapidly under mild conditions and any necessary solvent should be benign. These last two conditions can be difficult to meet. Reactions which proceed rapidly at room temperature in the solution phase are often sterically hindered in the polymer surface layers. Solvents which permeate into the polymer are likely to aid reaction but may, at the same time, give rise to surface reorganization, e.g. functional group migration into the bulk. This aspect of derivatization has been studied in some depth by Everhart and Reilley.[27,28]

9.5 Comparison of XPS with Other Techniques[29]

Before the advent of XPS, virtually the only technique available for studying polymer surfaces was reflection infra-red spectroscopy (either attenuated reflection (ATR) or multiple internal reflection (MIR) spectroscopy). This technique requires fairly large samples with flat or easily deformable surfaces and typically gives information pertaining to $\simeq 1$ μm into the material. In recent years Fourier transform infra-red (FT-IR) techniques have allowed surface sensitivity to be improved and sample size to be decreased, but even in the most favourable cases FT-IR cannot match the sensitivity of XPS. However, in systems with variable composition within the range 0–1 μm the two techniques are complementary and can effectively be used in tandem (see, for example, Ref. 30).

The application of AES in this field has commonly been discounted on the grounds of unacceptably high rates of radiation damage. However, this problem could not really be addressed because severe charging problems effectively prevented the acquisition of spectra. Very recently, van Ooij[31] has shown that small-area analysis can be achieved if the surface is first coated with a particular thickness of gold, followed by its removal by ion etching. This technique allows SEM features (observed *in situ* before sputtering) to be studied with respect to their surface elemental composition. It seems that surface graphetization results from this process so the technique has obvious limitations, but inorganic particulate analysis in polymeric matrices (an important area) is possible. Of course, in principle this overcomes the inability of XPS to give spatially resolved information. A similar goal was behind Briggs' study of the applicability of SIMS to polymer surface analysis.

Results of a systematic study[10] of practical problems gave grounds for optimism that spatial resolution of $\simeq 100 \ \mu$m would be possible, and molecular imaging at this resolution has recently been demonstrated.[32] Fast atom bombardment mass spectrometry (FABMS), in which the ion beam of SIMS is replaced by a neutral beam, gives equivalent data without the need for charge neutralization[33] but cannot, of course, be used for high spatial resolution work. Further studies[34] are now demonstrating the molecular fingerprinting capability of SIMS and FABMS and their potential for increased surface sensitivity over XPS. For ultimate surface sensitivity, also with the possibility of spatial resolution, ISS has potential. As with SIMS, this technique is just beginning to be applied to polymeric surfaces.

Another vibrational spectroscopy with much potential in this area is Raman spectroscopy. Spectra from thin polymer films have been described and with the Raman microprobe it is possible to analyse polymeric materials at high spatial resolution ($\simeq 1 \ \mu$m) whilst sampling the first few micrometres. The great advantage of the vibrational spectrocopies described in this section is their ability to comment on morphology (e.g. chain confrontation, crystallinity, orientation), which is a very important aspect of polymeric structure. The other techniques cannot do this.

Despite the lack of spatial resolution (this may not always be so!) XPS has had a dramatic impact on polymer surface analysis. This is undoubtedly because the technique gives elemental composition without major quantification problems and a fair degree of structural information, and does not suffer unduly from sample charging or radiation damage effects. In addition, sample shape is not restricted so that 'as received' materials of all types can usually be handled, subject only to overall dimensional considerations.

9.6   Application of XPS to Polymer Surface Analysis Problems

9.6.1   Some general considerations

The ability to detect small quantities of material on polymer surfaces merely through the appearance of characteristic core levels may seem to be a trivial point. The common polymers are comprised of a small number of elements and, furthermore, these have simple XP spectra (generally C $1s$ plus one or two peaks from O $1s$, N $1s$, F $1s$ and Cl $2s,2p$). However, common additives or contaminants contain additional elements such as S, P, Si, Al, Na, K, Br, Sn, Cr, Ni, Ti, Zn, Ca, Sb, Ge and the presence of these materials, even in very low concentrations, can therefore be detected very simply. The ability to detect such elements, especially on small or irregular 'as received' samples is invaluable in trouble-shooting or quality control operations which involve surface properties. These include optical properties (e.g. haze, gloss, stains), adhesive properties (e.g. wettability, printability, bondability, heat sealability and releaseability), electrical properties (e.g. static chargeability) and general processing and machine handling properties (e.g. friction and 'blocking' of film in reels). Specific contaminants can also lead to surface crazing and cracking of many plastics.

Additives in plastics and polymeric materials include agents to prevent oxidation (thermal and photochemical), to neutralize acidity, to promote fire

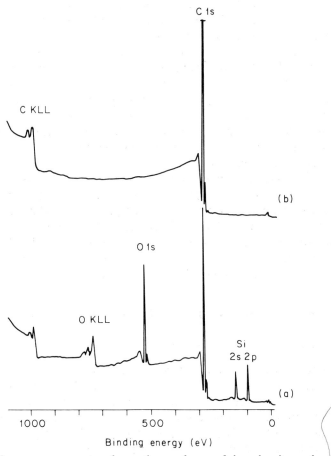

Figure 9.4 Survey-scan spectra from the surfaces of low-density polyethylene (LDPE) mouldings: (a) specimen exhibiting unacceptable surface cracking; (b) normal specimen. The cracked specimen has become contaminated on the surface with a silicone, probably of the type used as a mould-release aid, which is a stress corrosion agent for LDPE

retardancy and to aid processing. These are usually intended to be well distributed throughout the bulk but in certain circumstances surface segregation
may take place. Agents to lower surface friction, to increase surface conductivity and to prevent 'blocking' *are intended* to migrate to the surface. In some
cases they may not do so; in other cases they may migrate across interfaces

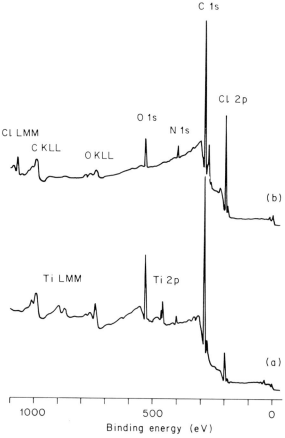

Figure 9.5   Survey-scan spectra from the surfaces of 'polypropylene' packaging films.
This material is coated on both sides with a vinylidenedichloride copolymer (which
improves the barrier properties of the film towards diffusion of oxygen and water) and
subsequently printed on one side. Both spectra are from the non-printed surface which
is required to heat-seal to itself in packaging operations: (a) specimen exhibiting
almost total loss of heat-sealing under normal conditions, (b) normal specimen. Heat-
seal failure is due to the presence of a titanium complex, an adhesion-promoting ink
component which has migrated from the 'cured' ink to the non-printed film surface
during contact in the stored reel. The degree of contamination is evidenced by the
marked reduction in intensity of the Cl 2p peak and the disappearance of the much
lower energy Cl *LMM* peak

through contact with another surface. Filler and pigment particles are often surface coated and material from this coating may pass into the polymer matrix and thence to the surface.

Surface segregation of emulsifier and stabilizer molecules used in some polymerization processes is a frequent occurrence. Common contaminants

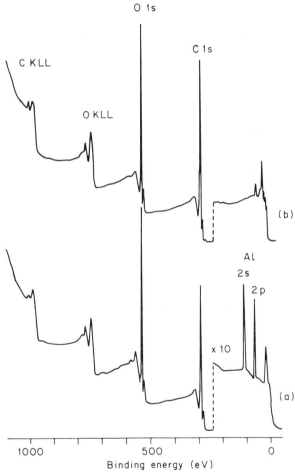

Figure 9.6   Survey-scan spectra from the surfaces of poly(ethyleneterephthalate) (PET) packaging films. This material has been vacuum-metallized on one side with a thin aluminium layer for improved gas-barrier and decorative effect. Both spectra are from the non-metallized surface: (a) specimen exhibiting reduced heat sealing and poor printability, (b) normal specimen. Abnormal behaviour is the result of enhanced adhesion between the two film surfaces (metal oxide and PET) during storage in a reel (mechanism not understood). On subsequent unwinding some aluminium containing material is transferred, contaminating the PET surface

generally are polymerization catalyst residues, lubricating oils and greases, mould release agents and dust. Typical examples of the power of simple element detection are shown in Figures 9.4 to 9.6.

An increased level of complexity is provided by thin layers of material which do not contain distinguishing elements. For example, Figure 9.7 shows C 1s and O 1s spectra from two sides of a poly(ethyleneterephthalate) (PET) film chemically modified on one side for adhesion enhancement. The modified layer thickness is too thin for analysis by reflection IR. No new core-level peaks are introduced and high resolution spectra need to be recorded. However, with a knowledge of appropriate relative elemental sensitivity factors stoichiometries can be obtained. The PET side is easily identified by the shake-up satellite (see Section 9.4.3).

Examples of this type of analysis by XPS have been described, amongst others, by Ratner and colleagues for polymers of importance in biomedical applications. Thus the surfaces of polyurethane materials (often used in the form of extruded tubes) show markedly varying compositions due to migration of oligomeric fractions rich in one component of the copolymer and to contamination pick-up.[35] Model surfaces prepared by casting purified polymers from solution also show variable composition depending on whether the surface was in contact with the substrate (e.g. cleaned glass) or air. Surface

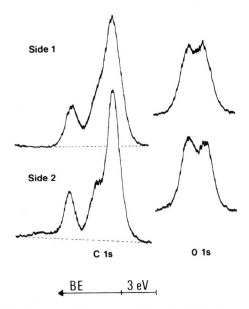

Figure 9.7   Core-level spectra from two sides of poly(ethyleneterephthalate) film chemically modified on one side for adhesion enhancement. Note the obvious $\pi \rightarrow \pi^*$ shake-up peak from the PET surface (side 2)

compositions derived from quantified XPS analyses are now revealing correlations with surface properties, e.g. blood platelet consumption.[35] In the same
field XPS has shown[36] that radiation-grafted hydrogels have surface compositions which may be drastically altered by changes in the degree of hydration
(these changes can, of course, be induced by the high vacuum of the instrument and special sample-cooling techniques have been described to overcome
the problems). Figure 9.8 illustrates this effect.

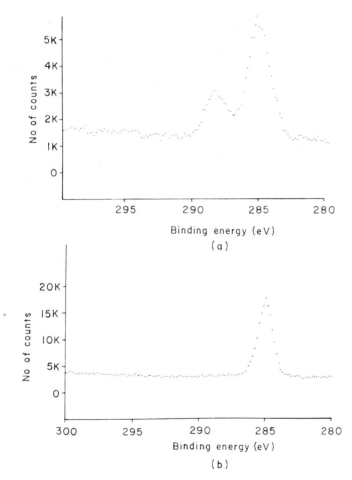

Figure 9.8   C 1s spectra from acrylamide (hydrogel) radiation-grafted onto silicone
rubber: (a) 160 K (hydrated), (b) 303 K (dehydrated). Dehydration led to disappearance of the N 1s signal (from acrylamide) and to a substantial increase in the Si 2p
signal (from the silicone substrate). The data point to an almost total migration of the
hydrogel into the substrate on dehydration. (After Ratner *et al.*[36])

It is not the intention in this chapter to catalogue the application of XPS in the analysis of polymer surfaces. This has previously been done in a review[37] by the present author covering the literature up to 1977 (including the important area of fibres and textiles) and the examples discussed therein are still entirely representative. A literature update is provided in the bibliography at the end of the chapter. Instead, activity in one particular (and large) field, that of polymer surface modification, will be discussed in some detail as a case history, in order to best illustrate the strengths and limitations of XPS.

9.6.2 Polymer surface modification

9.6.2.1 *Fluorinated systems*

The early days of XPS application to structural studies of polymers were dominated by the study of fluorinated systems for the simple reason that fluorine, as the most electronegative element, induces the largest chemical shift in C $1s$ BEs. Hence different structural units in polymers are the most easily identified from C $1s$ spectra in highly fluorinated materials. For the same reason surface modification studies were initially concerned with treatment of polytetrafluoroethylene (PTFE).

Adhesive bonding of PTFE can be achieved following pre-treatment in a number of ways, the most widely used treatments being sodium–liquid ammonia and sodium–naphthalene–THF. Independent studies of both treatments by von Brecht, Mayer and Bindel[38] and of the former treatment by Dwight and Riggs[39] led to the conclusion that the treatments lead to surface defluorination and the production of a highly unsaturated carbonaceous surface layer, which subsequently reacts with O_2–H_2O on exposure to air. Even at this early stage the inability of XPS to comment on the form of groups involving only carbon (or carbon bound to hydrogen) was appreciated and the reaction with Br_2 was used by both groups to gain some quantitative insight into unsaturation levels. (We now know that the oxygen-containing groups in the surface would complicate this interpretation.) Dwight and Riggs also observed[39] the further behaviour of the modified layer on abrasion, weathering. etc., as shown in Figure 9.9. This was also probably the first study which followed the relationship between contract angles and surface composition as determined by XPS. Collins, Lowe and Nicholas[40] also published at this time a study of the glow discharge treatment of PTFE in ammonia and, in less detail, in air. The five-component C $1s$ signal from the surface at equilibrium (NH_3 discharge) was not assigned but defluorination and incorporation of oxygen- and nitrogen-containing groups was obvious. The increased wettability and adhesion of this surface could be reversed by nitric-perchloric acid treatment. XPS showed this to be due to the removal of the modified layer from unmodified PTFE.

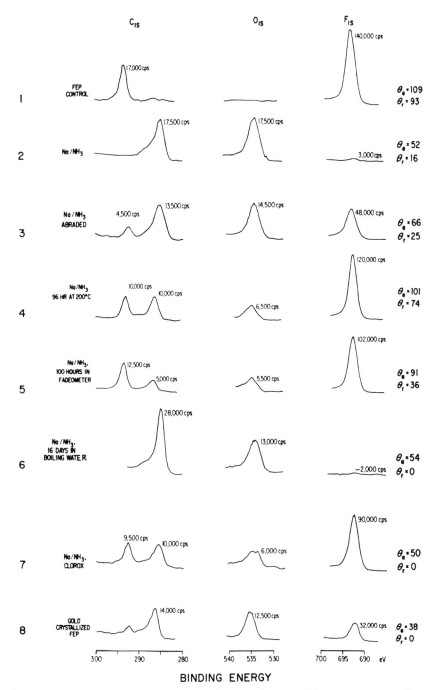

Figure 9.9   XPS and contact angle data from 'Teflon' FEP (copolymer of perfluoro-propylene with tetrafluoroethylene) before and after various surface treatments. (Reproduced from Dwight and Riggs[39] by permission of John Wiley & Sons, Inc.)

With the increased understanding derived from studies of simple fluoropolymers Clark *et al.*[41] were able, some two years after the above studies, to embark on the identification of species arising from the surface fluorination of high-density polyethylene. The complexity of the resulting C 1s profiles is illustrated by Figure 9.10, as is the reliance on curve resolution for detailed assignments. For the first time this study made a serious attempt to obtain a quantitative estimate of the distribution of modifying species in terms of structural type and depth distribution. The F 1s/F 2s peak intensity ratio was used for the purpose of investigating the latter, particularly as a function of reaction time.

These advantages of fluorinated systems were also used by Clark and Dilks in a model study[42] of the CASING (cross-linking by activated species of inert gases) process. The model polymer was an alternating random copolymer of ethylene and tetrafluoroethylene (52 per cent TFE) which gives a simple C 1s spectrum consisting of only two peaks due to —CH$_2$— and —CF$_2$— groups. In the argon plasma used the cross-linking of the surface could be followed through the loss of fluorine and formation of CF groups. Use of the F 1s/F 2s intensity ratio, angular variation technique and quantified peak intensity data allowed the kinetics of the process to be described in terms of both direct and radiative energy transfer to the polymer surface and subsurface, respectively.

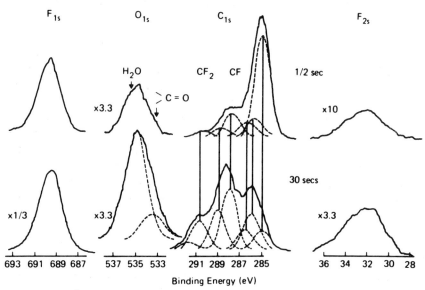

Figure 9.10   C 1s and F 1s spectra from the surface of high-density polyethylene after fluorination in F$_2$/N$_2$. (Taken from Clark *et al.*[41])

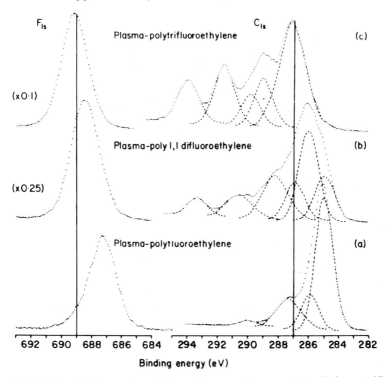

Figure 9.11   Core-level spectra of the plasma polymers derived from trifluoro-
ethylene; 1,1-difluoroethylene and fluoroethylene. (After Dilks[14])

More recently Dilks[9,14] has used XPS to study polymer synthesis in glow
discharges ('plasma polymerization'). Again the systems are highly fluorinated. The object of this work is to study the relationship between the structure of the monomer fed into the discharge, the composition of the resulting
polymer (which is deposited as a thin film) and parameters which characterize
the discharge and the system geometry. The kinetics of the polymerization
process can also be elucidated. Figure 9.11 illustrates the spectra obtained
and the deconvolutions required to quantitatively interpret the C 1s spectrum. Plasma polymerization onto wool fibre surfaces and plasma treatment
of textiles to modify surface properties, with XPS studies, has been discussed
by Millard.[43]

9.6.2.2 Polyolefins

Attractive as they are for model XPS studies, fluoropolymers fill only a very
small part of the total polymer spectrum either in terms of annual production
or range of end use. Polyolefins (low- and high-density polyethylene, poly-

propylene and a wide range of copolymers involving ethylene, propylene and higher α-olefins) are the dominant class, and surface modification of these materials is essential for their successful use in a wide range of applications. This is partly because the surfaces of the materials, being non-polar, have low surface energy and inherently poor wetting and adhesion characteristics, although the reasons have been the subject of a great deal of controversy for many years.[44]

XPS studies have cast much light on this subject. Since 1976, Briggs and colleagues have carried out a systematic study of commercial pre-treatments and other processes which lead to improved adhesion characteristics— chromic acid etching,[30,45] melting against aluminium,[17] thermal oxidation,[46] ozone treatment,[46] flame treatment[47] and electrical ('corona') discharge treatment.[21,22,48,49] Two of these studies will be discussed in some detail since they illustrate several important aspects of XPS application.

Chromic acid ($K_2Cr_2O_7/H_2O/H_2SO_4$) etching of polyolefinic surfaces[50] leads to severe roughening. Water contact angles on LDPE and HDPE fall rapidly as etching proceeds but in the case of PP a minimum is reached after a short time, followed by an increase to a plateau value close to that of the untreated surface. Reflection IR shows chemical modification (oxidation and sulphonation) in the case of LDPE but not HDPE or PP. This evidence might suggest that increased adhesion is the result of surface roughening or merely the removal of a 'weak' surface (or 'boundary') layer. Some of the XPS data from these polymer surfaces after treatment with chromic acid solutions and, for comparison, after concentrated sulphuric acid treatment are shown in Table 9.2.[30,44] The BE of the S $2p$ peak which appeared after treatment was consistent with the presence of $-SO_3H$ groups and a limited quantification of the data is shown assuming all sulphur is in this form. The adhesion data (lap shear strength of adhesive joints using an epoxy adhesive) show a clear corre- lation with the degree of oxidation (expressed in terms of the number of carbon–oxygen functions with an assumed constant average stoichiometry) but not with the $-SO_3H$ group concentration. For homogeneous samples the O $ls/O\ 2s$ intensity ratio (under the experimental conditions employed) is ≈10. Following the discussion in Section 9.4.1 it is clear from the O $1s/O\ 2s$ intensity ratio data that the thickness of the oxidized layer on PP is much less than that on LDPE after similar treatments (and for PP always less than the O $2s$ sampling depth, i.e. <100 Å) and that the degree of oxidation of the PP surface rapidly reaches an equilibrium. The XPS data suggests that a very thin equilibrium layer of oxidized polymer exists on PP during chromic acid treat- ment with continuous loss of material into solution, whereas LDPE continues to oxidize both in degree and in depth with exposure time. This accords with weight loss measurements. Angular dependent O $1s/C\ 1s$ intensity ratio data at low treatment levels (when roughness is not developed) further support this interpretation. Developing roughness does not affect the XPS intepreta-

Table 9.2 Surface composition and adhesion of etched polymers. (After Brewis and Briggs[44])

Polymer	Etching conditions	C/S atomic ratio	O/S atomic ratio	C atoms with SO_3H groups, %	O (not in SO_3H groups) to total C, % O/C	Lap shear strength, MN/m^2	Failure[†] type	O 1s O 2s[‡]
LDPE	None	—	—	—	0.25	0.55	I	—
LDPE	Conc. H_2SO_4 1 h at 70 °C	36.8	3.2	2.7	0.6	3.3	I	11.6
LDPE	Acid B* 5 s at 20 °C	198	9.1	0.5	3.1	4.8	I + M	—
LDPE	Acid A* 1 min at 20 °C	269	12.7	0.4	3.6	7.5	M	13.0
LDPE	Acid A 30 min at 70 °C	80.0	10.4	1.3	9.3	7.6	M	9.2
LDPE	Acid A 6 h at 70 °C	47.1	9.5	2.1	13.9	9.5	M	9.9
HDPE	None	—	—	—	0.52	0.38	I	—
HDPE	Conc. H_2SO_4 1 h at 70 °C	64.3	4.2	1.6	1.8	3.5	I	18.2
HDPE	Acid B 5 s at 20 °C	145	9.2	0.7	4.3	7.0	M	25.1
PP	None	—	—	—	0.25	0.28	I	—
PP	Conc. H_2SO_4 1 h at 70 °C	262	9.7	0.4	2.6	1.1	I	—
PP	Acid B 5 s at 20 °C	382	14.5	0.3	3.3	2.8	I	35.9
PP	Acid A 1 min at 20 °C	283	16.2	0.4	4.6	4.7	I	22.7
PP	Acid A 6 h at 70 °C	261	13.5	0.4	4.0	11.2	M	13.7

* Acid A 'normal' chromic–sulphuric acid ($K_2Cr_2O_7/H_2O/H_2SO_4$ = 7/12/150 by weight); acid B as acid A but 1/100th concentration with respect to $K_2Cr_2O_7$.

[‡] I, apparent interfacial failure; M, material failure.

[‡] For a homogeneous sample O 1s/O 2s ≈ 10.

Reproduced from Briggs et al.[30] with permission.

tion but markedly interferes with contact angle measurements as noted. The XPS data point to the overriding importance of surface oxidation and expose the inadequacy of reflection IR (as routinely used) for commenting on modification regimes of <100 Å.

A preferred etchant for treating PP mouldings prior to electroless metal plating (e.g. for decoration or polymer–metal lamination) is CrO_3–H_2O. Residual chromium has been shown to be important in achieving optimum metal adhesion to the polymer surface. Figure 9.12 shows XPS data from LDPE treated with this reagent.[30] Note the sharp maximum in chromium surface concentration and the close correlation of surface oxidation level with adhesion of the surface to an epoxide adhesive, as discussed for chromic–sulphuric acid treatment previously. The Cr $2p$ BE and the Cr $2p_{1/2}$–$2p_{3/2}$ spin–orbit splitting were found to be indicative of the Cr^{III} oxidation state. Figure 9.13 shows the corresponding high-resolution C $1s$ spectra for Figure 9.12. The high BE shoulder indicates domination by carbonyl groups initially but by carboxylate groups in the later stages of oxidation (confirmed by reflection IR), the transition point coinciding approximately with the maximum chromium concentration. The XPS data are consistent with likely oxidation mechanisms involving chromate ester intermediates. The three main sites for oxidative attack in LDPE are double bonds, tertiary carbons (branch points) and methylene groups. From the MIR evidence it appears that the

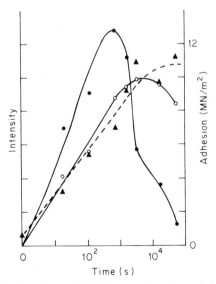

Figure 9.12 CrO_3–H_2O etching of LDPE as a function of time at 25 °C. ○ O $1s$ intensity (10^4 counts/s f.s.d.); ● Cr $2p_{3/2}$ intensity (10^3 counts/s f.s.d.); ▲ lap shear strength of adhesive joint (to epoxide adhesive). (After Briggs *et al.*[30])

first of these is relatively inert. Well-known mechanisms for reaction of the other two sites are as follows:

$$R_3CH + H_2CrO_4 \longrightarrow R_3C-OH + Cr^{IV}$$
$$R_2CH_2 + H_2CrO_4 \longrightarrow R_2CHOH + Cr^{IV}$$
$$Cr^{VI} + Cr^{IV} \longrightarrow 2Cr^{V}$$
$$R_2CHOH + Cr^{VI}/Cr^{V} \longrightarrow R_2C{=}O + Cr^{IV}/Cr^{III}$$

etc.

The final stable state of Cr is Cr^{III}. This could be present on the surface as a complex such as:

Further oxidation must result in chain cleavage and eventually in the production of carboxylic acids, but the mechanisms for these reactions are not well understood.

At the point where the chromium concentration at the surface is a maximum, calculation shows that there are ~3 chromium atoms per thousand carbon atoms. LDPEs typically contain about 20–30 branch points per thousand carbon atoms, but these are distributed between amorphous and less accessible crystalline phases. It is therefore not unreasonable to suggest that the chromium is bound as a Cr^{III} complex following attack at a tertiary carbon atom. Further oxidation at this site will lead to formation of a carboxylic acid with loss of chromium. This hypothesis is therefore in accord with the experimental observations. Furthermore, the acid hydrolysis which is required to remove the chromium still bound to the surface of etched LDPE would be explained by the reaction:

A tentative theory has been put forward[30] to explain the importance of surface chromium on final metal adhesion, but the XPS results clearly show why this adhesion is so dependent on exact treatment conditions; factors such as etchant recipe, solution temperature and number of times the etchant solu-

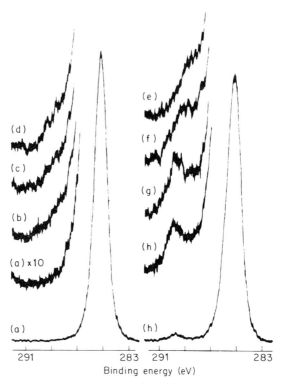

Figure 9.13   High-resolution C $1s$ spectra for LDPE etched with CrO_3–H_2O at 25 °C for the following times: (a) O s, (b) 20 s, (c) 40 s, (d) 2 min, (e) 10 min, (f) 1 h, (g) 5 h, (h) 16 h. Count rate $= 3 \times 10^3/s$ (f.s.d.). (After Briggs *et al.*[30])

tion previously used all affect the point in time during etching at which the maximum (optimum) surface chromium concentration is reached. This is a good example of the unique role which XPS could play in quality control.

In the preliminary stages of this work some irreproducibility of adhesion data for samples treated with CrO_3–H_2O was observed. XP spectra of these surfaces contained small peaks due to silicon. The species giving rise to these peaks were clearly confined to the outer surface of the polymer since angular variation measurements produced a marked change in the Si $2p$/C $1s$ intensity ratio. For example, a sample of LDPE treated for 1 min at 20 °C gave Si $2p$/C $1s$ intensity ratios of 7.3×10^{-3} ($\alpha = 80°$) and 1.6×10^{-2} ($\alpha = 15°$), a relative increase of $\times 2.2$ on moving to the lower take-off angle (with respect to the sample surface). By contrast the O $1s$/C $1s$ intensity ratio increased only marginally. The silicon-containing species were eliminated by redistilling the water until its surface tension was 72.5 mN/m. Table 9.3 indicates the effect of this contamination on adhesion values for oxidized surfaces.

Table 9.3 Adhesion of CrO_3–H_2O etched LDPE to an epoxide adhesive (lap shear strength). (After Briggs *et al.*[30])

	Adhesion, MN/m^2	
Etching conditions	Water A	Water B
1 min, 20 °C	1.32	3.45
20 s, 70 °C	2.50	9.51
1 min, 70 °C	5.21	11.06
5 min, 70 °C	7.82	11.38
15 min, 70 °C	9.42	11.38
30 min, 70 °C	9.92	11.20

Water A and B used in the etchant solution and for rinsing are ordinary distilled grade and redistilled (triply) to a surface tension of 72.5 mN/m, respectively.

Electrical ('corona') discharge treatment[50] of polyethylene and polypropylene film has been used for many years in order to render these surfaces printable and suitable for lamination or coating. In this process (Figure 9.14) the polymer film is passed over an earthed metal roller covered with a 'dielectric' (insulating) material. Separated from the film by $\simeq 2$ mm is an electrode bar to which a high voltage is applied (typically 15 kV at 20 kHz). Air in the film-electrode gap is ionized, the corona discharge thus formed is stable and this 'treats' the film surface.

Discharge treatment of low-density polyethylene (LDPE) has been studied by many workers, particularly with reference to the autoadhesion enhancement effect. LDPE autoadheres when two surfaces are contacted under pressure at temperatures above $\simeq 90$ °C. However, after fairly low levels of discharge treatment, treated surfaces will autoadhere at significantly lower temperatures ($\simeq 70$ °C). Two theories had been proposed to account for this effect. The first, due to Canadian workers,[51] suggested that electret formation

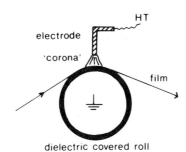

Figure 9.14   Schematic of electrical discharge treatment process

was involved—the resulting increased adhesion being electrostatic in nature. The evidence for this was that discharge treatment in both 'active' (air, oxygen) and 'inert' (nitrogen, argon, helium) gases gave the effect, its magnitude being related to the power dissipated in the discharge irrespective of which gas. Moreover, the maximum effect was achieved in oxidizing atmospheres before ATR-IR spectroscopy could detect any surface oxidation. The second, diametrically opposed theory due to Owens[52] suggested that hydrogen bonding between polar groups formed by the discharge treatment was responsible. However, only LDPE treated in air was studied.

In our work[8] discharge treatment was carried out in a model apparatus with static film samples and at low frequency (50 Hz). As Figure 9.15 demonstrates, XPS showed that treatment in air, nitrogen and argon (at atmospheric pressure) leads to surface chemical changes including oxidation in all cases. Clearly treatment in argon requires a longer time but power dissipation

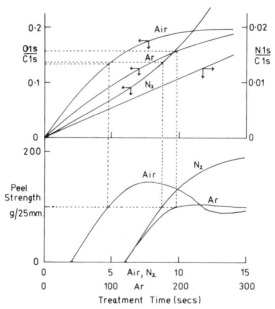

Figure 9.15 Comparison of autoadhesion (peel strength) and surface composition (from XPS) for LDPE discharge treated in air, nitrogen and argon (peak voltages 13.7, 13.7 and 2.2 kV, respectively, all at 50 Hz). Heat seals were made at 75 °C and 15 lb/in^2 with 2 s contact time. The O $1s$/C $1s$ intensity ratio is a qualitative measure of surface oxidation level. The N $1s$/C $1s$ ratios refer to surfaces treated in nitrogen . Note the similar surface oxidation levels for samples giving peel strengths of 100 gm/25 mm ---- (broken lines). Reproduced from Brewis, D. M., and Briggs, D. Polymer, 1981, **22**, 7, by permission of the publishers, Butterworth & Co (Publishers) Ltd. ©)

measurements confirmed the earlier result[51] that the power required to achieve a given level of autoadhesion (peel strength) was independent of the gas. It is also clear from Figure 9.15 that this corresponds roughly to the same degree of surface oxidation. Experiments were also carried out using a hydrogen discharge; this did not produce the autoadhesion enhancement effect and XPS did not detect surface oxidation. On the basis of these results the electret theory can be discounted, but the Owens' theory is given a firmer foundation. Figure 9.16 shows typical spectra for LDPE discharged treated in air.[48] A simple deconvolution of the high binding energy shoulder on the C 1s peak gives three peaks which can be ascribed to —CH$_2$O— (e.g. alcohol, ether, ester, hydroperoxide) at $\approx$286.5 eV, $>$C=O (e.g. aldehyde, ketone)

at $\approx$288.0 eV and $\overset{\displaystyle O}{\underset{\displaystyle -C-O-}{\|}}$ (carboxylic acid or ester) at $\approx$289.5 eV. The O

1s peak is even less informative. Most oxygen functional groups give O 1s

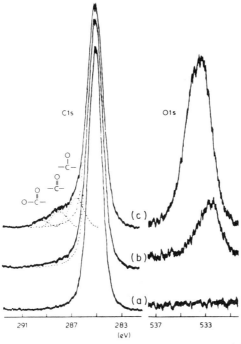

Figure 9.16 High-resolution C 1s and O 1s spectra for LDPE: (a) untreated; (b) and (c) discharge treated in air (13.7 kV peak voltage, 50 Hz) for 8 s and 30 s, respectively. Count rates are 3×10^3 counts/s f.s.d. (C 1s) and 10^3 counts/s f.s.d. (O 1s). Reproduced from Brewis, D. M., and Briggs, D. *Polymer*, 1981, **22**, 7, by permission of the publishers, Butterworth & Co. (Publishers) Ltd. ©)

BEs of ≈532 eV, the exception is the ester oxygen in carboxyl groups at ≈533.5 eV. The shift in the O $1s$ peak, shown in Figure 9.16, with increasing oxidation reflects the increasing relative concentration of carboxyl groups. Clearly this information is not specific enough to probe the detailed structure of the oxidized layer or to allow correlations to be made with the autoadhesion results. To overcome this problem a series of derivatization reactions have been devised to label specific functional groups. These are included in Table 9.1.[21,22]

The consensus view in the literature for the likely mechanism of oxidation during discharge treatment of LDPE is the following:

$$-CH_2-CH_2-CH_2- \quad \xrightarrow[-H\cdot]{I^+,\,e^-,\,M^*,\,h\nu}$$

$$-CH_2-\overset{\cdot}{C}H-CH_2-$$

$$\text{fast} \,\bigg|\, O_2$$

$$\xleftarrow[\text{H}^\cdot \text{ abstraction}]{\text{fast}} \quad -CH_2-\underset{\underset{O\cdot}{\overset{|}{O}}}{CH}-CH_2- \quad \xrightarrow{R^\cdot} \quad -CH_2-\underset{\underset{O-R}{\overset{|}{O}}}{CH}-CH_2-$$

$$-CH_2-\underset{\underset{OH}{\overset{|}{O}}}{CH}-CH_2-$$

$$\bigg\downarrow$$

Products ($-C{=}O$, $C-OH$, $C-O-R$, $-COOH$, $-COOR$, etc.)

both chain scission and cross-linking take place. The key intermediate is the hydroperoxide group, whose stability and decomposition has been the subject of much research.

The XPS data from the derivatized surfaces can be quantified to give the data in Table 9.4. The value for the population of $CH_2C{=}O$ groups assumes that on average two α–H atoms will be replaced during bromination. Since this group can tautomerize to give one enolic $-OH$ the population of $-CH_2C{=}O$ assessed by Br_2 uptake and enolic $-OH$ assessed by the CAC reaction should be comparable, as is observed.

The raw C $1s$ spectra tend to show broadly similar intensities for the $C-OH$ (etc.), $>C{=}O$ and $-COOH$ regions which is also borne out by these

Table 9.4 Quantification of functional groups in discharge treated LDPE surface. (Reproduced from Briggs and Kendall[21] by permission of Butterworth & Co (Publishers) Ltd)

Reaction	XPS ratio (core level/C 1s)	Atomic ratio (element/carbon)	Number of functional groups per original surface $-CH_2-$
PFPH	(F 1s) 0.205	5.5×10^{-2}	$>C=O$, 1.1×10^{-2}
Br$_2$–H$_2$O	(Br 3d) 3.6×10^{-2}	10.6×10^{-3}	$CH_2C=C$, 5.3×10^{-3}
CAC*	(Cl 2p) 1.3×10^{-2}	6.0×10^{-3}	$C-OH$, 6.0×10^{-3}
TAA†	(Ti 2p$_{3/2}$) 6.2×10^{-2}	1.5×10^{-2}	$C-OH$, 1.5×10^{-2}
NaOH	(Na 1s) 8.8×10^{-2}	1.1×10^{-2}	$-COOH$, 1.1×10^{-2}
SO$_2$	(S 2p) 7.6×10^{-3}	4.7×10^{-3}	$C-OOH$, 4.7×10^{-3}
None	(O 1s) 0.209	8.7×10^{-2}	

*Reacts apparently selectively with enolic $-OH$.
†Reacts apparently selectively with alcoholic $-OH$.
PFPH = pentafluorophenylhydrazine, CAC = chloroacetylchloride,
TAA = di-isopropoxytitanium bisacetylacetonate.

results. The total assay of $>C=O$, $C-OH$, $C-OOH$ and COOH groups would give an atomic O/C ratio of 5.7×10^{-2}, where 'C' is the carbon atoms in the original surface. This compares with the actual value of 8.7×10^{-2} from the discharge treated surface. Considering that ether, peroxide and ester groups are also likely to be present, in numbers comparable to the groups which have been derivatized, this assay is seen to be entirely reasonable. The apparent internal consistency of these results is additional evidence for the essential reliability of the derivatization procedures used. Hydroperoxides in polyethylene can have long lifetimes so if this mechanism is correct these groups should be detectable. The SO$_2$ reaction is positive identification and, we believe, the first direct evidence for this mechanism. Of the groups likely to be produced by hydroperoxide decomposition, derivatization techniques have therefore identified $-C=O$, $C-OH$ and $-COOH$.

These derivatization reactions also allow specific functional groups to be eliminated from the surface and so allow, with XPS as the built-in monitor, the study of specific interactions in adhesion to polymer surfaces. Thus the work described above allowed the Owens theory of discharge treatment induced autoadhesion to be completely confirmed, and other systems have been quantitatively studied.[21] This promises to be a fruitful extension of XPS capability in aiding the understanding of polymer surface behaviour.

Another process for modifying polyolefin surfaces is plasma discharge treatment, although this has yet to be commercialized, in which an electric discharge in a low-pressure gas (<1 torr) is employed. Clark and Dilks[5] have studied the modification of LDPE, PP and polystyrene by plasma oxidation. Angular variation studies show that compared with 'corona' treatment the modified layer is very thin and more highly oxidized (see Figure 9.1).

9.6.2.3   Poly(ethyleneterephthalate)

Although poly(ethyleneterephthalate) (PET) is a polar material with a relatively high surface energy, the achievement of good adhesion often necessitates surface treatment. 'Corona' discharge treatment is less successful than in the case of polyolefins due to the rather short timescale of the improved properties. For instance, Figure 9.17 shows the autoadhesive characteristics of discharge treated PET film as a function of time after treatment at which the treated surfaces are heat-sealed together (at the temperature used non-treated surfaces show no autoadhesion). This effect has been studied by XPS.[53] Figures 9.18 and 9.19 show the spectra from untreated and treated surfaces. Clearly additional oxygen is introduced into the surface and deconvolution of the C 1s envelope reveals the formation of two additional functional groups. On the basis of C 1s BEs from spectra of p-hydroxybenzoic acid and its methyl ester and other literature data[8] peaks 'd' and 'e' in Figure 9.18 are assigned to phenolic —OH and —COOH. Also the O 1s BE for —OH in phenol is in agreement with the BE of the main additional component to the O 1s spectrum of treated PET. By analogy with the free radical oxidation mechanism discussed previously for discharge treatment of LDPE, the introduction of phenolic —OH and free carboxylic acid end groups would indeed be expected. Formation of the latter group also implies chain scission of the PET backbone (i.e. a reduction in molecular weight).

Contact angle measurements show that the increase in the polar contribu-

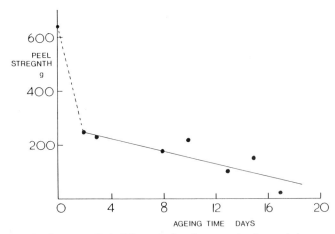

Figure 9.17   Peel strength (g/25 mm) for heat seals formed between discharge treated PET surfaces (130°, 2 s at 25 lb/in² contact) as a function of elapsed time between treating and sealing (ageing time). Reproduced from Briggs, D., *et al.* Polymer, 1980, **21**, 895, by permission of the publishers, Butterworth & Co. (Publishers) Ltd. ©)

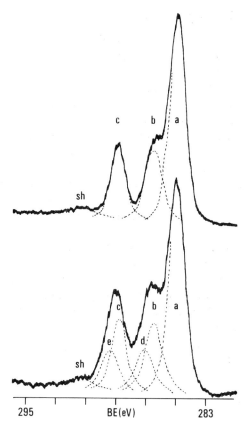

Figure 9.18   C 1s and O 1s spectra of PET surface before (upper) and after (lower) discharge treatment (40 s at 50 Hz with 15.9 kV peak voltage, equivalent to an energy input of 620 mJ/cm². Deconvolution was achieved using a Du Pont 310 curve resolver. Peak assignments are:

(a) ⟨phenyl ring⟩,   (b) —OCH₂—, (c) $$-\overset{\text{O}}{\overset{\|}{\text{C}}}-\text{O}-\text{CH}_2$$,   (d) —C—OH (phenolic),

(e) $-\overset{\text{O}}{\overset{\|}{\text{C}}}-\text{OH}$, (sh) $\pi \rightarrow \pi^*$ shake-up satellite. (Reproduced from Briggs, D., *et al.* Polymer, 1980, **21**, 895, by permission of the publishers, Butterworth & Co. (Publishers) Ltd. ©)

tion to the surface energy following treatment is rapidly reversed, paralleling the changes in adhesive properties. Following the surface changes with time by XPS gave the spectra in Figure 9.19; it was also shown that washing the surface led to the removal of most of the additional oxidized material. These results strongly suggest that most of the oxidized material produced by dis-

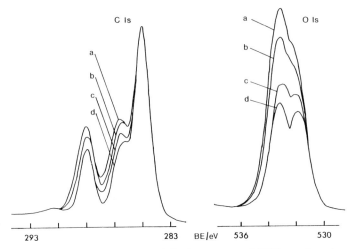

Figure 9.19   Changes in XPS spectra of discharge treated PET as a function of ageing time: (a) zero (corresponds to lower spectrum in Figure 9.18), (b) one day, (c) 23 days. Spectra (c) are virtually identical to those produced by washing the treated surface. Spectra (d) are for untreated film. Reproduced from Briggs, D., *et al*. Polymer, 1980, **21**, 895, by permission of the publishers, Butterworth & Co. (Publishers) Ltd. ©)

charge treatment of PET is of relatively low molecular weight. The initial increased adhesion is probably due in large part to the strong hydrogen-bonding capacity of the penolic —OH groups introduced by treatment. Following treatment the processes depicted schematically in Figure 9.20 probably occur, i.e. a fairly rapid migration of low molecular weight material into the bulk followed by a slower reorientation of the longer polymer chains, by which the newly introduced —OH groups are internally hydrogen-bonded and effectively rendered ineffective at the surface. XPS follows the first process but not the second since the latter changes take place within the sampling depth of the technique.

9.6.2.4   Other studies

As noted earlier it is not the intention of this chapter to catalogue applications of XPS in polymer technology. In an earlier review[37] this was attempted but the literature is now too extensive to repeat the exercise. Instead the following bibliography is divided into the most active areas and the individual titles are reasonably self-descriptive. A particular feature of the earlier review not covered herein was the (then) extant literature on the application of XPS to the study of fibres (man-made and natural) and textiles. A further review on this subject has recently appeared.[43] Carbon fibres are now receiving much

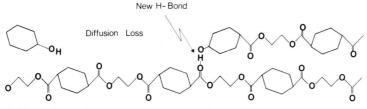

Figure 9.20   Schematic of molecular mechanisms of discharge treatment and ageing effects for PET

attention because of their importance in composite materials involving a variety of polymeric matrices. Optimization of the fibre–matrix interface via surface treatment is a key area of research. The changes taking place at polymer surfaces during weathering are an obvious clue to the mechanisms of photo-degradation (previously only studied using bulk analysis techniques). The polymer–metal interface is also the subject of intensive study both from the viewpoint of adhesion optimization and corrosion prevention. One aspect of this activity has already been described; other areas are plating of plastics for decoration, laminate composite manufacture, circuit board design and production, rubber tyre reinforcement and so on. A quite different application is the study of polyacetylenes in order to understand the mechanisms involved in the production of highly conducting systems when the polymer is doped (e.g. with halogens).

9.7 Conclusions

Although the rapid growth in the use of polymeric materials is largely due to the ease of processing compared with conventional materials (metals, wood, glass, etc.) many end uses demand non-inherent surface properties. Surface analysis is therefore of high importance in polymer technology and XPS has made a major contribution to its development. Already invaluable in the role of trouble-shooter, XPS is rapidly leading to a much more detailed understanding of crucial surface modification processes and promises to underpin the development of more efficient and effective processes in the future.

References

1. D. T. Clark, in *Photon, Electron and Ion Probes of Polymer Structure and Properties* (Eds D. W. Dwight, T. J. Fabish and H. R. Thomas), p. 255, ACS Symposium Series 162 (1981).
2. D. T. Clark, in *Handbook of X-ray and Ultraviolet Photoelectron Spectroscopy* (Ed. D. Briggs), p. 212, Heyden, London (1977).
3. D. R. Wheeler and S. V. Pepper, *J. Vac. Sci. Technol.*, **20**, 226 (1982).
4. D. T. Clark, in *Handbook of X-ray and Ultraviolet Photoselection Spectroscopy* (Ed. D. Briggs), p. 233, Heyden, London (1977).
5. D. T. Clark and A. Dilks, *J. Polym. Sci., Polym. Chem. Ed.*, **17**, 957 (1979).
6. D. T. Clark and Y. C. T. Fok, *J. Electron Spectrosc.*, **22**, 173 (1981).
7. R. F. Roberts, D. L. Allara, C. A. Pryde, D. N. E. Buchanan and N. D. Hobbins, *Surf. Interface Anal.*, **2**, 5 (1980).
8. See D. T. Clark and A. Harrison, *J. Polym. Sci., Polym. Chem. Ed.*, **19**, 1945 (1981), and previous papers in this series.
9. A. Dilks, in *Electron Spectroscopy—Theory, Techniques and Applications* (Eds C. R. Brundle and A. D. Baker), Vol. 4, Academic Press, London (1981).
10. D. Briggs and A. B. Wootton, *Surf. Interface Anal.*, **4**, 109 (1982).
11. D. T. Clark and A. Dilks, *J. Polym. Sci., Polym. Chem. Ed.*, **14**, 533 (1976).
12. J. J. O'Malley, H. R. Thomas and G. Lee, *Macromolecules*, **12**, 496 (1979).
13. R. D. Chambers, D. T. Clark, D. Kilcast and S. Partington, *J. Polym. Sci., Polym. Chem. Ed.*, **12**, 1647 (1974).
14. A. Dilks, in *Photon, Electron and Ion Probes of Polymer Structure and Properties* (Eds D. W. Dwight, T. J. Fabish and H. R. Thomas), p. 307, ACS Symposium Series 162 (1981).
15. J. J. Pireaux, J. Riga, R. Caudano and J. Verbist, in *Photon, Electron and Ion Probes of Polymer Structure and Properties* (Eds. D. W. Dwight, T. J. Fabish and H. R. Thomas), p. 169, ACS Symposium Series 162 (1981).
16. W. M. Riggs and D. W. Dwight, *J. Electron. Spectrosc.*, **5**, 447 (1974).
17. D. Briggs, D. M. Brewis and M. B. Konieczko, *J. Mater. Sci.*, **12**, 429 (1977).
18. H. Spell and C. Christenson, *TAPPI*, **1978**, 283 (1978).
19. D. S. Everhart and C. N. Reilley, *Anal. Chem.*, **53**, 665 (1981).
20. J. Hammond, J. Holubka, A. Durisin and R. Dickie, *Abstracts of Colloid and Interfacial Science Section*, ACS Miami Meeting, September 1978.
21. D. Briggs and C. R. Kendall, *Int. J. Adhesion, Adhesives*, **2**, 13 (1982).
22. D. Briggs and C. R. Kendall, *Polymer*, **20**, 1053 (1979).
23. C. D. Batich and R. C. Wendt, in *Photon, Electron and Ion Probes of Polymer Structure and Properties* (Eds. D. W. Dwight, T. J. Fabish and H. R. Thomas), p. 221, ACS Symposium Series 162 (1981).

24. A. Bradley and M. Czuha, Jr., *Anal. Chem.*, **47**, 1838 (1975).
25. M. Czuha, Jr. and W. M. Riggs, *Anal. Chem.*, **47**, 1836 (1975).
26. M. M. Millard and M. Marsi, *Anal. Chem.*, **46**, 1820 (1974).
27. D. S. Everhart and C. N. Reilley, *Surf. Interface Anal.*, **3**, 126 (1981).
28. D. S. Everhart and C. N. Reilley, *Surf. Interface Anal.*, **3**, 269 (1981).
29. D. Briggs, in *Surface Analysis and Pretreatment of Plastics and Metals* (Ed. D. M. Brewis), Chap. 4, Applied Science, London (1982), and references to individual techniques given therein.
30. D. Briggs, V. J. I. Zichy, D. M. Brewis, J. Comyn, R. H. Dahm, M. A. Green and M. B. Konieczko, *Surf. Interface Anal.*, **2**, 107 (1980).
31. W. van Ooij, *Abstracts of the Second International Conference on Quantitative Surface Analysis*, Teddington, November 1981; to be published in *Surface Interface Analysis*.
32. D. Briggs, *Surf. Interface Anal.*, **5**, 113 (1983).
33. D. Briggs, A. Brown, J. A. van der Berg and J. C. Vickerman, *Ion Formation from Organic Solids* (Ed. A. Benninghoven), Springer-Verlag (in press).
34. D. Briggs, *Surf. Interface Anal.*, **4**, 151 (1982).
35. B. D. Ratner, in *Photon, Electron and Ion Probes of Polymer Structure and Properties* (Eds. D. W. Dwight, T. J. Fabish and H. R. Thomas), p. 371, ACS Symposium Series 162 (1981).
36. B. D. Ratner *et al.*, *J. Appl. Polym. Sci.*, **22**, 643 (1978).
37. D. Briggs, in *Electron Spectroscopy : Theory, Techniques and Applications* (Eds C. R. Bundle and A. D. Baker), Vol. 3, Academic Press, London (1979).
38. H. von Brecht, F. Mayer and H. Binder, *Makromol. Chem.*, **33**, 89 (1973).
39. D. W. Dwight and W. M. Riggs, *J. Coll. Interface Sci.*, **47**, 650 (1974).
40. G. C. S. Collins, A. C. Lowe and D. Nicholas, *Euro. Polym. J.*, **9**, 1173 (1973).
41. D. T. Clark, W. J. Feast, W. K. R. Musgrave and J. Ritchie, *J. Polym. Sci., Polym. Chem. Ed.*, **13**, 857 (1975).
42. D. T. Clark and A. Dilks, *J. Polym. Sci., Polym. Chem. Ed.*, **15**, 2321 (1977).
43. M. M. Millard, in *Industrial Applications of Surface Analysis* (Ed. L. A. Casper and C. J. Powell), p. 143, ACS Symposium Series 199 (1982).
44. D. M. Brewis and D. Briggs, *Polymer*, **22**, 7 (1981).
45. D. Briggs, D. M. Brewis and M. B. Konieczko, *J. Mater. Sci.*, **11**, 1270 (1976).
46. D. Briggs, D. M. Brewis and M. B. Konieczko, *Euro. Polym. J.*, **14**, 1 (1978).
47. D. Briggs, D. M. Brewis and M. B. Konieczko, *J. Mater. Sci.*, **14**, 1344 (1979).
48. A. R. Blythe, D. Briggs, C. R. Kendall, D. G. Rance and V. J. I. Zichy, *Polymer*, **19**, 1273 (1978).
49. D. Briggs, C. R. Kendall, A. R. Blythe and A. B. Wootton, *Polymer*, **24**, 47 (1983).
50. D. Briggs, in *Surface Analysis and Pretreatment of Plastics and Metals* (Ed. D. M. Brewis), Chap. 9, Applied Science, London (1982), and references therein.
51. M. Stadal and D. A. I. Goring, *Can. J. Chem. Eng.*, **53**, 427 (1975), and references therein.
52. D. K. Owens, *J. Appl. Polym. Sci.*, **19**, 265 (1975).
53. D. Briggs, D. G. Rance, C. R. Kendall and A. R. Blythe, *Polymer*, **21**, 895 (1980).

Bibliography

Surface modification

D. T. Clark and A. Dilks, ESCA applied to polymers (XVIII), RF glow discharge modification of polymers in helium, neon, argon and krypton, *J. Polym. Sci., Polym. Chem. Ed.*, **16**, 911 (1978).

D. T. Clark and A. Dilks, ESCA applied to polymers (XXIII), RF glow discharge modification of polymers in pure oxygen and helium oxygen mixtures, *J. Polym. Sci., Polym. Chem. Ed.*, **17**, 957 (1979).

S-T. Huang, J-M. Chen and S-L. Chen, Application of photoelectron spectroscopy (ESCA) to study the surface fluorination of polypropyylene, *K'O Hsueh T'Ung Pao*, **24**, 831 (1979).

H. J. Leary and D. S. Campbell, ESCA studies of polyimide and modified polyimide surfaces. *ACS Symp. Ser.*, **162**, 419 (1981).

F. Yamamoto *et al.*, Surface grafting of polyethylene by mutual irradiation in methyl acrylate vapour (III), quantitative surface analysis by XPS, *J. Polym. Sci., Polym. Phys. Ed.*, **17**, 1581 (1979).

Photo-oxidation and weathering

A. Dilks and D. T. Clark, ESCA studies of natural weathering phenomena at selected polymer surfaces, *J. Polym. Sci., Polym. Chem. Ed.*, **19**, 2847 (1981).

J. Peeling and D. T. Clark, An ESCA study of the photo-oxidation of the surface of polystyrene film, *Polym. Degradation Stab.*, **3**, 97 (1981).

J. Peeling and D. T. Clark, ESCA study of the surface photo-oxidation of some nonaromatic polymers, *Polym. Degradation Stab.* **3**, 177 (1981).

J. Peeling and D. T. Clark, Photo-oxidation of the surfaces of polyphenylene oxide and polysulphone, *J. Appl. Polym., Sci.*, **26**, 3761 (1981).

Polymer–metal interactions

J. M. Burkstrand, Substrate effects on the electronic structure of metal overlayers—an XPS study of polymer–metal interfaces, *Phys. Rev.*, **B20**, 4853 (1979).

J. M. Burkstrand, Core-level spectra of chromium and nickel atoms on polystyrene, *J. Appl. Phys.*, **50**, 1152 (1979).

J. M. Burkstrand, Copper-poly(vinyl alcohol) interface: a study with XPS, *J. Vac. Sci. Technol.*, **16**, 363 (1979).

P. Cadman and G. M. Cossedge, The chemical nature of metal—PTFE tribological interactions as studied by XPS, *Wear*, **54**, 211 (1979).

E. Cernia and P. Ascarelli, Spectroscopic analysis of interactions between metallic surfaces and organic polymers, *Proc. Int. Conf. Org. Coat. Sci. Technol*, **1978**, 103 (1978).

M. Cini and P. Ascarelli, XPS applied to small metal particles and to metal–polymer adhesion: physical insight from initial and final-state effects, *NATO Adv Study. Inst. Ser.*, **C73**, 353 (1981).

K. Richter *et al.*, ESCA study of metal–polymer phase boundaries in composite materials, 2, *Chem.*, **18**, 390 (1978).

R. F. Roberts and M. Schonhorn, Surface modification of polymers by deposition of evaporated metals (I), Teflon FEP, *Polym. Prepr., Am. Chem. Soc., Div. Polym. Chem.*, **16**, 146 (1975).

J. P. Servais *et al.*, Examination by ESCA of the constitution and effects on lacquer adhesion of passivation films of tinplate, *Br. Corros. J.*, **14**, 126 (1979).

W. J. Van Ooij, A. Kleinhesselink and S. R. Leyenaar, Industrial applications of XPS : study of polymer-to-metal adhesion failure, *Surf. Sci.*, **89**, 165 (1979).

W. J. Van Ooij and A. Kleinhesselink, Application of XPS to the study of polymer–metal interface phenomena, *Appl. Surf. Sci.*, **4**, 324 (1980).

Fibres and textiles

R. R. Benerito *et al.*, Modifications of cotton cellulose surfaces by use of radio frequency cold plasmas and characterisation of surface changes by ESCA, *Text. Res. J.*, **51**, 224 (1981).

N. Chand, Surface studies of polyacrylonitrile yarn by XPS (ESCA), *Indian J. Text. Res.*, **6**, 133 (1981).

N. Chand, XPS study of drawn poly (ethylene terephthalate) fibres, *Pop. Plast. Rubber*, **26**, 6 (1981).

G. E. Hammer, Graphite fibre surface analysis by XPS and polar dispersive free energy analysis, *Appl. Surf. Sci.*, **4**, 340 (1980).

F. Hopfgarten, ESCA studies of carbon and oxygen in carbon fibres, *Fibre Sci. Technol.*, **12**, 283 (1979).

A. Ishitoni, Application of XPS to surface analysis of carbon fibre, *Carbon*, **19**, 269 (1981).

A. Proctor and P. M. A. Sherwood, XPS studies of carbon fibre surfaces (I). Carbon fibre surfaces and the effect of heat treatment, *J. Electron Spectrosc.*, **27**, 39 (1982).

A. Proctor and P. M. A. Sherwood, XPS studies of carbon fibre surfaces (III). The effect of electrochemical treatment, *Carbon.* **21**, 53 (1983).

A. Proctor and P. M. A. Sherwood, XPS studies of carbon fibre surfaces (III). Industrially treated fibres and the effect of heat and exposure to oxygen, *Surf. Interface Anal.*, **4**, 212 (1982).

F. J. Takeshige *et al.*, Fundamental studies on water repellency of textile fabrics by ESCA, *Teikoku Gakuen Kiyo*, **1981**, 83 (1981).

Polyacetylenes

S. K. Hsu *et al.*, Highly conducting iodine derivatives of polyacetylene: Raman, XPS and XRD studies, *J. Chem. Phys.*, **69**, 106 (1978).

I. C. Ikemoto *et al.*, XPS study of highly conductive bromine-doped polyacetylene, *Bull. Chem. Soc. Jpn.*, **55**, 721 (1982).

J. C. Ikemoto *et al.*, XPS study of highly conductive iodine-doped polyacetylene, *Chem. Lett.*, **10**, 1189 (1979).

W. R. Salanek *et al.*, Photoelectron spectra of arsenic (V) fluoride-doped polyacetylenes, *J. Chem. Phys.*, **71**, 2044 (1979).

G. C. Stevens, D. Bloor and P. M. Williams, Photoelectron valence band spectra of diacetylene polymers, *Chem. Phys.*, **28**, 399 (1978).

H. R. Thomas *et al.*, Photoelectron spectra of conducting polymers. Molecularly doped polyacetylenes, *Polymer*, **21**, 1238 (1980).

Miscellaneous

R. W. Bigelow *et al.*, Structure variation of sulphonated polystyrene surfaces, *Adv. Chem. Ser.*, **187**, 295 (1980).

D. Brion and G. Mavel, A study of the photoelectron spectroscopy (XPS) of mouldings based on PVC, *Eur. Polym. J.*, **16**, 159 (1980).

A. Casarini *et al.*, XPS and the problems of film producing industry, *Poliplasti Plast. Rinf.*, **27**, 80 (1979).

A. Dilks, The identification of peroxy-features at polymer surfaces by ESCA, *J. Polym. Sci., Polym. Chem. Ed.*, **19**, 1319 (1981).

D. W. Dwight *et al.*, ESCA analysis of polyphosphazene and poly (siloxane/carbonate) surfaces, *Polym. Prepr. An. Chem. Soc., Div. Polym. Chem.*, **20**, 702 (1979).

J. Gianelos and E. A. Grulke, Some studies of chlorinated PVC using XPS, *Adv. X-ray Anal.*, **22**, 473 (1979).

J. Knecht and H. Baessler, An ESCA study of solid 2,4-hexadiyne-1,6-diol bis (toluene sulphonate) and its constituents before and after polymerisation, *Chem. Phys.*, **33**, 179 (1978).

K. Knutson and D. J. Lyman, Morphology of black copolyurethanes (II). FTIR and ESCA techniques for studying surface morpholy, *Org. Coat Plast*. Chem., **42**, 621 (1981).

P. C. Lacaze and G. Tourillon, Spectroscopic study (XPS-SIMS) of the aging of polyacetonitrile thin films electrochemically deposited on a platinum electrode, *J. Chim. Phys., Phys. Chim. Biol.*, **76**, 371 (1979).

H. J. Leary and D. S. Campbell, Surface analysis of aromatic polymide films using ESCA, *Surf. Interface Anal.*, **1**, 75 (1979).

S. Nagarajan, Z. H. Stachurshi, M. E. Hughes and F. P. Larkins, A study of the PE-PTFE system (II). ESCA measurements, *J. Polym. Sci. Phys. Ed.*, **20**, 1001 (1982).

J. J. O'Malley and H. R. Thomas, Surface studies of multicomponent polymer solids, *Contemp. Top Poly. Sci.*, **3**, 215 (1979).

D. Shuttleworth *et al.*, XPS study of low molecular weight polystyrene-polydimethylsiloxane black copolymers, *Polym. Prepr., Am. Chem. Soc., Div. Polym. Chem.*, **20**, 499 (1979).

J. H. Stone-Masui and W. E. E. Stone, Characterisation of polystyrene latexes by photo-electron and infrared spectroscopy, *Polym. Colloids*, **2**, 331 (1980).

C. Sung, S. Paik and C. B. Hu, ESCA studies on surface chemical composition of segmented polyrethanes, *J. Biomed. Mater. Res.*, **13**, 161 (1979).

Y. Takai *et al.*, Photoelectron spectroscopy of poly-*p*-xylyene polymerised from the vapour phase, *Polym. Photochem.*, **2**, 33 (1982).

R. H. Thomas and J. J. O'Malley, Surface studies on multicomponent polymer studies by XPS : polystyrene/poly (ethyleoxide) homopolymer blends, *Macromolecules*, **14**, 1316 (1981).

Practical Surface Analysis
by Auger and X-ray Photoelectron Spectroscopy
Edited by D. Briggs and M. P. Seah
© 1983, John Wiley & Sons, Ltd

Chapter 10

Uses of Auger Electron and Photoelectron Spectroscopies in Corrosion Science

N. S. McIntyre

Surface Science Western, University of Western Ontario
London, Ontario, Canada N6A 5B7

10.1 Introduction

The processes of corrosion begin and terminate at very thin surface layers. Corrosive attack of a metal is initiated when a protective or 'passivating' oxide surface film is ruptured, allowing contact between the active metal and an invading atomic or molecular species. Re-passivation can result from coverage of the surface by as little as a few atomic layers of a chemically inert substance. Corrosion science involves understanding both aspects—the causes as well as the prevention of the chemical degradation of a metal. The analysis of such thin surface films using the many techniques available should therefore have been an early topic for exploitation by surface scientists.

The fact that this was not the case has had as much to do with the concern of most corrosion scientists with a vast array of practical materials problems as with the desire of many surface physicists or chemists to study less complex surface interactions. Contacts between the two disparate groups generally occurred when an actively corroding specimen was analysed for surface composition. The results of such investigations often told the corrosion scientists much about the surface in question, but less about the true origin of the corrosion. Thus, only in the past two or three years have there been concerted joint attempts to address the question of corrosion initiation and passivation using surface analysis techniques.

The scope of surface chemistry and physics related to corrosion processes is indeed broad. If the topic is defined as the understanding of the initiation or prevention of metal deterioration in the environment, the scope extends well beyond the traditional testing of metal durability in aqueous media. Electrochemistry has become one of the most important tools in controlling the number and rate of processes occurring at a metal interface. The combined use of electrochemistry and surface analysis is particularly promising and a number of examples are discussed in this chapter. Many corrosion-related problems involve as much metallurgy as chemistry. Corrosive attack of metals frequently occurs preferentially along boundaries between the metallic grains; heating and mechanical treatments of the metal have strong effects on such intergranular corrosion. Metallurgists, themselves, have made ample use of surface techniques, particularly Auger electron spectroscopy, to understand processes such as grain boundary segregation and alloy sensitization (see Chapter 7). However, few studies have yet used surface techniques to study corrosive invasion within metallic microstructure and fewer still have related surface effects to corrosive cracking under mechanical stress. The related microstructural phenomenon of surface pitting also has just begun to receive attention from surface scientists. The potential importance of such work will also be discussed in this chapter. The development of new methods and materials to provide surface protection is an area where analytical techniques are being more rapidly adopted. Some studies of organic and inorganic corrosion inhibitors are thus described in this chapter.

The earliest tool of the corrosion specialist used in surface studies was the optical microscope. This, coupled with a number of preferential chemical etchants, has allowed analysis of cross-sections cut through a corrosion film on a metal substrate. The surface distribution and homogeneity of corrosion films as thin as $5-10\ \mu m$ can be determined in this way. Crystallographic structure of such a film (if any) has sometimes been determinable by X-ray diffraction techniques. A major improvement to film analysis came in the 1960s with the general availability of electron probe microanalysis. With the electron microprobe an X-ray fluorescence elemental analysis could be made of a spacial region as small as $2\ \mu m$, thus allowing films to be analysed chemically in cross-section.

These techniques were, of course, only applicable to cases where corrosion was rather extensive and uniform and certainly could not provide information on the initial surface attack, on surface microstructure or on the detailed chemistry of any thin film playing a passivating role. The arrival of the electron spectroscopies in the early 1970s thus marks a watershed, beyond which a number of new approaches to corrosion science were possible.

Discussion of surface techniques in this chapter is limited to X-ray photoelectron spectroscopy (XPS or ESCA) and Auger electron spectroscopy (AES), in accord with the topics covered in this volume. Indeed, most of corrosion-related surface studies have used those techniques exclusively. Where limitations to these techniques exist, alternative approaches using other well-known surface techniques will be described.

In an attempt to provide information useful to specialists in either corrosion science or surface analysis, the first section discusses aspects of the electron spectroscopies which are particularly important for corrosion studies. The second section reviews past analytical use of AES and XPS in corrosion-related experiments. Supplemental information in both areas will be found in earlier reviews by Joshi[1] and Castle.[2, 3]

10.2 Special Aspects of XPS and AES for Corrosion Studies

10.2.1 Surface sensitivity

The 100–1000 eV electrons analysed in typical XPS and AES experiments have mean free paths ranging from 1 to 3 nm.[4] Thus, given the sensitivity of either technique, it is normally easy to detect a single monolayer of a passivating film on a substrate of different elemental composition. In certain cases, where chemical shifts are appreciable, XPS has been able to detect a top monolayer which differed only in *chemical* structure from the substrate.

Where lamellar layers of corrosion products are expected, the thickness of the uppermost layer on an infinitely thick substrate can be determined from the exponential relationship given by equation (5.27) in Chapter 5. It is possible to determine whether distribution of the top layer is lamellar or island-like by measuring the angular[5] or the kinetic energy dependence of the Auger electron or photo-electron intensities from the overlayer and the substrate. Castle[3] was able to use switch between two X-ray sources of different energy and compare the XPS intensity ratios obtained with results calculated for a lamellar model. If lamellar behaviour is confirmed, the uptake of as many as 10–20 monolayers should be measureable. In the future, changes in more complex surface distributions may be able to be followed quantitatively using scanning Auger microscopy (SAM).

Initial oxidation films fequently consist of several very thin layers of different composition; in such cases even very careful angle energy dependence studies may still yield an ambiguous model of the layers. An XPS study of an oxidized Alloy 600 surface showed nickel and iron oxides near the surface and chromium oxides nearer the metal interface. An even more surface-sensitive technique, low energy ion scattering (LEIS), was required to show that the top monolayer consisted uniquely of iron oxide.[6]

10.2.2 Elemental sensitivities

Both XPS and AES are sensitive to all elements of importance to corrosion scientists, excepting hydrogen. Access to the low atomic number elements (below $Z = 11$) has been particularly important for oxygen analysis of corrosion films; knowledge of the O 1s intensity and the binding energy have both been extremely valuable in passive film characterization.

The detection sensitivity in XPS is limited by the high background caused by the predominance of energy-degraded electrons in the spectrum. With most commercial X-ray sources and counting times of practical duration (approximately 1 h), elemental detection limits (signal/noise = 3/1) range between 1.0 and 0.1 per cent of the total composition. This means that many corrosion precursors such as chloride or phosphate are barely or not at all detected by XPS under many circumstances. Unless the count rate can be improved substantially (approximately an order of magnitude) from its present status, XPS will effectively remain a tool for studying only the major phases in a corrosion system.

With a few exceptions, equivalent AES detection limits tend to be somewhat poorer than XPS. There are several reasons for this (see Chapter 3), but for many corrosion specimens, electrical charging of insulating oxides and hydroxides is probably the most significant cause. Such charging is mainly controlled by using very low primary electron beam currents, but this, in turn, greatly increases counting times. The potential of AES for detecting some minor and trace constituents, such as chloride, is greater than for XPS since distribution of these is often quite localized.

The inability of XPS and AES to detect hydrogen is a definite limitation for corrosion research. Hydrogen is an important film constituent in the form of hydrates and hydroxides. In addition, metal hydrides which are formed in some systems are precursors to brittle fracture. Alternative methods for study of hydrogen are secondary ion mass spectrometry (SIMS) or nuclear microanalysis.

10.2.3　Quantitative analysis

A number of quantitative models for XPS and AES, with and without the use of standards, are discussed in Chapter 5. Several adaptations of these, particularly suited to corrosion film analysis have been developed. For XPS studies, probably the most complete 'first-principles' model was developed by Asami, Hashimoto and Shimodaira,[7,8] and is particularly suited to the analysis of thin passive oxides on alloy substrates. Equation (5.27) in Chapter 5 was modified to account for the attenuation effect of a carbon contamination overlayer and the differing densities of the oxide film and the metal substrate. The atomic fractions of chemically differentiated species were obtained in this way, even allowing for hydrogen content, estimated from the —OH and bound water detected (see below). This model has been successful in dealing with a surface oxide composition on a metal alloy substrate of quite a different composition.[9] In Figure 10.1, a comparison is made of the surface oxide compositions on a chromium–molybdenum steel as a function of anodic polarization potential in hydrochloric acid. The Fe $2p_{3/2}$, Cr $2p_{3/2}$ and Mo $3d$ XPS spectra were integrated, and were also curve resolved (see Appendix 3) to separate the lower binding energy metallic component from the oxide component of the

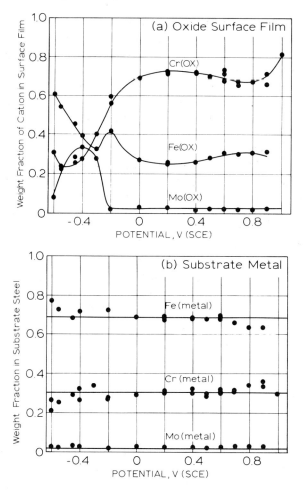

Figure 10.1   Quantitative measurement of compositions of oxide surface films and substrate alloys. Analysis of a chromium–molybdenum steel electrode polarized in $1M$ HC1 for 3600 s. (a) Weight fractions of cations in the oxide film are plotted as a function of applied potential. (b) Weight fractions of metal constituents in the stainless steel are plotted as a function of applied potential. The solid lines give the composition of the bulk steel. (Reproduced by permission of K. Hashimoto, K. Asami and K. Teramoto[9])

spectrum. The sums of these separated metal and oxide intensities were used to calculate the thickness of the oxide film, and this, in turn, was used in the calculation of the oxide and metal alloy compositions. The validity of the quantitative model is supported by the agreement between the known bulk

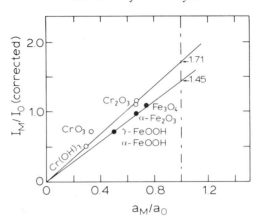

Figure 10.2 Quantitative XPS analysis of well-characterized metal oxides. Bulk oxygen/metal ratios are plotted against corrected O 1s/M 2p ratios for a number of chromium and iron oxides. (Reproduced by permission of K. Asami and K. Hashimoto[10])

alloy composition and the values consistently obtained for the metal substrate XPS analysis (see Figure 10.1b), despite major changes in the composition of the overlying oxide during the experiments.

Asami and Hashimoto have also calculated the compositions obtained from known oxide structures using this same XPS model.[10] Only one electron mean free path value was used for each photo-electron line, regardless of the chemical structure of the element. Figure 10.2 compares the known oxide/cation ratio for a given oxide with the O 1s/M 2p intensity ratio, corrected by the model for mean free path and contamination layer. The very good agreement for most oxides shows that it is valid to use a single mean free path value in such studies. Moreover, average O/M photo-electron cross-section ratios for chromium and iron oxides can be derived from the slopes in Figure 10.2.

The determination of the quantity of different types of oxygen present in the corrosion film is important, particularly where hydration and hydroxylation of the surface may affect its passivity. Asami *et al.*[11] have been able to account quantitatively for three different types of oxygen–metal oxide, metal hydroxide and bound water. Metal oxide and metal hydroxide bonding is detected on the basis of chemical shift; the bound water is determined as the difference between the total integrated O 1s intensity and the intensity accountable as bonded to cationic species detected over the entire XPS spectrum.

Other quantitative XPS studies of corrosion films have made use of standards to differentiate oxide and metal constituents.[6] Although this approach is less flexible than that described above, it also appears to achieve reasonable quantitative separation of major oxide and metal phases using a spectrum analysis method that can be automated.

Quantitative analysis by AES appears to have been carried out largely with standards, although a general 'first principles' approach is well documented (see Chapter 5). The problem of the quantitative contribution of underlying atomic layers to the Auger electron spectrum has been treated by Pons, Le Héricy and Longeron[12] using the multiple surface layer model. Quantitative AES depth–profile analysis of a concentration gradient through a 5–6 monolayer corrosion film is made more difficult by contributions from the underlying layers, themselves of somewhat different composition. The differential method uses composition information obtained further into the film during a depth profile to back-correct the preceding compositions. To be most effective, this method requires that new compositions be obtained every monolayer (~0.3 nm) in the depth profile. This method has been used to improve the depth resolution of AES composition profiles through passive films on Alloys 600 and 800.[13] Figure 10.3(a) shows the original depth profile made with AES line intensities, while Figure 10.3(b) shows the profiles of atomic fractions obtained after differential treatment of the data, as discussed above. Distinct maxima in chromium and iron concentrations within the surface oxide become evident in Figure 10.3(b) as a result of the removal of intensity contributions from underlying layers. The model of Pons, Le Héricy and Longeron has been extended by Mitchell[14] to include a simplified expression for the Auger electron absorption of each successive overlayer. In addition, the statistical nature of the sputtering process was taken into account using a modified Poisson distribution.

10.2.4 Spacial resolution

In XPS, the surface area sampled is of the order of several square millimetres, as defined by the solid acceptance angle of the electron optics. The irradiating X-ray flux normally covers a larger area. By contrast, the minimum area analysed by Auger electrons in the most recent scanning Auger microprobes is as small as 0.01 μm^2.

The low spatial resolution of XPS is a serious handicap in studying some corrosion problems. Corrosion processes by their nature are often localized; many reactions are initiated at surface kinks, grain boundaries or crevices. Unless such phenomena can be generated artificially, isolated from generalized surface corrosion, it is difficult to imagine the use of a standard XPS instrument to study localized corrosion chemistry. Recently, a prototype XPS system with a much-reduced X-ray beam diameter has been announced.* This development, as well as the use of powerful synchrotron photon sources with

---

*Surface Science Laboratories, Inc., Sunnyvale, California.

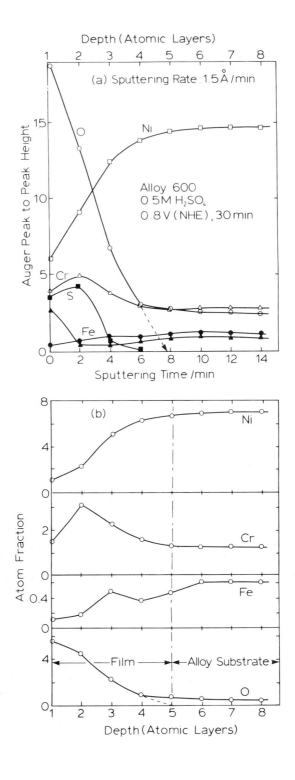

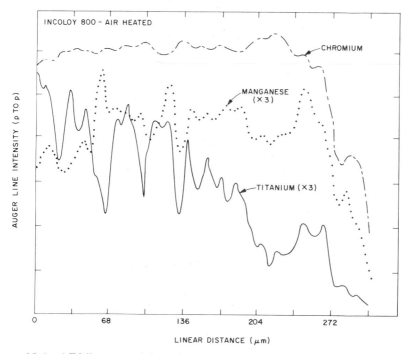

Figure 10.4   AES line scan of chromium, manganese and titanium distributions on an Alloy 800 surface oxidized at 600 °C for 1 min

reduced beam diameters, could lead to exciting new areas of research in corrosion chemistry.

The analysis of localized corrosion products by AES is well known to be a major advantage of the technique. Spot analyses of regions (e.g. pits and crevices) 1 μm apart are routinely made in studies of corrosion films. Line scans across a region changing in composition are particularly valuable, since a number of elements can be monitored simultaneously, and intensity references can be established. Figure 10.4 shows line scans of chromium, manganese and titanium made across an Alloy 800 surface oxidized at 600 °C in a flowing oxygen atmosphere.[15] Manganese and titanium distributions are highly localized and probably affected by the grain boundary intersections

---

Figure 10.3   Differential treatment of AES depth profile data for improved depth resolution. (a) AES intensity depth profile of an Alloy 600 surface passivated potentiostatically at 0.8 V(NHE) for 30 min in 0.5N H_2SO_4. (b) Differential composition profiles of data from (a) expressed in terms of atomic fractions for Ni, Cr, Fe and O. (Reproduced by permission of M. Seo and N. Sato[13])

with the surface. The distribution of titanium is, moreover, a strong function of the temperature which is altered with distance along the surface because of the flow of gas in that direction. The reconstruction of spacial distributions in the form of Auger images is qualitatively useful for corrosion studies. Auger maps showing chromium enrichment along grain boundaries has been obtained for 304 stainless steel heated to 1000 °C.[16] Many other examples of scanning Auger imaging may be found in the literature.*

10.2.5 Depth profiling

Analysis of a corrosion film to a depth greater than 2nm usually involves ion bombardment coupled with simultaneous analysis by AES or sequential analysis by XPS. Much has been written of the structural damage imparted by ion beams to oxide surfaces (see Chapter 4). Knowledge of this is important if oxide layers are to be chemically or quantitatively characterized during profiling. Some beam damage has been found to result in chemical alteration of metal oxides—severe in some cases and tolerable in others. Reports of the extent of chemical alterations on the same oxide system vary, suggesting that system geometry is important. For example, numerous workers have reported the decomposition of iron oxides to FeO under low-energy argon ion bombardment. Some[17, 18] find that FeO is reduced further to the metal, while others[19, 20] find that FeO is a stable product. If preservation of steady-state sputtering of an oxide structure does not result in its reduction to a metal, oxide and metallic phases can be distinguished quantitatively during a depth profile. Of the typical first-row 'corrosion' elements, McIntyre and others[6, 21] have found that chromium, cobalt, nickel and iron are not reduced to the metal by ion bombardment, under their experimental conditions.

Such a detailed analysis of the relationship between oxide and metal phases is only possible in XPS, where oxide metal contributions to the spectrum are well separated. Figure 10.5 shows cumulative presentations of oxide and metal phases on Alloy 600 surfaces, oxidized at 300 °C in 5 and 0.01 per cent oxygen gas mixtures. The major difference is the continuation of a nickel oxide phase deep into the substrate for the sample oxidized at higher oxygen partial pressure. Such effects would be unlikely to be detected using AES.

It goes without saying that differentiation of oxide chemical structures after ion bombardment is usually impossible. Higher oxidation states are reduced and dehydration has occurred. Other non-destructive methods for depth profiling may, at times, prove useful, if a bit tedious. Mechanical milling of corrosion surfaces using an *in situ* ball mill may be useful for determination of chemistry changes in oxide surfaces. The technique is now used with AES to profile rapidly through very thick films.[22] Chemical milling, which has been used for chemical depth profiles of semi-conductor surfaces,[23] may also be useful for analysis of some oxide films.

EFFECT OF PRESSURE

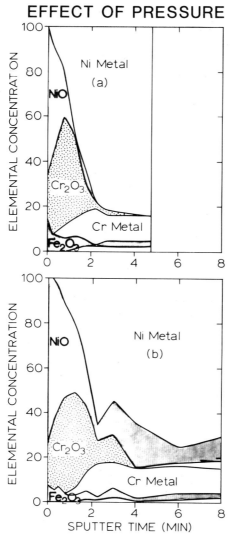

Figure 10.5   XPS depth profiles of surface films on Alloy 600 grown at 300 °C under two different oxygen pressures: (a) 0.01% O_2, 5 min; (b) 5% O_2, 5 min

10.2.6   Experimental techniques

10.2.6.1   *The corrosion experiment*

Over half of the surface analytical literature on corrosion describes analyses of specimens from contrived corrosion experiments, rather than the study of material from an actual corrosion problem. A wide range of tests have been documented in the corrosion literature to simulate conditions of pitting, crevice corrosion, stress corrosion or general surface attack. These tests are done in environments resembling 'working' conditions for the material (e.g. high-pressure steam autoclave, replenished aqueous autoclave, *in situ* applied tensile stress, etc.). The balance of studies involving corrosion, generally the more fundamental, are carried out under electrochemical control in a cell. In either situation, the question of specimen surface preparation is important, unless very thick film thicknesses are expected. Surfaces prepared by mechanical polishing (cold-working) exhibit different oxidation behaviour, depending on the degree of roughness of the surface,[6] while those polished electrochemically may lose or gain surface components.[3] Vacuum-annealed specimens are probably best for more basic studies of alloy oxidation; since, however, cold work is inevitably part of 'real-world' specimens, it is frequently desirable to study the effect itself.

10.2.6.2   Specimen transfer

Under many experimental conditions, and particularly for electrochemical studies, it is necessary to protect the specimen from further chemical alteration during transfer to the analytical chamber. The effect of atmospheric exposure will, of course, be greatest when an active metal or alloy is removed from the cell with a non-passive film on the surface. For certain relatively non-reactive films, the specimen can be stored in liquid nitrogen during a transfer.[13] A more elaborate method of transfer is to use an inert atmosphere glove box for the experimental cell and to transfer from this directly into the spectrometer via an air lock. This has proven to be satisfactory for the maintenance of $UO_{2.0}$ surfaces in a reduced state.[24] The most rigorous transfer systems, however, allow no contact with any atmosphere. Such a system for use with AES has been described,[25] and recently vacuum transfer chambers for use with combined electrochemical-XPS studies has been reported.[26, 27]

10.2.6.3   Vacuum effects

It is clear that the vacuum spectrometer results in the desorption of some hydrates and the decomposition of some hydroxides. Asami and Hashimoto[10] found that α-FeOOH decomposed slowly in vacuum at 50 °C, losing 3 per cent of its hydroxide every 100 minutes. Still, there is growing evidence that even water bound within the oxide lattice is being reproducibly detected in XPS O $1s$ spectra.[28,29] Water physically or chemically adsorbed on the surface is, of course, more readily desorbed and creates some scatter in the analytical measurements of XPS O $1s$ 'bound water' intensity.[9] The detection of bound water is an encouraging development for corrosion film characterization, since its presence has been linked to a degradation of the passive layer.[30] Further correlations of water content could be made using nuclear microprobe measurements before and after XPS analysis.

10.2.6.4 Beam effects

The X-ray source in XPS causes much less surface damage than an equivalent flux of electrons. However, X-ray-induced partial reductions of some metal oxide systems have been reported (e.g. $CuO \rightarrow Cu_2O$, $CoOOH \rightarrow CoO$), and evidence for these is likely to increase with the introduction of more powerful X-ray sources. Electron beams have been found to cause chemical decomposition,[31] desorption of water and adsorbed ions such as chloride[32] and reduction of oxides[33] under intense beam fluxes.

10.2.7 Chemical effects

In XPS the basis of the chemical shift of a core photo-electron peak is the change in electrostatic potential on the core electron when valence electron charge density is accepted or withdrawn from the atom (see Chapter 3). Thus, a relationship exists between the binding energy and the chemical state of the element. The presence of oxidized corrosion product elements can usually be clearly distinguished from the same elements as reduced metals, on the basis of their chemical shifts. Often, it is also possible to differentiate oxides of differing valence (e.g. $CrO_4{}^{2-}$ and Cr_2O_3) or oxide from hydroxide (e.g. NiO from Ni $(OH)_2$. The chemical structure also influences other photo-electron peak structures and many additional changes in oxidation states. Much of the detail of the spectroscopic lineshapes and chemical shifts, which are characteristic of different corrosion species, has not yet been tabulated in a form that is readily interpretable by corrosion researchers. The characteristic spectroscopic changes are generally not large, and the reported data in the literature often are in conflict or have error limits that are too large to be useful. This situation will improve as energy calibration procedures (Appendix 1) become more standardized and methods for a more accurate definition of peak shape and position are developed. Appendix 4 contains the most complete compilation of chemical shift data on compounds of interest to corrosion scientists (i.e. oxides of the structural elements, iron, nickel, chromium, cobalt, copper and manganese), and greatly extends earlier compilations in Ref. 34.

Well-characterized chemical shifts in AES are, in general, restricted to those between a metal and its oxide. It is quite likely that this situation will change in the future with the growing number of higher resolution energy analysers in AES systems.

A general review of XPS and AES chemical shifts is given elsewhere in this book. There are, however, a number of elements of particular interest to the corrosion specialist, where a detailed account of available oxide chemical shift data is useful. These are discussed below.

10.2.7.1 Oxygen

The O $1s$ line (approx. 530 eV) has been used extensively in the analysis of oxide surface species. The line has a relatively narrow width and a symmetric shape; this allows more accurate fitting of complex band combinations. An extensive review of O $1s$ binding energies of metal oxides has been reported by Johnson.[35] Although it appears that no simple correlation can be drawn between metal oxide bond character and O $1s$ binding energy, differences between experimental values for some oxides are sufficiently large to allow differentiation. One example of this is the shift between CuO and Cu_2O.[36,37]

Differences between hydroxyl oxygen (OH^-) and oxide oxygen (O^{2-}) are usually recognized by an O $1s$ shift of 1 to 1.5 eV. For example, the oxygen line associated with NiO is located at 529.6 ± 0.2 eV, while that for $Ni(OH)_2$ is located at a distinctly higher binding energy of 531.2 ± 0.2 eV.[37] Further, in the mixed oxide–hydroxide compound FeO(OH), contributions from each oxygen can be clearly distinguished in the O $1s$ spectrum.[19] The identification of water on surfaces from the O $1s$ spectrum is more difficult. Bonding of water molecules in different surface configurations apparently causes binding energy shifts over more than a 3 eV range. Norton[38] found that solid frozen water has a binding energy of ~533 eV, while Asami and coworkers find little difference between a hydroxide O $1s$ binding energy and that for water bound in the lattice (531 eV).[7,8] The broad envelope of O $1s$ intensity frequently detected in the range 531–534 eV suggests a multiplicity of forms of chemically and physically bound water on and within the surface. Unfortunately chemisorbed and physisorbed oxygen and hydroxyl are also found in this energy range.

Experimental Auger electron oxygen linewidths have been too broad for much use in chemical identification. Lattice oxide and chemisorbed oxygen have been identified in one structure.[39]

10.2.7.2 Titanium

XPS studies have shown that TiO_2 can be clearly distinguished from TiO in the Ti $2p_{3/2}$ spectrum.[34, 40, 41] TiN, a compound potentially formed within some high durability surfaces, can also be distinguished from TiO and TiO_2 on the basis of the Ti $2p_{3/2}$ chemical shift.[40]

The Auger $L_3M_{23}V$ spectra of TiO and TiO_2 are shifted by 2 eV, thus allowing their chemical identification by AES.[42] Lineshape changes in this transition have been used to differentiate metallic titanium and TiH_2.[43]

10.2.7.3 Vanadium

A chemical shift of about 1 eV has been detected between the oxides V_2O_5 and VO_2 using XPS. AES studies in this case have provided more information than those from XPS. A series of oxides of different vanadium–oxygen stoichiometries have been characterized using the $VL_{2,3}M_{2,3}V$ and $L_2M_{2,3}M_{2,3}$ Auger transitions.[44-46]

10.2.7.4 Chromium

Chromium oxide species are usually determined in XPS studies with the Cr $2p_{3/2}$ line. Chromium (III) and chromium (VI) oxides are fairly readily separated on the basis of the large (~2 eV) chemical shifts.[47,48] Chromium (III) oxide (Cr_2O_3) and the hydroxides ($Cr(OH)_3$ and $CrOOH$) can also be distinguished by a smaller shift (~0.5 eV) in binding energy.[10,49] Multiplet splitting of the Cr $2p_{3/2}$ line[10] may be useful in characterization of chromite structures.

10.2.7.5 Manganese

Several manganese oxides of different stoichiometries have been measured by XPS.[50] MnO and MnO_2 appear to be particularly distinguishable on the basis of their Mn $2p_{3/2}$ binding energies and peak shapes.[51]

10.2.7.6 Iron

Several XPS studies of iron oxides have been made.[19,52,53] Fe $2p$ photoelectron spectra of ferric and ferrous oxides are particularly complex, because of the large amount of coupling between the core hole created by photoemission and the high spin states of iron. Some of this complexity can be exploited for analytical purposes, since the spectral line shape is quite sensitive to chemical changes. This can be seen in Figure 10.6 where the Fe $2p_{3/2}$ spectra of α-FeOOH, α-Fe$_2$O$_3$, Fe$_3$O$_4$ and Fe(COOH)$_2$ are compared. The iron (III) hydroxide peak centre (Figure 10.2a) is shifted about 1 eV to higher binding energy than that for α-Fe$_2$O$_3$, α-Fe$_2$O$_3$ and γ-Fe$_2$O$_3$ (not shown) can be distinguished on the basis of the splitting in the main peak. Magnetite (Fe$_3$O$_4$) contains both Fe^{II} and Fe^{III}, both of which contribute to the Fe $2p_{3/2}$ spectrum with the two overlapping components in Figure 10.2(c). In real-life specimens, unfortunately, Fe$_3$O$_4$ surfaces are not usually as clearly identifiable since exposure to air causes partial oxidation of Fe$_3$O$_4$ to Fe$_3$O$_4$. Finally, some organo–iron compounds which are of interest to corrosion studies can be characterized separately from the oxides. In Figure 10.2(d), ferrous oxalate can be characterized on the bases of its prominent shake-up satellite (see arrow.

Several iron (II) oxides are also of interest to corrosion scientists. The Fe

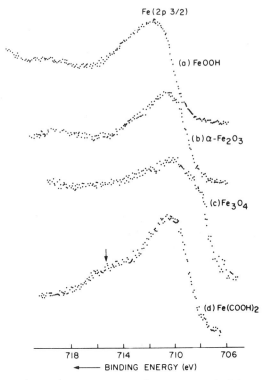

Figure 10.6 X-ray photo-electron spectra for some typical iron corrosion compounds. Fe $2p_{2/3}$ spectra are shown for (a) α-FeOOH; (b) α-Fe$_2$O$_3$; (c) Fe$_3$O$_4$; (d) Fe (COOH)$_2$

$2p_{3/2}$ spectrum of FeO has a prominent shake-up satellite,[18] as has the mixed oxide FeMoO$_4$.[54]

Sulphides of iron figure prominently in many corrosion studies. Surface iron sulphides frequently undergo hydrolysis to oxides in air, greatly complicating the spectra. However, FeS and FeS$_2$ are found to have quite different Fe $2p_{3/2}$ spectra, again resulting from differences in multiplet interaction. FeS$_2$ has a peak shape and position closely resembling that of metallic iron,[55] while FeS has a very broad peak centred nearly 2 eV higher in binding energy than the FeS$_2$ position.

Several iron oxides have also been characterized by low kinetic energy peaks in the AES spectrum. Ekelund and Leygraf[56] have noted two independent Fe *LVV* peaks at ~43 eV and 51 eV, respectively, and suggest that these are related to Fe^{2+} and Fe^{3+}, respectively. Seo *et al.*[57] also note a third peak at 46 eV and suggest that divalent iron is denoted by peaks at 46 and 51 eV, while trivalent iron is denoted by peaks at 43 and 51 eV.

10.2.7.7 Cobalt

Oxides of cobalt(II) and cobalt(III) are differentiated in XPS using their different magnetic properties. Paramagnetic cobalt(II) oxides have a strong shake-up satellite 6 eV above the Co $2p_{3/2}$ line, while the diamagnetic cobalt(III) oxides do not. Cobalt(II) hydroxide and cobalt(II) oxide are separated by a chemical shift of ~1.0 ± 0.2 eV.[18,37] The cobalt mixed oxides, $CoMoO_4$ and $CoAl_2O_4$, are shifted ~0.5 eV to a higher binding energy.[58]

10.2.7.8 Nickel

Unlike cobalt, only one oxidation state of nickel is believed to be present, for all practical purposes, on oxidized nickel surfaces. A chemical shift of approximately 2 eV is found between the Ni $2p_{3/2}$ line positions for NiO and $Ni(OH)_2$, thus facilitating identification of surface $Ni(OH)_2$.[37]

The Ni $2p_{3/2}$ spectrum of Ni^{2+} in different oxide lattices also changes significantly with the chemical structure, probably due to multiplet interaction.[6] In

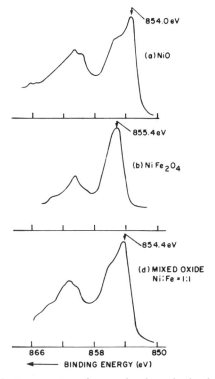

Figure 10.7   Photo-electron spectra of some simple and mixed nickel oxides. Ni $2p_{3/2}$ spectra are shown for (a) NiO; (b) $NiFe_2O_4$; (c) mixed iron–nickel oxide (Ni/Fe ratio = 1)

Figures 10.7(a) and (b) the Ni $2p_{3/2}$ peak maxima for NiO and the spinel NiFe$_2$O$_4$ are separated by 1.5 eV. In addition the NiO main peak is clearly split, while the peak for NiFe$_2$O$_4$ is not.

10.2.7.9 Copper

Copper has been one of the elements most extensively studied by XPS. Interpretation of oxide spectral structure is probably more straightforward for copper than for other transition metals. However, the oxides appear to transform more readily on heating or irradiation.

Two stable oxides of copper, Cu$_2$O and CuO, are found in solid-state films using XPS.[59] In addition, a cupric hydroxide has been characterized by XPS.[37]

All three compounds are likely surface products of the corrosion of copper metal, and their chemical differentiation from the metallic substrate is thus important. In the case of cuprous oxide, photo-electron spectra are identical to those for copper metal, within ± 0.1 eV. However, as initially described by Schoen,[60] the X-ray induced Auger spectra of copper metal and cuprous oxide (Cu $L_3M_{4,5}M_{4,5}$) are significantly different and allow a quantitative characterization[24] of oxide or metal. If contributions from the higher valence oxides are absent, the relative contributions of Cu$_2$O and metallic copper to the spectrum can be determined by ratioing peak intensities at two different energies. Copper $L_3M_{4,5}M_{4,5}$ Auger spectra of CuO and Cu(OH)$_2$ are both shifted to different kinetic energies, compared to Cu$_2$O. However, these peaks are broad and are somewhat more difficult to characterize because of the large number of discrete Auger lines under the envelope.

Characterization of CuO and Cu(OH)$_2$ is normally accomplished using the Cu $2p_{3/3}$ line. The principal Cu $2p_{3/2}$ peak maximum for CuO is shifted 1.3 ± 0.2 eV above that for Cu$_2$O, but this maximum is rather poorly defined with respect to the Cu$_2$O peak, because of significant broadening, probably associated with multiplet splitting. The Cu $2p_{3/2}$ peak for Cu(OH)$_2$ is shifted 2.5 ± 0.15 eV above that for Cu$_2$O. Cupric compounds are also characterized, in general, by the two strong shake-up peaks located 6 and 8 eV above the principal Cu $2p_{3/2}$ line.[37] The only difference noted between the shake-up spectra of CuO and Cu(OH)$_2$ is a change in the relative intensities of the two peaks under the shake-up envelope.

10.2.7.10 Molybdenum

The corrosion behaviour of molybdenum, as a component in many stainless steels, has been studied using XPS. Unlike the first-row transition elements, the spectral changes with oxidation state of second-row elements like molybdenum are more predictable and are not accompanied by satellite structure. Thus Mo^{IV} and Mo^{VI} oxides have been clearly characterized by chemical shifts of the Mo 3*d* line.[58] A chemical shift of ~0.8 eV is also noted between the Mo 3*d* line position for MoO_3 and $CoMoO_4$.

10.3 Review of XPS and AES Applications in Corrosion Science

The following section reviews the contributions XPS and AES to corrosion problems on a material-by-material basis. Studies not covered are those involving surface reactions of semi-conductor materials and reactions felt to be more related to metallurgy than corrosion. Surface reactions involving oxygen, water or other elements are considered, although many of the fundamental studies of surface reactions on metals are not considered if they do not appear to have a direct bearing on corrosion processes.

10.3.1 Light metals

Beryllium oxidation has been studied using AES by Zehner, Barbulesco and Jenkins[61] and the surface segregation of trace silicon has been detected using the energy loss peak below the Be *KLL* line. Anodic oxidation of single-crystal aluminium, studied using AES, has been found to result in stoichiometric Al_2O_3 layers.[62]

By contrast, XPS studies of naturally passivated aluminium foils have identified a number of hydroxides, as well as Al_2O_3.[63] Aluminium films formed on anodization in sulphuric acid have been shown by XPS to contain sulphate and sulphide ion,[64] as well as Al_2O_3. Scanning Auger microprobe studies combined with tensile stress tests on aluminium–zinc–magnesium alloys showed that the average grain boundary concentration of zinc, copper and magnesium could be indirectly correlated with the stress corrosion plateau crack velocity.[65]

10.3.2 First-row metals and their alloys

Almost all of the present surface literature is devoted to the study of alloys involving chromium, iron, cobalt, nickel and copper. These are described in the following sections.

10.3.2.1 Chromium

XPS and AES studies of the solid-state oxidation of chromium metal have been described by Conner.[66] The stability of chromate protective coatings on other metals has been the subject of several XPS investigations[67,68] which monitored, in particular, the oxidation state of chromium. The nature of passivation films on chromium surfaces in neutral and acidic solutions has been studied by AES[69]

10.3.2.2 Iron and low alloy steels

The air oxidation of pure iron has been studied by several authors using XPS and AES.[70, 71] Several electrochemical surface investigations have been made of the aqueous corrosion of pure iron in a borate buffer.[57, 72, 73] The consistent presence of boron in the outer portion of the passive film suggests that the inner portion of the oxide results from metal oxidation, while the outer portion results from precipitation of ferrous iron.[57] Depth profile evidence for two different oxide structures based on O/M ratios still requires confirmation that iron oxides are not being reduced by the ion beam. Other solution inhibitors such as CrO_4^{2-}, MoO_4^{2-} and $As_2O_5^{2-}$ were found to be incorporated into passive films to a greater extent than was boron.[74] The effect of chloride ion in penetrating iron oxide passive films is well known. To date, however, no chloride incorporation into the film can be detected by AES[75] or XPS.[76] Recent work[76] has suggested that hydrocarbon content of the iron oxide surface increases with increased chloride exposure; this is attributed to the loss of hydroxyl and bound water groups from the surface during depassivation. In addition, chloride attack was believed to result in an increased surface area of the oxide, as evidenced by the high O $1s$ binding energy (increased physisorbed gas).

The incorporation of iodide into a strained mild steel film has been followed as a function of polarization potential.[77] The measurement of iodine content of the film changes with potential (see Figure 10.8). The film is no longer stable to stress corrosion when iodine is not present.

An AES study of iron corrosion in nitrate solutions shows that the observed intergranular corrosion may be the result of non-passivation of the carbide phase in this solution.[78]

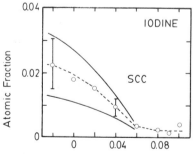

Figure 10.8 Relationship between iodine surface concentration and onset of passivity on an iron surface. (Reproduced by permission of K. Asami, N. Totsuka and M. Takano[77])

10.3.2.3 *Binary iron alloys and stainless steels*

Gas phase oxidation of iron–chromium alloys and stainless steels has been extensively investigated by AES and XPS.[79–88] For most temperatures studied, a thin chromium oxide layer is observed to form initially, followed by an overlying iron oxide layer. Baer[87] has carried out oxidations of 304 stainless steel at 800 °C using a range of oxygen partial pressures. With the use of scanning Auger microscopy, he showed that lower partial pressures favoured the initial growth of a pure chromium oxide, evenly distributed over the alloy grain surface. At high pressures the chromium oxide was concentrated near boundaries and a mixed oxide was found over the grain itself. In another study[88] the composition of the film formed on the grain surface of an Fe–26Cr alloy was analysed by AES and other methods following oxidation at 10^{-3} torr O_2 and 600 °C. After a few seconds of oxidation time, a duplex film was formed, the outer layer being identified as $\alpha-Fe_2O_3$ on the basis of the O/Fe AES intensity ratio.[88] An underlying layer of $\alpha-Cr_2O_3$ was identified, partly with the aid of AES, and the metal alloy substrate adjacent to this layer was shown to be depleted in chromium. Such information will assist efforts to grow protective films on alloys as part of their pre-treatment before service.[89]

The passivation of stainless steels during aqueous corrosion and the composition of its passive films have been one of the most studied subjects in corrosion science. Both AES and XPS studies have shown that the chromium concentration of the surface layer on Fe–Cr steels rises as the passivating potential is approached.[5, 8, 30, 89–93] The chromium enrichment is believed to result from a selective dissolution of iron from the oxide; in some cases, iron is also found to be depleted in the metal as well.[91] Castle and Clayton[5] and Asami, Hashimoto and Shimodaira[91] have examined the nature of the passive film in great detail. Castle and Clayton determined a duplex layer: the inner layer rich in chromium, the outer layer containing much hydroxide or bound water, held together with organic molecules. Asami, Hashimoto and Shimodaira found that both hydroxide and bound water reach a maximum in the film at the passivation potential, forming a hydrated chromium oxyhydroxide (see Figure 10.9).

The addition of molybdenum to stainless steels improves corrosion resistance, particularly from pitting attack. XPS studies[9, 94, 95] have helped to identify the role of molybdenum. It is proposed that hexavalent molybdenum reacts with active sites on a dissolving surface, where the oxyhydroxide cannot form. This leads to a decrease in activity at these sites and formation of a more uniform passive layer[9] (see Figure 10.9). A similar mechanism is proposed for the passivation of iron–molybdenum alloys. Two XPS studies[54,96] have shown that the major portion of passive oxide is an iron oxyhydroxide; only the active regions (i.e. pits) were found to contain molybdenum. In one

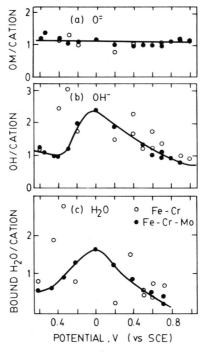

Figure 10.9   Surface concentrations of O^{2-}, OH^- and bound H_2O on steel surfaces measured as a function of polarization potential in 1 *M* HCl. For both Fe–30%Cr and Fe–30%Cr–2%Mo alloys the onset of passivation begins about -0.2 eV(VS SCE). The results here for Fe–30%Cr can be compared with the cation concentrations for the same alloy shown in Figure 10.1(a). Note that the better reproducibility of the analytical data for the iron–chromium–molybdenum alloy may be related to a more uniform passive layer. (After Asami, Hashimoto and Shimodaira[91])

of these studies[54] the molybdenum was identified as $FeMoO_4$. Another study of a molybdenum stainless steel, corroded in Ringers' solution, did not detect any significant molybdenum in the passive film.[97]

Reactions of Fe–Cr alloys with chloride solution have been studied and identification of surface chloride has been easier than for iron. Saniman[98] used XPS to measure the effect of solution chloride concentration on the surface uptake. Analysis of the film concentration and thickness showed that chloride intake resulted from exchange, not incorporation. Oxide film resistance to stainless steel crevice corrosion in chloride solution has also been studied.[99]

Anodic films formed on iron–nickel binary alloys have been compared with films on iron–nickel alloys also containing molybdenum and boron or phosphorus and boron.[100] This latter alloy had oxidized boron and phosphorus in the passive film. Using AES peak shapes, the phosphorus was identified tentatively as a phosphite; this form was believed responsible for reducing the nickel potential and hence its lower dissolution rate. Another AES–XPS study of the native oxide films on Fe–Ni–Cr–P–B alloys showed that boron and phosphorus are, in fact, enriched within the passive film, with most phosphorus distributed in a layer behind a chromium oxide enriched layer.[101]

AES studies of intergranular corrosion in 304 stainless steel[102] have revealed that many grain boundary inpurities play a role in increasing its susceptibility to attack. For example, sulphur in non-sensitized steels promotes intergranular corrosion in nitrate–dichromate solutions. However, chromium depletion at the grain boundary has been demonstrated by AES for specimens corroded in H_2SO_4–$CuSO_4$ solutions.

10.3.2.4   Cobalt and its alloys

An XPS study of the oxidation of cobalt in moist air[18] showed that $Co(OH)_2$ is an initial reaction product on the metal surface. The anodic oxidation of pure cobalt has been studied by AES and the passive film has been characterized.[103]

The corrosion of cobalt–chromium alloys in neutral or acidic solutions was found to proceed by selective dissolution of the cobalt.[67] McIntyre, Murphy and Zetaruk, using XPS, also showed that cobalt dissolves preferentially from a cobalt–chromium–tungsten alloy in alkaline solution at 300 °C. Gas phase oxidations done concurrently show that a cobalt-rich layer (Co_3O_4) forms in air; such outward migration of cobalt will enhance the preferential dissolution in the aqueous phase.

10.3.2.5   Nickel and its alloys

The oxidation of polycrystalline nickel has been followed by AES.[104–106] Müller *et al.*[106] have been able to identify the formation of NiO 'islands' early in the oxidation process.

Several investigations of Ni–Fe alloys have been reported, including AES analysis of oxidized permalloy surfaces (80%Ni–20%Fe)[107] and XPS analysis of Ni–20% Fe alloy oxidized at 500 °C.[108] An iron-rich oxide formed early on an annealed surface eventually gives way to a $NiFe_2O_4$ spinel.

Smith and Schmidt[109] have studied the oxidation of Ni–20%Cr alloy at 800 °C. A duplex Cr_2O_3–NiO was formed, except near grain boundary intersections. Similar XPS results were found for the short-term gas phase oxidation of Alloy 600 at 500 °C,[6] except that iron, a minor constituent in the bulk alloy, was found concentrated in the outermost layers of the oxide film. Oxygen partial pressure was shown to affect the nickel oxide content of the film.

A number of nickel-rich superalloy surfaces, oxidized near 1000 °C, have been analysed by XPS.[110] The surfaces were enriched in TiO_2, a minor constituent of all the alloys tested. High temperature reaction with Na_2SO_4 results in oxidation of most of the surface chromium to Cr^{VI}, apparently leading to accelerated corrosion.

XPS has been used to monitor the oxidation of a nickel electrode in contact with a solid electrolyte[27] in acid[111-113] and neutral[113] solutions and in liquid hydrogen fluoride.[114] Two XPS studies of the nickel electrode in acidic solution[111,112] show the passive film to be a NiO underlayer and $Ni(OH)_2$ on the surface. AES was used to measure the thickness of a NiO film during dissolution of nickel under transpassive conditions[115] but no information on film stoichiometry was forthcoming. MacDougall, Mitchell and Graham[113] analysed passive anodic films on polycrystalline nickel. Using AES to image the surface oxygen distribution, they showed that the passive film is one-third thinner on some grains than on others. Depth profiles of the film show a long interface between oxide and metal; the authors interpret this as evidence of a film high in defects which are important to the passivation process.

Anodic oxidation of Ni–25%Fe alloy in acidic solution[116] was shown by XPS to result in a passive film of NiO at the metal interface and nickel and iron hydroxides near the outer surface. By contrast, the duplex passive films found on Ni–Mo[117] and Hastalloy[31] had mixed nickel–molybdenum oxides in the inner film and only nickel oxide or hydroxide on the outer surface.

The surface films formed on Alloy 600 and Alloy 800 in NaOH have been measured by Hashimoto and Asami[118] using XPS. Primary passivation was ascribed to the formula of the chromium oxyhydroxide film and a secondary passive film was $Ni(OH)_2$. McIntyre, Zetaruk and Owen[119] analysed Alloy 600 surfaces following exposure to pressurized water reactor conditions and identified chromium oxyhydroxide under normal reducing conditions. Solution oxidizing conditions were varied to show a thick nickel-rich hydroxide film under highly oxidizing conditions and mixed chromium–nickel hydroxides under intermediate conditions. Passive films on Alloys 600 and 800 were depth-profiled using AES,[13] and the relative thermodynamic surface excess of chromium was found to decrease with increasing pH.

The aqueous corrosion of Ni–Cu alloy (66% Ni) under boiler conditions has been investigated by XPS,[120] and $Ni(OH)_2$ was shown to be the major surface species under active attack near 300 °C. Under reducing conditions, little or no surface oxide is found, in accordance with thermodynamic predictions.

10.3.2.6 Copper and its alloys

Copper and its oxidation have been studied using XPS and AES for almost ten years.[36, 121] Oxidation of a 70–30 Cu–Ni alloy results in a NiO-rich surface film, as deduced by XPS[122]

Anodic oxidation of copper metal in weakly alkaline or acidic solution was found by XPS to result in a duplex film of Cu_2O adjacent to the metal and a surface layer of $Cu(OH)_2$.[123] Recent XPS work of Shoesmith and coworkers has followed the growth of a similar layer in potassium hydroxide during the initial stages of formation.[24] The formation of films on copper–nickel alloys in saline solutions has been followed by XPS[124] and AES.[125] Duplex films were found in the latter work, which consisted of a mixed copper–nickel oxide over a thicker nickel-rich oxide. Oxidation of Cu–2%Be in ammoniacal solution results in a beryllium-rich surface layer.[126]

The chemistry of aluminium–brass condensor tube surfaces (copper 76%, zinc 22%, aluminium 2%) in seawater has been analysed by Castle, Epler and Peplow[127] using XPS. A layer high in magnesium and aluminium concentration is detected on the tubes and it is proposed that the magnesium from the seawater precipitates either by itself or in combination with aluminium from the alloy. In laboratory tests,[128] $Mg(OH)_2$ precipitation on aluminium brass was analysed by XPS as a function of pH. A higher pH at the interface compared with that in the bulk solution is required to rationalize the thickness of $Mg(OH)_2$ measured by XPS.

The surface analysis of organic protective films on copper has been carried out by several groups. The reaction of copper with benzotriazole has been studied by XPS[129,130] and its reaction with a cuprous oxide surface confirmed. The destructive oxidation of the cuprous benzotriazole to the cupric form has been monitored.[131] Mercaptobenzotriazole reactions with copper(I) surfaces to inhibit corrosion have also been investigated by XPS.[132]

10.3.3 Heavy metals

Corrosion-related surface analyses of elements heavier than those in the first row are much less extensive. Indeed, it is difficult to provide a coherent summary of studies of these elements because of the small number of results yet available.

Molybdenum electrochemistry and the oxidation of *molybdenum* have been studied by XPS by Ansell *et al.*[26] using a special inert atmosphere chamber. The cathodic decomposition of *niobium* oxide was shown by XPS to be a hydrated oxide of lower oxidation state.[133] The native oxides on the rare earth-like elements *cerium, ytterium* and *lanthanum* were determined by XPS[134] and *gadolinium* by AES.[135]

A number of studies of *tin* oxidation and corrosion have been reported. Ansell *et al.*[136] studied the electrochemical oxidation of tin in alkaline solution using XPS. In the passive region the surface oxide is determined to be Sn^{IV} rather than Sn^{II}, which grows initially. Investigations of the solid-state oxidation of tin by XPS have also been reported.[137] The passive film on a Sn–Ni alloy surface has been analysed also by XPS.[138] Servais *et al.*[139] examined chromium passivation films (tinplate) on tin surfaces and determined that lack of adhesion of such a film was caused by the growth of tin oxide.

Indium–gold and indium–lead–gold alloys were analysed for surface oxides by XPS.[140] The anodic oxidation of *gold*[141] in sulphuric acid was shown to produce Au_2O_3 as a surface film. *Lead* oxidation has been studied by XPS[142] and the electrochemical oxidation of *uranium oxide* has recently been characterized by Sunder *et al.*,[143] also using XPS.

10.3.4   Organic coatings

Organic protective coatings, applied to metal surfaces, are intended to provide a barrier between any invading species and the metal surface. Moreover, if penetration of the coating does occur, any corrosive attack will be limited to that immediate region and will not spread laterally, as on an uncoated surface. The loss of adhesion of such coatings on iron and steel surfaces has been studied using XPS.

Steel surfaces coated with expoxy-esters,[144] epoxy-urethane and epoxy-amines[145] have been corroded to de-adhesion in saline solutions. Each side of the film rupture was examined by XPS and subsequently the effects of additional chemical tests on the surface were monitored by XPS. Both organic and substrate sides of the rupture showed evidence of having undergone surface saponification to form sodium carboxylate groups. This is believed to result from corrosion-induced hydroxyl groups.

A highly sensitive method of detection of carboxylate groups involved their 'tagging' with silver—an ion with high selectivity for a carboxy group and a high XPS cross-section. Using this method a carboxylate surface concentration of about one per cent. can be detected (see also Chapter 9).

The effect of using a 'conversion' coating between the steel and the organic layer has been examined. The de-adhesion surfaces showed that the conversion coating, a zinc phosphate, remained adhered to the metal substrate, and showed evidence of some residual organics. Thus, de-adhesion takes place near the interface, but in the organic phase.

Hammond *et al.*[146] have analysed C 1s, N 1s and O 1s spectra of interfacial surfaces containing a number of organic coatings. They have detailed binding energy and intensity ratio data for several carboxylic anions and simple polymers.

The effects of mechanical and chemical de-adhesion of polybutadiene films have been compared by XPS.[147] Mechanical de-adhesion was shown to result from cohesive failure. As in earlier studies, chemical de-adhesion appears to result from ester hydrolysis at the metal interface.

10.4   Conclusion

It is clear that XPS and AES are beginning to be extremely effective in assessing the composition and chemistry of corrosion films. Of particular note is the degree to which both techniques have become truly quantitative over the past few years—not only in terms of elemental composition but also often in terms of the chemical entities present. Also, there is increasing evidence that even fragile structures, e.g. hydrates, can be preserved in the spectrometer vacuum. These two steps have made possible important discoveries about the nature of the passive film.

Significant quantities of data are beginning to accumulate in the subject area of iron–chromium–nickel corrosion. Thus, in some cases, corroborative evidence has been obtained from different types of experiments. This is the mark of a maturing field!

Electrochemical experiments have been particularly important since these allow corrosion conditions to be altered drastically within a convenient time frame. Many situations cannot be simulated in a cell; it is expected that surface analytical data can tie the result obtained in an electrochemical experiment with that obtained during 'open-circuit' corrosion, thus making corrosion in the electrochemical cell more relevant to corrosion in the field.

From this writer's viewpoint, there is a somewhat larger amount of data now coming from XPS than AES. This is related to the more benign excitation conditions and the better present-day knowledge of chemical shifts. Corrosion, however, is very much a microscopic process and the emphasis must again return to a microscopic surface technique such as AES, when XPS studies of a particular system have been exhausted. For this reason, attempts to understand AES electron beam effects are believed to be potentially important, as are Auger chemical shift studies. Microscopic surface studies will be particularly important for understanding the effects of organic and inorganic inhibition layers, as well as crevice and pitting corrosion.

It is hoped that the progress demonstrated by the usage of AES and XPS over the last five years will lead to the deserved inclusion of these techniques in a much larger number of future corrosion experiments.

Acknowledgement

The writer acknowledges the expert assistance of Miss T. Chan in preparing this manuscript.

References

1. A. Joshi, Investigation of passivity, corrosion and stress corrosion cracking phenomena by AES and ESCA, in *Corrosion 77,* Paper 16, National Association of Corrosion Engineers, Houston, Texas (1977).
2. J. E. Castle, *Surf. Sci.,* **68,** 583 (1977).
3. J. E. Castle, Application of XPS analysis to research into the causes of corrosion, in *Applied Surface Analysis,* (Eds T. L. Barr and L. E. Davis), ASTM STP 699, p. 182, American Society for Testing and Materials, Philadelphia, Pennsylvania (1980).

4. M. P. Seah and W. A. Dench, *Surf. Interfacial Anal.*, **1**, 2 (1979).
5. J. E. Castle and C. R. Clayton, *Corrosion Sci.*, **17**, 7 (1977).
6. N. S.McIntyre, D. G. Owen and D. C. Zetaruk, *Appl. Surf. Sci.*, **2**, 55 (1978).
7. K. Asami, K. Hashimoto and S. Shimodaira, *J. Jpn Inst. Metals*, **40**, 438 (1976).
8. K. Asami, K. Hashimoto and S. Shimodaira, *Corrosion Sci.*, **17**, 713 (1977).
9. K. Hashimoto, K. Asami and K. Teramoto, *Corrosion Sci.*, **19**, 3 (1979).
10. K. Asami and K. Hashimoto, *Corrosion Sci.*, **17**, 559 (1977).
11. K. Hashimoto, M. Naka, K. Asami and T. Masamoto, *Corrosion Sci.*, **19**, 165 (1979).
12. F. Pons, J. Le Héricy and J. P. Longeron, *Surf. Sci.*, **69**, 547 (1977); **69**, 565 (1977).
13. M. Seo and N. Sato, *Corrosion*, **36**, 334 (1980).
14. D. F. Mitchell, *Appl. Surf. Sci.*, **9**, 131 (1981).
15. N. S. McIntyre, Applications of surface analysis in the nuclear industry, in *Industrial Applications of Surface Analysis* (Eds C. J. Powell and L. Casper), ACS Symposium Series No. 99, p. 345, American Chemical Society, Washington, DC (1982).
16. D. R. Baer, *Appl. Surf. Sci.*, **7**, 69 (1981).
17. K. S. Kim, W. E. Baitinger, J. W. Amy and N. Winograd, *J. Elect. Spectrosc.*, **5**, 351 (1975).
18. T. J. Chuang, C. R. Brundle and K. Wandelt, *Thin Solid Films*, **53**, 19 (1978).
19. N. S. McIntyre and D. G. Zetaruk, *Anal. Chem.*, **49**, 1521 (1977).
20. D. F. Mitchell, G. I Sproule and M. J. Graham, *J. Vac. Sci. Technol.*, **18**, 690 (1981).
21. N. S. McIntyre, E. V. Murphy and D. G. Zetaruk, *Surf. Interfacial Anal.*, **2**, 151 (1979).
22. J. M. Walls, D. D. Hall and D. E. Sykes, *Surf. Interfacial Anal.*, **1**, 204 (1979).
23. C. J. Schmidt, P. V. Lenzo and E. G. Spencer, *J. Appl. Phys.*, **46**, 4080 (1975).
24. N. S. McIntyre, S. Sunder, D. W. Shoesmith and F. W. Stanchell, *J. Vac. Sci. Technol.*, **18**, 714 (1981).
25. R. W. Revie, B. G. Baker and J. L'M. Bockris, *Surf. Sci.*, **52**, 664 (1975).
26. R. O. Ansell, T. Dickinson, A. R. Povey and P. M. A. Sherwood, *J. Electroanal. Chem.*, **98**, 69 (1979).
27. C. Y. Yang and W. E. O'Grady, *J. Vac. Sci. Technol.*, **20**, 925 (1982).
28. K. Kudo, F. Shibata, G,. Okamoto and N. Sato, *Corrosion Sci.*, **18**, 809 (1978).
29. J. E. Castle, *Proc. of the Conference on Metal Coatings*, p. 435, Lehigh University (1978).
30. G. Okamoto, *Corrosion Sci.*, **13**, 471 (1973).
31. G. T. Burstein, *Materials Science and Engineering*, **42**, 207 (1980).
32. Z. Szklarska-Smialowska, H. Viefhaus and M. Hanik-Czachor, *Corrosion Sci.*, **16**, 649 (1976).
33. T. Smith, *Surf. Sci.*, **55**, 601 (1976).
34. C. D. Wagner, W. M. Riggs, L. E. Davis, J. F. Moulder and G. E. Muilenberg, *Handbook of X-ray Photoelectron Spectroscopy*, Physical Electronics Division, Perkin-Elmer Corporation, Eden Prairie, Minnesota (1979).
35. O. Johnson, *Chem. Scripta*, **8**, 162 (1975).
36. G. Schoen, *J. Electron. Spectrosc.*, **1**, 377 (1972).
37. N. S. McIntyre and M. G. Cook, *Anal. Chem.*, **47**, 2208 (1975).
38. P. R. Norton, *J. Catalysis*, **36**, 211 (1975).
39. T. Matsushima, D. B. Almy and J. M. White, *Surf. Sci.*, **67**, 89 (1977).
40. L. Ramqvist, K. Hamrin, G. Johansson, A. Fahlmann and C. Nordling, *J. Phys. Chem. Solids*, **30**, 1835 (1969).

41. J. F. Franzen, M. X. Umana, J. R. McCreary and R. J. Thorn, *J. Solid State Chem.,* **18,** 363 (1976).
42. J. S. Solomon and W. L. Baun, *Surf. Sci.,* **51,** 228 (1975).
43. M. L. Knotek and J. Houston, *Phys. Rev.,* **B15,** 4580 (1977).
44. G. A. Sawatsky and D. Post, *Phys. Rev.,* **B20,** 1546 (1979).
45. L. Fiermans and J. Vennik, *Surf. Sci.,* **35,** 42 (1973).
46. F. J. Szalkowski and G. A. Somorjai, *J. Chem. Phys.,* **61,** 2064 (1974).
47. G. C. Allen, M. T. Curtis, A. J. Hooper and P. M. Tucker, *J. Chem. Soc. (Dalton),* **1973,** 1675 (1973).
48. G. C. Allen and P. M. Tucker, *Inorg. Chim. Acta,* **16,** 41 (1976).
49. I. Ikemoto, K. Ishii, S. Kinoshita, H. Kuroda, M. A. A. Franco and J. M. Thomas, *J. Solid State Chem.,* **17,** 425 (1976).
50. M. Oku, K. Hirokawa and S. Ikeda, *J. Electron Spectrosc.,* **7,** 465 (1975).
51. A. Aoki, *Jpn J. Appl. Phys.,* **15,** 305 (1976).
52. G. C. Allen, M. T. Curtis, A. J. Hooper and P. M. Tucker, *J. Chem. Soc. (Dalton),* **1976,** 1526 (1976).
53. C. R. Brundle, T. J. Chuang and K. Wandelt, *Surf. Sci.,* **68,** 459 (1977).
54. D. A. Stout, J. B. Lumsden and R. W. Staehle, *Corrosion,* **35,** 141 (1979).
55. H. Binder, *Z. für Naturforsch.,* **B28,** 256 (1973).
56. S. Ekelund and C. Leygraf, *Surf. Sci.,* **40,** 179 (1973).
57. M. Seo, M. Sato, J. B. Lumsden and R. W. Staehle, *Corrosion Sci.,* **17,** 209 (1977).
58. T. A. Patterson, J. C. Carver, D. E. Leyden and D. M. Hercules, *J. Phys. Chem.,* **80,** 1702 (1976).
59. S. W. Gaarenstroom and N. Winograd, *J. Chem. Phys.,* **67,** 3500 (1977).
60. G. Schoen, *J. Electron Spectrosc.,* **7,** 377 (1972).
61. D. M. Zehner, M. Barbulesco and L. H. Jenkins, *Surf. Sci.,* **34,** 385 (1973).
62. S. Matsuzawa, N. Baba and S. Tajima, *Electrochim, Acta,* **24,** 1199 (1979).
63. T. L. Barr, *J. Vac. Sci. Technol.,* **14,** 660 (1977).
64. J. A. Treverton and N. C. Davies, *Electrochim. Acta,* **25,** 1571 (1980).
65. C. R. Shastry, M. Levy and A. Joshi, *Corrosion Sci.,* **21,** 673 ()1981).
66. G. R. Conner, *J. Vac. Sci. Technol.,* **15,** 343 (1978).
67. S. Storp and R. Holm, *Surf. Sci.,* **10,** 68 (1977).
68. D. R. Baer, Use of surface analytical techniques to examine metal corrosion problems, in *ACS Symposium Series, Industrial Application of Surface Analysis,* American Chemical Society, Washington, DC (1982).
69. M. Seo, R. Saito and N. Sato, *J. Electrochem. Soc.,* **127,** 1909 (1980).
70. C. R. Brundle, *Surf. Sci.,* **66,** 581 (1977).
71. J. K. Grimzewski, B. D. Padalia, S. Affrossman, L. M. Watson and D. J. Fabian, *Surf. Sci.,* **62,** 386 (1977).l
72. R. W. Revie, B. G. Baker and J. O'M. Bokris, *J. Electrochem. Soc.,* **122,** 1460 (1975).
73. D. H. Davies and G. T. Burstein, *Corrosion,* **36,** 416 (1980).
74. J. B. Lumsden, Z. Szklorska-Smialowska and R. W, Staehle, *Corrosion,* **34,** 169 (1977).
75. M. da Cunha Belo, B. Rondot, F. Pons and J. P. Lanyeron, *J. Electrochem. Soc.,* **124,** 1317 (1977).
76. D. L. Cocke, P. Nilsson, O. J. Murphy and J. O'M. Bokris, *Surf. Interfacial Anal.,* **4,** 94 (1982).
77. K. Asami, N. Totsuka and M. Takano, *Corrosion,* **35,** 208 (1979).
78. G. Tauber and H. J. Grabke, *Corrosion Sci.,* **19,** 793 (1979).
79. C. Leygraf, G. Hultqvist and S. Ekelund, *Surf. Sci.,* **51,** 409 (1975).

80. J. P. Coad and J. G. Cunningham, *J. Electron Spectrosc.,* **3,** 435 (1974).
81. G. Betz, G. K. Wehner, L. Toth and A. Joshi, *J. Appl. Phys.,* **45,** 5312 (1974).
82. I. Olefjord, *Corrosion Sci.,* **15,** 687 (1975).
83. C. Leygraf and G. Hultqvist, *Surf. Sci.,* **61,** 69 (1976).
84. R. K. Wild, *Corrosion Sci.,* **17,** 87 (1977).
85. T. Smith and L. W. Crane, *Oxid. Met.,* **10,** 135 (1976).
86. T. J. Driscoll and P. B. Needham, *Oxid. Met.,* **13,** 283 (1979).
87. D. R. Baer, *Appl. Surf. Sci.,* **7,** 69 (1981).
88. C. P. Jensen, D. F. Mitchell and M. J. Graham, *Corrosion Sci.* (in press).
89. B. Chattopodhyay and G. C. Wood, *J. Electrochem. Soc.,* **117,** 1163 (1970).
90. I. Olefjord and H. Fischmeister, *Corrosion Sci.,* **15,** 697 (1975).
91. K. Asami, K. Hashimoto and S. Shimodaira, *Corrosion Sci.,* **16,** 387 (1976).
92. K. Asami, K. Hashimoto and S. Shimodaira, *Corrosion Scie.,* **18,** 125 (1978).
93. C. Leygraf, G. Hultqvist, I. Olefjord, B. O. Elfström, V. M. Knyazheva, A. V. Plaskeyev and Y. M. Kolotyrkin, *Corrosion Sci.,* **19,** 343 (1979).
94. K. Sugimoto and Y. Sawada, *Corrosion,* **32,** 347 (1976).
95. K. Hashimoto and K. Asami, *Corrosion Sci.,* **19,** 251 (1979).
96. K. Hashimoto, M. Naka, K. Asami and T. Masumoto, *Corrosion Sci.,* **19,** 165 (1979).
97. J. R. Cahoon and R. Bandy, *Corrosion,* **38,** 299 (1982).
98. E. Saniman, MSc Thesis, University of Surrey, Guildford, Surrey (1977).
99. G. Hultqvist and C. Leygraf, *J. Vac. Sci. Technol.,* **17,** 85 (1980).
100. G. T. Burstein, *Corrosion,* **37,** 549 (1981).
101. D. R. Baer, D. A. Petersen, L. R. Pederson and M. T. Thomas, *J. Vac. Sci. Technol.,* **20,** 957 (1982).
102. A. Joshi and D. F. Stein, *Corrosion,* **28,** 321 (1972).
103. D. H. Davies and G. T. Burstein, *Corrosion Sci.,* **20,** 989 (1980).
104. S. H. Kulpa and R. P. Frankenthal, *J. Electrochem. Soc.,* **124,** 1588 (1977).
105. S. P. Sharma, *J. Electrochem. Soc.,* **125,** 2005 (1978).
106. K.-H. Müller, P. Beckmann, M. Schemmer and A. Benninghoven, *Surf. Sci.,* **80,** 325 (1979).
107. W. Y. Lee and J. J. Eldridge, *J. Electrochem. Soc.,* **124,** 1747 (1977).
108. I. Olefjord and P. Marcus, *Surf. Interfacial Anal.,* **4,** 23 (1982).
109. K. L. Smith and L. D. Schmidt, *J. Vac. Sci. Technol.,* **20,** 364 (1982).
110. S. R. Smith, W. J. Carter, G. D. Mateescu, F. J. Kohl, G. C. Fryburg and C. A. Stearns, *Oxidation of Metals,* **14,** 415 (1980).
111. T. Dickenson, A. F. Povey and P. M. A. Sherwood, *J. Chem. Soc., Faraday Trans. I,* **73,** 327 (1977).
112. P. Marcus, J. Oudar and I. Olefjord, *J. Microsc. Spectrosc. Electron,* **4,** 63 (1979).
113. B. MacDougall, D. F. Mitchell and M. J. Graham, *Corrosion,* **38,** 85 (1982).
114. N. Watanabe and M. Haruta, *Electrochim. Acta,* **25,** 461 (1980).
115. M. Datta, H. J. Mathieu and D. Landolt, *Electrochim. Acta,* **24,** 843 (1979).
116. I. Olefjord and P. Marcus, *Surf. Interfacial Anal.,* **4,** 29 (1982).
117. G. T. Burstein and T. P. Hoar, *Corrosion Sci.,* **17,** 939 (1977).
118. K. Hashimoto and K. Asami, *Corrosion Sci.,* **19,** 427 (1979).
119. N. S. McIntyre, D. G. Zetaruk and D. Owen, *J. Electrochem. Soc.,* **126,** 750 (1979).
120. N. S. McIntyre, T. E. Rummery, M. G. Cook and D. Owen, *J. Electrochem. Soc.,* **123,** 1164 (1976).
121. T. Robert, M. Bartel and G. Offergeld, *Surf. Sci.,* **33,** 128 (1972).
122. J. E. Castle, *Nature Physical Sciences,* **234,** 93 (1971).

123. H.-H. Strehblow and B. Titze, *Electrochim. Acta,* **25,** 839 (1980).
124. L. D. Hulett, A. L. Bacarella, L. LiDonnici and J. C. Griess, *J. Electron Spectrosc.,* **1,** 169 (1972).
125. G. E. McGuire, A. L. Bacarella, J. C. Griess, R. E. Clausing and L. D. Hulett, *J. Electrochem. Soc.,* **125,** 1801 (1978).
126. R. C. Newman and G. T. Brustein, *Corrosion Sci.,* **20,** 375 (1980).
127. J. E. Castle, D. C. Epler and D. B. Peplow, *Corrosion Sci.,* **19,** 457 (1979).
128. J. E. Castle and R. Tanner-Tremain, *Surf. Interfacial Anal.,* **1,** 49 (1979).
129. R. F. Roberts, *J. Electron Spectrosc.,* **4,** 273 (1974).
130. P. G. Fox, G. Lewis and P. J. Boden, *Corrosion Sci.,* **19,** 457 (1979).
131. D. Chadwick and T. Hashami, *Corrosion Sci.,* **18,** 39 (1978).
132. M. Ohsawa and W. Suetaka, *Corrosion Sci.,* **19,** 709 (1979).
133. K. Sugimoto, G. Belanger and D. L. Piron, *J. Electrochem. Soc.,* **126,** 535 (1979).
134. T. L. Barr, in *Quantitative Surface Analysis of Materials* (Ed. N. S. McIntyre), ASTM STP 643, p. 83, American Society for Testing and Materials, Philadelphia, Pennsylvania (1978).
135. A. J. Bevolo, B. J. Beaudry and K. A. Gschneidner, *J. Electrochem. Soc.,* **127,** 2556 (1980).
136. R. O. Ansell, T. Dickinson, A. F. Povey and P. M. A. Sherwood, *J. Electrochem. Soc.,* **124,** 1360 (1977).
137. C. L. Lau and G. K. Wertheim, *J. Vac. Sci. Technol.,* **15,** 622 (1978).
138. J. H. Thomas and S. P. Sharma, *J. Vac. Sci. Technol.,* **14,** 1168 (1977).
139. J. P. Servais, J. Lempereur, L. Renaud and V. Leroy, *Br. Corros. J.,* **14,** 126 (1979).
140. J. M. Baker, R. W. Johnson and R. A. Pollak, *J. Vac. Sci. Technol.,* **16,** 1534 (1979).
141. T. Dickinson, A. F. Povey and P. M. A. Sherwood, *J. Chem. Soc., Faraday Trans. I,* **71,** 298 (1975).
142. R. W. Hewitt and N. Winograd, *Surf. Sci.,* **78,** 1 (1978).
143. S. Sunder, D. W. Shoesmith, M. G. Bailey, F. W. Stanchell and N. S. McIntyre, *J. Electroanal. Chem.,* **130,** 163 (1981).
144. J. S. Hammond, J. W. Holubka and R. A. Dickie, *J. Coatings Technol.,* **51,** 655 (1979).
145. J. W. Holubka, J. S. Hammond, J. E. DeVries and R. A. Dickie, *J. Coatings Technol.,* **52,** 63 (1980).
146. J. S. Hammond, J. W. Holubka, J. E. DeVries and R. A. Dickie, *Corrosion Sci.,* **21,** 239 (1981).
147. R. A. Dickie, J. S. Hammond and J. W. Holubka, *Ind. Eng. Chem., Prod. Res. and Dev.,* **20,** 339 (1981).

Practical Surface Analysis
by Auger and X-ray Photoelectron Spectroscopy
Edited by D. Briggs and M. P. Seah
© 1983, John Wiley & Sons, Ltd.

Appendix 1

Spectrometer Calibration

M. T. Anthony

Division of Materials Applications,
National Physical Laboratory, Teddington, Middlesex, UK

A1.1 Calibration of XPS Instruments

The accurate calibration of X-ray photo-electron spectrometers has been an important and continuing objective of spectroscopists over the last ten years. Without an accurately defined spectral energy scale a full interpretation of the spectra gathered on different instruments is severely limited. Several workers[1-9] have reported energy calibrations by presenting binding energy tabulations for copper, silver and gold. These elements have the advantages of being easily cleaned and chemically inert as well as being stable conductors. For the purposes of calibration, binding energies are most accurately defined by referencing the zero to the Fermi level of conducting samples since this avoids errors arising from work function differences between spectrometers. The d-bands of palladium and nickel provide suitably intense and sharp Fermi edges in order to define this zero. The literature calibration values of binding energies for copper, silver and gold defined in this way are given in Table A1.1. It is evident that the accumulated data in this table shows a variability of about 0.3 eV which, surprisingly, has not reduced through the years.

The ASTM E-42 committee organized a round-robin[10] to report on binding energies obtained on a wide range of spectrometers. This highlighted the magnitude of errors arising in energy calibration. Analysis by means of Youden plots enables the contributions of random errors, zero shift and voltage scaling to be separated. The most significant error of around 0.3 eV is due to positioning of the zero point. The random error of about 0.1 eV shows the measurement repeatability. Finally, the voltage scaling of the spectrometer introduces errors of approximately 500 p.p.m. or 0.5 eV at 1000 eV binding energy.

Having regard to both the previous literature results[1-9] and those of the ASTM survey,[10] it was clear that accurate energy calibration would only be possible if there was full traceability of the calibration measurements. The

Table A1.1 XPS calibration binding energies, eV, from the literature

	Schon, 1972[2]	Johansson et al., 1973[3]	Asami, 1976[4]	Richter and Peplinski, 1978[5]	Wagner, Gale and Raymond, 1979[6]	Powell, Erickson and Madey, 1979[10]	Bird and Swift, 1980[7]	Fuggle and Mortensson, 1980[8]	Lebugle et al., 1981[9]
Cu $3p$	75.2 ± .1					75.1		75.1	75.13
Au $4f_{7/2}$	84.0	83.8 ± .2	84.07	84.0	83.8	84.0	83.98 ± .02	83.7	84.0
Ag $3d_{5/2}$	368.2	368.2 ± .2	368.23		367.9		368.21 ± .03	367.9	368.2
Cu $2p_{3/2}$	932.2 ± .1	932.8 ± .2	932.53	932.7	932.4	932.6	932.66 ± .06	932.5	932.57
Cu LMM, KE	919.0 ± .1	918.3 ± .2	918.65	918.35	918.6	918.7	918.64 ± .04		
BE	567.6 ± .1	568.35 ± .2	567.96	568.25	567.9	567.9	567.97 ± .04		
F.E. ref.	Pd	Pd	Pd	Pd	Au $4f_{7/2}$	E-42	Pd	—	—
Instrument	AEIES100	magnetic	AEIES200	AEIES200	Varian IEE-15	Survey	AEIES200B	VG ESCA3 Mk 1	HP 5950A

work undertaken at the NPL,[11] on a VG Scientific ESCA3 Mk II, gave particular attention to (a) the reproducibility of the calibration, (b) the correct establishment of the zero binding energy point and (c) the accurate measurement of the energy scale. The work determined in detail the effect of operating parameters on reproducibility and, in addition, concluded that the zero point was best defined by the nickel Fermi edge using Mg $K\alpha$ as the source radiation. The measurement of the energy scale was achieved by first checking the 1:1 energy-to-scan voltage equivalence and then by measuring the scan voltage as accurately as possible. The measurement chain was calibrated against a standard cell itself calibrated against the standard NPL volt. The NPL spectrometer energy scale therefore possessed full traceability to the primary volt standard. For electrons of energy 0–1550 eV a precision of measurement of around 2 p.p.m. was achieved with an overall accuracy of 11 p.p.m.

Binding energies of copper, silver and gold referenced to the nickel Fermi edge were determined on the NPL spectrometer and represent the first accurately traceable calibration of binding energies. The binding energies for both Al $K\alpha$ and Mg $K\alpha$ radiations are listed in Table A1.2.[11]

A comparison of previous literature values from Table A1.1 with the NPL tabulation shows that the recent careful work of Bird and Swift,[7] which quotes the smallest errors of the literature values, gives very close agreement. Other interlaboratory comparisons[11] have subsequently confirmed that where the appropriate methodology[12] is carried out, the calibration is repeatable to ±0.02 eV if some effort is taken and to ±0.05 eV with fair ease.

A1.2 Calibration of AES Instruments

The majority of electron spectrometers presently employed in Auger electron spectroscopy are cylindrical mirror analysers (CMA), as distinct from the

Table A1.2   XPS calibration binding energies, eV (Ni Fermi edge zero)

	Al $K\alpha$	Mg $K\alpha$
Cu $3p$	75.14 ± 0.02	75.13 ± 0.02
Au $4f_{7/2}$	83.98 ± 0.02	84.00 ± 0.01
Ag $3d_{5/2}$	368.27 ± 0.02	368.29 ± 0.01
Cu L_3MM	567.97 ± 0.02	334.95 ± 0.01
Cu $2p_{3/2}$	932.67 ± 0.02	932.67 ± 0.02
Ag M_4NN	1128.79 ± 0.02	895.76 ± 0.02

Notes
1 Al $K\alpha$–Mg $K\alpha$ = 233.02 eV
2 Au $4f_{7/2}$ Al $K\alpha$ BE lowered by Au $4f_{5/2}$ tail
3 Ag $3d_{5/2}$ Mg $K\alpha$ BE raised by Ag $3d_{3/2}$ X-ray satellite

spherical sector analysers (SSA) more common in the higher resolution instruments for XPS. The poorer resolution attainable using single-pass CMA analysers, together with the difficulties of accurate sample positioning, restrict the accuracy of the energy calibration compared with that for XPS. However, for chemical state information in AES, an accurate energy scale calibration is needed.

In contrast to XPS, where kinetic energies, E_K, are referenced to the Fermi level, in AES peaks are quoted for their kinetic energy, E_K, referenced to the vacuum level. As a result, the absolute energy scale will be sensitive to work function differences between spectrometers, although energy differences are not so dependent. In addition, for CMA instruments the precise positioning of the sample is critical, unlike the case for the SSA encountered in XPS instruments. Tests at the NPL[13] using a Varian CMA analyser showed that the position of a copper sample must be set to 15 μm to gain an experimental repeatability of 0.1 eV for the Cu L_3MM peak. This arises through the focal properties of the CMA where a change, Δl, in the focal length l of the analyser appears as a shift, ΔE, in the energy E. Generally l is of the order of 150 mm so that the above positioning accuracy cannot be avoided.

The recent ASTM E-42 round-robin[14] reports on the Auger peak positions for copper and gold reference metals; these peaks were recorded on a wide range of instruments. Peak energies showed substantial deviations about the median ranging from 1.9 eV for the low-energy peaks up to 7.4 eV for the high-energy gold $M_5N_{6,7}N_{6,7}$ transition. Clearly differences of this magnitude arise from calibration problems and make chemical state information unreliable.

As most Auger electron spectra are recorded in the derivative $(\mathrm{d}\{En(E)\}/\mathrm{d}E)$ mode with a modulation voltage applied to the analyser, peak energies are generally reported as the position of the high-energy negative excursion peak. Unfortunately, peak energies defined in this way are very sensitive to the modulation amplitude, as seen in Figure A1.1 for the 100 eV elastic peak, and also to analyser resolution. Accurate calibrations could be established for the copper, silver and gold peaks at a given modulation for various analyser resolutions; however, for most instruments the real modulation in terms of the contribution to the energy spectrum is not known with sufficient accuracy. For many instruments the modulations are not applied to all electrodes in phase in strict proportionality to the d.c. voltages present. This problem is discussed in detail by Seah,[15] who shows why this then makes it very difficult to calibrate the effective energy modulations of the spectrometer. The problem is most prevalent with XPS analysers used for AES but even the CMAs do not behave simply. Tests on two Varian CMAs show that the effective energy modulation of the spectrometer is 20 ± 1 per cent. lower than that predicted from the product of the measured voltage modulation of the mirror electrode and the spectrometer constant. In addition to this effect, which arises from the analyser design, will be the normal errors associated with ensuring that the voltage oscillator is kept in calibration. These problems are common to most AES instruments and not to any particular model.

In order to avoid the difficulties associated with the imprecision of the modulation and the need to correct for analyser resolution, each peak position in the differential mode may be replaced by the energy at which swing between the positive and negative excursions crosses the zero signal line. This zero crossing point is insensitive to the applied modulation and analyser resolution and can therefore be most accurately assigned. Figure A1.1 shows a series of different modulations for the 100 eV elastic peak and illustrates how the zero crossing point changes by less than 1 per cent. of the shift seen in the customary negative peaks for modulations up to 2 V. For small modula-

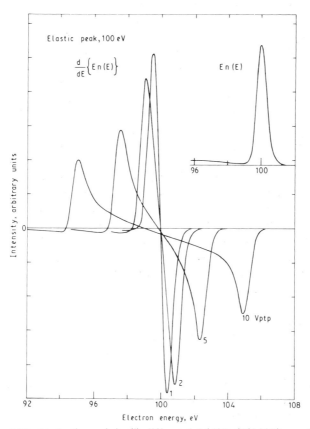

Figure.A1.1   100 eV elastic peak in (i) differential (d{$En(E)$}/dE), mode for 1, 2, 5 and 10 volts peak-to-peak modulation and (ii) direct, $En(E)$ mode

tions the zero crossing point corresponds to the peak positions as recorded in the direct *En(E)* spectrum (shown top right in Figure A1.1). Indeed, for the best possible accuracy of energy calibration, it is strongly advisable to use this *En(E)* mode, since only then is the peak energy entirely independent of the modulation and its associated errors in assessing its magnitude. The *En(E)* peak positions for copper, silver and gold reference metals, using a 5 kV electron beam, are given in Table A1.3.[13] The positions are the same as the zero crossing point in the differential mode for small modulations. The tabulated data were recorded at the NPL on a VG Scientific ESCA3 Mk II spectrometer modified for reference work with its energy scale traceable to the standard volt.[11] The peak positions thus have precise traceability to a primary standard, but, bearing in mind that kinetic energies referenced to the vacuum level (as customarily used) depend on the spectrometer 'work function', a final column in Table A1.3 is added with the kinetic energies referenced to the Fermi level. A further point that should be noted is that, with the varying energy in the deflector field, the effective mass of the electron is involved and not the rest mass, so that a relativistic effect occurs in which the true electron energy E_k is related to that using the low energy approximation for the analyser, E_k^0, by[16,17]

$$E_k = E_k^0 + \left(\frac{E_k^0}{1011R}\right)^2 + \cdots \text{SSA}$$

$$E_k = E_k^0 + \left(\frac{E_k^0}{1536}\right)^2 + \cdots \text{CMA}$$

where R is the retard ratio between the kinetic energy of the electron in the SSA deflector field and E_k. Thus, a CMA calibrated to give the correct value for the gold peak at 2020 eV will, if the energy scale is assumed to be linear measure, the copper 914.4 eV peak at 0.43 eV too high an energy, i.e. 914.8 eV.

Table A1.3   Auger electron peak kinetic energies, eV

	E_k	E_K
Cu $M_{2,3}M_{4,5}M_{4,5}$	59.2 ± 0.1	63.5 ± 0.1
Au $N_{6,7}O_{4,5}O_{4,5}$	68.2 ± 0.1	72.5 ± 0.1
Ag $M_4 N_{4,5}N_{4,5}$	353.6 ± 0.1	357.9 ± 0.1
Cu $L_3 M_{4,5}M_{4,5}$	914.4 ± 0.1	918.7 ± 0.1

Notes
1. Energy of peak recorded in *En(E)* mode, equivalent to the zero crossing point in the differential mode for small modulations.
2. Kinetic energies include the appropriate relativistic correction
   E_k = kinetic energy referenced to vacuum level
   E_K = kinetic energy referenced to Fermi level

Unlike the XPS case the use of Table A1.3 in calibrating AES analysers is not always trivial. For XPS instruments which can also provide AES, the spectrometer will already have been accurately calibrated using XPS. For XPS-style analysers used for AES and with no XPS present, the table can be used directly since the sample position is not critical. However, in CMA-based instruments where the sample will not be repositioned exactly to ± 15 μm an extra procedure must be used. Most CMAs are provided with a stabilized beam energy of 2000 eV so that the sample is positioned until the zero crossing of the elastic peak appears at 2000 eV on the energy scale. In general, the instrument needs to be left on this setting for 10 minutes or so for stability but, even then, the energy will not generally be 2000.0 eV. The spectrometer should be set up on copper or gold and the sample carefully moved until the energy difference of the peaks matches that in Table A1.3, as recorded by the energy DVM. The sample is now in the correct position and the true value of the nominal 2000 eV beam energy may be accurately established. This energy value is then used for all subsequent sample positionings and the instrument is calibrated as given in Table A1.3.

References

1. K. Siegbahn, C. Nordling, A. Fahlman, R. Nordberg, K. Hamrin, J. Hedman, G. Johansson, T. Bergmark, S-E. Karlsson, I. Lindgren and B. Lindberg, *ESCA—Atomic Molecular and Solid State Structure Studied by Means of Electron Spectroscopy,* Almqvist, Uppsala (1967).
2. G. Schon, *J. Electron Spectrosc.,* **1**, 377 (1972).
3. G. Johansson, J. Hedman, A. Berndtsson, M. Klasson and R. Nilsson, *J. Electron Spectrosc.,* **2**, 295 (1973).
4. K. Asami, *J. Electron Spectrosc.,* **9**, 469 (1976).
5. K. Richter and B. Peplinski, *J. Electron Spectrosc.,* **13**, 69 (1978).
6. C. D. Wagner, L. H. Gale and R. H. Raymond, *Anal. Chem.,* **51**, 466 (1979).
7. R. J. Bird and P. Swift, *J. Electron Spectrosc.,* **21**, 277 (1980).
8. J. C. Fuggle and N. Martensson, *J. Electron Spectrosc.,* **21**, 275 (1980).
9. A. Lebugle, U. Axelsson, R. Nyholm and N. Martensson, *Phys. Scrip.,* **23**, 825 (1981).
10. C. J. Powell, N. E. Erickson and T. E. Madey, *J. Electron Spectrosc.,* **17**, 361 (1979).
11. M. T. Anthony and M. P. Seah, (to be published).
12. M. P. Seah and M. T. Anthony, (to be published).
13. M. P. Seah and M. T. Anthony (to be published).
14. C. J. Powell, N. E. Erickson and T. E. Madey, *J. Electron Spectrosc.,* **25**, 87 (1982).
15. M. P. Seah, *Surf. Interface Anal.,* **1**, 91 (1979).
16. O. Keski-Rahkonen and M. O. Krause, *J. Electron Spectrosc.,* **13**, 107 (1978).
17. O. Keski-Rahkonen, *J. Electron Spectrosc.,* **13**, 113 (1978).

Practical Surface Analysis
by Auger and X-ray Photoelectron Spectroscopy
Edited by D. Briggs and M. P. Seah
© 1983, John Wiley & Sons, Ltd.

Appendix 2

Static Charge Referencing Techniques

P. Swift, D. Shuttleworth
*Shell Research Ltd., Thornton Research Centre, PO Box 1, Chester CH1 3SH
UK*

M. P. Seah
National Physical Laboratory, Teddington, Middlesex, UK

A2.1   Introduction

A problem commonly experienced in the interpretation of an XPS spectrum
is that of static charge referencing. Static charging arises as a consequence of
the build-up of a positive charge at the surface of non-conducting specimens
when the atoms lose electrons in the photo-emission process. This positive
charge produces a retarding field in front of the specimen such that the
photo-electrons have a kinetic energy, E_K, lower than that predicted by the
simple equation:

$$E_K = h\nu - E_B \qquad (A2.1)$$

where $h\nu$ is the photon energy of the X-ray source, E_B is the binding energy of
the appropriate core level and E_K is referenced to the Fermi level of the
spectrometer. Conductors in electrical contact with the spectrometer probe
do not exhibit such charging effects because the photo-electrons are replaced
by electrons flowing through the sample. However, for electrically isolated
conductors and insulating materials the rate of photo-electron loss is greater
than that of their replacement from within the specimen, and a charged
surface ensues.

Generally, in spectrometers using an achromatic X-ray source, this charge is
partially neutralized by a background of low-energy electrons (i.e. <5 eV)
produced by the bremsstrahlung X-rays striking the X-ray gun window and
the internal parts of the apparatus. If the surface charges positively, these
low-energy electrons are drawn to the sample so that an equilibrium steady-
state static charge results and the photo-electron peaks occur in the spectrum

with an energy now defined by

$$E_K = hv - E_B - C \qquad (A2.2)$$

where C is the steady-state static charge value, usually a positive number.

The steady-state static charge discussed above may be termed an equilibrium static charge, C, and is generally achieved within a few seconds of the irradiation of the specimen by X-rays from an achromatic source, unless the sample is of a changeable conductivity. The latter effect may arise from damage to a specimen by the X-rays, thereby changing its physical nature (e.g. reduction of an insulating compound to contain metallic sites).

The surface charging problem is accentuated if the X-ray source is monochromatized because of the reduced intensity of the photon flux and the loss of the bremsstrahlung X-rays. This leads to a significant decrease in the number of scattered low-energy electrons in the vicinity of the sample and, consequently, a steady-state static charge is seldom attained. If a monochromatized X-ray source is used it is necessary to direct a controlled beam of low-energy electrons from an electron gun at the specimen surface, as discussed in Section A2.6, and this produces a gun-controlled static charge, C_G.

In addition to the change of kinetic energy of the photo-electrons arising from the equilibrium static charge, spectral peak broadening may be observed for insulating materials. This is because, although the surface may attain an overall static charge at equilibrium with the photon flux, not all of the sites have equivalent C values. This effect is particularly noticeable for heterogeneous specimens where components vary in insulating nature (e.g. minerals) and, in extreme cases, spectral peaks arising from photo-electrons emitted from the same atomic level in different regions of the sample may be split by several electronvolts.

Although static charging can significantly shift the kinetic energy of the photo-electron peaks, it does not necessarily preclude the determination of their true binding energy provided the value of C is established. Several methods are discussed below which may be considered as supplementary or alternative to the use of the Auger parameter discussed in Chapter 3.

A2.2   Adventitious Surface Layers

Unless specimens are prepared for analysis under carefully controlled atmospheres, their surfaces generally suffer from adventitious contamination. Once in the spectrometer, further contamination can occur by the adsorption of residual gases, especially in instruments with oil diffusion pumps. These contamination layers can be used for referencing purposes if it is assumed that they truly reflect the steady-state static charge exhibited by the specimen surface and that they contain an element peak of known binding energy. Carbon is the element which is most commonly detected in contamination layers and the photo-electrons from the C 1s atomic energy level are those most generally adopted for referencing purposes. A binding energy of 284.8 eV is often used for the C 1s level of this contamination and the difference between its measured position in the energy spectrum and the above value gives the charging value, C, of equation (A2.2).

Siegbahn *et al*[1] first reported the presence of a carbon contamination layer and deduced that it originated from pump oil, since the C 1s peak increased in intensity with the exposure of the specimen to the spectrometer vacuum. Subsequently, it had generally been assumed that the C 1s electrons originate from saturated hydrocarbon components, or from carbon atoms in chemical domains of similar electronegativity to that in a saturated hydrocarbon, in the contamination layer. The experimental considerations and reliability of the use of adventitious carbon for energy referencing have been discussed in a review article by Swift.[2]

The main disadvantage of the technique is the uncertainty caused by the spread of values from 284.6 to 285.2 eV reported in the literature[2] for the C 1s electrons. Therefore, it is recommended that if the C 1s electrons from a contamination layer are to be used for referencing, their binding energy should be determined by the user on his own spectrometer. Ideally this determination should be carried out on a substrate closely similar in chemical and physical properties to the material to be analysed uniformly covered by only a thin contamination layer (i.e. of the order of a monolayer).

Factors that can affect the measured C 1s binding energy from adventitious carbon layers, and its subsequent use for energy referencing, include:

(a) The chemical state of the carbon in the contamination layer.[3]
(b) The thickness of the adventitious layer.[4–6]
(c) The chemical and physical nature of the substrate on which the measurements are made.[2,4]
(d) Specimen preparation and surface treatment effects.[4,7,8]
(e) The accuracy of the energy calibration of the spectrometer on which it is determined.[9]

Despite the apparent limitations and uncertainties associated with the use of adventitious carbon for static-charge referencing, it is the most convenient and commonly applied technique. However, for the reasons listed above and discussed in the review article by Swift,[2] the interpretation of binding energies must be regarded with caution. Furthermore, if results from different laboratories are to be compared, the absolute energy calibration of the spectrometers must be defined and a reliable reference binding energy (e.g. Au $4f_{7/2}$) reported. This information, together with the C 1s electron binding energy used and the substrate on which it was measured, should provide a reasonable basis to allow a meaningful comparison of results to be made.

A2.3   Deposited Surface Layers

In cases where a suitable adventitious overlayer is not present it is necessary
to introduce a calibrant by deliberate deposition. In general, organic compounds and metals are used for the calibrating species as they can often be
reliably condensed onto surfaces under UHV conditions. Unknown calibrants, such as pump oils, can be used if they re-evaporate without decomposition in UHV.

Care should be taken to ensure that the calibrant neither decomposes nor
reacts with the substrate upon condensation and that the binding energy
observed for the reference material, which may deposit either as a thin overlayer or in islands, is characteristic of the bulk. This can be done by taking
spectra as a function of time and amount of calibrant deposited. It may also be
advantageous to characterize the form of the deposition (i.e. as uniform overlayer or islands) by carrying out angular studies. This, of course, is not applicable to powder samples.

The deliberate deposition of oil vapour in diffusion pumped systems is frequently used for calibration purposes since it may be accomplished by allowing the liquid nitrogen cold traps to partially warm up. Alternatively,[10] in
certain types of spectrometer the cooled X-ray source window may be
allowed to warm to release condensed adventitious species directly onto the
specimen. In both cases condensation may be accelerated by cooling the
sample. When employing this method care should be exercised to avoid the
'memory effect' from previously introduced volatile material, which is not
intended to be used as a calibrant, condensing onto the sample and possibly
leading to an erroneous reference.

Another method employing organic materials as calibrants involves their
introduction from outside the vacuum chamber, as described by Connor.[11]
This little-used method allows the organic compound chosen to be compatible
with the sample under investigation, for instance in regard to wettability and
freedom from spectral overlap of core-level lines.

Metals are also frequently employed as calibrants in XPS and, because of
their general inertness and ease of evaporation, the noble metals are most
often used. In practice about a monolayer of, for example, gold is deposited,
preferably *in situ*. The method is one of the few charge-correction methods to
receive detailed study,[12–15] where it has been demonstrated[12,14,15] that the
conditions necessary to obtain electrical equilibrium, and therefore accurate
charge correction, depend on the surface coverage. An optimum thickness of
gold exists for effective charge correction, which for polyethylene and PTFE
is approximately 0.6 nm.[15] Such thicknesses should be regarded as equivalent
thicknesses since it is known[16] that gold deposits in the form of islands at these
low coverages. Care must be exercised here since there are indications that
small binding energy shifts ($\leqslant 0.25$ eV) for the Au $4f_{7/2}$ peak occurs at low

coverages.[17] Also, with certain types of substrate, notably halides and cyanides, reaction with the gold has been reported.[18-21] However, when the appropriate precautions are taken[13-15] the gold decoration technique may often be used to obtain charge-corrected binding energies using the calibrated binding energy of 84.0 eV for the Au $4f_{7/2}$ line.[9]

Should, of course, one be dealing with a sample that may safely evaporate without decomposition in UHV then referencing becomes relatively straight-forward. In this case a suitably thin layer of sample is first characterized by condensation onto a clean metal substrate (e.g. gold) and referenced by using a known core-level binding energy of the substrate. The reliability of the electrical contact should be verified by accumulating spectra at various sample thicknesses and by ensuring that all peaks in the spectrum shift by the same amount when an electrical bias is applied.

A2.4   Mixtures

The possibility of carrying out static charge energy referencing for insulating materials, by preparing finely ground mixtures with powders containing a component of known binding energy, has been investigated. Powdered graphite has been the most commonly used reference material for this technique,[4,22-24] whilst lithium fluoride,[25] potassium salts,[26,27] triplumbic tetroxide,[28] molybdenum trioxide[29,30] and gold[30] have also been considered. Generally, the experimental difficulty of obtaining intimate homogeneous mixtures (i.e. approaching atomic proportions) has been regarded as the main limitation of this technique.

Factors which must be considered when interpreting the data include:

(a) The particle size of the powders, which can affect the intimacy of the mixture.
(b) Particle–interface interaction between different components.
(c) Differential charging of the particles of the different components.

Results for mixed powders are usually very poor and Wagner[31] has recently shown that, for some systems, errors of up to 10 eV can be produced.

It is possible to achieve mixing at a molecular scale for a limited range of materials by fusion of a calibrant with an unknown compound; however, the data will rarely be characteristic of the original compounds. The co-condensation of a volatile calibrant with an unknown compound is also a method of energy referencing that has been extensively used by Connor,[11] who obtained good internal consistency leading to binding energies with a reproducibility of ±0.2 eV. The results were found to be independent of the relative proportions of the calibrant and unknown compound co-condensed and also of the specific calibrant molecule used.

Of the three mixture referencing techniques discussed above, the last has been shown to be the most reliable. However, this method is limited to stable volatile compounds. In general, the use of mixture techniques is not recommended as a means of energy referencing, either with or without static charge correction considerations.

A2.5   Internal Standards

Perhaps the most reliable method of spectral referencing lies in the use of an internal standard[32] which depends upon the invariance of the binding energy of a chosen chemical grouping in different molecules. It is attractive because the referencing specie is now 'locked' into the unknown material and must reflect the static charge of the system if it is uniform over the surface.

The method has found widespread use in the study of carbon-containing material[33] and, in particular, polymer systems, where it has found most reliable application in the study of chemical shift. The use of this method, of course, requires a knowledge of the molecular formula of the material being studied. The binding energy of the chemical grouping employed must be known to be invariant in a range of environments by some independent method (e.g. preparation of thin conducting films).

A novel variation on this method is in the use of bulk solvent as reference in quick-frozen solutions.[34] Although this is of limited application it is nonetheless valuable.

A2.6   Low-energy Electron Flood Gun

Low-energy electron flood guns may be used to stabilize the static charging of insulators examined by XPS[35] and, in particular, when monochromatized X-rays are employed. Recent work [36,37] into the design and operation of flood guns has shown that the best results are achieved with an electron beam of a low energy (<1 eV), with respect to the vacuum chamber at earth, from a source very close to the specimen. Optimum operating conditions, e.g. sample position, exist for particular configurations and must, in general, be determined by the user. Low electron energies should be used to maximize the neutralization effect and reduce the number of electron bombardment induced reactions.

A low-energy electron flood gun is potentially of use for three types of XPS application; these are:

(a) Studies using monochromatized X-rays where charging shifts for insulators are usually very large and not corrigible by the methods discussed previously. Flood guns compensate for the loss of bremsstrahlung-excited secondary electrons in changing from an achromatic to a monochromatic X-ray source. Operation in this manner normally overcomes the major problem but still leaves a small correction to be made using a method outlined in previous sections.

(b) Absolute calibration of spectra may be achieved by using a flood gun suitably calibrated for a particular instrument configuration.[36] In this way the value of C in equation (A2.2) can be reduced or even made negative. C_G, defined earlier in relation to equation (A2.2), is equal to the cathode voltage minus a gun constant. With a calibrated gun and for smooth samples the surface potential can be set to ± 0.01 eV, allowing a precise definition of binding energies. At the present time the design of flood guns does not enable this accuracy to be achieved with the type of powdered samples often studied with XPS. Here the absolute accuracy is reduced to about ± 0.50 eV.

(c) In situations where static charging is not homogeneous across the sample surface, peak broadening may occur and the electron flood gun can be used to minimize or even eliminate this broadening.[38]

Of the five energy referencing methods discussed in this appendix, the low-energy electron flood gun technique shows the greatest potential because of its applicability to all types of sample encountered in XPS.

References

1. K. Siegbahn, C. Nordling, A. Fahlman, R. Nordberg, K. Hamrin, J. Hedman, G. Johansson, T. Bergmark, S. Karlsson, I. Lindgren and B. Lindberg, *ESCA, Nova Acta Regiae, Soc. Sci. Ups.*, **IV**, 20 (1967).
2. P. Swift, *Surf. Interface Anal.*, **4**, 47 (1982).
3. E. S. Brandt. D. F. Untereker, C. N. Reilley and R. W. Murray, *J. Electron Spectrosc.*, **14**, 113 (1978).
4. G. Johansson, J. Hedman, A. Bendtsson, M. Klasson and R. Nilsson, *J. Electron Spectrosc.*, **2**, 295 (1973).
5. S. Kinoshita, T. Ohta and H. Kuroda, *Bull. Chem. Soc. Jpn,* **49**, 149 (1976).
6. D. T. Clark, A. Dilks and H. R. Thomas, *J. Polym. Sci., Polym Chem. Ed.*, **16**, 1461 (1978).
7. C. D. Wagner, *Applied Surface Analysis*, ASTM STP 699, American Society for Testing and Materials, Philadelphia (1980).
8. A. Jaegle, A. Kait, G. Nanse and J. C. Peruchetti, *Analysis,* **9**, 252 (1981).
9. R. J. Bird and P. Swift, *J. Electron Spectrosc.*, **21**, 227 (1980).
10. D. T. Clark, H. R. Thomas, A. Dilks and D. Shuttleworth, *J. Electron Spectrosc.*, **40**, 455 (1977).
11. J. A. Connor, in *Handbook of X-ray and Ultraviolet Photo-electron Spectroscopy* (Ed. D. Briggs), Chap. 5, p. 183, Heyden, London (1977).
12. D. J. Hnatowich, J. Hudis, M. L. Perlman and R. C. Ragnini, *J. Appl. Phys.*, **42**, 4883 (1971).
13. M. F. Ebel and H. Ebel, *J. Electron Spectrosc.*, **3**, 169 (1974).
14. C. R. Grinnard and H. M. Riggs, *Anal. Chem.*, **46**, 1306 (1974).
15. Y. Uwamine, T. Ishizuka and H. Yamatera, *J. Electron Spectrosc.*, **23**, 55 (1981).

16. D. W. Pashley, *Adv. Phys.*, **14**, 327 (1965).
17. I. Adams, Ph.D. Thesis, p. 137, University College of Wales, Aberystwyth (1972).
18. D. Betteridge, J. C. Connor and D. M. Hercules, *Electron Spectrosc.*, **2**, 327 (1973).
19. D. S. Urch and M. Webber, *J. Electron Spectrosc.*, **5**, 791 (1974).
20. L. I. Matiengo and S. O. Grim, *Anal. Chem.*, **46**, 2052 (1974).
21. V. I. Nefedov, Va. V. Salyn, G. Leonhardt and R. Scheibe, *J. Electron Spectrosc.*, **10**, 121 (1977).
22. R. Nordberg, H. Brecht, R. G. Albridge, A. Fahlman and J. R. Van Wazer, *Inorg. Chem.*, **9**, 2469 (1970).
23. V. I. Nefedov and M. M. Kakhana, *Zh. Anal. Khim.*, **27**, 2049 (1972).
24. G. Kumar, J. R. Blackburn, R. G. Albridge, W. E. Moddeman and M. M. Jones, *Inorg. Chem.*, **11**, 296 (1972).
25. W. Bremser and F. Linnemann, *Chem.*, **1971**, Fig. 95, 1011 (1971).
26. J. J. Jack and D. M. Hercules, *Anal. Chem.*, **43**, 729 (1971).
27. L. D. Hulett and T. A. Carlson, *Appl. Spectrosc.*, **25**, 33 (1971).
28. W. J. Stec, W. E. Morgan, R. G. Albridge, A. Fahlman and J. R. Van Wazer, *Inorg. Chem.*, **11**, 219 (1972).
29. W. E. Schwartz, P. H. Watts, J. P. Watts, J. W. Brasch and E. R. Lippincott, *Anal. Chem.*, **44**, 2001 (1972).
30. W. P. Dianis and J. E. Lester, *Anal. Chem.*, **45**, 1416 (1973).
31. C. D. Wagner, *J. Electron. Spectrosc.*, **18**, 345 (1980).
32. J. J. Ogilvie and A. Wolburg, *Appl. Spectrosc.*, **26**, 401 (1972).
33. D. T. Clark and H. R. Thomas, *J. Polym. Sci., Polym. Chem. Ed.*, **14**, 1671 (1976).
34. K. Burger, *J. Electron Spectrosc.*, **14**, 405 (1978).
35. D. A. Huchital and R. T. McKeon, *Appl. Phys. Lett.*, **20**, 158 (1972).
36. C. P. Hunt, C. T. H. Stoddart and M. P. Seah, *Surf. Interface Anal.*, **3**, 157 (1981).
37. C. P. Hunt, M. T. Antony, C. T. H. Stoddart and M. P. Seah, NPL Chem. Report 108, March 1980.
38. H. Windawi, *J. Electron Spectrosc.*, **22**, 373 (1981).

Practical Surface Analysis
by Auger and X-ray Photoelectron Spectroscopy
Edited by D. Briggs and M. P. Seah
© 1983, John Wiley & Sons, Ltd.

Appendix 3

Data Analysis in X-ray Photoelectron Spectroscopy

P. M. A. Sherwood

*Department of Inorganic Chemistry, The University,
Newcastle-upon-Tyne NE1 7RU, UK*

A3.1 Introduction

There is an increasing interest in the collection of data of high quality which can then be analysed by accurate data analysis methods. This has arisen because there is a growing awareness of the importance of being able to distinguish small changes in photo-electron spectra, and the availability of cheap microcomputers allows automation of data collection and many digital methods of data analysis to be applied without having to resort to large and expensive mainframe computers. This appendix will attempt to review the range of data-collection and analysis techniques, and provide examples from the work of the author which illustrate the application of the methods described.

A3.2 Data-collection Systems

The majority of X-ray photo-electron spectroscopic (XPS) data is still collected using analogue ratemeters and chart recorders. Analogue data are subject to problems of uncertainty in peak position and peak area due to dependence upon the instrument time constant and scanning speed. Analogue data give almost no idea of data quality due to the effective smoothing due to the choice of a high time constant. If smoothing of data is to be carried out it is much better to smooth digital data in a controlled way (to be discussed below) than to rely upon the correct selection of the time constant and scanning speed by the operator. Once the digital data have been collected it can be treated by various methods, but once analogue data have been collected it cannot be modified since it has already been altered by the analogue collection system. In addition the data analysis is often carried out using analogue methods such as curve fitting using a Du Pont curve resolver. This approach is highly subjective and, thus, often misleading. One of the most serious limitations of analogue data is that it prevents data accumulation to improve the signal-to-noise (S/N) levels in a spectrum.

Most commercial spectrometers can be purchased with a data system to control the spectrometer and to carry out some data analysis. Such data analysis systems may be fairly expensive, but the availability of cheap microcomputers allows data systems to be provided at a substantially reduced cost. The microcomputers have the advantage that even though they are generally single-task machines their cost allows one to have a number of them performing particular tasks. In addition, the programs can always be adjusted to accommodate the latest developments and improvements in data analysis and collection which would not be possible with dedicated apparatus such as multichannel analysers.

The way in which a microcomputer-based data analysis system might be constructed is illustrated in Figure A3.1, which shows the way in which micro- and mainframe computers are associated with the spectrometer in the author's laboratory. This system uses two microcomputers and five terminals associated with the mainframe computer. Printers and a plotter are all linked together through a local distribution board, which allows devices such as the printers to be shared by the microcomputers and the mainframe computer

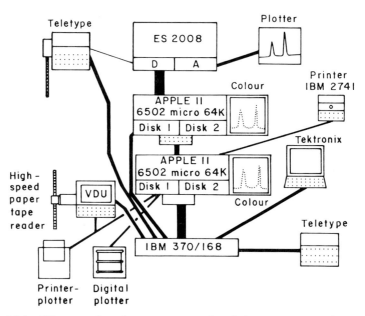

Figure A3.1   Diagram of a microcomputer and mainframe computer data system

terminals, and the microcomputers and the terminals to talk to the mainframe computer through six hard-wired lines.

The data are collected simultaneously by an analogue ratemeter (A) and a digital ratemeter(D). The analogue ratemeter gives an X–Y analogue plot which is useful for collection of an overall spectrum and for monitoring any changes in the spectrum with time (which might be caused by decomposing samples or a drifting sample potential). The digital ratemeter can also be used to monitor any changes in the spectrum with time by an additional output which allows data, when required, to be punched out on paper tape simultaneously with the main output which is transferred to the Apple II microcomputer system.

The Apple II microcomputer system has 64K of random access memory available which gives good space for complex data-collection programs with full graphics facilities. This micro is programmed to communicate with the operator to ensure that the machine is being operated in the correct manner, as well as to collect all the relevant spectral information. The program is so designed that an inexperienced operator is provided with a considerable amount of information on an 80×24 upper and lower case screen. The experienced operator can ignore the information since it is printed at a very fast rate. The spectrum is then displayed while it is collected on an additional graphics screen displayed simultaneously with the text screen. This allows simultaneous display of the program instructions with the generation of graphics, allowing instruction manuals to be eliminated and making it possible to easily construct self-teaching programs. This is of particular value when the microcomputer system is being used to analyse the data. The data are stored on floppy disc, which can be transferred to another 64K Apple II system for data analysis, while the first Apple II system is used to collect another spectrum, thus eliminating any dead time.

The spectra are stored permanently on floppy disk (a 5.25 inch floppy disk can store approximately seventy reasonable length spectra), but are also transferred to the mainframe computer where they are stored on magnetic tape. The mainframe computer is used for data analysis and for large calculations (such as molecular orbital calculations) for data interpretation. Mainframe computing power is needed for some data analysis methods, but the microcomputers are being increasingly used for this purpose in the author's laboratory.

A3.3   Simple Data Operations

Many data systems control most of the spectrometer functions, such as starting kinetic energy, range, number of scans, scanning to a predetermined S/N ratio, X-ray gun voltage and current settings, and the operation of an argon ion gun. The data collection may be displayed as it is collected or after the spectrum is terminated. Some caution has to be exercised in the way in which the data collection is carried out, since some manufacturers pre-treat the data (by smoothing and background subtraction) as an automatic part of their data system. This is a very regrettable practice since it should be clear from this appendix that there is no 'correct' way for carrying out such operations, and thus this practice can lead to misleading results. In general, then, all the data treatment methods must be treated with caution, and the original untreated data should always be retained as a permanent record.

In particular, some very simple operations can be carried out, and these are discussed below. All these operations can easily be carried out on a microcomputer system. Such a system has the value that the altered spectrum can be instantly displayed after the process has been applied.

A3.3.1 Spectral display and expansion

The display is generally made as a series of points representing each digital data point collected for a particular time at a specific kinetic energy, or sometimes at a particular time over a small lineraly varied kinetic energy range. This dot display of points is in the author's view to be preferred over a display which links the points by a line to give a continuous spectrum. The spectrum often has cross wires superimposed upon it so that individual points in the spectrum can be identified in terms of their x (kinetic energy or binding energy) and y (counts) position which is generally printed out at the bottom of the spectrum. Simple x and y expansion of the scale is generally made available. A facility to normalize the spectrum so that the point with maximum counts is at the graphics display maximum is generally useful, especially if there is a facility to remove part of a spectrum in order to concentrate upon particular spectral features.

A3.3.2 Integration and area measurement

Some facility for area measurement is useful, and this generally takes the form of a facility where the area between two points (identified by positioning the cross wires at these points) is calculated as the area between a line joining the points and the spectrum. The whole spectrum is sometimes integrated by adding all the data points together (after the removal of some horizontal background value) and displaying the integrated curve as the total sum at that point counting from left to right.

A3.3.3 Spike removal

Spikes can sometimes be generated in a spectrum as a result of power surges or the switching on of apparatus, and a facility to remove a particular point and replace it by another point is valuable. Obviously this process should only be applied to a limited number of points!

It is very important to remove spikes before further data analysis is carried out since only one or two such spurious points can very seriously effect data analysis operations such as smoothing and non-linear least squares curve fitting.

A3.3.4 Satellite subtraction

Some commercial data systems provide for the X-ray radiation satellites to be removed, and even claim that such an approach is as good as using an X-ray monochromator, which of course removes such satellites. Satellite subtraction must be carried out with great care. Firstly, the background in the region of interest must be accurately subtracted, and it will be seen below that this is by no means an operation that can be performed with complete accuracy. Secondly, any Auger peaks present in the region of interest must be identified (since these, of course, will have no X-ray satellite contribution) and their spectral contribution accurately calculated, so that this intensity can be removed from the intensity that contributes to the X-ray radiation satellite intensity. Thirdly, any satellites due to other peaks outside the region of interest must be identified, e.g. X-ray satellites, satellites of Al $K\alpha_{12}$ radiation from magnesium anode systems with aluminium X-rays windows and cross-over in two anode systems, and discrete energy loss satellites from photo-electron peaks at a higher binding energy than the region of interest (see Chapter 3).

A3.3.5 Baseline removal

The removal of the background of a spectrum is a non-trivial operation which will be discussed below. The removal of a horrizontal baseline is a trivial operation, however, which is a convenient method for the visual improvement of a spectrum which consists of small spectroscopic changes superimposed over a large background.

A3.3.6 Addition and subtraction of spectra

For comparative purposes two spectra may be added together or substracted one from the other. Many data systems provide such a facility for direct addition and subtraction. If the subtraction technique is to be used in a more thorough manner the subtraction technique must be carried out in a non-trivial manner, such methods being discussed below.

A3.3.7 Peak maximum location

A procedure to locate the maximum count, i.e. peak maximum, in a spectrum is easily done. Such a procedure generally allows a genuine peak to be distinguished from a stray point.

A3.4 Smoothing

Nearly all data systems provide a facility to smooth the data. Such an operation must be carried out with care and this section is concerned with the effect that smoothing has upon data and suggests how smoothing might be carried out in the most effective and least distorting way. In general it is always much better to use unsmoothed data, since any smoothing procedure is bound to introduce some sort of distortion to the spectra. Nevertheless, there are many cases where some smoothing is required, such as when very weak signals are recorded with a prohibitively long time needed to record a spectrum with a good signal-to-noise ratio or when a sample is decomposing and it is necessary to record a spectrum over a very small period of time. In addition it is often necessary to smooth data with a fairly good signal-to-noise ratio as a prerequisite to further analysis involving processes such as difference or derivative spectra.

Smoothing is a process that attempts to increase the correlation between points while suppressing uncorrelated noise. Smoothing is achieved by convolution of the data with a suitable smoothing function in an appropriate way. Two different types of smoothing process can be used, namely the least squares central point smoothing techniques proposed by Savitsky and Golay[1] and related methods, and the Fourier transform approach. The first method is the most common, and in its simplest form is found on most data systems.

A3.4.1 Least squares approach

This method involves convoluting the data with a convoluting function such that

$$y(0) = \sum_{t=-m}^{m} \frac{C_t f(t)}{\text{NORM}} \tag{A3.1}$$

where $y(0)$ is the smoothed point, which corresponds to the centre of an odd number interval of points, P $(p = 2m + 1)$; $f(t)$ are experimental data points in the given interval, which are convoluted with the appropriate integers, C_t; and NORM is a normalizing factor.

In the simplest case, that of a moving average, $C_t = 1$ for all t and NORM = P. A least squares fit to the data over the interval P using a polynomial of degree n can be achieved exactly by using the above formula since discrete integers, C_i, exist for distinct values of P and n. Hildebrand[2] and Proctor and Sherwood[3] quote a general formula for the necessary set of polynomials p.

In general a number of points concerning the use of this type of smoothing can be identified:[3]

(a) Smoothing increases with the number of points used in the smoothing interval. The optimum value for the smoothing interval is 0.7 (peak width at half maximum).

(b) Smoothing decreases as the degree of the smoothing polynomial increases. In general, fits to quadratic or cubic polynomials are more effective than those to higher order polynomials.

(c) Normal methods of smoothing cause a loss of points at each end of the spectrum. If the smoothing interval is of N points then the number of points lost from each end of the spectrum is $(N-1)/2$; thus the effect becomes more serious as the smoothing interval is increased. This effect can be eliminated by estimating the points lost by a suitable equation (equation 8 in Ref. 3). This allows the smoothing operation to be repeated as many times as desired.

(d) Smoothing can be repeated (especially when no points are lost when carried out as described in point c) in order to increase the amount of smoothing, though the largest amount of smoothing occurs during the first passage through the smoothing process. Repeat smoothing has the effect of generating a new smoothing polynomial which gives more emphasis to positive values, is broader than the original, has a height that falls as the function gets broader and causes ripples to occur at the extreme ends of the functions. This wide range of smoothing functions makes repeat smoothing a valuable process.[3] It is found that the smallest interval possible, repeated as many times as possible, is a good way to perform such a repeat smoothing operation, though the choice of smoothing interval and repeat number is not critical provided that the interval chosen is not too large with respect to the peak width.

Figure A3.2 and Table A3.1 show how the methods described above can be successfully applied to a typical core photo-electron spectrum. Thus a noisy O $1s$ spectrum (which can be fitted to three peaks) can be smoothed by a twenty-one-point interval repeated a hundred times (21 is $0.7 \times$ peak width at half maximum), and the result compared with the same spectrum run for a long period of time. The smoothed spectrum contains no new information, of course, since smoothing is a cosmetic process, but the smoothed spectrum makes some spectral features more evident to the observer and so is more useful than the unsmoothed spectrum. Thus the smallest of the three peaks is more clearly present visually and in terms of the fitted information (Table

(a)

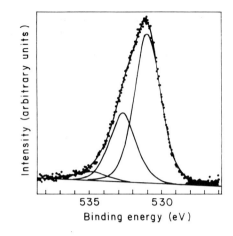

(b)

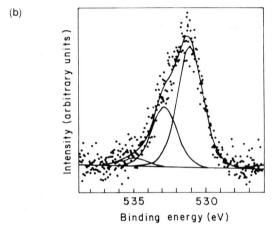

(c)

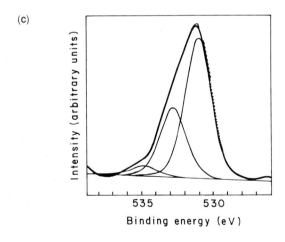

Table A3.1   Non-linear least squares curve fitting results for and O 1s spectrum obtained with good statistics, poor statistics, and poor statistics smoothed

Parameter	Good statistics (Figure A3.2a)	Poor statistics (Figure A3.2b)	Poor statistics smoothed (Figure A3.2c)
Binding	534.81(16)	534.79(76)	534.73(12)
Energy, eV	532.66(04)	532.74(18)	532.71(03)
	530.95(02)	530.99(08)	530.99(01)
FWHM, eV	2.10(04)	2.04(15)	2.10(03)
Area ratio	1/7.77/16.90	1/7.44/14.94	1/7.08/14.34
Chi-squared	584.4	15636.9	321.97

Note. Figures in brackets refer to 95 per cent confidence limits.

A3.1) in the smoothed spectrum. Thus the chi-squared value (*vide infra*) falls to 42 per cent. of the original value when the third peak is added to the smoothed spectrum, but to only 97 per cent. of the original value in the unsmoothed spectrum.

A3.4.2   Fourier transform approach

Smoothing can be carried out by a very different approach which involves the use of Fourier transform analysis.[4-7] Fourier transformation of a spectral array containing noise results in a spatial frequency distribution for both the signal and the noise. It is possible to truncate the noise contribution in the spatial frequency distribution by multiplication with a suitable weighting function whereupon subsequent transformation back will result in a smoothed spectrum.

The problem with this approach is that a suitable weighting function must be chosen. Various approaches have been used[7-11] but much depends upon the degree of smoothing desired. An optimum S/N filter has been suggested by Turin,[11] but as the S/N ratio is increased so the peak resolution decreases.

Any smoothing methods involves many subjective factors and all methods must be used with great caution.

Figure A3.2   Oxygen 1s spectra illustrating the effect of smoothing. In all cases the spectra are fitted to three peaks, the details being given in Table A3.1. (a) Data for the same material obtained with good statistics by running the spectrum for a long period, (b) data for the same material with bad statistics by running the spectrum for a short period and (c) the result of smoothing the spectrum (b) 100 times with a smoothing interval of 21

A3.5   The Analysis of Overlapping Spectral Features

In many cases the information provided by photo-electron spectroscopy is contained in a spectrum that consists of a number of overlapping peaks, often of different peak shapes and intensities. In the core region the peaks will be a series of chemically shifted peaks, satellites, energy loss features and Auger peaks. In the valence region the spectra will reflect all the complex features of the ground-state valence band, together with complications due to the need to include the excited state (i.e. joint density of states) at low photon energies. In both cases no spectrum can be unambiguously analysed and there is no definite way to proceed with the analysis. Experimentally it is best to employ a monochromatized photon source, especially in the X-ray region, but this is not always available, and in any case the substantial loss of intensity may lead to unacceptable spectral collection times. There are two main ways to try to unscramble this information, namely deconvolution and curve fitting. Neither method can give a unique solution. Most attention has been given to curve fitting, though deconvolution has been applied with success to a number of spectra. Curve fitting is very dependent upon the initial 'guess', i.e. the number of peaks and their peak parameters, and deconvolution can provide valuable assistance in making the most suitable initial guess. Suitable initial guesses can also be assisted by derivative methods and curve synthesis. Certainly the application of as many approaches as possible will provide the best chance of a reasonable spectral interpretation.

A3.5.1   Deconvolution methods

A photo-electron spectrum has an observed spectral width that includes a broadening due to instrumental factors. These factors include the resolving power of the spectrometer and the line width of the photon source used (see Chapter 3). If one knew the exact contribution due to these factors and could construct a function (B) that described them, then it would be known that the observed spectrum was a convolution of the 'true' spectrum (f_t) and this instrumental function. The 'true' spectrum could then be obtained by the deconvolution of the observed spectrun (f_o) that arises from the way in which the experiment is constructed. The observed spectrum is related to the desired 'true' spectrum by a 'convolution' equation:

$$f_o = f_t B$$

The retrieval of f_t from f_o is referred to as 'deconvolution'. Random high-frequency noise leads to the addition of a noise term to equation (A3.2) which accounts for noise present in all real data to some extent. This noise term means that the deconvolution process is not unique, which has the effect of possibly introducing noise and spurious oscillations into the deconvoluted spectrum. This serves to highlight the fact that such resolution improvement or enhancement is only achieved at the expense of the signal-to-noise (S/N) ratio in the same way that improvement in S/N ratio by smoothing is generally accompanied by a loss of resolution.[12,13]

Several methods have been used to deconvolute the effects of data distortion. These include differentiation techniques (*vide infra*). However, the three main methods used are:

(a) Direct solution of the convolution equation.
(b) Fourier transform analysis.
(c) Iterative techniques.

It is found that the success of the process depends less upon the method of deconvoltion chosen than upon the quality of the initial data. Data with very good statistics are an important starting point.

The whole subject has been very effectively reviewed by Carley and Joyner[14] and the reader is strongly advised to read this paper for a full discussion of this subject.

A3.5.2   Derivative spectra

Derivative spectra provide a useful method for peak location in appropriate cases. A number of workers[15-21] have pointed out the usefulness of derivative spectra. Automatic peak searching routines, based upon the properties of ideal derivatives, have been suggested, but in XPS these are only of limited use. Second-derivative spectra have been used[15,16] as a quick and easy method for the prediction of peak positions as well as providing useful information in themselves. In second-derivative spectra negative peaks occur corresponding approximately to overlapping peak positions in the original spectrum.

Derivative spectra cannot give accurate peak positions since one is examining the envelope of overlapping peaks so that the observed maximum of each peak is always shifted by the presence of other peaks. Nevertheless derivative spectra provide useful information. In general the most accurate peak position is observed when the differentiating convoluting interval[15] is similar to the FWHM of the component peaks, though there is a need to balance convoluting interval with resolution.

Figure A3.3 illustrates how second-derivative spectra can be used to predict the position and relative intensity in a simple two-component test spectrum based upon two overlapping O 1*s* peaks. The second-derivative spectrum shows two peaks of varying intensity, corresponding well to the amounts of the two overlapping peaks in the original spectrum. In addition to slight errors in peak position due to the factors discussed above the relative intensity of the two peaks is only approximate. This is due to the cancellation

Original spectrum   Second derivative

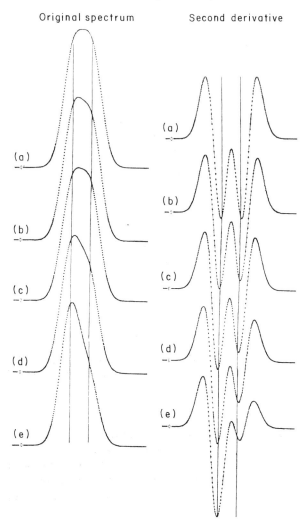

Figure A3.3 Second-derivative spectra of original model spectra consisting of two overlapping peaks of varying intensity: (a) 1 : 1, (b) 1 : 0.95, (c) 1 : 0.90, (d) 1 : 0.75, (e) 1 : 0.5. Convoluting interval = 21, FWHM = 60, separation = 50 channels

effects that occur when the positive lobe of one component peak overlaps with the negative lobe of the other peak. In general this effect will depend upon the number of component peaks, their widths and intensities.

Figure A3.5 compares the smoothed second derivative of an O 1s spec- a carbon fibre (Figure A3.4a) which consists of overlapping peaks. The ratio $Q = A/B$ gives some indication of the asymmetry of the C 1s peak, in the same way that measurement of peak width could indicate peak asymmetry.

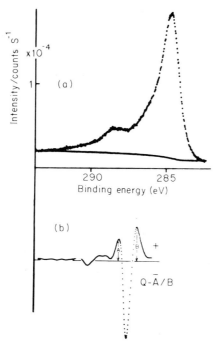

Figure A3.4   Second-derivative of a C 1s spectrum of a carbon fibre. (a) original spectrum, (b) second-derivative spectrum showing the way in which the Q factor is evaluated

However, accurate determination of Q is much simpler than any such width measurement, once the second derivative is obtained.

Figure A3.5 compares the smoothed second derivative of an O 1s spectrum with the curve fitted spectrum clearly illustrating the value of second-derivative spectra in providing a suitable guess to peak positions that can then be used in curve fitting.

Two relatively simple methods for calculating derivative spectra are available.

A3.5.2.1   Fourier methods

In this approach,[5,22] if $y(x)$ has the Fourier transform $F(s)$, then dy/dx has the Fourier transform $(i2\pi s)F(s)$. Extending this to the general nth derivative:

$$y(x) \rightleftarrows F(s) \qquad \frac{d^n y}{dx^n} \rightleftarrows (i2\pi s)^n F(s) \qquad (A3.3)$$

Thus, by applying the relevant weighting function in the Fourier domain,

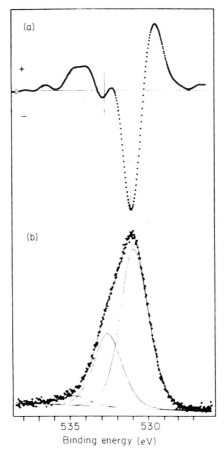

Figure A3.5   Comparison of peak prediction in an O 1s spectrum by means of (a) a second-derivative spectrum and (b) a non-linear least squares curve fitting to the original spectrum

the subsequent transformation will result in the generation of the nth derivative of the original data. However, because more weight is given to higher spectral frequencies (large S) the resultant derivatives will have increased noise associated with them.

A3.5.2.2   *Polynomial methods*

The derivative of a spectral point is simply the derivative of an initially fitted nth-order polynomial at that point. Thus the whole derivative spectrum can be obtained by sliding convolution processes exactly analogous to the extended least squares smoothing approach.[3] The examples discussed here have all been calculated using this method.

As in the Fourier transform case, differentiation using this method also enhances noise characteristics[15] and so even spectra with high S/N characteristics require some prior smoothing.

Repeating the derivative process does produce a fairly satisfactory higher derivative (i.e. first-derivative procedure repeated n times gives an approximation to the nth derivative) and so initial convoluting functions of greater complexity than first or second need not be known. However, practical use of derivatives greater than $n = 2$ is very limited.

A3.5.3   Curve synthesis

A spectrum can be synthesized by using digital or analogue methods to sum a series of functions representing individual peaks in order to produce a final function that closely represents the experimental spectrum. The peak function is generally designed to be a function of appropriate peak variables such as position, intensity, width, function type and peak tail characteristics. Simple curve synthesis can readily be carried out on a microcomputer system or an analogue curve-fitting device such as a Du Pont curve analyser. The microcomputer system has the advantage that a wider range of functions can be used, and the result of the fit can be rapidly displayed graphically together with the appropriate statistical information to allow the quality of the fit to be evaluated. Such a curve synthesis provides a useful initial guess for the more thorough process of non-linear least squares curve fitting which is described below.

A number of types of function have been used for this purpose, the most common being Gaussian or Lorentzian functions. In a work of the author and others[23] a mixed Gaussian/Lorentzian function is used, the one found to be most effective in practice being a product function:

$$f(x) = \frac{\text{peak height}}{[1 + M(x - x_0)^2/\beta^2]\exp\{(1 - M)[\ln 2(x - x_0)^2]/\beta^2\}} \quad \text{(A3.4)}$$

where x_0 is the peak centre and β is a parameter that is nearly 0.5 (FWHM). The actual FWHM is calculated from β using an iterative method. M is the mixing ratio and takes the value 1 for a pure Lorentzian peak and the value 0 for a pure Gaussian peak.

Curve fitting of this type assumes that a particular peak profile (Gaussian or Lorentzian) is uniquely characterized once its peak width at half maximum (PWHM) has been fixed, and cannot be resolved into subcomponents. In 1968 Perram[24] showed that a single Gaussian profile ($y = \exp(-x^2/2.05)$) could be represented by the sum of two separate Gaussian profiles ($y = \pm 0.515359 \exp[-(x^{\pm}0.244622)^2/2(0.988937)^2]$), suggesting that the numerical decomposition of a structureless contour is not unique. However Baruya and Maddams[25] have shown that this result is atypical since it relates to two component peaks of equal intensity and half width, symmetrically separated by −10 per cent of their half widths on either side of the initial Gaussian peak. They conclude that in most practical situations, Gaussian and Lorentzian profiles are unique and curve fitting may be undertaken. The scope and limitations of curve fitting in general have been discussed by Maddams.[13]

A3.5.3.1 Chi-square (X^2)

The quality of the curve fit obtained can usefully be evaluated by evaluation of the weighted χ^2, which is defined as

$$\chi^2 = \sum_{r=1}^{n} w_r [y_r - f(x_r/q)]^2 \qquad (A3.5)$$

where y_r is the observed count at $x = x_r$, $F(x/q)$ is the fitted peak envelope, k is the total number of points in the spectrum and w_r is a weighting function which in this case is chosen as y_r^{-1}, thus making χ^2 equal to the χ^2 statistic.[26-28]

A3.5.3.2 Goodness of fit

The value of χ^2, or more particularly $\Delta\chi^2$ from one fit to another on the same data, provides the statistical information about the 'goodness of fit'.

Strictly speaking, only χ^2 values with the same number of degrees of freedom f ($f = k$ − number of free parameters) can be directly compared. Athough f may vary slightly from fit to fit this is generally insignificant. Also by the very nature of χ^2, the larger the spectral intensity, the larger the associated χ^2 values for a similar quality fit. Thus, although a change in χ^2 from fit to fit for the same data is meaningful, comparison of χ^2 values for different spectra is meaningless unless the peak intensities are equal.

A3.5.3.3 Tail information

Three tail parameters have been used by the author, namely a constant tail ratio(CT), an exponential tail ratio (ET) and the tail mixing ratio (TM. The tail function (T) used then becomes

$$T = \text{TM} \cdot \text{CT} + (1 - \text{TM})\exp(-D_x \cdot \text{ET}) \qquad (A3.6)$$

where D_x is the separation from the peak centre in channels. The peak function to the right of the peak centre is chosen to have no tail and can be represented as $H \cdot GL$, where H is the peak height and GL the peak function.

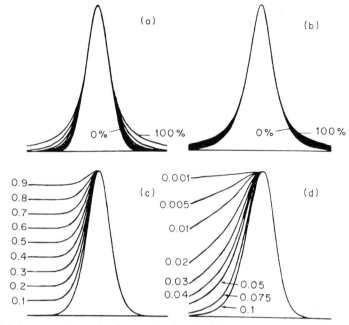

Figure A3.6   Possible peak shapes obtained with a Gaussian/Lorentzian product function by varying the mixing ratio and tail parameters. (a) Variation of the mixing ratio in steps of 10 per cent from 10 to 100 per cent Lorentzian character. (b) As (a) but the range is in steps of 1 per cent from 88 to 100 per cent. (c) Variation of constant tail parameters. (d) Variation of exponential tail parameters.

The peak function with the tail may be represented as

$$Y = H[GL + (1 - GL)\ T] \qquad (A3.7)$$

Figure A3.6 shows the way in which peak shapes can be synthesized by the variation of the peak parameters in the Gaussian/Lorentzian product function combined with the tail parameters.

A3.5.4   Non-linear least squares curve fitting

This attempts to optimize the curve synthesis process by using the method of non-linear least squares, which recognizes that the process concerned may be complex, in that the appropriate parameters enter into the algebraic expression that describes the process in a non-linear manner. Such is the case for photo-electron spectral data which can be represented as a function $F(x/g)$ (see equation A3.5 above) which depends in a non-linear manner upon the parameters (q) described above. The process to minimize χ^2 is carried out computationally, rather than manually, by the operator of the curve synthesis process by a procedure of guessing. A number of possible non-linear least squares methods are available. The author uses the Gauss–Newton method. In this method the requirement for the minimization of χ^2 allows one to construct the equations:

$$0 = \frac{\partial \chi^2}{\partial q_i} = -2 \sum_r w_r [y_r - F(x_r/q)] \left(\frac{\partial F}{\partial q_i} \right)_{x=x_r} \qquad (A3.8)$$

which gives k equations where k is the number of parameters q. The problem is a non-linear one, since q values enter non-linearly. If one puts

$$q_i = q_i^0 + \delta_i \qquad (A3.9)$$

where q_i is the value of q which minimizes χ^2, q_i^0 is some initial guess and δ_i is the correction requires to q_i^0 to give q_i. If

$$F_r = F(x_r/q^0) \qquad \text{and} \qquad F_r^i = \left(\frac{\partial F}{\partial q_i} \right)_{x=x_r, q=q^0} \qquad (A3.10)$$

then one can expand $F(x_r/q)$ as a Taylor series where the series is truncated after the linear terms, i.e.:

$$F(x_r/q) = F_r + \sum \delta_j F_r^j \qquad (A3.11)$$

where $j = 1, 2, \ldots, k$. Then equations (A3.8) can be rewritten to give k linear equations:

$$0 \approx \sum_r w_r [y_r - F_r - \sum_j \partial_j F_r^j] F_r^i \qquad (A3.12)$$

These equations can be solved by an iterative process, the process being stopped when the δ_i values (equation A3.9) vary by an insignificant amount for all δ_i values from one iteration to the next. The process does not always move steadily to convergence and problems may arise such as the matrices calculated being singular or having a negative determinant. The process has the advantage that it allows a certain amount of valuable statistical information, such as the standard deviations of all the calculated parameters, to be calculated. Such information allows some quantitative significane to be attached to the calculated parameters. In addition to the final χ^2 value, the probability that χ^2 is less than χ^2 (calculated) can be evaluated.[24] Normally a value less than 95 per cent. is considered statistically significant, though it must be remembered that the stringency of this test is greated when the number of electron counts is large. In general the most useful approach is to compare how χ^2 or f changes for different fits to the same spectrum rather than to use the information in any absolute sense. All this information may well assist in the overall decision of the 'best fit'. Such a decision must rest upon the statistical information consistent with chemical and spectroscopic sense. Often no unique solution is provided, but the spectrum can be reduced

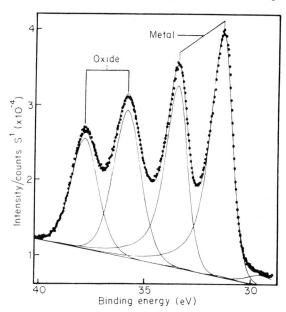

Figure A3.7   Curve-fitted W 4*f* spectrum

to a number of slightly different possibilities allowing the amount of significant information provided by a particular spectrum to be assessed. An example
of such a case has been provided.[29]

The process as carried out in practice can be illustrated with the aid of the
W 4*f* spectrum illustrated in Figure A3.7. This spectrum contains four clearly
resolvable peaks, which for the purposes of the fit are split into two separate
groups of two peaks: one group for the oxide (WO_3) and the other for the
metal. Such separation into different groups is necessary when peaks are
known to have different widths and tail parameters.

Each peak has even parameters associated with it:

(a) Peak centre
(b) Peak height.
(c) Peak width.
(d) Gaussian/Lorentzian peak shape mixing ratio.
(e) Constant tail height.
(f) Exponential tail slope.
(g) Constant/exponential tail mixing ratio.

Thus for *n* peaks are 7*n* peak parameters. In addition there are also the
linear background slope and intercept. Thus a maximum of 7*n* + 2 parameters can be allowed to vary or 'float' as desired. However it is impossible to

let this happen in practice and so some parameters (generally parameters d to g) are fixed.

The spectrum is fitted to the Gaussian/Lorentzian product function described above with a mixing ratio M of 0.5 for all peaks and a spin-orbit splitting intensity ratio of each doublet fixed at the expected value for $4f_{5/2}$: $4f_{7/2}$ of 1/1.333 (see Chapter 3). The tail parameters are different for the oxide and metal peaks. The asymmetry in the metal peak shape is due to conduction band interaction effects,[30-32] and in this case is accounted for by adding some exponential tail characteristic to the high binding energy side of the metal peaks (the amount being determined from a previous fit to a pure metal spectrum). We have found that using such a method is satisfactory for most purposes, though it is different from the method used by Hufner and Wertheium,[31] who first deconvoluted the instrumental broadening function from the observed data and then fit the deconvoluted data with the Doniach and Sunjic[30] lineshape. The results, shown in Table A3.2, show that the asymmetric tail in the calculation of peak intensity (area) for the metal peaks is clearly very significant.

The fit calculates a linear background slope and intercept which is often satisfactory though removal of a non-linear background can be useful (*vide infra*). X-ray satellite contributions were also taken into account, the full statellite peaks being added to each peak in the fit. The satellite contribution is very noticeable at the lower binding energy.

It should be clear that non-linear least squares curve fitting is preferable to analogue curve synthesis as it provides a quantitative handle with which to gauge the quality of the fit and, unlike analogue methods, peak parameters are perfectly reproducible. Most important is that operator subjectivity is largely eliminated.

Table A3.2 Results of curve fit to W $4f$ spectrum ($\chi^2 = 2855, f = 298$) (Figure A3.7)

	Oxide		Metal	
	5/2	7/2	5/2	7/2
Binding energy*, eV	37.66(0.01)	35.75(0.01)	33.5(0.01)	31.50(0.01)
Separation, eV	1.91(0.01)		2.00(0.01)	
PWHM†, eV	1.43(0.01)		1.01(0.01)	
Intensity × 10^{-7}	1.127(1%)	1.503	31.720(1%)	2.293
(peak area)			21.323(1%)	1.764
Relative intensity	1	1.333	1	1.333
Exponential tail slope	20.0	20.0	0.032	0.032

*Reference C $1s$ = 284.6 eV.
†Excluding exponential tail contribution.
‡Including exponential tail contribution.

A3.6 Background Removal

The accurate removal of the background contribution to a spectrum is a process that must be carried out with care since, except in the case of the trival removal of a horizontal background discussed above, it may involve distortion of the data when incorrectly carried out. Any background removal will alter absolute peak intensities and will cause problems with any quantification model which must be defined with respect to clear background conditions. There is no definite way to remove a background and the whole process is still controversial (see Chapter 5).

A rather crude removal of the background may be achieved by removal of a linear background drawn as a straight line between the first and last set of points in a spectrum. The correct choice of a first and last set of points, or a suitable average over them, has a critical effect upon the resulting background. Such a process is best not carried out directly but may be included in the curve-fitting process as described above and illustrated in Figure A3.7. This means that the original data are not altered and if a non-linear background has been removed, as described below, then such a linear background would be expected to be horizontal, as shown in Figure A3.9 below.

The removal of a non-linear background requires a complete understanding of the processes that give rise to background electrons. Such a process will involve inelastic electron ejection processes, but the background must be distinguished from any peaks, which may be broad and difficult to distinguish, that may arise from specific elastic processes such as satellite peaks of various sorts and Auger electron peaks. In addition any angular dependence of the background (which may be machine dependent since β factors will differ for different machines; see Chapter 5) and inhomogeneities in the material may cause problems. The problems that may arise from plasmon losses and asymmetric broadening being confused with the inelastic background has been pointed out.[33] The most common method of non-linear background subtraction, often called the Shirley method, considers the background at any point due to inelastically scattered electrons, is assumed to arise solely from the scattering of electrons of higher kinetic energy and is thus proportional to the integrated photo-electron intensity to higher kinetic energy. The method has been used by a number of workers[34-38] and various refinements have been introduced. Bishop[39] has modified this model to include a linearly falling contribution to the background of inelastically ejected electrons as the energy falls from high to low kinetic energy. In general the non-linear background removal becomes more and more prone to error as the energy range over which the background is subtracted is increased, so while modified models such as that due to Bishop may help, problems due to features such as negative peaks may appear when the energy range is large. It is important to ensure that the beginning and end of the spectral region that is to be back-

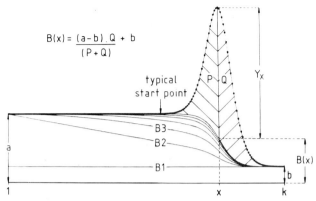

Figure A3.8   Inelastic background determination. $B(x)$ is the background at point x in the spectrum which contains k equally spaced points

ground subtracted represents points on the background profile (rather than a part of some specific peak) or else the background removal process will remove significant peak intensity.

Non-linear background subtraction may be performed in the way used by the author,[15] which is illustrated in Figure A3.8. In such a spectrum the value of the background at a point x in a spectral array of k equally spaced points of separation h is

$$B(x) = \frac{(a - b)Q}{P + Q} + b \qquad (A3.13)$$

where a is the average start point, b the average end point, $(P + Q)$ the total background subtracted (BS) peak area and Q the (BS) peak area from point x to point k. Using the trapezoidal rule:

$$Q = h\left[\left(\sum_{i=x}^{k} y_i\right) - 0.5(y_x + y_k)\right] \qquad (A3.14)$$

the (BS) areas are calculated initially by choosing a linear constant background of magnitude b, line B1 in Figure A3.8. Substitution in equation (A3.13) leads to the background B2 which is then used to calculate new (BS) areas resulting in the background B3. The process is repeated until $(P = Q)$ remains essentially unchanged on successive iterations. Typically the start point is chosen close to the peak to minimize the number of required iterations. In Figure A3.8 the start point is somewhat removed from the peak to illustrate the iterative process; however the final result will be the same.

The term $0.5(y_x + y_k)$ is the correction introduced by the trapezoidal rule to the simple sum of points over the range x to k. In determining $(P = Q)$,

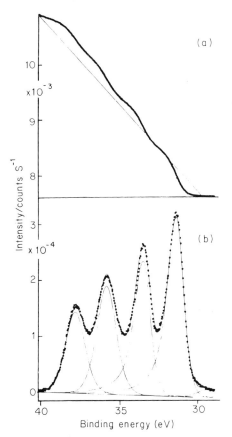

Figure A3.9 W $4f$ spectral region. (a) Inelastic integral background spectrum in comparison with linear background. (b) Spectrum with integral background subtracted and curve fitted

where $x = 1$, y_1 and y_k both zero, the latter correction is insignificant. As x increases the correction does become more significant since y_x increases.

Figure A3.9(a) shows the calculated background for the W $4f$ region of Figure A3.7 compared with the computer-calculated linear background from the fit. It is generally true that a linear background underestimates the 'true background' to the high binding energy side of a peak, the converse being true at the lose binding energy side. This could be of major significance in determination of peak areas and subsequent peak area ratios. In this case, as shown by the fit in Figure A3.9(b), the effect is minimal. A point to note, however, is that at the low binding energy side of the peak there is still intensity due to an X-ray satellite. The satellite is accounted for by allowing the fitted background slope to vary such that a sloping linear background is calculated, but it is much less than before. A horizontal background could have been obtained by estimating the position of the true background intensity at the low binding energy side of the peak prior to subtraction.

A3.7   Difference Spectra

Difference spectra can be a very useful method for data analysis of a range of similar samples subjected to different amounts of chemical treatment. The simple subtraction of one spectrum from another is a trivial operation, but the proper use of difference spectra requires this subtraction process to be carried out with care and according to clearly defined criteria. The application of this technique to XPS has recently been discussed[15] and it has been shown that there are a number of points, specific to XPS, that need to be considered.

A3.7.1   Alignment

The first step in any difference spectra process is to ensure that the two spectra have been correctly aligned, i.e. that their data points correspond directly in kinetic energy value. This generally requires that the two spectra contain the same number of data points. Careful checks of calibration must then be carried out to ensure this correct alignment. It has been shown[15] that small differences in spectral alignment can cause large differences in the resulting difference spectra, stressing the importance of carrying out this operation with care and accuracy. Methods for using difference spectra to obtain good spectral alignment have been discussed.[15]

A3.7.2   Normalization

When the spectra have been correctly aligned it is then necessary to carry out a normalization process before the difference spectra is calculated. This requires that all the points in one spectrum be multiplied by this factor before the spectra are subtracted. The process of normalization requires that one knows the exact proportion of one spectrum in the spectrum from which it is to be subtracted. When one is attempting to subtract two spectra, each of which contains clearly separated chemically shifted peaks, the process is trivial, but such cases would hardly benefit from difference spectra. The typical difference spectrum is obtained from spectra that contain a number of overlapping peaks which makes normalization non-trivial and central to the difference spectral process. Thus one needs to know the correct answer before carrying out the process! Normally the two spectra are treated by appropriate preliminary methods such as smoothing (using the same smoothing parameters) and background subtraction (bearing in mind the problems discussed above). Then the maximum points in the two spectra are identified and the spectra adjusted so that these maximum points are the same (ten thousand in the spectra illustrated here). After this preliminary process normalization can be carried out and it is possible to identify three types of normalization process that can be carried out in a clearly defined manner:

(a) Height normalization.
(b) Optimal normalization.
(c) Area normalization.

Height normalization and area normalization are processes that can be carried out easily. Optimal normalization represents an attempt to obtain the 'correct' normalization factor by an iterative procedure. The normalization process can be illustrated (Figure A3.10) using model C 1s spectra A and B, where spectrum B consists of a number of overlapping chemically shifted peaks B, B1, B2 and B3 and spectrum A is a single peak. Spectrum B is then subtracted from spectrum A to hopefully reveal the weaker peaks B1, B2 and B3.

A3.7.2.1   Height normalization

Having carried out the preliminary procedures on the aligned spectra, described above, then a normalization factor x can be defined. If the height of

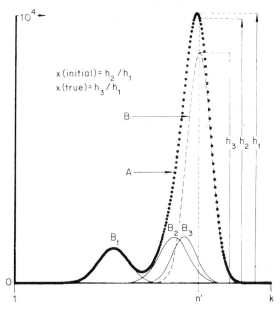

Figure A3.10   Model C 1s 'oxidized' spectrum, A, comprising 'unoxidized' component B, and chemically shifted components B1, B2 and B3

the peak maximum of the most intense peak in a spectrum to be subtracted (which will be called spectrum B) from another spectrum composed of a number of peaks (which will be called spectrum A) is h_3, and the height in spectrum A at this peak maximum is h_1, then

$$x = h_3/h_1$$

as illustrate in Figure A3.10. Unfortunately x is not generally known though it must be less than unity. Simple height normalization is then carried out with x = unity. This type of height normalization can be applied to assist in alignment of the two spectra.[15]

A3.7.2.2 Optimal normalization

Optimal normalization is based upon the idea that it might be possible to obtain a better value for x than unity by using iterative methods. The method used by the author[15] decreases x below unity and then follows certain properties of the resulting difference spectrum until x is considered to be at its optimum value. The most successful test criterion in model spectra was found to be that the difference spectrum should maintain a negative slope in the region bounded by the peak maximum of spectrum B and the right-hand side of spectrum A (points n' and k in Figure A3.10), i.e.:

$$(A - B)(n) > (A - B)(n - 1) \qquad\qquad (A3.16)$$

where n varies from n' to $(k - 1)$.

Starting with a value of $x = x(\text{initial}) = h_2/h_1$ (thus ensuring that the calculated value of x will always be less than or equal to 1.0) the procedure then tests expression (A3.16).[16] If the criterion is not satisfied for a particular pair of points the value of x is reduced by an amount $0.01x(\text{initial})$, and the test begins again. When (A3.16) is satisfied over the whole region, n' to k, the value of x is $x(\text{optimal})$ and the spectra are optimally normalized (ON). Using this process with model spectra it was possible to obtain a value of x of 0.854, very close to the correct value of 0.855 for the example chosen. In real spectra it has been found that the range of the test must be reduced from the right-hand side of spectrum A by C channels (Figure A3.10) or else the optimal value of x may fall to zero!

A3.7.2.3 Area normalization

The value of x is chosen to be the ratio of the total area of the spectrum of interest(spectrum A) and the total area of the reference spectrum(spectrum B). In contrast to the height and area normalized spectra this normalization factor will be greater than unity. The resulting difference spectrum will contain negative peaks corresponding to the fact that spectrum A contains other features than just spectrum B; the less of spectrum B in spectrum A then the larger will be the negative peak. If x(optimal) is obtained as above and x(area normalization) is also evaluated, these two normalization factors can give useful quanitities. Thus the fraction of the reference spectrum(B) in the spectrum of interest(A) will be x(area normalization)$/x$(optimal) and thus the difference between this fraction and unity is the fraction of the total area due to spectral features other than the reference spectrum. Thus if the reference spectrum were that of a pure metal and the spectrum of interest contained metal and oxide, such a calculation would allow the amount of oxide in the spectrum to be evaluated. This might usefully be compared with curve-fitting results.

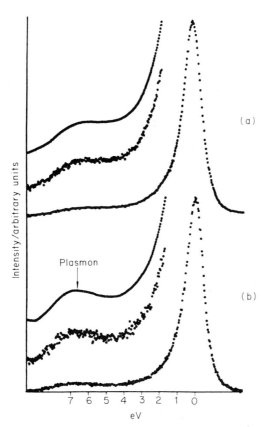

Figure A3.11   C 1s spectra obtained from untreated type II carbon fibres: (a) B1, at ambient temperature, (b) B2, after heat treatment to 1400 °C. In addition to the original spectrum, the plasmon region is shown magnified and smoothed ($P = 21, 15$ times)

A3.7.3 Examples of difference spectra

Figure A3.12 illustrates the difference spectrum obtained when the two spectra illustrated in Figure A3.11 are subtracted. The two spectra show only small visual differences, but the difference spectra highlight these differences. Thus the area normalized spectrum shows that there are more chemically shifted C $1s$ groups in B1 than in B2, since there is a large negative peak in the spectrum corresponding to the 'graphitic' carbon (chemical shift shown as zero). The height normalized spectrum is shown, together with other spectra corresponding to smaller values of x. The optimal normalization factor is satisfied with $C = 30$ for $x = 0.95$. As expected, the value of x(optimal) varies as C is varied, as shown in Figure A3.13, but 0.95 corresponds to a

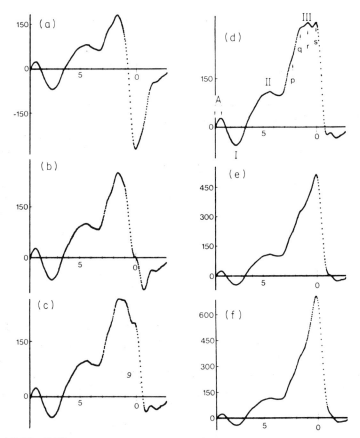

Figure A3.12   Difference spectrum B1 − B2 (Figure A3.11). Various normalization factors: (a) area normalized $z = 1.03$, (b) height normalized $x = 1.0$, (c) $x = 0.98$, (d) $x = 0.97$, (e) optimally normalized $x = 0.95$, (f) $x = 0.93$

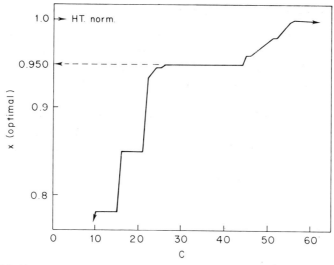

Figure A3.13   Variation of x (optimal) with C for the difference B1(3)−B2(1)

plateau in the graph which makes it seen a suitable value. In fact, the spectrum with $x = 0.97$ seems to contain more information, which may mean that it represents a more suitable choice of x than the optimal value which may mean that the test is too severe. Region I is thought to correspond to a difference in the plasmon intensity in the two spectra, and regions II and III to differences in surface C/O groups. Clearly the choice of x(optimal) in real spectra requires subjective decisions and is not uniquely defined as are area and height normalized spectra. However, real information is present and, when combined with other data analysis methods, can lead to useful chemical information.

A3.8   Conclusions

Data analysis clearly has a vital role to play in the interpretation of X-ray photo-electron spectroscopy. This appendix has attempted to present the main techniques that are currently available and provide suitable illustrations of their application. Much time and effort can be spent upon data analysis, and one needs to be careful not to over-interpret the data. However, there are many systems where spectral changes are very small but highly significant, and in such cases careful data analysis may provide the only means of extracting useful results. The ready availability of cheap data-collection systems should encourage more workers to collect digital data, and hopefully use such techniques to improve spectral accuracy. Great care must be taken in data analysis that changes the original data or attempts to provide automated information on peak positions, peak intensities or even elemental compositions. Such an analysis may be based upon approximations and assumptions about the sample that makes these methods incorrect and highly misleading when used in certain situations. Notwithstanding the need for care and discretion in their application, the wider application of data analysis methods will clearly make an important contribution to the future useful development of photo-electron spectroscopy.

Acknowledgement

I would like to thank the American Chemical Society for permission to reproduce many of the figures in this appendix. I would also like to thank Dr Andrew Proctor for his contribution, while a research student and postdoctoral fellow, to a number of the data analysis methods described here.

References

1. A. Savitsky and M. J. E. Golay, *Anal. Chem.*, **36**, 1627 (1964).
2. F. B. Hildebrand, *Introduction to Numerical Analysis*, Chap. 7, McGraw-Hill, New York (1956).
3. A. Proctor and P. M. A. Sherwood, *Anal. Chem.*, **52**, 2315 (1980).
4. W. F. Maddams, *Appl. Spec.*, **34**, 245 (1980).
5. J. O. Lephardt, *Transform Techniques in Chemistry* (Ed. P. R. Griffiths), Chap. 11, p. 285, Plenum Press, New York (1978).
6. G. Horlick, *Anal. Chem.*, **44**, 943 (1972).
7. K. R. Betty and G. Horlick, *Appl. Spec.*, **30**, 23 (1976).
8. T. A. Maldacker, J. E. Davis and L. B. Rogers, *Anal. Chem.*, **46**, 637 (1974).
9. D. W. Kirmse and A. W. Westerberg, *Anal. Chem.*, **43**, 1035 (1971).
10. C. A. Bush, *Anal. Chem.*, **46**, 890 (1974).
11. G. L. Turin, *IRE Trans. Inform. Theory*, **IT-6**, 311 (1960).
12. R. R. Ernst, *Advances in Magnetic Resonance* (Ed. J. S. Waugh), Vol. 2, p. 1, Academic Press, New York (1966).
13. W. F. Maddams, *Appl. Spec.*, **34**, 245 (1980).
14. A. F. Carley and R. W. Joyner, *J. Electron Spectrosc.*, **16**, 1 (1979).
15. A. Proctor and P. M. A. Sherwood, *Anal. Chem.*, **54**, 13 (1982).
16. H. P. Yule, *Anal. Chem.*, **38**, 103 (1966).
17. T. Inouye, T. Harper and N. C. Rasmussen, *Nucl. Instr. Methods*, **67**, 125 (1969).
18. A. W. Westerberg, *Anal. Chem.*, **41**, 1770 (1969).
19. J. R. Morrey, *Anal. Chem.*, **40**, 905 (1968).
20. A. E. Panlath and M. M. Millard, *Appl. Spec.*, **33**, 502 (1979).
21. J. J. Pireaux, *Appl. Spec.*, **30**, 219 (1976).
22. D. C. Champery, *Fourier Transforms and Their Physical Applications*, p. 17, Academic Press, New York (1973).
23. R. O. Ansell, T. Dickinson, A. F. Povey and P. M. A. Sherwood, *J. Electroanal Chem.*, **98**, 79 (1979).
24. J. W. Perram, *J. Chem. Phys.*, **49**, 4245 (1968).
25. A. Baruya and W. F. Maddams, *Appl. Spec.*, **32**, 563 (1978).
26. E. Caulcott, *Significance Tests*, Routledge and Kegan Paul, London (1973).

27. J. S. Bendat and A. G. Piersol, *Random Data: Analysis and Measurement Procedures*, Wiley-Interscience, New York (1971).
28. F. E. Fisher, *Fundamental Statistical Concepts*, Canfield Press, San Francisco (Harper and Row) (1973).
29. A. Proctor and P. M. A. Sherwood, *Surf. Interface Anal.*, **2**, 191 (1980).
30. S. Doniach and M. Sunjic, *J. Phys.*, **C3**, 285 (1970).
31. S. K. Hufner and G. K. Wertheim, *Phys. Rev.*, **B11**, 678 (1975).
32. G. K. Wertheim and D. N. E. Buchanan, *Phys. Rev.*, **B16**, 2613 (1975).
33. G. K. Wertheim and S. Hufner, *Phys. Rev. Lett.*, **35**, 53 (1975).
34. D. A. Shirley, *Phys. Rev.*, **B5**, 4709 (1972).
35. M. O. Krause, T. A. Carlson and R. D. Dismukes, *Phys. Rev.*, **170**, 37 (1968).
36. D. W. Fischer, *Advan. X-ray Anal.*, **13**, 159 (1969).
37. A. Barrie and F. J. Street, *J. Electron Spectrosc.*, **7**, 1 (1975).
38. N. S. McIntyre and D. G. Zetaruk, *Anal. Chem.*, **49**, 1521 (1975).
39. H. E. Bishop, *Surf. Interface Anal.*, **3**, 272 (1981).

Practical Surface Analysis
by Auger and X-ray Photoelectron Spectroscopy
Edited by D. Briggs and M. P. Seah
© 1983, John Wiley & Sons, Ltd.

Appendix 4

Auger and Photoelectron Energies and the Auger Parameter: A Data Set

C. D. Wagner

29 Starview Drive, Oakland CA 94618, USA

This is a comprehensive survey of $N(E)$-type Auger line energies from the literature to 1982. It includes data on the sharpest Auger line and, where available, the most intense photo-electron line. From these are calculated the modified Auger parameter, which is the difference in the line energies plus the energy of the exciting radiation or, more simply,

$$\alpha' = KE(Auger) - KE(photo\text{-}electron) + h\upsilon = KE(Auger)$$
$$+ BE(photo\text{-}electron)$$

where the zero reference for both the Auger and photo-electrons is the Fermi edge. This quantity is useful because it is not subject to problems with determination of steady state charge, and because its chemical shifts reflect changes in screening energy (75–7). Compilations similar to this have appeared in earlier versions, including plots of Auger energy versus photo-electron energy, termed chemical state plots (77–21, 78–8, 79–11, 79–12).

There are some problems, of course, in selecting the lines to be included. We are guided by emphasizing those lines that can be observed in conventional ESCA or XPS spectra. It was decided not to attempt to include valence-type or cvv Auger lines because they are usually composed of broad bands, and vary greatly in line distribution with the chemical state. Thus, we do not include the elements through $Z = 7$ (nitrogen) and compilations already exist for available data for oxygen (80–21) and fluorine (77–21) so these data need not be repeated here. The rule against including cvv lines is bent with inclusion of some fragmentary data on the wide LVV lines of sulphur and chlorine, the $LM_{23}V$ lines of Ti, V, Cr, and Mn and the LVV lines of Fe, Co and Ni. Similarly, there are included data on the $M_{45}VV$ lines of Mo, Ru, Rh, and Pd. Finally, in the *NOO* series, there are data on the *NVV* of Pt and Au. Data on Auger lines for transitions higher in energy than that accessible by Al *K* X-rays are supplied for Al, Si, P, S, Cl, Ar, Br, Kr, fragmentary data for the $L_3M_{45}M_{45}$ lines of Sr, Y, Zr, Nb and Mo, and the higher energy $M_5N_{67}N_{67}$ series for the heavy metals, Ta, W, Os, Pt, Au, Hg, Tl, Pb and Bi. There are no data for Rb or for any of the rare earths. With the latter the Auger lines are too broad to be analytically useful in this way. There are also no data available on the actinides. In summary, there are entries for all of the stable elements except $Z<10$, Rb, the rare earths, Hf, Re, Ir, and the actinides. Values for both a high and low energy Auger series are shown for S, Cl, Ar, Pt, Au and Pb. In the $M_{45}N_{45}N_{45}$ and $N_{67}O_{45}O_{45}$ series, the $M_4N_{45}N_{45}$ and $N_6O_{45}O_{45}$ components are the ones cited, because in insulating compounds they appear sharper and more easily measured accurately than the M_5 and N_7 counterparts.

The selection of the companion photo-electron line is simpler. It is the $1s$ line through Na, the $2p_{3/2}$ line through Zn, the $3d_{5/2}$ line through Ba, and the $4f_{7/2}$ line through Hf–Bi. In the overlapping regions this is somewhat arbitrary. Thus, the $1s$ energy is also supplied where it was available in the data for Mg and the following elements through Ar, even though they may require higher energy sources than Mg or Al X-rays to be produced. Data on $2p_{3/2}$ line energies are supplied for Ga, Ge, As, Se, and Br. Some data on $4d$ energies are included for antimony and tellurium. Data on the $3d_{5/2}$ line are supplied for tungsten compounds.

In this tabulation a serious attempt has been made to assemble the line energy data in as self-consistent a manner as possible. To this end we emphasize the following:

(a) Data on gas phase materials are assumed to have good accuracy, usually inherent to using mixtures with rare gas standards. These data are presented without change. A comprehensive review of gas phase photo-electron line energies has appeared, and information there on gas phase referencing may be useful (80–22). All data from gas phase are referenced to the vacuum level and are indicated here by the symbol 'v'.

(b) References 80–4*, 82–1*, and 82–2* are all recent efforts to obtain accurate line energies of easily prepared solids relative to the Fermi level. The values shown in this tabulation are corrected to Au $4f_{7/2} = \underline{84.00}$; their absolute values are actually, respectively, 83.98, 84.1 and 84.00.

(c) For all of the other data on solids, we have made the assumption of a common energy scale with a spectrometer work function such that Au $4f_{7/2} = \underline{84.0}$, Ag $3d_{5/2} = \underline{368.2}$ and Cu $2p_{2/2} = \underline{932.6}$. Data on any of these elements afford the opportunity to adjust the position of the voltage scale of the reference on this common basis. Similarly, data on insulators accompanied by a valid method of charge reference, such as gold decoration, use of adventitious hydrocarbon or use of a hydrocarbon moiety of the sample, assuming hydrocarbon $C1s = \underline{284.8}$ eV, permits adjustment also. When adjustment to the data is done for these reasons, the reference is followed by 'r'. References supplying no single natural line energy, or no referral to the Fermi level, are designated 'n'. Often, though the articles may not state it, it is assumed that papers from a given laboratory closely spaced in time have the same reference for one end of the voltage scale (usually Au $4f_{7/2}$). The data for insulators not charge referenced, or of doubtful charge reference validity, are not included, but in many cases the Auger parameters are included because they are independent of charge correction.

References including $1s$ and $2p$ lines for Na, Mg, Al and Si, and those with $2p_{3/2}$ and $3d$ lines for Zn, Ga, Ge, As and Se afford the possibility to check on the magnitude of the voltage scale, because these line differences for the elements should be equal to the X-ray energy. This was possible for a few references, and minor corrections in high binding energy lines were made and noted by 'c', using X-ray energy data from (67–1). It is not possible to make corrections for non-linearity in the voltage scales.

Older data from a given laboratory supplanted by later data have been omitted in favour of the later data. Some data appearing to be clearly inconsistent with those of most other workers have been omitted. These are usually older studies, of the order of ten years, before instruments were developed to their present state. Some data hitherto unpublished by the author are included as reference 82–5. Most are data obtained in conjunction with reference (79–12), and some for reference (79–11), and were obtained in the same way as data in those references.

Some studies emphasize the changes in line energy, where lines from both chemical or physical states are present in the same spectrum. There are included in the data as, e.g. Mg → Mg ox, with values of line energy changes indicated by Δ. Where there is a change of phase, e.g. Zn → Zn (g), the reference level is of course the same.

The order of presentation of compounds for each element is roughly in decreasing Auger parameter, but with regard to clustering of similar compounds.

References

66–1   H. Körber and W. Mehlhorn, *Z. Physik*, **191**, 217 (1966).
66–2   A. Fahlman, R. Nordberg, C. Nordling and K. Siegbahn, *Z. Physik*, **192**, 476 (1966).
67–1   J. A. Bearden, *Rev. Mod. Phys.*, **39**, 78 (1967).
69–1   K. Siegbahn, C. Nordling, G. Johansson, J. Hedman, P. F. Heden, K. Hamrin, U. Gelius, T. Bergmark, L. O. Werme, R. Manne and Y. Baer, *ESCA Applied to Free Molecules*, North Holland Publishing Company, Amsterdam (1969).
69–2   L. Ramqvist, K. Hamrin, G. Johansson, A. Fahlman and C. Nordling, *J. Phys. Chem. Solids*, **30**, 1835 (1969).
70–1   D. W. Langer and C. J. Vesely, *Phys. Rev.*, **B2**, 4885 (1970).

70–2 R. Spohr, T. Bergmark, N. Magnusson, L. O. Werme, C. Nordling and K. Siegbahm, *Phys. Scr.*, **2**, 31 (1970).
70–3 S. Aksela, M. Pessa and M. Karras, *Z. Physik*, **237**, 381 (1970).
71–1 S. Aksela, *Z. Physik*, **244**, 268 (1971).
72–1 L. O. Werme, T. Bergmark and K. Siegbahn, *Phys. Scr.*, **6**, 141 (1972).
73–1 G. Schön, *Acta Chem. Scand.*, **27**, 2623 (1973).
73–2 G. Johansson, J. Hedman, A. Berndtsson, M. Klasson and R. Nilsson, *J. Electron Spectrosc. Relat. Phenom.*, **2**, 295 (1973).
73–3 S. P. Kowalczyk, L. Ley, F. R. McFeely, R. A. Pollak and D. A. Shirley, *Phys. Rev.*, **B8**, 3583 (1973).
73–4 W. B. Perry and W. L. Jolly, *Chem. Phys. Lett.*, **23**, 529 (1973).
73–5 S. P. Kowalczyk, R. A. Pollak, F. R. McFeely, L. Ley and D. A. Shirley, *Phys. Rev.* **B8**, 2387 (1973).
73–6 G. Schön, *J. Electron Spectrosc.*, **2**, 75 (1973).
73–7 G. Schön, *Surf. Sci.*, **35**, 96 (1973).
73–8 C. D. Wagner and P. Biloen, *Surf. Sci.*, **35**, 82 (1973).
73–9 L. O. Werme, T. Bergmark and K. Siegbahn, *Phys. Scr.*, **8**, 149 (1973).
74–1 H. Aksela and S. Aksela, *J. Phys. B. Atom. Mol. Phys.*, **7**, 1262 (1974).
74–2 S. Aksela and H. Aksela, *Phys. Lett.*, **48A**, 19 (1974).
74–3 B. Breukmann and V. Schmidt, *Z. Physik*, **268**, 235 (1974).
74–4 J. E. Castle and D. Epler, *Proc. Roy. Soc. London*, **A339**, 49 (1974).
74–5 N. E. Erickson, *J. Vac. Sci. Technol.*, **11**, 226 (1974).
74–6 H. Hillig, B. Cleff, W. Mehlhorn and W. Schmitz, *Z. Physik*, **268**, 225 (1974).
74–7 M. Klasson, A. Berndtsson, J. Hedman, R. Nilsson, R. Nyholm and C. Nordling, *J. Electron Spectrosc.*, **3**, 427 (1974).
74–8 S. P. Kowalczyk, L. Ley, F. R. McFeely, R. A. Pollak and D. A. Shirley, *Phys. Rev.*, **B9**, 381 (1974).
74–9 T. D. Thomas and R. W. Shaw, *J. Electron Spectrosc.*, **5**, 1081 (1974).
74–10 P. Larson, *J. Electron Spectrosc.*, **4**, 213 (1974).
74–11 W. B. Perry and W. L. Jolly, *Inorg. Chem.*, **13**, 1211 (1974).
75–1 L. Fiermans, R. Hoogewijs and J. Vennik, *Surf. Sci.*, **63**, 390 (1977).
75–2 J. C. Fuggle, L. M. Watson, D. J. Fabian and S. Affrossman, *J. Phys. F. Metal. Phys.*, **5**, 375 (1975).
75–3 H. C. Halder, J. Alonso and W. E. Swartz, *Z. Naturforsch.*, **30a**, 1485 (1975).
75–4 C. K. Jørgenson and H. Berthou, *Chem. Phys. Lett.*, **36**, 432 (1975).
75–5 L. Ley, F. R. McFeely, S. P. Kowalczyk, J. G. Jenkin and D. A. Shirley, *Phys. Rev.*, **B11**, 600 (1975).
75–6 E. D. Roberts, P. Weightman and C. E. Johnson, *J. Phys. C. Sol. State Phys.*, **8**, 1301 (1975).
75–7 C. D. Wagner, *Faraday. Disc. Chem. Soc.*, **60**, 291 (1975).
76–1 L. Asplund, P. Kelfve, H. Siegbahn, O. Goscinzki, H. Fellner-Feldegg, K. Hamrin, B. Blomster and K. Siegbahn, *Chem. Phys. Lett.*, **40**, 353 (1976).
76–2 M. K. Bahl, R. O. Woodall, R. L. Watson and K. J. Irgolic, *J. Chem. Phys.*, **64**, 1210 (1976).
76–3 G. Dufour, J.-M. Mariot, P.-E. Nilsson-Jatko and R. C. Karnatak, *Phys. Scr.*, **13**, 370 (1976).
76–4 H. C. Halder, J. Alonso and W. E. Swartz, *Phys. Rev.*, **B13**, 2418 (1976).
76–5 K. S. Kim, S. W. Gaarenstroom and N. Winograd, *Chem. Phys. Lett.*, **41**, 503 (1976).
76–6 O. Keski-Rahkonen and M. O. Krause, *J. Electron Spectrosc.*, **9**, 371 (1976).
76–7 N. S. McIntyre, T. E. Rummery, M. G. Cook and D. Owen, *J. Electrochem. Soc.*, **123**, 1165 (1976).

76–8   R. Reisfeld, C. D. Jørgenson, A. Bornstein and H. Berthou, *Chimia*, **30**, 451 (1976).

76–9   P. Weightman, *J. Phys. C. Sol. State Phys.*, **9**, 1117 (1976).

76–10 J. D. Nuttall and T. E. Gallon, *J. Phys. C. Sol. State Phys.*, **9**, 4063 (1976).

76–11 J. E. Castle, L. B. Hazell and R. D. Whitehead, *J. Electron Spectrosc.*, **9**, 247 (1976).

77–1   H. Aksela, S. Aksela, J. S. Jen and T. D. Thomas, *Phys. Rev.*, **A15**, 985 (1977).

77–2   L. Asplund, P. Kelfve, B. Blomster, H. Siegbahn, K. Siegbahn, R. L. Lozes and U. I. Wahlgren, *Phys. Scr.*, **16**, 273 (1977).

77–3   L. Asplund, P. Kelfve, B. Blomster, H. Siegbahn and K. Siegbahn, *Phys. Scr.*, **16**, 268 (1977).

77–4   M. K. Bahl and R. L. Watson, *J. Electron Spectrosc.*, **10**, 111 (1977).

77–5   M. K. Bahl, R. L. Watson and K. J. Irgolic, *J. Chem. Phys.*, **66**, 5526 (1977).

77–6   N. R. Armstrong and R. H. Quinn, *Surf. Sci.*, **67**, 451 (1977).

77–7   T. A. Carlson, W. B. Dress and G. L. Nyberg, *Phys. Scr.*, **16**, 211 (1979).

77–8   J. C. Fuggle, *Surf. Sci.*, **69**, 581 (1977).

77–9   J. C. Fuggle, E. Källne, L. M. Watson, and D. J. Fabian, *Phys. Rev.*, **B16**, 750 (1977).

77–10 S. W. Gaarenstroom and N. Winograd, *J. Chem. Phys.*, **67**, 3500 (1977).

77–11 R. Hoogewijs, L. Fiermans and J. Vennik, *J. Electron Spectrosc.*, **11**, 171 (1977).

77–12 R. Hoogewijs, L. Fiermans and J. Vennik, *J. Microsc. Spec. Electron*, **1**, 109 (1977).

77–13 J. Haber and L. Ungier, *J. Electron Spectrosc.*, **12**, 305 (1977).

77–14 A. W. C. Lin, N. R. Armstrong and T. Kuwana, *Anal. Chem.*, **49**, 1228 (1977).

77–15 J. M. Mariot and G. Dufour, *Chem. Phys. Lett.*, **50**, 219 (1977).

77–16 J. F. McGilp and P. Weightman, *J. Phys. C. Sol. State Phys.*, **9**, 3541 (1977).

77–17 J. H. Fox, J. D. Nuttall and T. E. Gallon, *Surf. Sci.*, **63**, 390 (1977).

77–18 V. I. Nefedov, A. K. Zhumadilov and T. Y. Konitova, *J. Struct. Chem.*, **18**, 692 (1977).

77–19 R. B. Shalvoy, G. B. Fisher and P. J. Stiles, *Phys. Rev.*, **B15**, 1680 (1977).

77–20 J. Väyrynen, S. Aksela and H. Aksela, *Phys. Scr.*, **16**, 452 (1977).

77–21 C. D. Wagner, in *Handbook of X-ray and Ultra-violet Photoelectron Spectroscopy*, (Ed. D. Briggs). Heyden and Sons, 1977, Chap. 7 and Appendix 3, pages 387–392.

77–22 E. Antonides, E. C. Janse and G. A. Sawatzky, *Phys. Rev.*, **B15**, 1669 (1977).

77–23 A. Barrie and F. J. Street, *J. Electron Spectrosc.*, **7**, 1 (1977).

77–24 S. Aksela, H. Aksela, M. Vuontisjarvi, J. Väyrynen and E. Lähteenkorva, *J. Electron Spectrosc.*, **11**, 137 (1977).

78–1   M. K. Bahl, R. L. Watson and K. J. Irgolic, *J. Chem. Phys.*, **68**, 3272 (1978).

78–2   J. Haber, T. Machej, L. Ungier and J. Ziolkowski, *J. Sol. State Chem.*, **25**, 207 (1978).

78–3   J. M. Mariot and G. Dufour, *J. Electron Spectrosc.*, **13**, 403 (1978).

78–4   Y. Mizokawa, H. Iwasaki, R. Nishitani and S. Nakamura, *J. Electron Spectrosc.*, **14**, 129 (1978).

78–5   R. Romand, M. Roubin and J. P. Deloume, *J. Electron Spectrosc.*, **13**, 229 (1978).

78–6   P. Steiner, F. J. Reiter, H. Höchst, S. Hüfner and J. C. Fuggle, *Phys. Lett.*, **66A**, 229 (1978); *Phys. Stat. Solidi*, **90**, 45 (1978).

78–7   P. M. Th. M. Van Attekum and J. M. Trooster, *J. Phys. F. Metal Phys.*, **8**, L169 (1978).

78–8   C. D. Wagner, *J. Vac. Sci. Technol.*, **15**, 518 (1978).

79-1   A. J. Ashe, M. K. Bahl, K. D. Bomben, W. T. Chan, J. Gimzewski, P. A. Sitton and T. D. Thomas, *J. Am. Chem. Soc.*, **101**, 1764 (1979).

79-2   H. Aksela, J. Väyrynen and S. Aksela, *J. Electron Spectrosc.*, **16**, 339 (1979).

79-3   C. R. Brundle and D. Seybold, *J. Vac. Sci. Technol.*, **16**, 1186 (1979).

79-4   R. G. Cavell and R. Sodhi, *J. Electron Spectrosc.*, **15**, 145 (1979).

79-5   S. M. Barlow, P. Bayat-Mokhtari and T. E. Gallon, *J. Phys. C. Sol. State Phys.*, **12**, 5577 (1979).

79-6   H. Kuroda, T. Ohta and Y. Sato, *J. Electron Spectrosc.*, **15**, 21 (1979).

79-7   R. Kumpula, J. Väyrynen, T. Rantala and S. Aksela, *J. Phys. C. Sol. State Phys.*, **12**, L809 (1979).

79-8   M. Pessa, A. Vuoristo, M. Villi, S. Aksela, J. Väyrynen, T. Rantala and H. Aksela, *Phys. Rev.*, **B20**, 3115 (1979).

79-9   A. C. Parry-Jones, P. Weightman and P. T. Andrews, *J. Phys. C. Sol. State Phys.*, **12**, 1587 (1979).

79-10 T. Rantala, J. Väyrynen, R. Kumpula and S. Aksela, *Chem. Phys. Lett.*, **66**, 384 (1979).

79-11 C. D. Wagner, L. H. Gale and R. H. Raymond, *Anal. Chem.*, **51**, 466 (1979).

79-12 C. D. Wagner, W. M. Riggs, L. E. Davis, J. F. Moulder and G. E. Muilenberg, *Handbook of X-Ray Photoelectron Spectroscopy*, Perkin-Elmer Corporation, Physical Electronics Division, Eden Prairie, Minnesota (1979).

79-13 D. M. Zehner and H. H. Madden, *J. Vac. Sci. Technol.*, **16**, 562 (1979).

79-14 L. Hilaire, P. Légaré, Y. Holl and G. Maire, *Solid State Comm.*, **32**, 157 (1979).

79-15 J. E. Castle and R. H. West, *J. Electron Spectrosc.*, **16**, 195 (1979).

79-16 G. D. Nichols and D. A. Zatko, *Inorg. Nucl. Chem. Lett.*, **15**, 401 (1979).

79-17 J. E. Castle, L. B. Hazell and R. H. West, *J. Electron Spectrosc.*, **16**, 97 (1979).

79-18 V. Y. Young, R. A. Gibbs and N. Winograd, *J. Chem. Phys.*, **70**, 5714 (1979).

80-1   E. J. Aitken, M. K. Bahl, K. D. Bomben, J. K. Gimzewski, G. S. Nolan and T. D. Thomas, *J. Am. Chem. Soc.*, **102**, 4874 (1980).

80-2   S. Aksela, J. Väyrynen, H. Aksela and S. Pennanen, *J. Phys. B. Atom Molec Phys.*, **13**, 3745 (1980).

80-3   C. W. Bates and L. E. Galan, *Proc. Ninth IMEKO Symp. on Photon Detectors*, Budapest, Hungary, 9–12 September, 1980 pp. 100–129.

80-4* R. J. Bird and P. Swift, *J. Electron Spectrosc.*, **21**, 227 (1980).

80-5   M. K. Bahl, R. L. Watson and K. J. Irgolic, *J. Chem. Phys.*, **72**, 4069 (1980).

80-6   J. E. Castle and R. H. West, *J. Electron Spectrosc.*, **18**, 355 (1980).

80-7   S. Evans, *Proc. Roy. Soc. London*, **A370**, 107 (1980).

80-8   R. A. Gibbs, N. Winograd and V. Young, *J. Chem. Phys.*, **72**, 4799 (1980).

80-9   J. Hedman and N. Martensson, *Phys. Scr.*, **22**, 176 (1980).

80-10 P. Kelfve, B. Blomster, H. Siegbahn, K. Siegbahm, E. Sanhueza and O. Goscinski, *Phys. Scr.*, **21**, 75 (1980).

80-11 L. L. Kazmerski, P. J. Ireland, P. Sheldon, T. L. Chu, S. S. Chu and C. L. Lin, *J. Vac. Sci. Technol.*, **17**, 1061 (1980).

80-12 H. Van Doveren and J. A. Th. Verhoeven, *J. Electron Spectrosc.*, **21**, 265 (1980).

80-13 J. A. Th. Verhoeven and H. Van Doveren, *Appl. Surf. Sci.*, **5**, 361 (1980).

80-14 J. Väyrynen, S. Aksela, M. Kellokumpu, and H. Aksela, *Phys. Rev.*, **A22**, 1610 (1980).

80-15 C. D. Wagner, *J. Electron Spectrosc.*, **18**, 345 (1980).

80-16 C. D. Wagner and J. A. Taylor, *J. Electron Spectrosc.*, **20**, 83 (1980).

80-17 P. Weightman and P. T. Andrews, *J. Phys. C. Sol. State Phys.*, **13**, L815, L821 (1980).

80–18 P. Weightman and P. T. Andrews, *J. Phys. C. Sol. State Phys.*, **13**, 3529 (1980).

80–19 S. J. Yang and C. W. Bates, *Appl. Phys. Lett.*, **36**, 675 (1980).

80–20 N. H. Turner, J. S. Murday and D. E. Ramaker, *Anal. Chem.*, **52**, 84 (1980).

80–21 C. D. Wagner, D. A. Zatko and R. H. Raymond, *Anal. Chem.*, **52**, 1445 (1980).

80–22 A. A. Bakke, H.-W. Chen and W. L. Jolly, *J. Electron Spectrosc.*, **20**, 333 (1980).

81–1 G. Van der Laan, C. Westra, C. Haas and G. A. Sawatzky, *Phys. Rev.*, **B23**, 4369, (1981).

81–2 N. S. McIntyre, S. Sunder, D. W. Shoesmith and F. W. Stanchell, *J. Vac. Sci. Technol.*, **18**, 714 (1981).

81–3 M. Schärli and J. Brunner, *Z. Physik*, **B42**, 285 (1981).

81–4 J. A. Taylor, *Appl. Surf. Sci.*, **7**, 168 (1981).

81–5 J. A. Taylor and J. W. Rabalais, *J. Chem. Phys.*, **75**, 1735 (1981).

81–6 J. Väyrynen, *J. Electron Spectrosc.*, **22**, 27 (1981).

81–7 S. Aksela, M. Kellokumpu, H. Aksela and J. Väyrynen, *Phys. Rev.*, **A23**, 2374 (1981).

81–8 C. D. Wagner, H. A. Six, W. T. Jansen and J. A. Taylor, *Appl. Surf. Sci.*, **9**, 203 (1981).

82–1* C. J. Powell, N. E. Erickson and T. Jach, *J. Vac. Sci. Technol.*, **20**, 625 (1982)

82–2* M. P. Seah and M. T. Anthony, Appendix 1 of this book.

82–3 P. Swift, *Surf. Inter. Anal.*, **4**, 47 (1982).

82–4 J. A. Taylor, *J. Vac. Sci. Technol.*, **20**, 751 (1982).

82–5 C. D. Wagner, unpublished data.

82–6 R. H. West and J. E. Castle, *Surf. Inter. Anal.*, **4**, 68 (1982).

82–7 C. D. Wagner, D. E. Passoja, H. F. Hillery, T. G. Kinisky, H. A. Six, W. T. Jansen and J. A. Taylor, *J. Vac. Sci. Technol.*, **21**, 933 (1982).

82–8 C. D. Wagner and J. A. Taylor, *J. Electron Spectrosc.*, **28**, 211 (1982).

82–9 R. M. Henry, T. A. B. Fryberger and P. C. Stair, *J. Vac. Sci. Technol.*, **20**, 818 (1982).

82–10 M. Polak, *J. Electron Spectrosc.* (in press).

82–11 S. Aksela and J. Sivonen, *Phys. Rev.*, **A23**, 1243 (1982).

82–12 L. Pederson, *J. Electron Spectrosc.*, **28**, 203 (1982).

NEON	$1s$	$KL_{23}L_{23}$	α'	Ref.
Ne (implanted in Fe)	863.4	818.0	1681.4	75-7 r
Ne (implanted in diamond)	863.1	818.7	1681.8	80-7
Ne (g)	870.31 v	804.56 v	1674.87	74-9
Ne (g)	870.37 v	804.52 v	1674.89	73-2
Ne (g)	870.2 v	804.8 v	1675.0	69-1
Ne (g)	870.0 v	804.15 v	1674.15	66-1

SODIUM				
Na	1071.8 c	944.3	2066.1 c	77-23 r
Na	1071.5 c	994.2	2065.7 c	73-3
Na		994.5		78-6 n
NaI	1071.6	991.2	2062.8	75-7 r
NaI			2062.2	74-8
NaI			2063.7	79-6
NaBr	1071.7	990.6	2062.3	75-7 r
NaBr			2063.1	79-6
Na → Na ox	Δ +0.7	−4.5	−3.8	77-23
Na ox	1072.7	989.8	2062.5	77-23 r
Na ox			2062.0	74-8
Na ox		998.8		66-2 n
Na_3Sb			2064.8	80-3
Na_2KSb			2065.1	80-3
$NaBiO_3$	1071.3	990.9	2062.2	79-11 r
Na_2MoO_4	1070.9	991.0	2061.9	79-11 r
Na_2CrO_4	1071.2	990.9	2062.1	79-11 r
$Na_2Cr_2O_7$	1071.6	990.4	2062.0	79-11 r
Na_2PdCl_4	1071.8	990.2	2062.0	75-7 r
NaCl	1071.6	990.3	2061.9	75-7 r
NaCl			2062.8	79-6
NaSCN	1071.3	990.5	2061.8	75-7 r
Na_2SeO_3	1070.8	991.0	2061.8	75-7 r
$Na_2S_2O_4$	1071.2	990.6	2061.8	75-7 r
$Na_2S_2O_3$	1071.6	990.1	2061.7	75-7 r
Na_2WO_4	1071.3	990.4	2061.7	79-11 r
Na_2SO_3	1071.4	990.2	2061.6	79-11 r
Na_2SO_3			2062.6 c	80-20 r
Na thioglycollate	1071.2	990.4	2061.6	79-11 r
Na_2TeO_4	1071.1	990.5	2061.6	75-7 r
$NaAsO_2$	1070.9	990.7	2061.6	75-7 r
Na_2HPO_4	1071.6	989.9	2061.5	79-12 r
Na_2HPO_4	1071.5	989.7	2061.2	82-3
$Na_2SnO_3 \cdot 3H_2O$	1071.1	990.3	2061.4	79-11 r
$NaNO_2$	1071.6	989.8	2061.4	75-7 r
Na_2CO_3	1071.5	989.8	2061.3	79-11 r
$Na_2C_2O_4$	1070.8	990.5	2061.3	79-11 r

SODIUM—*cont.*	1s	$KL_{23}L_{23}$	α'	Ref.
Na_3PO_4	1071.1	990.2	2061.3	82-3
Na salt of EDTA*	1070.8	990.4	2061.2	79-11 r
$NaHCO_3$	1071.3	989.8	2061.1	79-11 r
$Na_2IrCl_6 \cdot 6H_2O$	1071.9	989.2	2061.1	75-7 r
albite $(NaAlSi_3O_8)$	1072.2	988.9	2061.1	82-7 r
NaH_2PO_4	1072.0	989.1	2061.1	82-3
$NaPO_3$	1071.7	989.3	2061.0	75-7 r
$NaPO_3$	1071.6	989.4	2061.0	82–3
Na_2SO_4	1071.2	989.8	2061.0	75-7 r
Na_2SO_4			2062.1c	80-20 r
Na_2SO_4			2061.9	79-6
NaOAc	1071.1	989.9	2061.0	75-7 r
Na benzenesulphonate	1071.3	989.7	2061.0	79-11 r
NaH_2PO_2	1071.1	989.8	2060.9	82-3
NaOOCH	1071.1	989.8	2060.9	79-11 r
natrolite $(Na_2Al_2Si_3O_{10} \cdot 2H_2O)$	1072.4	988.5	2060.9	82-7 r
$NaNO_3$	1071.4	989.4	2060.8	79-11 r
$NaNO_3$			2062.2	79-6
Na zeolite A $(NaAlSiO_4)$	1071.7	988.9	2060.6	82-7 r
hydroxysodalite	1070.5	989.8	2060.3	82-7 r
Na_2ZrF_6	1071.5	988.7	2060.2	75-7 r
Na_2TiF_6	1071.6	988.5	2060.1	75-7 r
Na_3AlF_6	1071.9	988.2	2060.1	79-11 r
NaF	1071.2	988.6	2059.8	75-7 r
NaF			2061.0	79-6
$NaBF_4$	1072.7	987.1	2059.8	75-7 r
Na_2GeF_6	1071.7	988.1	2059.8	75-7 r
Na_2SiF_6	1071.7	987.7	2059.4	75-7 r
Na (g)	1078.6 v	976.7 v	2055.3	74-6

*Ethylenediaminetetracetic acid

MAGNESIUM	2p	$KL_{23}L_{23}$	α'	1s	Ref.
Mg	49.95	1185.5	1235.45		79-12 r
Mg		1185.9			78-7 n
Mg	49.5*	1185.5	1235.0	1303.2 c	78-6 n
Mg	49.6	1185.7	1235.3		77-11 r
Mg	49.6	1185.6	1235.2	1303.2 c	77-8
Mg		1184.9			77-7 r
Mg	49.4	1186.5 c	1235.9 c	1303.0 c	75-3, 76-4
Mg	49.4	1185.3	1234.7	1303.0	75-5
Mg	50.0	1185.6	1235.6		75-7 r
Mg		1183.0 v			77-20

*From $2s = 88.6$, assuming $2s - 2p = 39.1$.

MAGNESIUM—*cont.*	2p	$KL_{23}L_{23}$	α'	1s	Ref.
Mg$_2$Cu	49.5	1186.0	1235.5	1302.6	75-2*
Mg$_3$Au	49.4	1185.7	1235.1	1302.7	75-2*
Mg$_3$Bi	50.3	1184.9	1235.2	1303.6	75-2*
Mg → Mg ox	Δ +1.3	−6.1	−4.8		79-11
Mg → Mg ox	Δ +1.4	−5.9	−4.5		77-11
Mg → Mg ox	Δ +0.8	−5.1	−4.3	+1.3	77-8
Mg → Mg ox	Δ +1.2	−5.2	−4.0	+1.5	77-7 r
Mg → Mg ox	Δ +1.5	−6.2	−4.7	+0.6	75-3, 76-4
Mg → Mg ox	Δ	−6.2			73-8
MgO	50.4	1180.4	1230.8	1304.0	79-11 r
MgO (single crystal)			1231.6		77-11
Mg acetylacetonate	50.1	1180.5	1230.6	1304.0	79-11 r
Mg cyclohexanebutyrate	50.7	1180.0	1230.7	1304.2	79-11 r
Mg erucate	50.7	1180.2	1230.9	1304.4	79-11 r
MgSeO$_4$	51.1	1180.7	1231.8	1304.7	79-11 r
MgSO$_4$·7H$_2$O	51.6	1178.8	1230.4	1305.2	79-11 r
talc, Mg$_3$Si$_4$O$_{11}$·H$_2$O	50.46	1180.3	1230.76		82-7 r
MgF$_2$	50.95	1178.15	1229.1	1305.0	80-15 r
MgF$_2$	50.9	1178.0	1228.9	1304.8	79-11 r
Mg (g)		1167.0 v			77-20
Mg (g)		1167.1 v			74-3
Mg (s) → Mg (g)		Δ −16.0			77-20

*Line energies from 75-2 corrected by +0.3 eV kinetic energy, based upon later work from same laboratory, 77-8.

ALUMINIUM	2p	$KL_{23}L_{23}$	α'	1s	α'	Ref.
Al	72.92	1393.21	1466.13			81-8 r, 82-7 r
Al	72.85	1393.29	1466.14			82-4
Al		1393.2				78-7 n
Al		1393.2				77-7
Al		1393.0				76-3 r
Al			1466.2			79-15
Al				1558.2	2953.2	79-17
AlAs	73.6	1391.2	1464.8			82-4
AlN	74.0	1388.9	1462.9			81-5

ALUMINIUM—*cont.*	2p	$KL_{23}L_{23}$	α'	1s	α'	Ref.
Al ox	74.15*	1387.48	1461.63			81-8 r
Al → Al ox	Δ +2.7	−7.4	−4.7			82-4
Al → Al ox	Δ +2.8	−7.4	−4.6	+3.0		77-7
Al → Al ox		−6.5		+3.0	−3.5	76-11
Al_2O_3 sapphire	74.10	1387.87	1461.97			81-8 r
Al_2O_3 sapphire,						
heated 450°	74.32	1387.68	1462.00			81-8 r
α-Al_2O_3	73.85	1388.24	1462.09			81-8 r
Al_2O_3 corundum					2948.5	82-6
Al_2O_3			1461.9			79-15
γ-Al_2O_3	73.72	1387.83	1461.55			81-8 r
AlO(OH)						
boehmite	74.22	1387.60	1461.82			81-8 r
$Al(OH)_3$						
bayerite	73.90	1387.62	1461.52			82-7 r
$Al(OH)_3$						
bayerite	74.3	1387.7	1462.0			82-4
$Al(OH)_3$ gibbsite	74.00	1387.43	1461.43			81-8 r
AlF_3			1460.7			79-15
kaolinite	74.68	1386.73	1461.41			82-7 r
kaolinite					2948.6	79-17
pyrophyllite	74.71	1386.75	1461.46			82-7 r
pyrophyllite					2948.9	82-6
muscovite mica	74.25	1387.06	1461.31			82-7 r
muscovite mica					2948.8	82-6
albite	74.34	1386.47	1460.81			82-7 r
albite					2947.7	82-6
natrolite	74.25	1386.53	1460.78			82-7 r
					2947.8	82-6
spodumene	74.32	1387.13	1461.45			82-7 r
sillimanite	74.58	1386.86	1461.44			82-7 r
andalusite					2948.5	82-6
almandine					2948.9	82-6
anorthite					2948.5	82-6
beryl					2949.2	82-6
cordierite					2948.7	82-6
epidote					2948.6	82-6
kyanite					2948.85	82-6
microcline					2946.9	82-6
plagioclase					2948.0	82-6
staurolite					2949.1	82-6
stilbite					2946.9	82-6
sodalite					2947.9	82-6

*From C 1s charge reference

ALUMINIUM—*cont.*	2p	$KL_{23}L_{23}$	α'	1s	α'	Ref.
molecular sieve type A	73.66	1386.90	1460.56			81-8 r
molecular sieve type X	74.13	1386.25	1460.38			81-8 r
molecular sieve type Y	74.45	1385.85	1460.30			81-8 r
H Zeolon	74.82	1385.52	1460.34			82-7 r
hydroxysodalite	73.95	1486.35	1460.3			82-7 r
SILICON						
Si	99.44	1616.68	1716.12			81-8 r
Si	99.6	1616.3	1715.9			81-4
Si	99.6	1616.4	1716.0			80-6 n
Si vapour deposited	99.7	1616.2	1715.9	1839.3 c	3455.5 c	74-7 r
Si		1616.4				77-7
Si					3456.3	79-17
$PdSi_{0.6}$	99.8	1617.4	1717.2			80-16 r
$MoSi_2$	99.56	1617.20	1716.76			81-8 r
$MoSi_x$	99.4	1617.4	1716.8			81-8 r
SiC			1714.1			80-6
SiC					3453.7	82-6
Si_3N_4	101.9*	1612.2	1714.1			81-4
Si_3N_4			1713.7		3454.15	80-6, 82-6
phenylsilicone resin	102.74	1609.96	1712.70			82-7 r
methylsilicone resin	102.92	1608.80	1711.72			82-7 r
poly-dimethyl-silicone	102.40	1609.38	1711.78			82-7 r
Si ox	103.43	1608.27	1711.70			81-8 r
Si ox (on Si100)	103.4	1608.6	1712.0			81-4
SiO_2 (on Si100)			1711.9			81-4
SiO_2 vapour deposited	103.4	1608.8	1712.2	1842.7 c	3451.5 c	74-7 r
SiO_2 α-cristobalite	103.25	1608.64	1711.89			81-8 r
SiO_2 α'-quartz	103.65	1608.6	1712.25			82-7 r
SiO_2 quartz			1712.2		3452.4	80-6, 82-6
SiO_2 Vycor	103.5	1608.5	1712.0			78-8 r

*Formed by N_2^+ bombardment of Si(100) wafer.

SILICON—*cont.*	$2p$	$KL_{23}L_{23}$	α'	$1s$	α'	Ref.
SiO$_2$ gel	103.59	1607.87	1711.46			82-7 r
SiO$_2$ gel	104.1	1607.4	1711.5			78-8 r
SiO$_2$ gel			1711.3			80-6
ZnSiO$_3$			1711.8			80-6
hemimorphite	101.96	1610.52	1712.48			82-7 r
					3452.7	82-6
wollastonite	102.36	1609.99	1712.35			82-7 r
pseudo-						
wollastonite	102.16	1610.27	1712.43			82-7 r
talc	103.13	1608.93	1712.06			82-7 r
talc			1712.3		3453.6	80-6, 82-6
kaolinite	102.98	1609.03	1712.01			82-7 r
kaolinite			1711.9		3451.5	80-6, 79-17
pyrophyllite	102.88	1609.20	1712.08			82-7 r
pyrophyllite			1712.1		3453.1	80-6, 82-6
muscovite mica	102.36	1609.64	1712.00			82-7 r
muscovite			1712.0		3452.5	80-6, 79-17
sillimanite	102.64	1609.48	1712.12			82-7 r
spodumene	102.46	1609.59	1712.05			82-7 r
almandine			1712.4		3453.0	80-6, 82-6
anorthite			1712.3		3452.4	80-6, 82-6
biotite			1712.15			80-6
bentonite			1712.1			80-6
lepidolite			1712.0			80-6
microcline			1711.95		3452.0	80-6, 82-6
beryl			1711.7		3452.1	80-6, 82-6
stilbite			1711.7		3451.9	80-6, 82-6
andalustite					3452.6	82-6
staurolite					3453.6	82-6
epidote					3452.7	82-6
uvarovite					3452.2	82-6
kyanite					3452.6	82-6
cordierite					3452.7	82-6

SILICON—*cont.*	$2p$	$KL_{23}L_{23}$	α'	$1s$	α'	Ref.
olivine					3452.2	82-6
enstatite					3452.6	82-6
asbestos					3452.5	82-6
zircon					3452.65	82-6
serpentine					3453.1	82-6
soda glass	102.95	1608.72	1711.67			82-7 r
albite	102.63	1609.26	1711.89			82-7 r
albite					3452.3	82-6
natrolite	102.22	1609.62	1711.84			82-7 r
natrolite					3452.4	82-6
hydroxysodalite	101.65	1610.7	1712.35			82-7 r
sodalite					3452.35	82-6
plagioclase					3452.4	82-6
molecular sieve type A	101.43	1610.09	1711.52			81-8 r
molecular sieve type X	102.16	1609.40	1711.56			81-8 r
molecular sieve type Y	102.84	1608.63	1711.47			81-8 r
H Zeolon	103.28	1608.40	1711.68			82-7 r
$SiCl_4$ (g)	110.17 v	1600.16 v	1710.33			80-10
$Si(OMe)_4$ (g)	107.70 v	1601.81 v	1709.51			80-10
$Si(OEt)_4$ (g)	107.56 v	1602.27 v	1709.83			80-10
$SiMe_4$ (g)	105.94 v	1603.74 v	1709.68			80-10
$SiEt_4$ (g)	106.03 v	1604.3 v	1710.33			80-10
$SiCl_3Ph$ (g)	108.81 v	1601.95 v	1710.76			80-10
$SiCl_3C_3H_5$ (g)	108.92 v	1601.51 v	1710.43			80-10
$SiCl_3Et$ (g)	108.97 v	1601.34 v	1710.31			80-10
$SiCl_3C_2H_3$ (g)	109.05 v	1601.29 v	1710.34			80-10
$SiCl_3Me$ (g)	109.15 v	1600.96 v	1710.11			80-10
$SiCl_3H$ (g)	109.44 v	1600.3 v	1709.74			80-10
$SiCl_2MeC_2H_3$ (g)	108.07 v	1602.08 v	1710.15			80-10
$SiCl_2Et_2$ (g)	107.85 v	1602.52 v	1710.37			80-10
$SiCl_2Me_2$ (g)	108.10 v	1601.82 v	1709.92			80-10
$SiCl_2MeH$ (g)	108.53 v	1601.17 v	1709.70			80-10
$SiClMe_2C_3H_5$ (g)	106.98 v	1603.18 v	1710.16			80-10
$SiClEt_3$ (g)	106.6 v	1603.8 v	1710.4			80-10
$SiClMe_3$ (g)	107.06 v	1602.80 v	1709.86			80-10
$SiMe_3CH_2Cl$ (g)	106.23 v	1603.62 v	1709.85			80-10
$SiMe(OEt)_3$ (g)	107.09 v	1602.70 v	1709.79			80-10
$SiMe_2(OEt)_2$ (g)	106.69 v	1603.00 v	1709.69			80-10
$SiMe_3OEt$ (g)	106.29 v	1603.29 v	1709.58			80-10
SiF_4 (g)	111.70 v	1595.34 v	1707.04			80-10
SiH_4 (g)	107.1 v	1601.2 v	1708.3	1847.0	3448.2	74-11, 79-4

PHOSPHORUS	$2p$	$KL_{23}L_{23}$	α'	$1s$	Ref.
Gap	128.7	1858.9	1987.6		80-16 r
P (red)	130.2	1857.0	1987.2		81-3 r
$NaPO_3$	134.7	1848.3	1983.0		78-8 r
Na_2HPO_4	133.1	1850.8	1983.9		80-16 r
$SPCl_3$ (g)	141.15 v	1842.6 v	1983.75		79-4
$OPCl_3$ (g)	141.35 v	1841.3 v	1982.65		79-4
PCl_3 (g)	140.15 v	1842.4 v	1982.55		79-4
SPF_3 (g)	142.85 v	1839.1 v	1981.95		79-4
C_5H_5P, phosphazene (g)	136.1 v	1845.3 v	1981.4		79-1
PF_5 (g)	144.65 v	1836.2 v	1980.85		79-4
OPF_3 (g)	143.25 v	1836.9 v	1980.15		79-4
PF_3 (g)	142.05 v	1837.4 v	1979.45		79-4
PH_3 (g)	137.35 v	1842.0 v	1979.35	2150.5 v	79-4
PH_3 (g)	137.3 v	1841.4 v	1978.7		79-1
SULPHUR					
WS_2	162.8	2115.6	2278.4		78-8 r
$NiWS_2$	162.6	2115.9	2278.5		78-8 r
NiS	162.8	2116.1	2778.9		80-16 r
$Na_2S_2O_3$ (central S)	168.6	2107.8	2276.4		78-8 r
(peripheral S)	162.5	2112.5	2275.0		78-8 r
Na_2SO_3	166.6	2108.5	2275.1		82-8 r
$CuSO_4$	169.1	2108.0	2277.1		80-16 r
SF_6	174.4	2100.45	2274.85		82-8 r
SF_6 (g)	180.4 v	2092.6 v	2273.0	2490.1 v	76-6
SF_6 (g)	180.28 v	2092.52 v	2272.80		77-2
CS_2	163.6	2111.65	2275.25		82-8 r
CS_2 (g)	170.03 v	2101.40 v	2271.43		76-1
SO_2	167.4	2106.2	2273.6		82-8 r
SO_2 (g)	174.8 v	2095.5 v	2270.3	2483.7 v	76-6
SO_2 (g)	174.84 v	2095.40 v	2270.24		77-2
COS (g)	170.8 v	2099.2 v	2270.0		76-1
H_2S (g)	170.2 v	2098.7 v	2268.9	2478.5 v	76-6
H_2S (g)	170.44 v	2098.42 v	2268.86		77-2
H_2S (g)		2099.1 v		2477.7 v	79-4
		LLV			
S (s)	164.25	152	316		82-5

CHLORINE	$2p_{3/2}$	*LVV*	α'	$1s$	Ref
poly(vinyl chloride)	200.1	182.3	382.4		79-12 r
KCl	199.3	181.0	380.3		77-18 r
$KClO_3$	206.5	181.0	387.5		77-18 r
$KClO_4$	208.7	180.7	389.4		77-18 r

CHLORINE—*cont.*	$2p_{3/2}$	$KL_{23}L_{23}$	α'	$1s$	Ref.
CCl_4 (g)	207.04 v	2375.72 v	2582.76		80-1
$CHCl_3$ (g)	206.86 v	2375.52 v	2582.38		80-1
CCl_3F (g)	207.20 v	2374.93 v	2582.13		80-1
t-C_4H_9Cl (g)	205.38 v	2376.64 v	2582.02		80-1
i-C_3H_7Cl (g)	205.62 v	2376.17 v	2581.79		80-1
n-C_3H_7Cl (g)	205.81 v	2375.77 v	2581.58		80-1
CH_2Cl_2 (g)	206.62 v	2375.15 v	2581.77		80-1
CCl_2F_2 (g)	207.47 v	2374.18 v	2581.66		80-1
C_2H_5Cl (g)	205.92 v	2375.46 v	2581.38		80-1
Cl_2 (g)	207.82 v	2373.72 v	2581.54		80-1
$CClF_3$ (g)	207.83 v	2373.30 v	2581.13		80-1
CH_3Cl (g)	206.26 v	2374.51 v	2580.77		80-1
ClF (g)	209.18 v	2370.73 v	2579.91		80-1
HCl (g)	207.38 v	2371.98 v	2579.36		80-1
HCl (g)		2372.2 v		2829.2 v	79-4

ARGON	$2p_{3/2}$	$L_3M_2M_3^*$	$L_2M_2M_3^*$	α'	$1s$	$KL_{23}L_{23}$	Ref.
Ar (implanted in Fe)	241.7	216.9*		458.6			75-7 r
Ar (implanted in Be)		211.0 v	212.8 v				79-13 n
Ar (adsorbed on Ag)		210.6 v	212.7 v				76-10
Ar (multilayer on Ag)		207.6 v					76-10
Ar (g)	248.62 v	203.49 v	205.61 v	453.17[†]			73-2
Ar (g)					3205.9 v	2660.51 v	69-1, 77-3

*Centre of unresolved peaks.
[†]Calculated for mean of Auger peaks listed.

POTASSIUM

K				547.9[†]	80-3
K_3Sb				546.8[†]	80-3
Na_2KSb				547.0[†]	80-3

KI	292.8		250.8*	543.6	75-7 r
KBr	293.1	248.3	250.7	542.6	79-12 r
KCl	292.5		250.4*	542.9	82-5

KNO_3	292.9		249.3*	542.2	75-7 r
K_2SO_4				542.5	80-20
K_2SO_3				542.3	80-20
KF	292.5		249.6*	542.1	75-7 r
$KSbF_6$	293.7		248.6*	542.3	75-7 r
K (g)	300.7 v	236.67 v	239.51 v	540.2[†]	81-7

*Peaks composed mainly of $L_3M_{23}M_{23}{}^1D_2$ and $L_2M_{23}M_{23}{}^1D_2$ respectively. Values centred between these columns are for the unresolved doublet.
[†]Auger parameter based upon $L_2M_{23}M_{23}$. The others are based upon Auger energies of the unresolved doublet.

CALCIUM	$2p_{3/2}$	$L_3M_2M_3^*$	$L_2M_2M_3^*$	α'	Ref.
Ca	345.9	298.2	644.1		80-12 r
CaCl$_2$	348.3	291.9	640.2		75-7 r
CaO	346.5	291.9	638.4		82-5
CaO	347.3	292.5	639.8		80-12 r
Ca → CaO	Δ +1.4	−5.7	−4.3		80-12
CaSO$_4$	347.6	291.2	638.8		82-5 r
CaCO$_3$	347.0	291.8	638.8		79-12 r
wollastonite					
(Ca silicate)	347.0	291.5	638.5		82-7 r
CaF$_2$	347.9	288.9	636.8		82-5
SCANDIUM					
Sc$_2$O$_3$	401.9	334.9	736.8		79-12 r
Sc oxalate	403.3	333.5	736.8		82-5
Sc acetylacetonate	402.2	333.4	735.6		82-5
ScF$_3$	405.0	329.8	734.8		82-5

*Peaks composed mainly of $L_3M_{23}M_{23}{}^1D_2$ and $L_2M_{23}M_{23}{}^1D_2$, respectively. Values centred between these columns are for the unresolved doublet.

TITANIUM	$2p_{3/2}$	$L_3M_{23}M_{23}$	α'	$KL_{23}L_{23}$	α'	Ref.
Ti			836.6			77-6
TiO$_2$			840.3			77-6
		$L_3M_{23}V$				
Ti	454.0	419.1	873.1			79-12 r
TiN	455.7	420.0	875.7			79-11 r
TiC	454.6	418.2	872.8			79-11 r
TiO$_2$	458.7	414.9	873.6			82-5
TiO$_2$	458.5	414.7	873.2			79-11 r
TiO acetylacetonate	458.4	414.8	873.2			79-11 r
titanocene dichloride	457.2	414.9	872.1			79-11 r
Na$_2$TiF$_6$	462.6	409.8	872.4			79-11 r
K$_2$TiF$_6$	462.1	409.4	871.5			79-11 r
Ti → TiC	Δ +1.3			−1.5	−0.2	69-2
Ti → TiN	Δ +1.5			−1.9	−0.4	69-2
Ti → TiO	Δ +1.0			−1.6	−0.6	69-2
Ti → TiO$_2$	Δ +4.9			−5.8	−0.9	69-2
VANADIUM						
V	512.15	472.0	984.15			79-12 r
V → VC	Δ + 1.8			−2.4	−0.6	69-2
CHROMIUM						
Cr	574.3	527.2	1101.5			79-12 r

MANGANESE	$2p_{3/2}$	$L_3M_{23}V$	α'	Ref.
Mn	639.0	586.4	1225.4	79-12 r
Mn		582.8 v		81-6
MnO_2	642.4	584.9	1227.3	82-5
MnO_2	642.3	583.9	1226.2	79-11 r
$MnCl_2$	642.8	580.9	1223.7	79-11 r
$MnSiO_3$	642.3	582.4	1224.7	79-11 r
$K_3Mn(CN)_6$	641.7	582.0	1223.7	79-11 r
$Mn(C_{24}H_{27}N_7)(PF_6)_2$*	640.8	582.2	1223.0	79-11 r
Mn (g)		561.6 v		81-6
Mn (s) $\rightarrow$ Mn (g)		Δ -21.2		81-6

*The nitrogen ligand contains three pyridine rings.

IRON	$2p_{3/2}$	L_3VV*	α'	Ref.
Fe	706.95	702.4	1409.35	79-12 r
Fe_2O_3	711.0	703.1	1414.1	79-11 r
Fe_2O_3	710.9	702.0	1412.9	82-5
$FeWO_4$	711.5	703.1	1414.6	79-11 r
$Fe_2(WO_4)_3$	711.1	702.5	1413.6	79-11 r
FeS	710.4	703.2	1413.6	79-11 r
FeS_2	707.4	702.7	1410.1	79-11 r
Fe^{II} acetylacetonate	711.5	700.8	1412.3	79-11 r
Fe^{III} acetylacetonate	711.8	700.3	1412.1	79-11 r
Fe cyclohexanebutyrate	712.0	700.8	1412.8	79-11 r
Fe dithiodibutylcarbamate	711.3	701.2	1412.5	79-11 r
$FeSO_4 \cdot 7H_2O$	711.0	700.4	1411.4	79-11 r
$Fe(C_{10}H_8N_2)_3(PF_6)_2$†	708.2	699.6	1407.8	79-11 r
$K_3Fe(CN)_6$	709.9	698.4	1408.3	79-11 r
$K_4Fe(CN)_6$	708.5	698.9	1407.4	79-11 r
K_3FeF_6	713.8	698.6	1412.4	79-11 r

*The Auger line for all but Fe and FeS_2 is very broad, ca. 8 eV wide, and accuracy of the line
energy is therefore limited.
†Ligand is $C_5H_5NCH=NCH_3$.

COBALT	$2p_{3/2}$	L_3VV	α'	Ref.
Co	778.1	773.2	1551.3	79-12 r
Co	778.2	773.0	1551.2	77-13
Co_3O_4	780.0	773.6	1553.6	82-5
Co_3O_4	780.3	773.9	1554.2	79-11 r
Co_3O_4	779.3	773.6	1552.9	77-13
CoO	780.2	773.6	1553.8	82-5

COBALT—*cont.*	$2p_{3/2}$	L_3VV	α'	Ref.
$CoSiO_3$	781.5	770.4	1551.9	79-11 r
Co_2SiO_4	781.3	770.6	1551.9	79-11 r
Co cyclohexanebutyrate	781.6	770.0	1551.6	79-11 r
Co dibutyldithiocarbamate	779.5	770.2	1549.7	79-11 r
$Co(NH_3)_6Cl_3$	781.7	768.6	1550.3	79-11 r
$Co(N_4$-tetramethylethylenediamine$)(NO_3)_2$	780.0	770.1	1550.1	79-11 r
$K_3Co(CN)_6$	781.9	766.8	1548.7	79-11 r
$Co(C_{24}H_{27}N_7)(PF_6)_2$*	780.5	773.4	1553.9	79-11 r
$CoSiF_6$	783.6	768.2	1551.8	79-11 r

*The nitrogen ligand contains three pyridine rings.

NICKEL	$2p_{3/2}$	L_3VV	α'	Ref.
Ni	852.7*	846.2*	1698.9*	82-1* r
Ni	852.5	846.2	1698.7	79-12 r
Ni	852.6	845.9	1698.5	76-5
Ni	852.7	845.8	1698.5	74-4 r
Ni	852.9	846.1	1699.0	80-8
Ni (s) → Ni (implanted in C)	Δ +0.9	−0.7	+0.2	80-8
NiO	853.5	846.0*	1699.5	82-5
NiO	853.5	846.4*	1699.9	79-11 r
NiO	854.1	845.8	1699.9	74-4
Ni cyclohexanebutyrate	856.3	842.5	1698.8	79-11 r
Ni acetylacetonate	855.7	842.9	1698.6	79-11 r
KNi biuret	856.8	841.9	1698.7	79-11 r
Ni trifluoroacetate	856.9	841.8	1698.7	79-11 r
$NiSiO_3$	856.9	841.4	1698.3	79-11 r
Ni dimethylglyoxime	854.8	842.4	1697.2	79-11 r
$Ni(C_{24}H_{27}N_7)(PF_6)_2$†	855.4	842.1	1697.5	79-11 r
$K_2Ni(CN)_4$	855.4	840.3	1695.7	79-11 r
NiF_2	857.4	842.4	1699.8	79-11 r
$NiSiF_6$	858.7	840.4	1699.1	79-11 r

*Taken as the centre of a broad square line.
†The nitrogen ligand contains three pyridine rings.

COPPER	$2p_{3/2}$	$L_3M_{45}M_{45}$	α'	Ref.
Cu	932.67*	918.65*	1851.32*	82-2*
Cu	932.68*	918.62*	1851.30*	80-4*
Cu	932.6*	918.8*	1851.4*	82-1*
Cu	932.6	918.4	1851.0	79-12 r

COPPER—*cont.*	$2p_{3/2}$	$L_3M_{45}M_{45}$	α'	Ref.
Cu	932.8	918.5	1851.3	77-9 r
Cu	932.8	918.3	1851.1	73-2
Cu	932.6	918.6	1851.2	75-7 r
Cu	932.6	918.8	1851.4	77-10
Cu	932.7 c	918.6	1851.3	76-5
Cu	932.2	919.0	1851.2	73-7
Cu	932.4	919.0	1851.4	76.7
Cu	933.0	918.4	1851.4	78-5
Cu	933.1	918.2	1851.3	78-2
Cu	932.8	918.0	1850.8	73-5
Cu	933.0	918.1	1851.1	73-2
Cu	932.6	918.9	1851.5	81-2 r
Cu		918.9		70-3
Al_2Cu	933.9	918.0	1851.9	77-9 r
CuAgSe	932.3	917.6	1849.9	78-5
Cu_2Se	932.3	917.5	1849.8	78-5
CuSe	932.4	918.3	1850.7	78-5
Cu_2S	932.5	917.4	1849.9	75-7 r
$Cu \rightarrow Cu_2S$	$\Delta + 0.07$	-1.37	-1.30	82-5
$Cu \rightarrow Cu_2S$	$\Delta + 0.1$	-1.8	-1.7	74-10
CuS	932.6	917.8	1850.4	78-5
Cu_2O	932.6	916.6	1849.2	82-5
Cu_2O	932.4	917.2	1849.6	77-10
Cu_2O	932.2	917.6	1849.8	76-7 r
Cu_2O	933.1	916.2	1849.3	78-2 r
$Cu \rightarrow Cu_2O$	$\Delta -0.11$	-2.00	-2.11	82-5
$Cu \rightarrow Cu_2O$	$\Delta + 0.1$	-2.3	-2.2	74-10
$Cu \rightarrow Cu_2O$	Δ	-2.3		81-2
$Cu \rightarrow Cu_2O$	$\Delta\ 0.0$	-2.2	-2.2	73-5
CuO	933.8	917.9	1851.7	82-5
CuO	933.6	918.1	1851.7	77-10
CuO	933.8	917.8	1851.6	78-2 r
CuO	933.5	917.9	1851.4	76-7 r
CuO	933.0	917.9	1850.9	73-7
$Cu \rightarrow CuO$	$\Delta +0.96$	-0.88	$+0.08$	82-5
$Cu \rightarrow CuO$	$\Delta +1.2$	-1.0	$+0.2$	74-10
$Cu \rightarrow CuO$	$\Delta +1.3$	-0.8	$+0.5$	81-2
$Cu \rightarrow Cu(OH)_2$	$\Delta +2.5$	-2.7	-0.2	81-2
CuCl	932.4	915.6	1848.0	77-10
CuCl	932.6	915.0	1847.6	75-7 r
$CuCl_2$	934.4	915.5	1849.9	77-10

COPPER—cont.	$2p_{3/2}$	$L_3M_{45}M_{45}$	α'	Ref.
$CuCl_2$	935.2	915.1	1850.3	79-11 r
$CuCl_2$	934.8	915.3	1850.1	81-1 r
$CuBr_2$	933.3	916.9	1850.2	81-1 r
CuCN	933.1	914.5	1847.6	75-7 r
$Cu_2Mo_3O_{10}$	932.0	916.5	1848.5	78-2 r
$CuSO_4$	935.5	915.9	1851.4	82-5
$CuSO_4$	935.5	915.6	1851.1	77-18 r
$Cu(NO_3)_2$	935.5	915.3	1850.8	77-18 r
$CuCO_3$	935.0	916.3	1851.3	79-11 r
$CuMoO_4$	934.5	916.6	1851.1	78-2 r
$CuSiO_3$	934.9	915.2	1850.1	79-11 r
$CuC(CN)_3$	933.2	914.5	1847.7	77-18 r
$Cu(C_{24}H_{27}N_7)(PF_6)_2$*	934.0	915.9	1849.9	79-11 r
CuF_2	937.0	914.8	1851.8	79-11 r
CuF_2	936.1	916.0	1852.1	77-10
CuF_2	936.8	914.4	1851.2	81-1 r
Cu → Cu atoms in SiO_2	Δ 0.7	−4.1	−3.4	79-18
Cu → Cu (g)	Δ 2.5	−13.2	−10.7	82-11

*The nitrogen ligand contains three pyridine rings.

ZINC	$2p_{3/2}$	$L_3M_{45}M_{45}$	α'	Ref.
Zn	1021.65	992.2	2013.85	79-12 r
Zn	1021.6	992.3	2013.9	77-15 r
Zn	1021.7	992.5	2014.2	73-6
Zn	1021.4 c	992.0	2013.4 c	76-5
Zn	1022.0	991.8	2013.8	74-8
Zn	1021.7 c	992.0	2013.7 c	77-10
Zn	1021.9	992.4	2014.3	74-4 r
Zn	1021.4 c	992.3	2013.7 c	77-12 n
Zn		988.4 c		77-17
Zn		988.2 v		77-20
Zn		991.7		70-3
ZnTe	1021.6 c	991.3	2012.9 c	77-12 n
ZnTe			2012.2	70-1
ZnSe	1022.0 c	989.5	2011.5 c	77-12 n
ZnSe			2010.2	70-1
ZnS	1021.6 c	989.7	2011.3 c	77-10
ZnS	1021.9 c	988.2	2010.1 c	77-12 n
ZnS			2011.9	70-1

ZINC—*cont.*	$2p_{3/2}$	$L_3M_{45}M_{45}$	α'	Ref.
ZnI_2	1022.5 c	988.7	2011.2 c	77-10
ZnI_2			2011.1 c	77-12
$ZnBr_2$			2010.3 c	77-12
$ZnBr_2$	1023.4	987.3	2010.7	75-7 r
ZnO	1021.7	988.2	2009.9	73-6
ZnO	1022.1 c	987.6	2009.7 c	77-12 n
ZnO	1022.1 c	987.7	2009.8 c	77-10
ZnO			2010.3	70-1
ZnO	1021.9	988.6	2010.5	82-5
ZnO	1022.2	987.4	2009.6	74-8
Zn → Zn ox	Δ + 0.3	−4.2	−3.9	79-11
Zn → Zn ox	Δ	−4.2		77-17
Zn → Zn ox	Δ + 0.2	−3.5	−3.3	74-4
$ZnCl_2$			2009.2 c	77-12
Zn acetylacetonate	1021.4	987.7	2009.1	75-7 r
$Zn(C_{24}H_{27}N_7)(BF_4)_2$*	1021.3	988.3	2009.6	79-11 r
hemimorphite (Zn silicate)	1021.96	987.30	2009.26	82-7 r
ZnF_2	1021.8 c	986.2	2008.0 c	77-10
ZnF_2			2008.2 c	77-12 n
ZnF_2	1022.8	986.7	2009.5	75-7 r
ZnF_2			2007.4	74-8
Zn (g)		973.3 v		74-2
Zn (g)		974.5 v		77-20
Zn → Zn (g)	Δ 3.15	−13.1	−9.9	79-7

*The nitrogen ligand contains three pyridine rings.

GALLIUM	3d	$L_3M_{45}M_{45}$	α'	$2p_{3/2}$	Ref.
Ga	18.7	1068.1	1086.8		79-12 r
Ga	18.7	1068.1	1086.8	1116.6 c	82-5, 75-7 r
Ga	18.6	1068.3	1086.9	1116.5 c	73.6
Ga	18.4	1068.2	1086.6		77-22
Ga	18.5	1069.0	1087.5	1116.4 c	74-4 r
Ga	18.5	1068.1	1086.6		78-4 r
GaAs (cleaved)	19.2	1066.4	1085.6		79-3 n
GaAs (cleaved)	19.4	1066.2	1085.6		78-4 r
GaAs (sputtered)	19.0	1067.1	1086.1		78-4 r
GaAs (chem etch)	19.3	1066.2	1085.5		78-4 r
GaP	19.3	1066.2	1085.5		82-5
GaN	19.54	1064.5	1084.04		80-9 r
Ga → Ga ox	Δ +2.0	−6.3	−4.3		82-5
Ga → Ga ox	Δ +2.6	−6.4	−3.8		74-4

GALLIUM—*cont.*	3d	$L_3M_{45}M_{45}$	α'	$2p_{3/2}$	Ref.
Ga_2O_3	21.0	1061.9	1082.9		78-4 r
Ga_2O_3	20.3	1062.8	1083.1	1117.9	73-6 r
GaAs → Ga ox	Δ +1.0	−3.5	−2.5		79-3

GERMANIUM	3d	$L_3M_{45}M_{45}$	α'	$2p_{3/2}$	α'†	Ref.
Ge	29.0	1145.4	1174.4			77-19 r
Ge	29.15	1145.2	1174.35	1217.2 c	2362.4 c	79-12 r
Ge	29.5*	1145.0*	1174.5	1217.4*	2362.4	77-16
Ge	29.0	1145.0	1174.0	1217.0 c	2362.0 c	74-4 r
Ge	29.2	1144.9	1174.1			77-22
$GeTe_2$	30.0	1144.8	1174.8			77-19 r
$GeSe_2$	30.9	1143.8	1174.7			77-19 r
GeS_2	30.5	1143.7	1174.2			77-19 r
Ge → Ge ox	Δ +4.1	−8.0	−3.9	+3.8	−4.2	74-4
Ge → Ge ox	Δ +3.8	−7.9	−4.1	+3.7	−4.2	82-5, 75-7
						75-7 r
GeO_2	32.7	1137.7	1170.4	1220.6	2358.3	82-5,
						75-7 r
GeO_2					2358.9	76-8
Na_2GeF_6	33.3	1135.7	1169.0	1221.3	2357.0	82-5,
						75-7 r
K_2GeF_6					2357.15	76-8
$GeBr_4$ (g)	38.95 v	1130.32 v	1169.27			73-4
$GeCl_4$ (g)	39.6 v	1129.01 v	1168.61			73-4
$GeMe_4$ (g)	35.63 v	1132.64 v	1168.27			73-4
GeH_3Br (g)	37.65 v	1129.81 v	1167.46			73-4
GeH_3Cl (g)	37.77 v	1129.4 v	1167.17			73-4
GeH_3Me (g)	36.44 v	1130.55 v	1166.99			73-4
GeH_4 (g)	36.9 v	1129.5 v	1166.4			73-4
GeF_4 (g)	41.55 v	1124.28 v	1165.83			73-4

*Vacuum-level referenced values corrected by a 4.3 eV work function.
†Based upon $2p_{3/2}$ and $L_3M_{45}M_{45}$.

ARSENIC	3d	$L_3M_{45}M_{45}$	α'	$2p_{3/2}$	Ref.
As	41.8	1225.0	1266.8	1323.4 c	76-2
As	41.5	1226.1	1267.6		75-6 r
As	41.5	1225.2	1266.7	1323.1	82-5, 79-11 r
As	42.1	1225.0	1267.1	1323.7 c	74-4 r
As	41.6	1225.4	1267.0	1323.3	82-4
GaAs (cleaved)	41.3	1224.5	1265.8		79-3 n
GaAs (chem etch)	41.2	1225.0	1266.2	1323.0	82-4

ARSENIC—*cont.*	$3d$	$L_3M_{45}M_{45}$	α'	$2p_{3/2}$	Ref.
GaAs (sputtered)	41.0	1225.5	1266.5	1322.7	82-4
GaAs (sputtered)	40.9	1225.4	1266.3	1322.8	79-12 r
NbAs	40.8	1226.0	1266.8		76-2
As$_2$Te$_3$		1225.0			76-2
As$_2$Se$_3$	43.0	1223.3	1266.3	1324.7	76-2
As$_2$S$_3$	43.5	1222.0	1265.5	1325.6	76-2 r
As$_4$S$_4$	43.1	1222.7	1265.8	1325.1	76-2 r
Ph$_3$AsS	44.1	1220.0	1264.1	1325.9	76-2 r
Me$_3$AsS	44.0	1219.3	1263.3	1325.8	76-2 r
AsI$_3$	43.5	1222.9	1266.4	1325.6	76-2 r
MeAsI$_2$	43.5	1222.3	1265.8	1325.1	76-2 r
AsBr$_3$	45.3	1218.1	1263.4	1327.4	76-2 r
Ph$_3$As	42.4	1221.1	1263.5	1324.3	76-2 r
As$_2$O$_3$	44.4	1218.9	1263.3	1326.7	76-2
As$_2$O$_3$	45.0	1218.8	1263.8	1326.4	79-11 r
As$_2$O$_3$	44.9	1218.7	1263.6	1326.6	82-4 r
As$_2$O$_5$	46.2	1217.4	1263.6	1328.1	76-2 r
As$_2$O$_5$	44.9	1218.6	1263.5	1328.8	74-4 r
NaAsO$_2$	44.2	1219.4	1263.6	1325.6	79-11 r
NaAsO$_2$	44.3	1219.6	1263.9		82-4 r
Na$_2$HAsO$_4$	45.5	1217.1	1262.6	1326.8	79-11 r
Ph$_3$AsO	44.3	1219.5	1263.8	1325.5	76-2
PhAsO(OH)$_2$	45.2	1218.4	1263.6	1326.8	76-2
Ph$_2$AsOOH	44.4	1219.0	1263.4	1326.8	76-2
Me$_2$AsOOH	44.6	1218.4	1263.0	1326.3	76-2
(C$_{10}$H$_{21}$)$_2$AsOOH	44.0	1219.0	1263.0	1325.5	76-2
BuAsO(OH)$_2$	45.1	1218.3	1263.4	1327.0	76-2
KAsF$_6$	47.8	1213.8	1261.6	1330.0	79-11 r
GaAs → As$_2$O$_3$	Δ 3.1	−5.9	−2.8		79-3

	$3s$	$L_2M_{45}M_{45}$			
C$_5$H$_5$As (arsabenzene) (g)	211.2	1248.2			79-1
AsMe$_3$ (g)	211.1	1247.9			79-1
AsH$_3$ (g)	212.4	1245.1			79-1

SELENIUM	$3d_{5/2}$	$L_3M_{45}M_{45}$	α'	$2p_{3/2}$	$\alpha'\dagger$	Ref.
Se	55.5	1307.0	1362.5			80-5 r
Se	55.7	1306.7	1362.4			79-11 r
Se	55.5	1306.6*	1362.1*	1434.2 c	2740.8 c	79-12 r
Se	56.3	1305.8	1362.1	1435.0 c	2740.8 c	74-4 r

SELENIUM—*cont.*	$3d_{5/2}$	$L_3M_{45}M_{45}$	α'	$3p_{3/2}$	$L_3M_{23}M_{45}$	Ref.
Se				161.9	1178.2	77-18 r
$USe_{1.88}$				161.1	1180.0	77-18 r
Ph_2Se	56.0	1303.8	1359.8			80-5 r
Ph_2Se_2	56.0	1304.1	1360.1			80-5 r
$Se\!=\!C(NH_2)_2$	55.0	1305.2	1360.2			80-5 r
Ph_2SeI_2	58.3	1301.9	1360.2			80-5 r
Ph_2SeCl_2	57.9	1302.7	1360.6			80-5 r
$C_7H_8SeCl_3$	58.1	1302.6	1360.7			80-5 r
SeO_2	59.0	1301.4	1360.4			80-5 r
H_2SeO_3	59.2	1300.8	1360.0			80-5 r
H_2SeO_4	61.2	1297.9	1358.1			80-5 r
Ph_2SeO	57.8	1301.7	1359.5			80-5 r
Na_2SeO_3	58.5	1301.2	1359.7			79-11 r
Na_2SeO_3				164.9	1173.1	77-18 r
Na_2SeO_4	60.6	1298.9	1359.5			79-11 r
$MgSeO_4$	59.5	1299.2	1358.7			79-11 r
$(NH_4)_2SeO_4$	59.2	1300.1	1359.3			79-11 r
K_2SeO_4				165.9	1173.1	77-18 r

$^*\pm 0.5$ eV.   †Based upon $2p_{3/2}$ and $L_3M_{45}M_{45}$.

BROMINE	$3d_{5/2}$	$L_3M_{45}M_{45}$	α'	$2p_{3/2}$	Ref.
KBr	68.7	1388.0	1456.7		80-16 r
LiBr	68.9	1389.2	1458.1	1548.8	78-8 r
NaBr	68.9	1388.3	1457.2	1549.0	78-8 r
$C_{16}H_{33}NMe_3Br$	67.5	1390.1	1457.6	1547.5	78-8 r
$KBrO_3$	74.8	1384.4	1459.2	1553.9	78-8 r
tetrabromophenol-sulphonphthalein*	70.5	1387.9	1458.4	1550.5	78-8 r
CBr_4 (g)	76.7 v	1378.9 v	1455.6		70-2
$CHBr_3$ (g)	76.8 v	1378.6 v	1455.4		70-2
CH_2Br_2 (g)	76.6 v	1378.1 v	1454.7		70-2
CH_3Br (g)	76.3 v	1377.6 v	1453.9		70-2

KRYPTON
Kr (g)	93.8 v	1460.4 v	1554.2		73-2, 72-1

STRONTIUM
SrF_2		1640.6			78-8 r

YTTRIUM
Y_2O_3		1736.5			78-8 r

ZIRCONIUM
Zr ox		1831.0			78-8 r

NIOBIUM
Nb ox		1919.7			78-8 r

MOLYBDENUM	$3d_{5/2}$	$L_3M_{45}M_{45}$	α'			
				$M_{45}VV$	α'	
Mo	227.9			222.8	450.7	79-12 r
Mo	228.0	2038.8	2266.8			80-16 r
Mo ox	232.7	2032.2	2264.9			80-16 r
Mo → Mo ox	Δ +4.7	−6.6	−1.9			80-16
MoSi$_2$	227.7	2039.0	2266.7			82-7 r

*The indicator, bromophenol blue.

RUTHENIUM	$3d_{5/2}$	$M_{45}VV$	α'	Ref.
Ru	280.2	274.2	554.4	79-12 r
Ru$_3$(CO)$_{12}$	280.9	271.8	552.7	82-5

RHODIUM				
Rh	307.2	301.3	608.5	79-12 r
Rh acetylacetonate	310.0	298.8	608.8	79-11 r
Rh(NO$_3$)$_3$·2H$_2$O	310.7	297.7	608.4	79-11 r
RhCl$_3$·3H$_2$O	310.0	298.2	608.2	79-11 r
Na$_3$RhCl$_6$	310.0	297.7	607.7	79-11 r
Rh$_6$(CO)$_{16}$	308.6	298.7	607.3	79-11 r
Rh(NH$_3$)$_5$Cl$_3$	310.0	297.1	607.1	79-11 r
(PH$_3$P)$_3$RhBr	309.5	297.5	607.0	79-11 r
(PH$_3$P)$_3$RhHCO	309.5	297.4	606.9	79-11 r
(PH$_3$P)$_3$RhCl	307.5	297.3	604.8	79-11 r

PALLADIUM	$3d_{5/2}$	$M_4N_{45}N_{45}$	α'	Ref.
Pd	335.1	327.8	662.9	79-12 r
Pd	334.6	328.5	663.1	79-14 n
Pd	335.1	327.8	662.9	80-17
Mg$_{75}$Pd$_{25}$	336.2	326.4	662.6	80-17
Al$_{80}$Pd$_{20}$	337.4	325.5	662.9	80-17
Ag$_{30}$Pd$_{50}$	334.6	328.8	663.4	80-17
Ag$_{80}$Pd$_{20}$	334.9	329.8*	664.7	80-17
Ag$_{90}$Pd$_{10}$	334.9	329.7*	664.6	80-17
Pt$_{50}$Pd$_{50}$	334.6	327.5	662.1	79-14 n
Au$_{40}$Pd$_{60}$	334.5	327.5	662.0	79-14 n
Pd ox	336.4	326.3	662.7	79-14 n
PdSO$_4$	338.7	324.8	663.5	79-11 r
Pd acetylacetonate	338.1	324.9	663.0	79-11 r
Pd(NO$_3$)$_2$	338.2	324.7	662.9	79-11 r
PdBr$_2$	337.7	324.9	662.6	79-11 r
Pd(CN)$_2$	338.7	323.7	662.4	79-11 r
PdCl$_2$	338.0	325.2	662.2	79-11 r
(cyC$_8$H$_{12}$)$_2$PdCl$_2$	338.5	323.8	662.3	79-11 r
Pd(NH$_3$)$_4$Cl$_2$	338.4	323.8	662.2	79-11 r
(Ph$_3$P)$_2$PdCl$_2$	338.0	323.6	661.6	79-11 r
Na$_2$PdCl$_4$	338.0	323.4	661.4	79-11 r
K$_2$PdCl$_4$	337.9	323.1	661.0	79-11 r
(Ph$_3$P)$_4$Pd	335.1	324.4	659.5	79-11 r

*Line distribution is unusual.

SILVER	$3d_{5/2}$	$M_4N_{45}N_{45}$	α'	Ref.
Ag	368.28*	357.84*	726.12*	82-2*
Ag	368.1	357.9	726.0	79-12 r
Ag	368.2	358.1	726.3	77-9 r
Ag	368.3	358.0	726.3	77-10
Ag	368.3	358.0	726.3	73-1
Ag	368.4	357.5*	725.9	78-5
Ag	368.2	357.6*	725.8	77-10
Ag	368.2	357.9	726.1	79-9
Ag		357.8		78-3 r
Ag		359.3		71-1
Ag		353.7 v		76-10
$Mg_{21}Ag_{79}$	368.3	358.1*	726.4	80-18
$Mg_{30}Ag_{50}$	368.7	357.9*	726.6	80-18
$Mg_{97}Ag_3$	368.8	358.2*	727.0	80-18
$AlAg_2$	368.7	357.7	726.4	77-9 r
$Al_{40}Ag_{60}$	368.8	357.7*	726.5	80-18
$Al_{95}Ag_5$	369.0	357.5*	726.5	80-18
Ag_2S	368.2	356.8*	725.0	78-5
Ag_2S			724.7	80-20
Ag_2Se	367.9	357.0*	724.9	78-5
$AgCuSe$	367.9	356.9*	724.8	78-5
Ag_2O	368.5	356.0*	724.5	78-5
Ag_2O	367.9	356.6	724.5	73-1
Ag_2O	367.8	356.7*	724.5	77-10
AgO	367.6	357.2	724.8	73-1
AgO	367.4	356.6*	724.0	77-10
AgI	368.0	356.1*	724.1	77-10
$AgOOCCF_3$	368.8	355.1	723.9	75-7 r
Ag_2SO_4	368.3	354.2	722.5	75-7 r
Ag_2SO_4	367.9	355.1*	723.0	77-10
Ag_2SO_4			722.5	80-20
AgF	367.7	355.3*	723.0	77-10
AgF_2	367.3	355.6*	722.9	77-10
Ag (g)		341.8*		80-14

*6.0 eV added to value for $M_5N_{45}N_{45}$ to give value for $M_4N_{45}N_{45}$.

CADMIUM	$3d_{5/2}$	$M_4N_{45}N_{45}$	α'	Ref.
Cd	405.0	383.6	788.6	79-12 r
Cd	404.9	383.7	788.6	75-7 r
Cd	405.0	384.0*	789.0	77-10
Cd	405.3	383.5*	788.8	79-9

CADMIUM—*cont.*	$3d_{5/2}$	$M_4N_{45}N_{45}$	α'	Ref.
Cd		385.6		71-1
Cd		380.5 v*		77-20
Cd		380.4 v		77-24
Cd		380.1 v		76-9
$CdSe_{0.65}Te_{0.35}$	404.9	382.2	787.1	82-10 r
CdTe	405.2	382.4	787.6	82-10 r
CdTe	405.0	382.6*	787.6	77-10
CdSe	405.3	381.5*	786.8	77-10
CdSe	405.0	381.7	786.7	82-10 r
CdS	405.3	381.3*	786.6	77-10
CdI_2	405.4	381.2*	786.6	77-10
CdO	404.2	382.4*	786.6	77-10
$Cd(OH)_2$	405.1	380.0	785.1	79-11 r
CdF_2	405.8	378.8	784.6	75-7 r
CdF_2	405.9	379.0	784.9	77-10
$Cd \rightarrow Cd$ (g)	Δ +2.95	−11.8	−8.85	79-7
Cd (g)		368.2 v		77-24
Cd (g)		368.3 v*		77-20
Cd (g)		367.9 v		74-1

*6.8 eV added to value for $M_5N_{45}N_{45}$ to give value for $M_4N_{45}N_{45}$.

INDIUM

	$3d_{5/2}$	$M_4N_{45}N_{45}$	α'	Ref.
In	443.84	410.41	854.25	79-8
In	443.8	410.4	854.2	79-12 r
In	444.0	410.5	854.5	77-14 r
In	444.3	410.1	854.4	75-7 r
In	444.0	410.2	854.2	79-9
In	443.6	410.5	854.1	80-11 n
In	443.4	411.5	854.9	79-16 n
In		411.1		71-1
InTe	444.3	409.2	853.5	79-11 r
In_2Te_3	444.5	408.9	853.4	79-11 r
InSe	445.0	408.0	853.0	79-11 r
In_2Se_3	444.8	408.3	853.1	79-11 r
InS	444.5	408.3	852.8	79-11 r
In_2S_3	444.7	407.3	852.0	79-11 r
In_2S_3	444.3	408.0	852.3	79-16 n
In_2S_3	444.4	408.9	853.3	80-11 n
InP	444.0	410.0	854.0	80-11 n
In → In ox	Δ +1.2	−3.6	−2.4	75-7
In_2O	444.3	406.8	851.1	77-21 r
In_2O	444.0	406.6	850.6	80-11 n

INDIUM—*cont.*	$3d_{5/2}$	$M_4 N_{45} N_{45}$	α'	Ref.
In$_2$O$_3$	444.3	406.4	850.7	77-21 r
In$_2$O$_3$	444.9	406.3	851.2	77-14 r
In$_2$O$_3$	444.9	407.2	852.1	80-11 n
In$_2$O$_3$	443.7	407.2	850.9	79-16 n
In(OH)$_3$	445.0	405.0	850.0	79-11 r
In(OH)$_3$	445.8	404.8	850.6	80-11 n
In(OH)$_3$	444.6	405.4	850.0	79-16 n
InI$_3$	445.8	405.8	851.6	77-21 r
InI$_3$ (red)	445.3	406.5	851.8	79-16 n
InI$_3$ (yellow)	445.7	405.8	851.5	79-16 n
InBr$_3$	446.0	404.8	850.8	77-21 r
InBr$_3$	445.7	405.2	850.9	79-16 n
InCl	444.8	405.7	850.5	77-21 r
InCl$_3$	446.0	404.6	850.6	77-21 r
InCl$_3$	445.8	403.8	849.6	80-11 n
InCl$_3$	445.7	404.8	850.5	79-16 n
InF$_3$	446.2	403.7	849.9	75-7 r
InF$_3$	445.9	403.7	849.6	79-16 n
(NH$_4$)$_3$InF$_6$	445.6	404.1	849.7	77-21 r
(NH$_4$)$_3$InF$_6$	445.3	404.0	849.3	79-16 n
InCl$_3$ (g)		394.1 v *		80-2
InCl (g)		393.95 v *		80-2
In (g)		393.25 v *		80-2

*7.6 eV added to value for $M_5 N_{45} N_{45}$ to give value for $M_4 N_{45} N_{45}$

TIN	$3d_{5/2}$	$M_4 N_{45} N_{45}$	α'	Ref.
Sn	484.87	437.27	922.14	79-8
Sn	484.85	437.6	922.45	79-12 r
Sn	485.0	437.4	922.4	75-7 r
Sn	484.9	437.5	922.4	77-14 r
Sn	484.8	437.4*	922.2	79-9
Sn		434.0 v		79-5
SnS	485.6	435.7	921.3	75-7 r
Sn → Sn ox	Δ +1.5	−4.7	−3.2	75-7
Sn → Sn ox	Δ	−4.1		79-5
SnO	486.8	432.15	918.95	82-5
SnO	486.7	432.0	918.7	79-11 r
SnO$_2$	487.3	431.8	919.1	79-11 r
SnO$_2$	486.6	432.4	919.0	82-5
SnO$_2$	486.6	432.6	919.2	77-14 r

Tin—*cont.*	$3d_{5/2}$	$M_4N_{45}N_{45}$	α'	Ref.
Na_2SnO_3	486.7	431.7	918.4	75-7 r
$(R_3Sn)_2O$	485.9	432.4	918.3	79-11 r
$NaSnF_3$	487.4	430.8	918.2	75-7 r
$SnCl_2$ (g)		420.24 v		79-8

*8.5 eV added to value for $M_5N_{45}N_{45}$ to give value for $M_4N_{45}N_{45}$.

ANTIMONY	$3d_{5/2}$	$M_4N_{45}N_{45}$	α'	$4d$	Ref.
Sb	528.02	464.29	992.31	31.94	79-8
Sb	528.25	463.9	992.15		79-12 r
Sb	528.2	464.2	992.4		75-7 r
Sb		465.7			71-1
Cs_3Sb			991.5		80-3
K_3Sb			990.7		80-3
Na_2KSb			990.6		80-3
Na_3Sb			990.6		80-3
Sb_2S_3	529.5	462.1	991.6		75-7 r
Sb_2S_5	529.3	462.2	991.5		75-7 r
Sb_2O_5	530.0	459.7	989.7		75-7 r
$KSbF_6$	532.9	454.4	987.3		75-7 r
Sb_4 (g)		452.95 v			78-9, 79-2

TELLURIUM					
Te	572.85	492.13	1064.98	40.26	79-8
Te	572.9	492.2	1065.1		79-12 r
Te	573.1	491.8	1064.9	40.5	77-4
Te		492.2			71-1
Te → Te ox	Δ +3.7	−5.2	−1.5		75-7
CdTe	572.7	490.8	1063.5		82-10 r
$CdSe_{0.65}Te_{0.35}$	572.6	491.1	1063.7		82-10 r
TeO_2	576.1	487.1	1063.2	43.4	77-5
TeO_3	577.3	485.5	1062.8	44.6	77-5
$Te(OH)_6$	577.1	485.1	1062.2	45.0	77-5 r
Ph_2Te_2	573.9	488.5	1062.4	42.8	77-5* r
$PhTeI_3$	575.8	488.2	1064.0	44.8	77-5 r
Ph_2TeI_2	575.4	487.8	1063.2	44.4	77-5 r
Et_2TeI_2	575.3	487.6	1062.9	44.2	77-5 r
Me_2TeI_2	575.6	487.6	1063.2	42.8	77-5 r
$TeBr_4$	576.7	487.3	1064.0	44.0	77-5
$PhTeBr_3$	576.6	486.8	1063.4	43.9	77-5 r
Ph_2TeBr_2	576.2	486.7	1062.9	43.5	77-5 r
$FC_6H_4TeBr_3$	576.3	487.0	1063.3	43.7	77-5 r

TELLURIUM—cont.	$3d_{5/2}$	$M_4N_{45}N_{45}$	α'	$4d$	Ref.
BuTeBr$_3$	576.6	486.5	1063.1	43.9	77-5 r
MePhTeBr$_2$	576.0	486.6	1062.6	43.2	77-5 r
TeCl$_4$	576.9	486.1	1063.0	44.3	77-5
p-MeOC$_6$H$_4$TeCl$_3$	576.7	485.9	1062.6	44.4	77-5 r
Ph$_2$TeCl$_2$	576.2	486.3	1062.5	43.8	77-5 r
(NH$_4$)$_2$TeCl$_6$	576.9	486.4	1063.3	45.3	77-5 r
Te(thiourea)$_2$Cl$_2$	574.7	488.3	1063.0	42.0	77-5 r
p-MeC$_6$H$_4$TeOOH	576.1	486.6	1062.7	43.7	77-5 r
Na$_2$TeO$_4$	576.8	485.5	1062.3		75-7 r
Te$_2$ (g)		479.20 v			79-2, 79-8

*All 77-5 references with 'r' are assumed to be referenced to adventitious carbon C $1s$, necessitating a correction.

IODINE	$3d_{5/2}$	$M_4N_{45}N_{45}$	α'	Ref.
LiI	619.7	517.0	1136.7	79-12 r
NaI	618.9	517.2	1136.1	79-11 r
KI	618.7	517.0	1135.7	79-11 r
NiI$_2$	619.0	518.8*	1137.8	77-10
CuI	619.0	518.6*	1137.6	77-10
ZnI$_2$	619.8	517.5*	1137.3	77-10
AgI	619.4	518.3*	1137.7	77-10
CdI$_2$	619.2	518.5*	1137.7	77-10
InI$_3$	619.9	517.2	1137.1	79-11 r
KIO$_4$	624.1	513.3	1137.4	79-11 r
I$_2$ (g) → I (g)		Δ −3.25		79-10

*11.5 eV added to the value for $M_5N_{45}N_{45}$ to provide value for $M_4N_{45}N_{45}$.

XENON

	$3d_{5/2}$	$M_4N_{45}N_{45}$	α'	Ref.
Xe (implanted in graphite)	669.65	545.2	1214.85	79-12 r
Xe (implanted in diamond)	668.9	545.4*	1214.3	80-7 n
Xe (implanted in Fe)	670.2	544.8	1215.0	75-7 r
Xe (adsorbed on Pt)	669.5	544.2	1213.7	74-5 n
Xe (g) → Xe (adsorbed on Mo)	Δ −1.2	+6.3	+5.1	82-9
Xe (g) → Xe (adsorbed on MoO$_2$)	Δ −3.5	+7.1	+3.6	82-9
Na$_4$XeO$_6$	674.1	541.4	1215.5	77-21
Na$_4$XeO$_6$			1216.3*	75-4
Xe (g)	676.4 v	532.7 v	1209.1	69-1, 72-1
Xe (g)		532.8 v		73-9

*12.7 eV added to $M_5N_{45}N_{45}$ value to give value for $M_4N_{45}N_{45}$.

CESIUM	$3d_{5/2}$	$M_4N_{45}N_{45}$	α'	Ref.
Cs			1296.6	80-3
Cs			1297.1	80-19
Cs$_3$Sb			1296.1	80-3

CESIUM—*cont.*	$3d_{5/2}$	$M_4N_{45}N_{45}$	α'	Ref.
CsI			1294.0	80-3
Cs_2O_4			1293.1	80-19
CsCl	724.9	568.7	1293.6	82-5
CsOH	724.15	568.7	1292.85	79-12 r
Cs_2SO_4	723.9	568.4	1292.3	77-21 r
BARIUM				
Ba	779.3*	602.0*	1381.3	80-12 r
Ba ox	779.1	598.4	1377.5	80-12 r
Ba → Ba ox	Δ −0.2	−3.6	−3.8	80-12
BaO	779.85	597.5	1377.35	79-12 r
$BaSO_4$	780.8	596.1	1376.9	77-12 r
Ba erucate	780.2	596.2	1376.4	82-5
Ba cyclohexanebutyrate	780.1	596.1	1376.2	82-5
Ba chloranilate	779.6	596.7	1376.3	82-5
BaF_2	779.7	597.2	1376.9 ♦	82-5

*These data are assumed to supersede those in 80-13.

TANTALUM	$4f_{7/2}$	$M_5N_{67}N_{67}$	α'	$M_4N_{67}N_{67}$	$3d_{5/2}$	Ref.
Ta	21.9	1674.65	1696.55			80-16 r
TUNGSTEN						
WC	32.5	1729.1	1761.6	1791.5	1807.7	78-8 r
WS_2	33.0	1728.5	1761.5	1790.8	1807.7	78-8 r
$NiWS_2$	32.8	1728.7	1761.5	1791.1	1807.7	78-8 r
$K_4W(CN)_8$	32.8	1725.6	1758.4	1787.9	1807.5	78-8 r
$(C_5H_5)_2WCl_2$		1725.9		1787.9	1808.5	78-8 r
WO_3	36.1	1723.8	1759.9	1786.3	1810.6	78-8 r
H_2WO_4	36.1	1723.9	1760.0	1786.3	1810.5	78-8 r
$W(OPh)_6$	37.3	1723.8	1761.1	1786.2	1811.3	78-8 r
$CoWO_4$	36.0	1725.0	1761.0	1787.2	1810.7	78-8 r
$CuWO_4$	36.1	1725.3	1761.4	1786.9	1810.7	78-8 r
$NiWO_4$	36.0	1724.3	1760.3	1786.6	1811.2	78-8 r
$Fe_2(WO_4)_3$	36.3	1723.8	1760.1	1786.3	1810.8	78-8 r
$Na_2WO_4 \cdot 2H_2O$	36.4	1722.8	1759.2	1785.3	1810.4	78-8 r
Li_2WO_4	35.7	1722.8	1758.5	1785.1	1810.6	78-8 r
OSMIUM						
K_2OsCl_6	56.2			1907.7		78-8 r
$K_4Os(CN)_6$	57.2			1909.8		77-12 r

PLATINUM					$N_{67}W$	
Pt	71.1				63.4	82-5
Pt	71.3	1960.7	2032.0	2041.1		78-8 r
K_2PtCl_4	73.4			2035.2		75-7 r

GOLD

Au	84.0*	2015.9*	2099.9*			82-1* r
Au	84.0	2015.7	2099.7			80-16 r
Au	84.0				69.4*	82-5
Au	84.0	2015.8	2099.8	2101.2		78-8 r
Au		2015.9		2101.3		81-3 r

*Probably mainly component N_7VV.

MERCURY	$4f_{7/2}$	$M_4N_{67}N_{67}$	α'	$N_6O_{45}O_{4\,5}^*$	α'	$N_7O_{45}O_{45}$	Ref.
Hg	99.9			80.55	180.45	77.75	82-5
Hg (g)						63.5 v	77-1

THALLIUM							
Tl	117.8					85.1	82-5

LEAD

Pb	137.0	2180.5	2317.5				80-16 r
Pb	136.8			96.25	233.05	92.95	82-5
Pb	136.8			96.25	233.05		82-12
PbTe	137.25			95.45	232.7		82-12
PbSe	137.6			94.75	232.35		82-12
PbS	137.5			94.55	232.05		82-12
PbO	137.25			92.85	230.1		82-12
PbO_2	137.4			93.05	230.45		82-12
PbI_2	138.35			93.35	231.7		82-12
$PbBr_2$	138.8			92.6	231.4		82-12
$PbCl_2$	138.9			92.1	231.0		82-12
PbF_2	138.5			90.6	299.1		82-12
$Pb(OH)_2$	137.95			91.95	229.9		82-12
$Pb(NO_3)_2$	138.5			91.7	230.2		82-12
$PbSiO_3$	138.65			91.1	229.75		82-12
$PbSO_4$	140.0			90.1	230.1		82-12
$PbTiO_3$	138.0			92.6	230.6		82-12
$PbCrO_4$	138.3			92.75	231.05		82-12
$PbZrO_3$	138.5			91.7	230.2		82-12
$Pb(IO_4)_2$	138.2			92.7	230.9		82-12
$PbWO_4$	138.7			91.8	230.5		82-12
PbNCN	137.5			94.0	231.5		82-12
$Pb(OAc)_2$	138.5			91.45	229.95		82-12

*This component, made up chiefly of $N_6O_{45}O_{45}$ lines, is believed to be most easily measurable in most compounds (cf. Ref. 77-1).

BISMUTH	$4f_{7/2}$	$N_6O_{45}O_{45}{}^*$	α'	$N_7O_{45}O_{45}$	Ref.
Bi	157.0	103.7	260.7	100.1	82-5

*This component, made up chiefly of $N_6O_{45}O_{45}$ lines, is believed to be most easily measurable in most compounds (cf. Ref. 77-1).

Practical Surface Analysis
by Auger and X-ray Photoelectron Spectroscopy
Edited by D. Briggs and M. P. Seah
© 1983, John Wiley & Sons, Ltd

Appendix 5

Empirically Derived Atomic Sensitivity Factors for XPS

The following empirically derived set of atomic sensitivity factors, relative to
F $1s$ = 1.00, is obtained from a combination of data from the Varian IEE and
Physical Electronics (Perkin-Elmer) 550 spectrometers. These spectrometers
utilize scanning by varying the retarding voltage applied to the emitted electrons, with the analyser operated at constant-pass energy. This gives a transmission function for the spectrometer varying with the inverse of the electron
kinetic energy. The factors therefore should be applicable to other spectrometers with the same transmission characteristics (cf. M. P. Seah, *Surf.
Interface Anal.*, **2**, 222, 1980), but will not be applicable to those operating in
a different mode. These data are reproduced from C. D. Wagner, L. E. Davis,
M. V. Zeller, J. A. Taylor, R. M. Raymond and L. H. Gale, *Surf. Interface
Anal.*, **3**, 211 (1981).

	Strong line		Secondary line‡	
	Area $1s$	Height† $1s$	Area $2s$	Height $2s$
Li	0.020	0.020		
Be	0.059	0.059		
B	0.13	0.13		
C	0.25	0.25		
N	0.42	0.42		
O	0.66	0.66	0.025	0.025
F	1.00	1.00	0.04	0.04
Ne	1.5	1.5	0.07	0.07
Na	2.3	2.3	0.13	0.12
Mg	3.5*§	3.3	0.20	0.15

512 Practical Surface Analysis

	Strong line			Secondary line‡	
	Area		Height†	Area	Height†
	$2p_{3/2}$	$2p$	$2p_{3/2}$	$2s$	$2s$
Mg		0.12	0.12	0.20	0.15
Al		0.185	0.18	0.23	0.17
Si		0.27	0.25	0.26	0.19
P		0.39	0.36	0.29	0.21
S		0.54	0.49	0.33	0.24
Cl		0.73	0.61	0.37	0.25
Ar		0.96	0.75	0.40	0.26
K	0.83	1.24	0.83	0.43	0.26
Ca	1.05	1.58	1.05	0.47	0.26
Sc	(1.1)	(1.65)	(1.1)	0.50	0.26
Ti	(1.2)	(1.8)	(1.2)	0.54	0.26
				$3p$	$3p$
Ti	(1.2)¶	(1.8)	(1.2)	0.21	0.15
V	(1.3)	(1.95)	(1.3)	0.21	0.16
Cr	(1.5)	(2.3)	(1.5)	(0.21)	(0.17)
Mn	(1.7)	(2.6)	(1.7)	(0.22)	(0.19)
Fe	(2.0)	(3.0)	(2.0)	(0.26)	(0.21)
Co	(2.5)	(3.8)	(2.5)	(0.35)	(0.25)
Ni	(3.0)	(4.5)	(3.0)	(0.5)	(0.3)
Cu	(4.2)	(6.3)	(4.2)	(0.65)	(0.4)
Zn	4.8		4.8	0.75	0.40
Ga	5.4		5.4	0.84	0.40
Ge	6.1*		6.0*	0.92	0.40
As	6.8*		6.8*	1.00	0.43

	Area		Height	Area		Height
	$3d_{5/2}$	$3d$	$3d_{5/2}$	$3p_{3/2}$	$3p$	$3p_{3/2}$
Ga		0.31	0.31		0.84	0.40
Ge		0.38	0.37		0.91	0.40
As		0.53	0.51		0.97	0.42
Se		0.67	0.64		1.05	0.48
Br		0.83	0.77		1.14	0.54
Kr		1.02	0.91	0.82	1.23††	0.60
Rb		1.23	1.07	0.87	1.30	0.67
Sr		1.48	1.24	0.92	1.38	0.69
Y		1.76	1.37	0.98	1.47	0.71
Zr		2.1	1.5	1.04	1.56	0.72

	Strong line				Secondary line‡		
	Area		Height		Area		Height
	$3d_{5/2}$	$3d$	$3d_{5/2}$		$3p_{3/2}$	$3p$	$3d_{5/2}$
Nb	1.44	2.4	1.57		1.10		0.72
Mo	1.66	2.75	1.74		1.17		0.73
Tc	1.89	3.15	1.92		1.24		0.73
Ru	2.15	3.6	2.15		1.30		0.73
Rh	2.4	4.1	2.4		1.38		0.74
Pd	2.7	4.6	2.7		1.43		0.74
Ag	3.1	5.2	3.1		1.52		0.75
Cd	3.5		3.5		1.60		0.75
In	3.9		3.9		1.68		0.75
Sn	4.3		4.3		1.77		0.75
	Area		Height†		Area	$4d$	Height $4d$
Sb	4.8		4.8			1.00	0.86
Te	5.4		5.4			1.23	0.97
I	6.0		6.0			1.44	1.08
Xe	6.6		6.6			1.72	1.16
Cs	7.2		7.0			2.0	1.25
Ba	7.9		7.5			2.35	1.35
La		(10)¶				(2)	
Ce		(10)				(2)	
Pr		(9)				(2)	
Nd		(7)				(2)	
Pm		(6)				(2)	
Sm		(5)				(2)	
Eu		(5)				(2)	
Gd	(3)*					(2)	
Tb	(3)*					(2)	
		$4d$			$4p_{3/2}$		
Dy		(2)¶			(0.6)¶		
Ho		(2)			(0.6)		
Er		(2)			(0.6)		
Tm		(2)			(0.6)		
Yb		(2)			(0.6)		
Lu		(2)			(0.6)		

	Strong line			Secondary line‡		
	Area		Height $4f_{7/2}$	Area		Height $4d_{5/2}$
	$4f_{7/2}$	$4f$		$4d_{5/2}$	$4d$	
Hf		2.05	1.70	1.42	2.35	0.90
Ta		2.4	1.89	1.50	2.50	0.90
W		2.75	2.0	1.57	2.6	0.90
Re		3.1	2.1	1.66	2.75	0.90
Os		3.5	2.2	1.75	2.9	0.90
Ir	2.25	3.95	2.4	1.84		0.90
Pt	2.55	4.4	2.55	1.92		0.90
Au	2.8	4.95	2.8	2.05		0.90
Hg	3.15	5.5	3.15	2.15		0.95

	Area		Height $4f_{7/2}$	Area $4d_{5/2}$	Height $4d_{5/2}$	Area		Height $5d_{5/2}$
	$4f_{7/2}$	$4f$				$5d_{5/2}$	$5d$	
Tl	3.5	6.15	3.5	2.25	0.95		0.9	0.55
Pb	3.85	6.7	3.82	2.35	1.00		1.0	0.6
Bi	4.25	7.4	4.25	2.5	1.00		1.1	0.65
Th	7.8		7.8	3.5	1.2	0.9	1.5	0.9
U	9.0		9.0	3.85	1.3	1.0	1.6	1.0

†Height sensitivity factors based on line widths for strong lines of 3.1 eV, typical of lines obtained in survey spectra on insulating samples. When spin doublets are unresolved, data are for the convoluted peak height.

‡Factors for the strong lines are insensitive to the radiation source (Mg or Al). Factors for the secondary lines ($2s$, $3p$, $4d$ and $5d$) are dependent to an extent upon the photon energy. Values shown are average for Al and Mg. For more accurate results, multiply the factors by 0.9 when Mg radiation is used and by 1.1 when Al radiation is used.

§Starred data are for peaks obtained only by using Al X-rays.

¶Data in parentheses indicate great variability with chemical state, because of the prevalence of multielectron processes. Data shown for the series Ti–Cu are for diamagnetic forms; data for paramagnetic forms will be lower in general. Data for the rare earths are based on few experimental points, and should be regarded only as a rough approximation.

††Many of the area data are supped for spin doublets for $3p$ and $4d$ because of the considerable width of many of those lines. Data for combined spin doublets in the $2p$ series for transition metals and the $3d$ for the rare earths are supplied because of the prevalence of shake-up lines, which make it desirable to deal with the doublet as a whole.

Practical Surface Analysis
by Auger and X-ray Photoelectron Spectroscopy
Edited by D. Briggs and M. P. Seah
© 1983, John Wiley & Sons, Ltd.

Appendix 6

(a) Line Positions* from Mg X-rays, by Element (BE Scale)

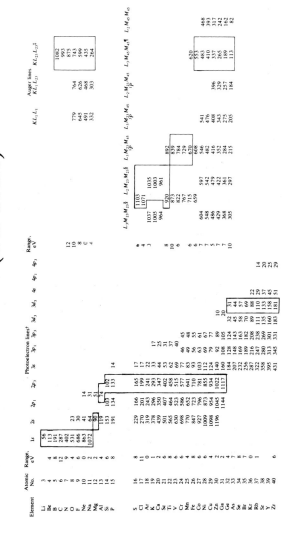

Photoelectron lines†

Element	Atomic No.	Range eV	1s	2s	2p₁	2p₃	3s	3p₁	3p₃	3d₃	3d₅	4s	4p₁	4p₃
Li	3	4	56											
Be	4	4	113											
B	5	8	191											
C	6	12	287											
N	7	9	402											
O	8	6	531	23										
F	9	6	686	30										
Ne	10	2	863	41	14	14								
Na	11	4	1072	64	31	31								
Mg	12	2		90	51	51								
Al	13	6		119	74	74								
Si	14	8		153	102	102	14							
P	15			191	134	133	14							
S	16	8		229	166	165	17							
Cl	17	11		270	201	199	17							
Ar	18	0		319	243	241	22							
K	19	1		378	296	293	33							
Ca	20	2		439	350	347	44	25						
Sc	21	1		501	407	402	53	31						
Ti	22	8		565	464	458	62	37						
V	23	6		630	523	515	69	40						
Cr	24	6		698	586	577	77	46	45					
Mn	25	4		770	652	641	83	49	48					
Fe	26	8		847	723	710	93	56	55					
Co	27	6		927	796	781	103	63	61					
Ni	28	4		1009	873	855	112	69	67					
Cu	29	2		1098	954	934	124	79	77					
Zn	30	2		1196	1045	1022	140	92	89	10				
Ga	31				1144	1117	160	108	105	20				
Ge	32						184	128	124	32	31			
As	33	8					207	148	143	45	44			
Se	34	7					232	169	163	58	57			
Br	35	5					256	189	182	70	69			
Rb	37	0					322	247	238	111	110	29		14
Sr	38	1					358	280	269	135	133	37		20
Y	39						395	313	301	160	158	45		25
Zr	40	6					431	345	331	183	181	51		29

Auger lines (light elements, KLL)

Element	KL₁L₁	KL₁L₂,₃	KL₂,₃L₂,₃‡
B			1082
C			993
N			875
O	779	764	743
F	645	626	599
Ne	491	468	435
Na	332	303	264

Auger lines (L₃M₄,₅M₄,₅ / L₂M₄,₅M₄,₅)

Element	L₃M₄,₅M₄,₅	L₂M₄,₅M₄,₅
Mn	620	
Fe	553	468
Co	483	393
Ni	410	317
Cu	337	242
Zn	265	162
Ga	189	82
Ge	113	

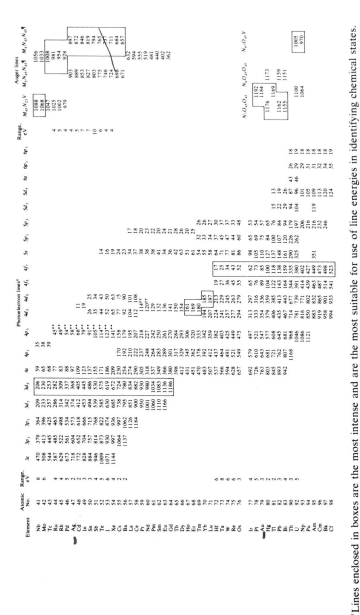

*Lines enclosed in boxes are the most intense and are the most suitable for use of line energies in identifying chemical states.

†For brevity, $2p_{3/2}$, $3d_{5/2}$ equals $3d_5$, etc.

‡Includes KVV designation when L_{23} is not a core level.

§Designation is oversimplified.

¶Includes LVV when M levels are not in core, and MVV when N levels are not in core.

**No simple $4p_{1/2}$ line exists for this group of elements.

††The $4d$ doublet for these elements is complex and is variable with chemical state because of multiplet splitting and multielectron process.

Source: *Handbook of X-ray Photoelectron Spectroscopy*, C. D. Wagner, W. M. Riggs, L. E. Davis, J. F. Moulder and (ed.) G. E. Muilenberg, Perkin-Elmer Corporation, Eden Prairie (1979).

(b) Line Positions* from Al X-rays, by Element (BE Scale)

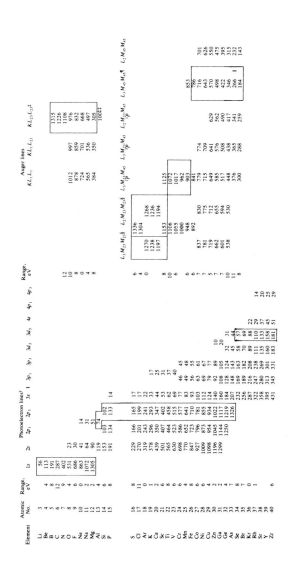

Photoelectron lines[†]

Element	Atomic No.	Range, eV	$1s$	$2s$	$2p_1$	$2p_3$	$3s$	$3p_1$	$3p_3$	$3d_3$	$3d_5$	$4s$	$4p_1$	$4p_3$
Li	3	4	56											
Be	4	4	113											
B	5	8	191											
C	6	12	287											
N	7	9	402											
O	8	6	531	23										
F	9	6	686	30	14	14								
Ne	10	0	863	41	31	31								
Na	11	2	1072	64	31	31								
Mg	12	2	1305	90	51	51								
Al	13	4		119	74	73								
Si	14	6		153	103	102	14							
P	15	8		191	134	133								
S	16	8		229	166	165	17							
Cl	17	11		270	201	199	17							
Ar	18	0		319	243	241	25	17	17					
K	19	1		378	296	294	31	22	22					
Ca	20	2		439	350	347	44	33	33					
Sc	21	6		501	407	402	53	44	44					
Ti	22	8		565	464	458	62	53	53					
V	23	6		630	523	515	69	62	62					
Cr	24	4		698	586	577	77	69	69	46	45			
Mn	25	8		770	652	641	83	77	77	49	48			
Fe	26	6		847	723	710	93	83	83	56	55			
Co	27	6		927	796	781	103	93	93	61				
Ni	28	6		1009	873	855	112	103	103	67				
Cu	29	4		1098	954	934	124	112	112	77				
Zn	30	2		1196	1045	1022	140	124	124	89				
Ga	31	2		1299	1144	1117	160	148	143	108	105			
Ge	32	4			1250	1326	184	169	163	128	124	32		
As	33	7					207	189	182	148	143	45		
Se	34	8					232	216	208	169	163	58		
Br	35	0					256	247	238	189	182	70		
Kr	36	1					287	280	269	216	208	89		14
Rb	37	—					322	313	301	247	238	110	133	20
Sr	38	—					358	358	280	280	269	135	158	25
Y	39	—					395	395	313	301	160	181		29
Zr	40	6					431	345	331	183	181			

Auger lines

Element	Atomic No.	Range, eV	$L_3M_{23}M_{23}\beta$	$L_2M_{23}M_{23}$	$L_3M_{23}M_{45}$	$L_3M_{45}M_{45}$	$L_3M_{45}M_{45}$ §	$L_2M_{45}M_{45}$	KL_1L_1	KL_1L_{23}	$KL_{23}L_{23}$ ‡
B	5	12									1315
C	6	10									1226
N	7	8									1108
O	8	0							1012	997	976
F	9	4							878	859	832
Ne	10	8							724	701	668
Na	11								565	536	497
Mg	12								384	350	305
Al	13										100 ‡
Si	14	6	1270	1336	1125						
P	15	4	1238	1304	1072						
S	16	0	1197		1017	1153					
Cl	17			1268	962	1166					
Ar	18	8		1236	903	1055					
K	19	10		1194	841	1000					
Ca	20	6			779	948					
Sc	21	6			715	892					
Ti	22	7	837	830	649						
V	23	5	781	775	585	774					
Cr	24	5	719	712	517	709	853	701			
Mn	25	7	662	655	448	641	786	626			629
Fe	26	10	601	594	376	576	716	550			562
Co	27	11	538	530	300	508	643	475			490
Ni	28	8				438	570	395			417
Cu	29					365	498	315			341
Zn	30					288	422	232			259
Ga	31						346	143			
Ge	32						266				
As	33						184				
Se	34	6					—				

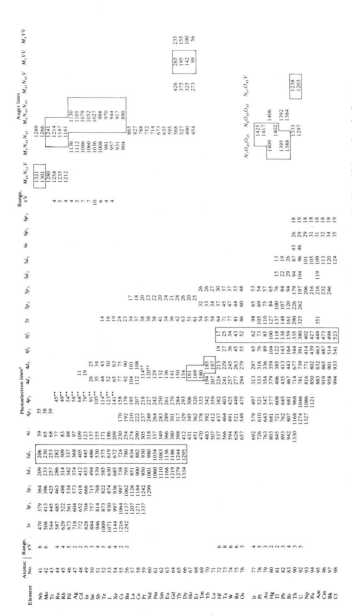

*Lines enclosed in boxes are the most intense and are the most suitable for use of line energies in identifying chemical states.

†For brevity, $2p_3$ equals $2p_{3/2}$, $3d_5$ equals $3d_{5/2}$, etc.

‡Includes KVV designation when L_{23} is not a core level.

§Designation is oversimplified.

¶Includes LVV when M levels are not in core and MVV when N levels are not in core.

**No simple $4p_{1/2}$ line exists for this group of elements.

††The $4d$ doublet for these elements is complex and is variable with chemical state because of multiplet splitting and multielectron processes.

‡‡Often observable, induced by bremsstrahlung.

Source: *Handbook of X-ray Photoelectron Spectroscopy*, C. D. Wagner, W. M. Riggs, L. E. Davis, J. F. Moulder and (ed.) G. E. Muilenberg, Perkin-Elmer Corporation, Eden Prairie (1979).

Practical Surface Analysis
by Auger and X-ray Photoelectron Spectroscopy
Edited by D. Briggs and M. P. Seah
© 1983, John Wiley & Sons, Ltd

Appendix 7

(a) Line Positions from Mg X-rays, in Numerical Order

17 Hf $4f_7$	102 Si $2p_3$	206 Nb $3d_5$	359 Lu $4p_3$	575 Te $3d_5$	863 Ne $1s$
23 O $2s$	105 Ga $3p_3$	208 Kr $3p_3$	359 Hg $4d_5$	577 Cr $2p_3$	872 Cd (A)
25 Ta $4f_7$	108 Ce $4d_5$	213 HF $4d_5$	362 Gd (A)	594 Ce (A)	875 N (A)
30 F $2s$	110 Rb $3d_5$	229 S $2s$	364 Nb $3p_3$	599 F (A)	882 Ce $3d_5$
31 Ge $3d_5$	113 Be $1s$	229 Ta $4d_5$	368 Ag $3d_5$	618 Cd $3p_3$	897 Ag (A)
34 W $4f_7$	113 Ge (A)	230 Mo $3d_5$	378 K $2s$	619 I $3d_5$	920 Sc (A)
40 V $3p$	114 Pr $4d$	238 Rb $3p_3$	380 U $4f_7$	632 La (A)	928 Pd (A)
41 Ne $2s$	118 Tl $4f_7$	241 Ar $2p_3$	385 Tl $4d_5$	641 Mn $2p_3$	930 Pr $3d_5$
43 Re $4f_7$	119 Al $2s$	245 W $4d_5$	396 Mo $3p_3$	657 Ba (A)	934 Cu $2p_3$
44 As $3d_5$	120 Nd $4d$	263 Re $4d_5$	402 N $1s$	666 In $3p_3$	954 Rh (A)
45 Cr $3p_3$	124 Ge $3p_3$	264 Na (A)	402 Eu (A)	670 Mn (A)	961 Ca (A)
48 Mn $3p_3$	132 Sm $4d$	265 Zn (A)	402 Sc $2p_3$	672 Xe $3d_5$	970 U (A)
50 I $4d_5$	133 P $2p_3$	269 Sr $3p_3$	405 Cd $3d_5$	677 Th $4d_5$	980 Nd $3d_5$
51 Mg $2p$	133 Sr $3d_5$	270 Cl $2s$	410 Ni (A)	684 Cs (A)	981 Ru (A)
52 Os $4f_7$	136 Eu $4d$	279 Os $4d_5$	413 Pb $4d_5$	686 F $1s$	993 C (A)
55 Fe $3p_3$	138 Pb $4f_7$	282 Ru $3d_5$	435 Ne (A)	710 Fe $2p_3$	1003 K (A)
56 Li $1s$	143 As $3p_3$	284 Tb $4p_3$	439 Ca $2s$	711 Xe (A)	1005 Th (A)
57 Se $3d_5$	150 Tb $4d$	287 C $1s$	440 Sm (A)	715 Sn $3p_3$	1022 Zn $2p_3$
61 Co $3p_3$	153 Si $2s$	293 Dy $4p_3$	443 Bi $4d_5$	724 Cs $3d_5$	1035 Ar (A)
62 Ir $4f_7$	154 Dy $4d$	293 K $2p_3$	445 In $3d_5$	729 Cr (A)	1071 Cl (A)
63 Xe $4d_5$	158 Y $3d_5$	297 Ir $4d_5$	458 Ti $2p_3$	737 I (A)	1072 Na $1s$
64 Na $2s$	159 Bi $4f_7$	301 Y $3p_3$	463 Ru $3p_3$	739 U $4d_5$	1082 B (A)
67 Ni $3p_3$	161 Ho $4d$	306 Ho $4p_3$	483 Co (A)	743 O (A)	1083 Sm $3d_5$
69 Br $3d_5$	163 Se $3p_3$	309 Rh $3d_3$	486 Sn $3d_5$	765 Te (A)	1088 Nb (A)
73 Pt $4f_7$	165 S $2p_3$	316 Pt $4d_5$	498 Rh $3p_3$	768 Sb $3p_3$	1103 S (A)
74 Al $2p$	169 Er $4d$	319 Ar $2s$	501 Sc $2s$	780 Ba $3d_5$	1117 Ga $2p_3$
75 Cs $4d_5$	180 Tm $4d$	320 Er $4p_3$	515 V $2p_3$	781 Co $2p_3$	1136 Eu $3d_5$
77 Cu $3p_3$	181 Zr $3d_5$	331 Zr $3p_3$	519 Nd (A)	784 V (A)	1155 Bi (A)
85 Au $4f_7$	182 Br $3p_3$	333 Tm $4p_3$	530 Sb $3d_5$	794 Sb (A)	1162 Pb (A)
87 Zn $3p_3$	185 Yb $4d_5$	335 Th $4f_7$	531 O $1s$	819 Sn (A)	1169 Tl (A)
88 Kr $3d_5$	189 Ga (A)	336 Au $4d_5$	534 Pd $3p_3$	822 Te $3p_3$	1176 Hg (A)
90 Ba $4d_5$	191 B $1s$	337 Pd $3d_5$	553 Fe (A)	834 La $3d_5$	1184 Au (A)
90 Mg $2s$	191 P $2s$	337 Cu (A)	555 Pr (A)	839 Ti (A)	1186 Gd $3d_5$
100 Hg $4f_7$	197 Lu $4d_5$	342 Yb $4p_3$	565 Ti $2s$	846 In (A)	1192 Pt (A)
101 La $4d_5$	199 Cl $2p_3$	347 Ca $2p_3$	573 Ag $3p_3$	855 Ni $2p_3$	

An A in parentheses denotes Auger line. The sharpest Auger line and the two most intense photo-electron lines per element are included in the table. For brevity, $2p_3$ equals $2p_{3/2}$, $3d_5$ equals $3d_{5/2}$, etc. All lines are on the binding energy scale.

Source: Handbook of X-ray Photoelectron Spectroscopy, C. D. Wagner, W. M. Riggs, L. E. Davis, J. F. Moulder and (ed.) G. E. Muilenberg, Perkin-Elmer Corporation, Eden Praire (1979).

(b) Line Positions from Al X-rays, in Numerical Order

17 Hf $4f_7$	110 Rb $3d_5$	229 Ta $4d_5$	385 Tl $4d_5$	667 Th $4d_5$	1072 Na $1s$
23 O $2s$	113 Be $1s$	230 Mo $3d_5$	396 Mo $3p_3$	686 F $1s$	1072 Ti (A)
25 Ta $4f_7$	114 Pr $4d$	238 Rb $3p_3$	402 N $1s$	710 Fe $2p_3$	1079 In (A)
30 F $2s$	118 Tl $4f_7$	241 Ar $2p_3$	402 Sc $2p_3$	715 Sn $3p_3$	1083 Sm $3d_5$
34 W $4f_7$	119 Al $2s$	245 W $4d_5$	405 Cd $3d_5$	716 Co (A)	1105 Cd (A)
40 V $3p$	120 Nd $4d$	263 Re $4d_5$	413 Pb $4d_5$	724 Cs $3d_5$	1108 N (A)
41 Ne $2s$	124 Ge $3p_3$	265 Tb (A)	422 Ga (A)	739 U $4d_5$	1117 Ga $2p_3$
43 Re $4f_7$	132 Sm $4d$	266 As (A)	439 Ca $2s$	752 Nd (A)	1130 Ag (A)
44 As $3d_5$	133 P $2p_3$	269 Sr $3p_3$	443 Bi $4d_5$	768 Sb $3p_3$	1136 Eu $3d_5$
45 Cr $3p_3$	133 Sr $3d_5$	270 Cl $2s$	445 In $3d_5$	780 Ba $3d_5$	1153 Sc (A)
48 Mn $3p_3$	136 Eu $4d$	279 Os $4d_5$	458 Ti $2p_3$	781 Co $2p_3$	1161 Pd (A)
50 I $4d_5$	138 Pb $4f_7$	282 Ru $3d_5$	463 Ru $3p_3$	786 Fe (A)	1186 Gd $3d_5$
52 Os $4f_7$	141 Gd $4d$	287 C $1s$	486 Sn $3d_5$	788 Pr (A)	1187 Rh (A)
55 Fe $3p_3$	142 Ho (A)	293 K $2p_3$	497 Na (A)	822 Te $3p_3$	1194 Ca (A)
56 Li $1s$	150 Tb $4d$	297 Ir $4d_5$	498 Zn (A)	827 Ce (A)	1205 U (A)
57 Se $3d_5$	153 Si $2s$	301 Y $3p_3$	498 Rh $3p_3$	832 F (A)	1214 Ru (A)
61 Co $3p_3$	154 Dy $4d$	305 Mg (A)	501 Sc $2s$	834 La $3d_5$	1219 Ge $2p_3$
62 Ir $4f_7$	158 Y $3d_5$	306 Ho $4p_3$	515 V $2p_3$	855 Ni $2p_3$	1226 C (A)
63 Xe $4d_5$	159 Bi $4f_7$	309 Rh $3d_5$	530 Sb $3d_5$	863 Ne $1s$	1230 Th (A)
64 Na $2s$	161 Ho $4d$	316 Pt $4d_5$	531 O $1s$	865 La (A)	1236 K (A)
67 Ni $3p_3$	163 Se $3p_3$	319 Ar $2s$	534 Pd $3p_3$	882 Ce $3d_5$	1244 Tb $3d_5$
69 Br $3d_5$	165 S $2p_3$	320 Er $4p_3$	565 Ti $2s$	890 Ba (A)	1268 Ar (A)
73 Pt $4f_7$	169 Er $4d$	331 Zr $3p_3$	570 Cu (A)	903 Mn (A)	1295 Dy $3d_5$
74 Al $2p$	180 Tm $4d$	333 Tm $4p_3$	573 Ag $3p_3$	917 Cs (A)	1301 Mo (A)
75 Cs $4d_5$	181 Zr $3d_5$	335 Th $4f_7$	575 Te $3d_5$	930 Pr $3d_5$	1304 Cl (A)
77 Cu $3p_3$	182 Br $3p_3$	336 Au $4d_5$	577 Cr $2p_3$	934 Cu $2p_3$	1305 Mg $1s$
85 Au $4f_7$	184 Se (A)	337 Pd $3d_5$	595 Gd (A)	944 Xe (A)	1315 B (A)
87 Zn $3p_3$	185 Yb $4d_5$	342 Yb $4p_3$	618 Cd $3p_3$	962 Cr (A)	1321 Nb (A)
88 Kr $3d_5$	191 B $1s$	346 Ge (A)	619 I $3d_5$	970 I (A)	1326 As $2p_3$
90 Ba $4d_5$	191 P $2s$	347 Ca $2p_3$	635 Eu (A)	976 O (A)	1336 S (A)
90 Mg $2s$	195 Dy (A)	359 Lu $4p_3$	641 Mn $2p_3$	980 Nd $3d_5$	1388 Bi (A)
99 Er (A)	197 Lu $4d_5$	359 Hg $4d_5$	643 Ni (A)	998 Te (A)	1395 Pb (A)
100 Hg $4f_7$	199 Cl $2p_3$	364 Nb $3p_3$	666 In $3p_3$	1017 V (A)	1402 Tl (A)
101 La $4d_5$	206 Nb $3d_5$	368 Ag $3d_5$	668 Ne (A)	1022 Zn $2p_3$	1409 Hg (A)
102 Si $2p_3$	208 Kr $3p_3$	378 K $2s$	672 Xe $3d_5$	1027 Sb (A)	1417 Au (A)
105 Ga $3p_3$	213 Hf $4d_5$	380 U $4f_7$	673 Sm (A)	1052 Sn (A)	1425 Pt (A)
108 Ce $4d_5$	229 S $2s$				

An A in parentheses denotes Auger line. The sharpest Auger line and the two most intense photo-electron lines per element are included in the table. For brevity, $2p_3$ equals $2p_{3/2}$, $3d_5$ equals $3d_{5/2}$, etc. All lines are on the binding energy scale.

Source: *Handbook of X-ray Photoelectron Spectroscopy*, C. D. Wagner, W. M. Riggs, L. E. Davis, J. F. Moulder and (ed.) G. E. Muilenberg, Perkin-Elmer Corporation, Eden Praire (1979).

Practical Surface Analysis
by Auger and X-ray Photoelectron Spectroscopy
Edited by D. Briggs and M. P. Seah
© 1983, John Wiley & Sons, Ltd

Appendix 8

Kinetic Energies of Auger Electrons: Experimental Data from Spectra Acquired by X-ray Excitation

C. D. Wagner

29 Starview Drive, Oakland, CA 94618, USA

The following table is believed to represent the best available data on the most prominent Auger lines in X-ray excited spectra. Lines presented are mostly of the core type, with no attempt made to include the valence-type lines such as *KVV* for Li–O, *LVV* for Mg–Cl, *MVV* for Cr–Y and *NVV* for Ir–Hg. The low energy *NOO* lines for Tl–Bi and the rather prominent Coster–Kronig lines for the heavy metals are not included. Certain lines of very low intensity, such as *KLV*, *LMV* and *MNV* lines, and shoulders on the intense lines are also not included.

Data are presented mainly for elemental conductive states. Gas phase data are supplied when the data are significantly more detailed than the solid phase. Data on oxides are included as well for certain elements where the spectra are significantly different for element and oxide. Other tabulations should be consulted for data on chemical shifts.

Most of the data are derived from information used to produce reference 79–1. Other references of particular value for the distributions of line energies are noted. Data are Fermi level referenced unless otherwise noted. For many lines, the energies are close to those observed with conventional electron-excited, vacuum-referenced dN/dE data, customarily used in Auger electron spectroscopy, because the work function of 4–5 eV approximately offsets the difference between the peak position and the inflection point on the high-energy side. Some notations used are:

> (g) = gas phase
> v = referenced to vacuum level
> i = interpolated or extrapolated value
> b = broad line
> l = low intensity

KLL	KL_1L_1 ${}^1S$	KL_1L_{23} ${}^1P$	KL_1L_{23} ${}^3P$	$KL_{23}L_{23}$ ${}^1S$	$KL_{23}L_{23}$ ${}^1D$	Ref.
F alkali fluorides	611	630	638	653	656	68-1
Ne implanted in Fe	762	785	798 i	814	818	75-2
(g)	749 v	772 v	783 v	801 v	805 v	66-1
Na	926	955	967	989	994	74-2
Mg	1107	1140	1155	1181	1186	75-1, 78-1
Al	1302	1341	1357	1387	1393	76-1, 78-1
Si	1514	1559 i	1576	1610	1617	77-5
P GaP	1745	1794 i	1813	1852	1859	80-2
S WS$_2$	1991	2045	2065	2108	2116	77-3
Cl CCl$_4$ (g)	2245 v	2304 v	2326 v	2374 v	2383 v	69-1
Ar (g)	2508 v	2576 v	2600 v	2651 v	2661 v	77-2

LMM	$L_3M_1M_{23}$ ${}^1P$	$L_3M_1M_{23}$ ${}^3P$	$L_3M_{23}M_{23}$ ${}^1D$	$L_3M_{23}M_{23}$ ${}^3P$	$L_3M_{23}M_{45}$ ${}^1F$	$L_3M_{23}M_{45}$ ${}^3D{}^3P$	$L_2M_{23}M_{45}$ ${}^1F$	$L_3M_{45}M_{45}$ ${}^1G$	$L_2M_{45}M_{45}$ ${}^1G$	Ref.
Ar in Fe			217							
(g)			204 v	206 v*						75-2
K KBr			248	250*						73-1
KBr (g)			237 v	239 v						75-2
Ca element				298*						81-1
CaCO$_3$			285	288*	363					80-1
Sc Sc$_2$O$_3$	307	315	383	335*	419					
Ti element	346	355	382	389	415			452		71-2
TiO$_2$										

Element										Ref
V element	400	411	432	438	472				510	71-2
V$_2$O$_3$			428	432	468					71-2
Cr element	446	460	1	489	527				570	71-2
Cr$_2$O$_3$					528					71-2
Mn element	500 i	514 i	1	543	586	635				71-2
MnO$_2$				541	585	634				
Fe element	548	563	591	598	647	651	702	1	715	71-2
Fe$_2$O$_3$				597	647		703			71-2
Co element	605	618	648	655	710	715	723	773	789	71-2
CoO	604	617	647	654	709			773	789	71-2
Ni element	661	674	709	715	775	781	1	846	863	71-2
NiO			710		775			846		
Cu element	716	729	768	775	840	847	859	919	939	77-1
CuO			764		837	844	858	918		
Zn element	774	787	826	834	905	913	928	992	1015	77-1, 6, 73-2, 77-1
Ga element	831	846	888	897	973	983	1000	1068	1095	
Ge element	890 i	895 i	953	962	1043	1054	1074	1145	1176	77-1
As element	950	966	1020	1029	1116	1127	1152	1225	1261	77-7
Se element	1013	1033	1086	1096	1189	1201	1230	1307	1348	
Br bromo-methanes (g)	1143 v	1155 v		1253 v	1266 v	1300 v		1378 v	1424 v	70-1
Kr (g)	1210 v	1223 v		1327 v	1342 v	1380 v		1460 v	1513 v	72-1
Rb										
Sr SrF$_2$								1641		78-2
Y Y$_2$O$_3$								1737		78-2
Zr Zr oxide								1831		78-2
Nb Nb oxide								1930		78-2
Mo element								2039	2144	80-2

*Includes $L_2M_{23}M_{23}$ (^1D). Omitted because of low intensity are $L_2M_{23}M_{23}$, $L_3M_1M_{45}$ and $L_2M_{23}M_{45}$ (^3D).

MNN	$M_{45}N_1N_{23}$	$M_{45}N_{23}N_{23}$	$M_{45}N_1N_{45}$	$M_{45}N_{23}N_{45}$	$M_5N_{45}N_{45}$ ^1G	$M_5N_{45}N_{45}$ $^3F_{2,3}$	$M_4N_{45}N_{45}$	Ref.
Zr	94	119	1	150	200			
Nb	114 b	136	145	168	223			
Mo	122	1	163	188				
Tc								
Ru	152		201	231	274			
Rh			223	253 b	302			
Pd			243	274 b	328		333	
Ag			262	296 b	352	353	358	71-1
Cd			280	b	377	379	384	71-1, 74-1
In			1	b	403	405	411	71-1
Sn			1	b	429	431	438	79-2
Sb					455	457	464	71-1
Te					482	484	492	77-4
I					506	509	517	71-1
LiI					532	536	545	
Xe in carbon (g)					520 v	524 v	533 v	72-1

MNN	$M_5N_{45}N_{45}$ ^1G	$M_5N_{45}N_{45}$ $^3F_{2,3}$	$M_4N_{45}N_{45}$	$M_5N_{45}N_{67}$	$M_4N_{45}N_{67}$	$M_5N_{67}N_{67}$	$M_4N_{67}N_{67}$	Ref.
Cs CsOH	555	559	569	630	643			
Ba element			602					80-1
BaO	585		598	669	683			
La La_2O_3	622		638	711 i	728 i			
Ce CeO_2	660			755	771			
Pr	695 i				795 i			
Nd	733 i				840 i			
Pm	770 i				885 i			
Sm Sm_2O_3	808 b				950 ib	1068	1094	
Eu	846 i				980 i	1120 i	1150 i	

		$M_{45}N_{45}N_{67}$	$M_5N_{67}N_{67}$	$M_4N_{67}N_{67}$†	Ref.
Gd	884 i	1020 i	1170 i	1202 i	
Tb element	920 b	1068 b	1223	1256	
Dy	960 i	1115 i	1280 i	1318 i	
Ho	998 i	1165 i	1332 i	1372 i	
Er element	1035	1218 b	1387	1428	
Er$_2$O$_3$	1037	1221 b	1386	1429	
Tm	1080 i	1270 i	1440 i	1487 i	
Yb		1320 i	1500 i	1549 i	
Lu		1370 i	1560 i	1615 i	
Hf		1420 i	1615 i	1669 i	80-2
Ta		1462	1675	1733	78-2
W			1730	1792	
Re			1790 i	1856 i	78-2
Os K$_2$OsCl$_6$			1837	1908	
Ir			1900 i	1975 i	
Pt			1961	2041	78-2
Au			2016	2102	
Hg			2070 i	2160 i	80-2
Tl			2128 i	2223 i	
Pb			2181	2282	80-2
Bi			2235 i	2343 i	
Th					
U					

†The difference in the binding energies of the $3d_{5/2}$ and $3d_{3/2}$ levels was used to calculate many of the $M_4N_{67}N_{67}$ energies from the $M_5N_{67}N_{67}$ energies.

Acknowledgement

Unpublished data from spectra obtained at Physical Electronics Division, Perkin-Elmer Corporation in connection with references 79-1 and 80-2 were used in assembling these data tables.

References

66–1 H. Körber and W. Mehlhorn, *Z. Phys.*, **191**, 217 (1966).
68–1 R. G. Albridge, K. Hamrin, G. Johansson and A. Fahlman, *Z. Phys.*, **209**, 419 (1968).
69–1 B. Cleff and W. Mehlhorn, *Z. Phys.*, **219**, 311 (1969).
70–1 R. Spohr, T. Bergmark, N. Magnusson, L. O. Werme, C. Nordling and K. Siegbahn, *Phys. Scr.*, **2**, 31 (1970).
71–1 S. Aksela, *Z. Phys.*, **244**, 268 (1971).
71–2 J. P. Coad, *Phys. Lett.*, **37A**, 437 (1971).
72–1 R. Spohr, T. Bergmark, N. Magnusson, L. O. Werme, C. Nordling and K. Siegbahn, *Phys. Scr.*, **2**, 31 (1970).
73–1 G. Johansson, J. Hedman, A. Berndtsson, M. Klasson and R. Nilsson, *J. Electron Spectrosc.*, **2**, 295 (1973).
73–2 G. Schön, *J. Electron Spectrosc.*, **2**, 75 (1973).
74–1 H. Aksela and S. Aksela, *J. Phys.*, **B7**, 1262 (1974).
74–2 A. Barrie and F. J. Street, *J. Electron Spectrosc.*, **7**, 1 (1977).
75–1 J. C. Fuggle, L. M. Watson, D. J. Fabian and S. Affrossman, *J. Phys.*, **F5**, 375 (1975).
75–2 C. D. Wagner, *Faraday Disc. Chem. Soc.*, **60**, 291 (1975).
76–1 G. Dufour, J.-M. Mariot, P.-E. Nilsson-Jatko and R. C. Karnatak, *Phys. Scr.*, **13**, 370 (1976).
77–1 E. Antonides, E. C. Janse and G. A. Sawatzky, *Phys. Rev.*, **B15**, 1669 (1977).
77–2 L. Asplund, P. Kelfve, B. Blomster, H. Siegbahn and K. Siegbahn, *Phys. Scr.*, **16**, 268 (1977).
77–3 L. Asplund, P. Kelfve, B. Blomster, H. Siegbahn, K. Siegbahn, R. L. Lozes and U. I. Wahlgren, *Phys. Scr.*, **16**, 273 (1977).
77–4 M. K. Bahl and R. L. Watson, *J. Electron Spectrosc.*, **10**, 111 (1977).
77–5 T. A. Carlson, W. B. Dress and G. L. Nyberg, *Phys. Scr.*, **16**, 211 (1977).
77–6 J.-M. Mariot and G. Dufour, *Chem. Phys. Lett.*, **50**, 219 (1977).
77–7 E. D. Roberts, P. Weightman and C. E. Johnson, *J. Phys. C. Sol. State Phys.*, **8**, 1301 (1975).
78–1 P. M. Th. M. Van Attekum and J. M. Trooster, *J. Phys.*, **F8**, L169 (1978).
78–2 C. D. Wagner, *J. Vac. Sci. Technol.*, **15**, 518 (1978).
79–1 C. D. Wagner, W. M. Riggs, L. E. Davis, J. F. Moulder and G. E. Muilenberg, *Handbook of X-ray Photoelectron Spectroscopy*, Physical Electronics Division, Perkin-Elmer Corporation, Eden Prairie, Minnesota, 1979.
79–2 S. M. Barlow, P. Bayat-Mokhtari and T. E. Gallon, *J. Phys. C. Sol. State Phys.*, **12**, 5577 (1979).
80–1 H. Van Doveren and J. A. Th. Verhoeven, *J. Electron Spectrosc.*, **21**, 265 (1980).
80–2 C. D. Wagner and J. A. Taylor, *J. Electron Spectrosc.*, **20**, 83 (1980).
81–1 S. Aksela, M. Kellokumpu, H. Aksela and J. Väyrynen, *Phys. Rev.*, **A23**, 2374 (1981).

Index

527

emission induced by protons (PIXE),
  11
energies, **50**
fluorescence, 94, 398
ghosts, **126**
induced AES, 293
lines, **50**
linewidths, **50**
monochromator, **53**, 361, 366, 438,
  442, 449
penetration depth, 135

satellites, **126**, 361
  subtraction of, 449
sources, **48**, 128
spectroscopy (XES), 7
window, **53**, 118

Yield, Auger electron, **182**
  sputtering, **212**

Zeolites, 287, 295, 322, **325**